AN

INTRODUCTION

TO

NATURAL PHILOSOPHY;

DESIGNED AS A

TEXT-BOOK

FOR THE USE OF

STUDENTS IN COLLEGE.

BY DENISON OLMSTED, LL.D.,
LATE PROFESSOR OF NATURAL PHILOSOPHY AND ASTRONOMY IN YALE COLLEGE.

SECOND REVISED EDITION

BY E. S. SNELL, LL.D.,
PROFESSOR OF MATHEMATICS AND NATURAL PHILOSOPHY IN AMHERST COLLEGE.

NEW YORK:
COLLINS & BROTHER,
106 LEONARD STREET.
1871.

Electrotyped by SMITH & McDOUGAL, 82 Beekman Street.

PREFACE.

THE object kept in view in the present revision is the same as heretofore—to prepare a book suitable for use in the College recitation room. The work does not aim to be a handbook of Physics, giving information on all points relating to the subjects treated of, but merely a book of first principles, accompanied by a sufficient number of illustrative statements and diagrams to render those principles clear, and to impress them on the memory.

The more prominent departures from the former revision are the following:

1. Most of the second part of Mechanics is omitted; and what is retained is introduced in appropriate connections in Part I. The second part was originally intended as a sort of substitute for a course of experimental lectures. But colleges are now so generally supplied with apparatus for illustration, that it seems unnecessary to encumber the volume with information which can be presented so much more satisfactorily by the lecturer.

2. Instead of the brief Part entitled "Electro-magnetism," first presented in the work when the former revision was made, the subject of "Dynamical Electricity" is now discussed as fully as the other branches.

3. The subject of "Heat" for the first time forms one Part of the work, although the course of instruction in many colleges still retains it in the chemical department.

4. Some additions are made to the applications of the differential and integral calculus, and all the discussions of this character are brought together in an "Appendix" at the close of the volume.

5. More than three-fourths of the engravings are new, most of which were drawn expressly for this revision.

Besides these more apparent alterations, it should be added that a large part of the whole book has been carefully rewritten, and additions and improvements made in almost every page of it.

The author of the revision wishes to express his indebtedness to Professor JOSEPH FICKLIN, of the University of Missouri, who has furnished much valuable material for the Part on "Mechanics," and has critically examined the mathematical portions of the book, and rendered essential aid in correcting and improving them. The Part on "Dynamical Electricity" is almost entirely the work of Professor CHARLES H. SMITH, of Cincinnati, Ohio, formerly Tutor of Natural Philosophy in Yale College. Much credit is due to him for presenting the principles of that extensive and complex branch of physics in a clear and systematic form. Dr. B. JOY JEFFRIES, of Boston, Lecturer on Optical Phenomena and the Eye in Harvard University, has kindly assisted in preparing the description of the eye and its adjustments, and has allowed his own original drawings to be copied in the engravings relating to this subject.

E. S. SNELL.

AMHERST COLLEGE, *September*, 1870.

CONTENTS.

INTRODUCTION.

PART I.—MECHANICS.

CHAPTER I.

CHAPTER II.

CHAPTER III.

CHAPTER IV.

CHAPTER V.

CHAPTER VI.

CHAPTER VII.

CHAPTER VIII.

CHAPTER IX.

PART II.—HYDROSTATICS.

CHAPTER I.

CHAPTER II.

PART III.—PNEUMATICS.

CHAPTER I.

CHAPTER II.

CHAPTER III.

PART IV.—SOUND.

CHAPTER I.

CHAPTER II.

CHAPTER III.

CHAPTER IV.

PART V.—MAGNETISM.

CHAPTER I.

CHAPTER II.

PART VI.—FRICTIONAL OR STATICAL ELECTRICITY.

CHAPTER I.

CHAPTER II.

CHAPTER III.

CHAPTER IV.

CHAPTER V.

PART VII.—DYNAMICAL ELECTRICITY.

CHAPTER I.

CHAPTER II.

CHAPTER III.

CHAPTER IV.

PART VIII.—HEAT.

CHAPTER I.

CHAPTER II.

CHAPTER III.

CHAPTER IV.

CHAPTER V.

PART IX.—LIGHT.

CHAPTER I.

CHAPTER II.

CHAPTER III.

CHAPTER IV.

CHAPTER V.

CHAPTER VI.

CHAPTER VII.

CHAPTER VIII.

CHAPTER IX.

CHAPTER X.

APPENDIX.

APPLICATIONS OF THE CALCULUS.

I. Fall of Bodies.

II. Centre of Gravity.

III. Centre of Oscillation.

IV. Centre of Hydrostatic Pressure.

V. Angular Radius of the Rainbow and the Halo.

NATURAL PHILOSOPHY.

INTRODUCTION.

Art. 1. Classification of Physical Sciences.—The material world consists of two parts—the *organized*, including the animal and vegetable kingdoms; and the *unorganized*, which comprehends the remainder. Organized matter is treated of in *Physiology*, and in those branches of science usually called *Natural History*. Unorganized matter forms the subject of *Natural Philosophy* and *Chemistry*. *Chemistry* considers the internal constitution of bodies, and the relations of their smallest parts to each other. *Natural Philosophy* deals principally with the external relations of bodies and their action upon one another. If, however, the bodies are so large as to constitute *worlds*, of which the earth itself is one, this science takes the name of *Astronomy*.

The word *Physics* is much used to include both Natural Philosophy and Chemistry; but sometimes it is applied to the branches of Natural Philosophy, except Mechanics. According to this use of the word, Natural Philosophy is divided into two general subjects, Mechanics and Physics.

2. Definitions relating to Matter.—

A *body* is a separate portion of matter, whether large or small.

An *atom* is a portion of matter so small as to be indivisible.

A *particle* denotes the smallest portion which can result from division by mechanical means, and consists of many atoms united together.

The word *molecule* signifies a very small portion of matter, either atom or particle.

Mass is the quantity of matter in a body, and is usually measured by its weight.

Volume signifies the space occupied by a body.

Density expresses the relative mass contained within a given volume. Thus, if one body has twice as great a mass within a certain volume as another has, it is said to have twice the density.

Pores are the minute portions of space within the volume of a body, which are not filled by the material of that body. All matter is porous, some kinds in a greater and some in a less degree.

Force is the name of any cause, whatever it may be, which gives motion to matter, or which changes its motion.

3. Properties of Matter.—

(1.) *Extension.*—Every portion of matter, however small, has length, breadth, and thickness, and thus occupies space. This is its extension.

(2.) *Impenetrability.*—While matter occupies space, it excludes all other matter from it, so that no two atoms can be in exactly the same place at the same time. This property is called impenetrability.

The two foregoing are often called *essential* properties, because we cannot conceive matter to exist without them.

(3.) *Divisibility.*—Matter is *divisible* beyond any known limits. After being divided, as far as possible, into particles by mechanical methods, it may be still further reduced by chemical action to atoms, which are too small to be in any way recognized by the senses.

(4.) *Compressibility.*—Since pores exist in all matter, it may be compressed into a smaller volume. Hence all matter is *compressible,* though in very different degrees.

(5.) *Elasticity.*—After a body has suffered compression, it shows, in some degree at least, a tendency to restore itself to its former volume. This property is called elasticity. A body is said to be *perfectly elastic* when the force by which it recovers its size is equal to that by which it was before compressed. The word elasticity is used generally in a wider sense than is given in the above definition, namely, the tendency which a body has to recover its original *form,* whatever change of form it may have previously received. Thus, if a body is stretched, bent, twisted, or distorted in any other way, it is called *elastic,* if it tends to resume its form as soon as the force which altered it has ceased. *Torsion* is the name of the elastic force which tends to untwist a thread or wire when it has been twisted.

(6.) *Attraction.*—This is the general name used to express the universal tendency of one portion of matter towards another. It receives different names, according to the circumstances in which it acts. The attraction which binds together atoms of different kinds, so as to form a new substance, is called *affinity,* and is discussed in Chemistry; that which unites particles, whether simple or compound, so as to form a body, is called *cohesion;* the clinging of two kinds of matter to each other, without forming a new substance, is called *adhesion;* and the tendency manifested by masses of matter toward each other, when at sensible distances, is called *gravity.*

(7.) *Inertia.*—This is also a universal property of matter, and signifies its tendency to continue in its present condition as to motion or rest. If at rest, it cannot move itself; if in motion, it cannot stop itself or change its motion, either in respect to direction or velocity.

4. Branches of Natural Philosophy.—Natural Philosophy is generally divided into *Mechanics, Hydrostatics, Pneumatics, Sound, Magnetism, Electricity, Heat,* and *Light.*

Mechanics treats of the motion and equilibrium of bodies, caused by the application of force. Since there are three conditions of matter, solid, liquid, and gaseous, it is convenient to divide the general subject of Mechanics into three branches.

1st. The mechanics of solids, also called *Mechanics.*

2d. The mechanics of liquids, called *Hydrostatics.*

3d. The mechanics of gases, called *Pneumatics.*

All the other branches of Natural Philosophy (often called Physics) treat of various phenomena caused by *minute vibrations* in the particles of matter. These vibrations are excited in different ways, and when transmitted to us, affect one or more of our senses. Thus, *sound* consists of such vibrations as affect the sense of hearing; and *light* is another mode of vibration, that affects only the sense of vision.

It was formerly customary to regard magnetism, electricity, heat, and light, as so many kinds of *imponderable* matter, that is, matter having no sensible weight, and thus distinguished from solids, liquids, and gases, which are the different forms of *ponderable* matter. But it is now known that when forces are applied to matter, they not only produce the visible forms of motion, but may be made to develop either sound, magnetism, electricity, heat, or light; and that most of these modes of motion may be transformed into others, and each may be made a measure of the force which is employed to produce it.

PART I.

MECHANICS.

CHAPTER I.

MOTION AND FORCE.

5. Classification of Motions.—Motion is change of place, and is either *uniform* or *variable*. In *uniform* motion equal spaces are passed over in equal times, however small the times may be. In *variable* motion the spaces described in equal times are unequal. Such motion may be either *accelerated* or *retarded*. In *accelerated* motion the spaces described in equal times become continually greater; in *retarded* motion they become continually less. Motion is said to be *uniformly accelerated* if the increments of space in equal times (however small) are equal; and *uniformly retarded* if the decrements are equal.

Velocity is the space described in the *unit* of time. In Mechanics, one second is much used as the unit of time, and one foot as the unit of space; hence, velocity is the number of feet described in one second.

6. Uniform Motion.—When motion is uniform, the number of feet described in one second, multiplied by the number of seconds, obviously gives the whole space. Let s = space, t = time, and v = velocity; then $s = t\,v$; $\therefore t = \frac{s}{v}$, and $v = \frac{s}{t}$. If this space is compared with another, s', described in the time t', with the velocity v', then $s : s' :: t\,v : t'\,v'$; or briefly, in the form of a variation, $s \propto t\,v$. In like manner $t \propto \frac{s}{v}$, and $v \propto \frac{s}{t}$.

If two bodies, moving uniformly, describe *equal* spaces, then $s = s'$; $\therefore t\,v = t'\,v'$; $\therefore t : t' :: v' : v$. That is, in order that two bodies may describe equal spaces, their velocities must vary inversely as the times during which they move.

7. Questions on Uniform Motion.—

1. A ball was rolled on the ice with a velocity of 78 feet per second, and moved uniformly 21 seconds; what *space* did it describe? *Ans.* 1638 feet.

2. A steamboat moved uniformly across a lake 17 miles wide, at the rate of 20 feet per second; what *time* was occupied in crossing? *Ans.* 1*h.* 14*m.* 4*s.*

3. On the supposition that the earth describes an orbit of 600 millions of miles in $365\frac{1}{4}$ days, with what *velocity* does it move per second? *Ans.* 19 miles, nearly.

4. Three planets describe orbits which are to each other as 15, 19, and 12, in times which are as 7, 3, and 5; what are their *relative velocities?* *Ans.* 225, 665, and 252.

8. Momentum.—The momentum of a body signifies its *quantity of motion,* and is reckoned according to the *mass,* or quantity of matter, which is moving, and the *velocity* with which it moves. The momentum, therefore, varies as the product of the mass and the velocity.

Let the momentum of a body $= m$, its mass $= q$, and its velocity $= v$; then $m = q\,v$, $q = \frac{m}{v}$, and $v = \frac{m}{q}$. In order to compare the momentum of one body with that of another, let m', q', v', represent the momentum, mass, and velocity, respectively of the second body; then $m : m' :: q\,v : q'\,v'$; or $m \propto q\,v$; $\therefore\ q \propto \frac{m}{v}$, and $v \propto \frac{m}{q}$.

If the momentum of one body equals that of another, then, since $m = m'$, $q\,v = q'\,v'$, $\therefore q : q' :: v' : v$. That is, in order that the momenta of two bodies should be equal, their masses must vary inversely as their velocities.

Since there are two elements entering into the momentum of a body—namely, its *mass,* usually expressed in pounds, and its *velocity,* expressed in feet per second—therefore momentum cannot be measured either in pounds or in feet, being in nature unlike either. The word *foot-pound* is employed for the unit of momentum whenever the unit of mass is a pound and the unit of velocity is a foot per second.

9. Questions on Momentum.—

1. A ship weighing 336,000 lbs. is dashed against the rocks in a storm, with a velocity of 16 miles per hour; with what momentum did she strike? *Ans.* 7,884,800 foot-pounds.

2. A ball weighing 1 oz. is fired into a log weighing 53 lbs., suspended so as to move freely, and imparts a velocity of 2 ft. per second. Assuming that the log and ball have a momentum equal to the previous momentum of the ball alone, required the velocity of the ball. *Ans.* 1,698 ft. per sec.

3. Suppose a comet, whose velocity is 1,000,000 miles per hour,

has the same momentum as the earth, whose velocity is 19 miles per second; what is the ratio of their masses? *Ans.* 1 : 14.6.

4. Two railway cars have their quantities of matter as 7 to 3, and their momenta as 8 to 5; what are their relative velocities? *Ans.* As 24 to 35, or nearly 5 to 7.

5. The momentum of a cannon-ball was 434 foot-pounds; what must be the velocity of a half-ounce bullet, in order to have the same momentum? *Ans.* 13,888 feet.

10. Classification of Forces.—The principal forces in nature are the following:

1. *Attraction* in its several forms. *Cohesion* and *chemical affinity* are the forces which bind together the particles and atoms of bodies, and *gravity* is that which everywhere near the earth causes bodies to fall toward it, or to press upon it.

2. *Elasticity.*—This is a force which, in many kinds and conditions of matter, tends to repel the particles from each other.

The forces, whether attraction or repulsion, which exist among the atoms or molecules of a body, are called *molecular* forces.

3. *Muscular force.*—All living beings are endowed with this force, by which they put in motion bodies around them, and by acting upon other bodies, are enabled also to move themselves from place to place.

4. *Matter in motion.*—If a body which some force has put in motion impinges on another body, it imparts motion to it, and is therefore itself a force. This is true not only of ordinary visible motions, but of those small and often invisible vibrations, which manifest themselves as sound, heat, &c. Gravity, or any other force, may cause heat, and heat may cause light and electricity. Thus, any form of motion is a force, and it can be employed to produce other forms.

11. Impulsive and Continued Forces and their Effects.—An *impulsive* force is one which has no sensible continuance, as the blow of a hammer. A *continued* force is one which acts during a perceptible length of time. Continued forces are subdivided into *constant* and *variable.* A *constant* force has the same intensity during the whole time of its action; a *variable* force is one whose intensity changes.

Keeping in mind the property of inertia, we associate different kinds of motion with the forces which produce them, as follows:

1. An *impulsive* force causes *uniform* motion.
2. A *continued* force, *accelerated* motion.
3. A *constant* force, *uniformly accelerated* motion.
4. A *variable* force, *unequally accelerated* motion.

If the force is applied in a direction opposite to that in which the body has a previous uniform motion, the connection is the following:

5. An *impulsive* force causes *uniform* motion, or *rest.*

6. A *continued* force, *retarded* motion.

7. A *constant* force, *uniformly retarded* motion.

8. A *variable* force, *unequally retarded* motion.

In cases 1 and 5, it is obvious that, the impulse being given, the body is left to itself, and cannot change the state of motion or rest impressed on it.

In 2, 3, and 4, it must be considered that the force at each instant adds a new *increment* to the uniform motion which the body would have had if the force had ceased; and if the force is constant, those increments are equal; if variable, they are unequal.

In 6, 7, and 8, the same statements may be made in regard to decrements. It is also plain that in these three last cases, if the force continues to act indefinitely, the motion will be retarded until the body comes to a state of momentary rest, and then is accelerated in the direction of the force.

12. Measure of Force.—The intensity of an *impulsive* force is measured by the *momentum* which it will produce or destroy; that is, $f \propto m$. But $m \propto q\,v$; $\therefore f \propto q\,v$. Hence, if q is constant, $f \propto v$. If, then, an impulse is applied to a given *mass*, the intensity of that impulse is measured by the *velocity* which it imparts or destroys.

But in the case of a *constant* force, the momentum depends not only on the intensity of the force, but on the time during which it is applied; that is, $f\,t \propto m$, and $f \propto \frac{m}{t}$. If the mass of the body is given, then, as in the case of an impulsive force, q being constant, $f\,t \propto v$, and $f \propto \frac{v}{t}$.

To express the measure of a *variable* force, let t be a constant and infinitely small portion of time; then the force varies as the mass multiplied by the increment of velocity imparted in that time.

13. The Three Laws of Motion.—All the phenomena of motion in Mechanics and Astronomy are found to be in accordance with three first principles, which Newton announced in his Principia, and which are to be regarded as forming the basis of mechanical science. They may be named and defined as follows:

1. The law of *inertia.*—A body at rest tends to remain at rest; and a body in motion tends to move forever, in a straight line, and uniformly.

2. The law of the *coexistence of motions.*—If several motions are communicated to a body, it will ultimately be in the same position, whether those motions are simultaneous or successive.

3. The law of *action and reaction.*—If any kind of action takes place between two bodies, it produces equal momenta in opposite directions; or, every action is accompanied by an equal and opposite reaction.

The truth of these laws cannot be established, except approximately, by direct experiments, because gravity, friction, and the resistance of air, interfere more or less with every possible experiment. They are to be learned rather by a careful study of the phenomena of motion in general. We see an approximation to the *first* law, in rolling a ball on a horizontal surface; first, on the earth, then on a floor, and again on smooth ice, the motion approaching toward uniformity as obstructions are diminished, and gravity producing no direct effect, because acting at right angles to the line of motion. The discussion of the *second* law is reserved for Chapter III. The *third* law is illustrated by a variety of cases in collision, attraction, and repulsion. Suppose that a body *A*, being in motion, strikes directly against *B*, which is at rest; it is found that *B* acquires a certain momentum, and that *A loses* (that is, *acquires* in an opposite direction) an equal amount. The same is true if *B* is in motion, and *A* either overtakes or meets it. In the collision of two railroad trains, it is immaterial as to the effects which they will respectively suffer, whether each is moving towards the other, or whether one is at rest, provided that in the latter case the moving train has a momentum equal to the momenta of the two trains in the former case. When a magnet attracts a piece of iron, each moves towards the other with the same momentum. A spring between two bodies *A* and *B* drives *A* from *B* with as much momentum as *B* from *A*; and the sudden expansion of burning gunpowder, which propels the balls when a broadside is fired, causes an equal amount of motion of the ship in the opposite direction.

14. Force of Gravity.—Every mass of matter near the earth, when free to move, pursues a straight line towards its centre. The force by which this motion is produced is called *gravity;* either the gravity of the body or the gravity of the earth; for the attraction is mutual and equal, in accordance with the third law of motion. It is easy to understand why a small mass should attract a large one, as much as the large mass attracts the small one. Let *A* consist of *one* atom of matter, and *B*, at any distance from it, consist of *ten* atoms. If it be admitted that *A* attracts *one* atom of *B* as much as *that one* atom attracts *A*, then the above conclusion follows. For A attracts *each* of the ten atoms of B as much as *each* of the same ten attracts A; so that A exerts ten units of attraction on B, while B exerts ten units of attraction on A. The same reasoning obviously applies to the earth in relation to the small bodies on its surface.

15. Relation of Gravity and Mass.—At the same distance from the centre of the earth, *gravity varies as the mass.* This is because it operates equally on every atom of a body; hence the greater the number of atoms in a body, the greater in the same ratio is the attraction exerted upon it. That gravity varies as the mass is also proved from the observed fact, that in a vacuum it gives the same velocity, in the same time, to every mass, however great or small, and of whatever species of matter. For a constant force, acting for a given time, is measured by the *momentum* which it produces (Art. 12), and that momentum, if the velocity is the same, varies as the mass: therefore the force also varies as the mass to which it imparts the given velocity.

If a body is not free to move, its tendency towards the earth causes *pressure;* and the measure of this pressure is called the *weight* of the body. Weight is usually employed as a measure of the mass in bodies. The foregoing relations are embodied in the following expressions: $g \propto q$; and $w \propto q$.

16. Relation of Gravity and Distance.—At different distances from the earth, *gravity varies inversely as the square of the distance from the centre.* The demonstration of this proposition is reserved for astronomy, where it is shown by the movements of the bodies in the solar system that this law applies to them all.

The moon is 60 times as far from the earth's centre as the distance from that centre to the surface: therefore the attraction of the earth upon the particles of the moon is 3600 times less than upon particles at the surface of the earth. At the height of 4000 miles above the earth, gravity is four times less than at the surface. But the heights at which experiments are commonly made upon the weights of bodies bear so small a ratio to the radius of the earth, that this variation is commonly imperceptible. At the height of *half a mile*, the diminution does not amount to more than about $\frac{1}{4000}$th part of the weight at the surface. For, let r = the radius of the earth = 4000 miles, nearly; and let x be the height of the body, w its weight at the earth's surface, and w' its weight at the height x. Then,

$$w : w' :: (r+x)^2 : r^2 :: r^2+2rx+x^2 : r^2.$$

$$w : w-w' :: r^2+2rx+x^2 : 2rx+x^2 \therefore w-w' = \frac{w\,(2rx+x^2)}{r^2+2rx+x^2} \;(A).$$

But when x is a small fraction of r, x^2 may be neglected, and the formula becomes $w - w' = \frac{w \times 2x}{r + 2x}$ . . . . (B).

Let x be *half a mile;* then $\frac{w \times 1}{4000 + 1} = \frac{1}{4001}$th part of the whole weight; or, a body would weigh so much less at the height of half a mile than at the surface of the earth. But if the height were as great as 100 miles above the earth, the loss should be calculated by formula (A), since the other would give a result too small by one per cent. or more, according to the height.

What loss of weight would a body sustain by being elevated 500 miles above the earth? *Ans.* $\frac{17}{81}$, or more than $\frac{1}{5}$ of its weight.

The relation of gravity to distance is expressed by the formula $g \propto \frac{1}{d^2}$; and as $g \propto q$ also, it varies as the product of the two; that is, $g \propto \frac{q}{d^2}$; or *gravity towards the earth varies as the mass of the body directly, and as the square of the distance from the earth's centre inversely.*

17. Gravity within a Hollow Sphere.—A particle situated *within a spherical shell* of uniform density, is equally attracted in all directions, and *remains at rest.* This is true, because, in every direction from the body, the mass varies at the same rate as the square of the distance, so that attraction increases for one reason, as much as it diminishes for the other; which is proved as follows:

Let the particle P (Fig. 1) be at any point within the spherical shell $A\ B\ C\ D$. Let two opposite cones of revolution, of very small angle, have their vertices at P, and suppose the figure to be a section through the centre of the sphere and the axis of the cones. Then $A\ B$ and $a\ b$ will be the major axes of the small ellipses, which are the bases of the cones, and which may be considered as plane figures. By geometry, $A\ P : P\ B :: P\ b : P\ a$; and the angles at P being equal, the triangles are similar; hence the angles B and a are equal. Therefore, the bases of the cones are similar ellipses, being sections of similar cones, equally inclined to the sides. By similar triangles, $A\ P^2 : P\ b^2 : A\ B^2 : a\ b^2$. Let q and q' represent the masses of the thin laminæ which form the bases; then, since similar ellipses are to each other as the squares of their major axes, we have

FIG. 1.

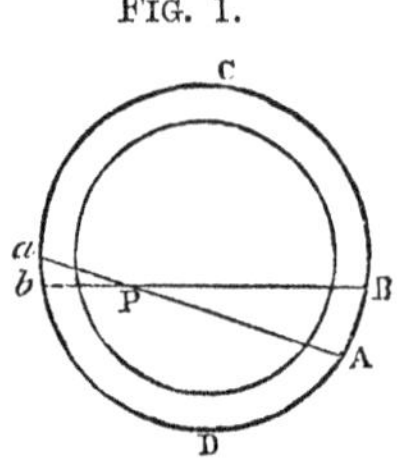

$$q : q' :: A\ P^2 : P\ b^2, \text{ or } \frac{q}{A\ P^2} = \frac{q'}{P\ b^2}.$$

But $\frac{q}{A\ P^2}$ and $\frac{q'}{P\ b^2}$ represent the attractions of the bases respectively on the particle (Art. 16); and since these are equal, the particle is equally attracted by all the opposite parts of the spherical shell.

18. Gravity within a Solid Sphere.—Within a *solid sphere* of uniform density, weight varies directly *as the distance from the centre.*

FIG. 2.

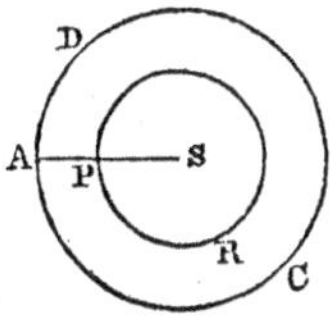

Let a particle P (Fig. 2) be within the solid sphere $A\ D\ C$; and call its distance from the centre d. Now, by the preceding article the shell exterior to it, $A\ D\ R$, exerts no influence upon it, and it is attracted only by the sphere $P\ R\ S$. Let q represent the quantity of this sphere; then gravity varies as $\frac{q}{d^2}$. But $q \propto d^3$; $\therefore g \propto \frac{d^3}{d^2} \propto d$. Hence, in the earth (if it be supposed spherical and uniformly dense, though it is neither exactly), a body at the depth of 1000 miles weighs *three-fourths* as much as at the surface, and at 2000 miles it weighs half as much, while at the centre it weighs nothing.

Comparing this proposition with Art. 16, we learn that just at the surface of the earth a body weighs more than at any other place without or within. Within, the weight diminishes *nearly* as the distance from the centre diminishes; without, it diminishes as the square of the distance from the centre increases.

At the surface of spheres having the same density, weight varies as the radius of the sphere. Let r be the radius of the sphere, and q its mass; then, since $g \propto \frac{q}{r^2}$, in this case it varies as $\frac{r^3}{r^2} \propto r$. Therefore, if two planets have equal densities, the weight of bodies upon them is as their radii or their diameters. If a ball two feet in diameter has the same density as the earth, a particle of dust at its surface is attracted by it nearly 21 millions of times less than it is by the earth.

19. Questions for Practice.—

1. How much weight would a rock that weighs ten tons (22,400 lbs.) at the level of the sea, lose if elevated to the top of a mountain five miles high? *Ans.* 55.8952 lbs.

2. If the earth were a hollow sphere, and if, through a hole bored through the centre, a man were let down by a rope, would the force required to support him be increased or diminished as he descended through the solid crust, and where would it become equal to nothing?

3. How much would a 44-pound shot weigh at the centre of

the earth; how much at a point half way from the centre to the surface; and how much 100 miles below the surface?

4. If a hole were bored through the centre of the earth, and a stone were dropped into it, in what manner would the stone move in its way to the centre and after it reached the centre?

5. Suppose a 32-pound cannon-ball, fired with the velocity of 2,000 feet per second, to have the same momentum as a battering-ram whose weight is 5760 pounds; find the velocity of the latter.

Ans. 11.11 ft. per sec.

6. Suppose light to have weight, and one grain of it moving at the rate of 192,000 miles per second, to impinge directly against a mass of ice moving at the rate of 1.45 feet per second, and to stop it; requireh the weight of the ice.

Ans. 99877.832 lbs., or nearly $44\frac{1}{2}$ tons, reckoning 7000 gr. = 1 lb.

7. If a ball of the same density with the earth, $\frac{1}{10}$th of a mile in diameter, were to fall through its own diameter toward the earth, what space would the earth move through to meet the ball, the diameter of the earth being taken at 8,000 miles?

Ans. $\frac{1}{80000000000}$ inch, nearly.

8. Two men are pulling a boat ashore by a rope, one at each end, *A* being in the boat and *B* on the shore; how will the time of bringing the boat ashore compare with the time in which *A* would pull it ashore alone, were the other end of the rope fixed to an immovable post?

9. Suppose the rope to pass from *A* in one boat to *B* in another equal boat; how fast will *B's* boat move? will *A's* boat have the same velocity as when *B* was on the shore?

CHAPTER II.

VARIABLE MOTION. FALLING BODIES.

20. Uniform Motion represented Geometrically.—When a body moves uniformly for a given time, the space described equals the time multiplied by the velocity (Art. 6). Therefore, if one side of a rectangle represents the *time* of motion, and an adjacent side the *uniform velocity*, the area of the rectangle will represent the *space* described in that time, because the area equals the product of two adjacent sides. Thus, let *A B*, *B C*, &c. (Fig. 3), represent any equal portions of time, and *A F*, *B G*, &c., the uniform velocity;

Fig. 3.

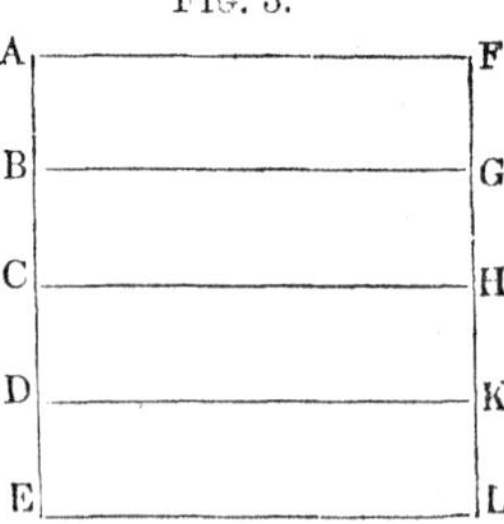

then $A\ G$, $B\ H$, &c., may be used to represent the spaces described, and the rectangle $A\ L$ may represent the space passed over with the velocity $A\ F$, in the time $A\ E$.

21. Velocity Increased at Finite Intervals.—Suppose the body to receive equal impulses at the beginning of all the equal portions of time, $A\ B$, $B\ C$, &c. (Fig. 4). Then, $A\ F$ being the velocity given by the first impulse, $G\ H$, $K\ L$, &c., the increments of velocity, will each be equal to $A\ F$; and $B\ H$, the velocity during the second portion of time, equals $2\ A\ F$; $C\ L$, that of the third, equals $3\ A\ F$, &c. Therefore, $B\ G$, $C\ K$, $D\ M$, &c., are as 1, 2, 3, etc. But $A\ B$, $A\ C$, $A\ D$, &c., are as 1, 2, 3, &c. Hence the triangles $A\ B\ G$, $A\ C\ K$, &c., are similar, and the same straight line $A\ O$ passes through the angles of all the rectangles. Now these rectangles represent the successive spaces described in the equal times, and their sum represents the whole space described in the time $A\ E$. This exceeds the triangle $A\ O\ E$ by the sum of the small equal triangles $A\ F\ G$, $G\ H\ K$, &c.

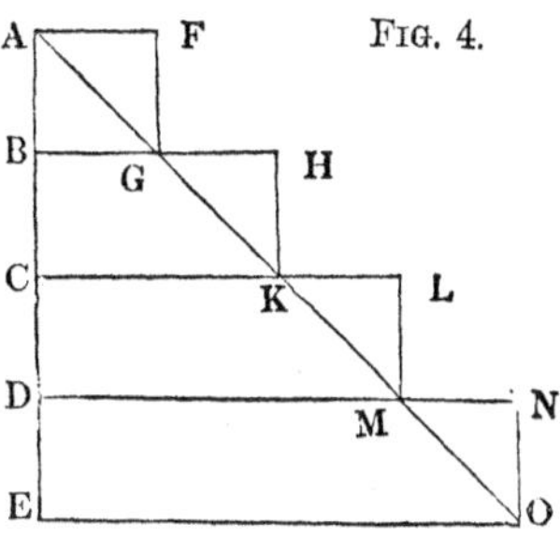

Fig. 4.

22. Uniformly Accelerated Motion Represented.—Let the increments of time and velocity (Fig. 5) be half as great as before. Then the sum of all the rectangles, or the whole space described, exceeds the triangle $A\ O\ E$ by the sum of the triangles $A\ f\ g$, $g\ F\ G$, &c. These triangles are one-half the sum of those in Fig. 4. Therefore, by continually halving the increments of time and velocity, the sum of the rectangles continually approaches the area of the triangle; and when these increments become infinitely small, the first velocity becomes *zero*, and the sum of the rectangles equals the triangle. Therefore, the space described by a body which begins to move from rest by the action of a *constant* force may be represented by a right-angled triangle, as $A\ O\ E$, whose side $A\ E$ represents the time, and the side $E\ O$ the last acquired velocity.

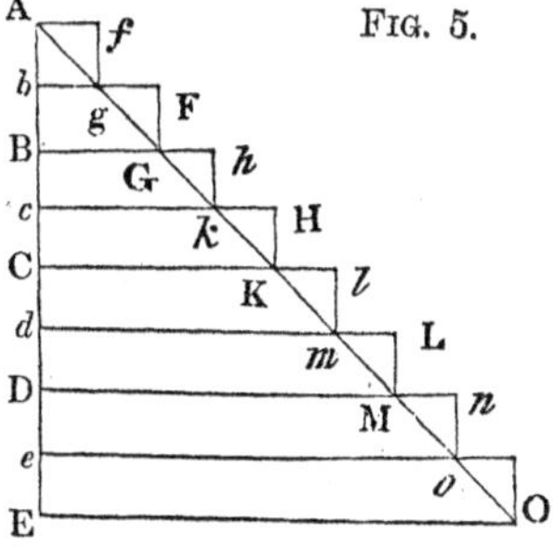

Fig. 5.

Such motion is said to be uniformly accelerated (Art. 11). An example of this is found in the *fall* of a body in a vacuum. For gravity acts *incessantly*, and within the range of our experiments it may be considered as acting with *equal intensity.* The properties of the triangle enable us to ascertain very readily the laws of the fall of a body.

23. Laws of the Fall of Bodies.—When bodies fall from rest by the force of gravity, and unobstructed by the air, the following relations exist between the space, time, and velocity:

1. *The spaces vary as the squares of the times.*
2. *The spaces vary as the squares of the acquired velocities.*
3. *The times vary as the acquired velocities.*

For let s be the space described, v the velocity acquired by a body falling from rest for the time t, s' the space described, v' the velocity acquired at any other period t' of its fall; then, from what has already been demonstrated, if t and t' be represented by the lines $A\ B$ and $A\ D$ (Fig. 6), and v and v' by the lines $B\ C$ and $D\ E$, drawn at right angles to them, s and s' will be represented by the triangles $A\ B\ C$, $A\ D\ E$. Now,

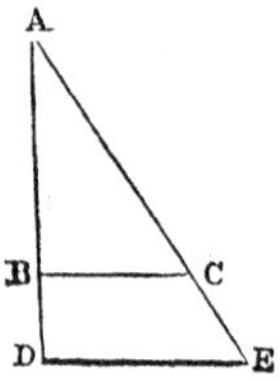

$A\ B\ C : A\ D\ E :: A\ B^2 : A\ D^2$; or, as $B\ C^2 : D\ E^2$;

hence, $s : s' :: t^2 : t'^2$, or as $v^2 : v'^2$.

Fig. 6.

As equal increments of velocity are generated in equal times, it is farther evident that the velocity acquired varies as the time; the same conclusion may also be deduced from the similar triangles $A\ B\ C$, $A\ D\ E$; for $B\ C : D\ E :: A\ B : A\ D$, i. e. $v : v' :: t : t'$.

Since the spaces described are as the squares of the times, if a body falls from rest during times which are represented by the numbers 1, 2, 3, 4, 5, &c., the spaces described in those times will be as the square numbers 1, 4, 9, 16, 25, &c.; and the spaces described in equal successive portions of time will be as the odd numbers 1, 3, 5, 7, 9, &c., as exhibited in the following table:

Times.	Spaces described.	Spaces described in equal successive portions of time.
1	1	In 1st portion of time 1
2	4	2d " " . . . 4 − 1 = 3
3	9	3d " " . . . 9 − 4 = 5
4	16	4th " " . . . 16 − 9 = 7
5	25	5th " " . . . 25 − 16 = 9
&c.	&c.	&c. . . . . &c. = &c.

The odd number expressing the space described in any unit of time, and which is found in the above table by taking the difference of the squares of successive numbers, may also be obtained by subtracting *one* from twice the number of units in the time. Thus, in the table, $3 = 2 \times 2 - 1$; $5 = 2 \times 3 - 1$, &c. This is true to any extent. Let n represent any whole number; the number next less is $n - 1$. The space described in n units of time is represented by n^2, and that described in $n - 1$ units, by $(n - 1)^2$. Therefore, the space described in the nth unit is represented by $n^2 - (n - 1)^2 = 2n - 1$. This is an odd number, and it equals twice the given number, less one.

24. Uniformly Retarded Motion.—If a body be projected perpendicularly upward in a vacuum, with the velocity which it has acquired in falling from any height, it will rise to the point from which it fell, before it begins to descend again, and the motion will be uniformly retarded. As the force of gravity adds equal velocities in equal times to a descending body, so it destroys equal velocities in equal times in a body which is ascending. The spaces described in successive units of time, by a body thus ascending, reckoning from the beginning of its motion, will be the same as those stated in the foregoing table, but in an inverted order: thus, if the time be divided into four equal parts, then the spaces described in the descent of the body during these equal times are as the numbers 1, 3, 5, 7, but in its ascent they will be as 7, 5, 3, 1; that is, the space described in the first portion of time, in its ascent, will be the same as that described in the last, in its descent, and so on till the body arrives at its highest point.

25. Acquired Velocity.—If a body moves uniformly with the acquired velocity, it will pass over twice as great a space, in the same time, as it falls through to acquire it.

FIG. 7.

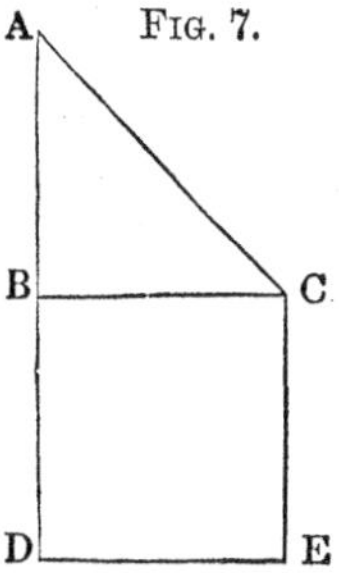

A B C D E

Let the triangle $A\ B\ C$ (Fig. 7) represent the space described by gravity in the time $A\ B$, and $B\ C$ the last acquired velocity; produce $A\ B$ to D, making $B\ D$ equal to $A\ B$, and complete the rectangle $B\ E$; then, if a body moves during the time $B\ D$ with the uniform velocity represented by $B\ C$, the space described in that time will be represented by the rectangle $B\ E$; but the triangle $A\ B\ C$ is half $B\ E$; hence the space described with the velocity $B\ C$ continued uniformly is twice that which would be described in the same time $A\ B$, falling from rest.

Since the space described by a body falling from rest is half that which it would describe in the same time with its greatest velocity continued uniformly, and since a body projected perpendicularly upward rises to the same height as that from which it must fall to acquire the velocity of projection, the whole space described by a body projected perpendicularly upward is *half* that which it would describe in the same time with its first velocity continued uniformly.

26. Projection Downward.—The space described in any time by a body *projected downward* with a given velocity is equal to the space which would be described with that velocity continued uniformly during that time, together with the space through which a body would fall from rest by the action of gravity in the same time.

Let $A\ D$ (Fig. 8) represent the given velocity of projection, and $A\ B$ the given time, and complete the rectangle $A\ E$; produce $B\ E$ to C, and let $E\ C$ represent the velocity generated by gravity in the time $A\ B$ or $D\ E$, and join $D\ C$. Then the body, moving by projection alone (Art. 20), would describe the rectangle $A\ E$ in the time $A\ B$; but, by gravity alone, it would describe the triangle $D\ E\ C$ (Art. 22). Hence, by the coexistence of both motions (Art. 13), it would describe the trapezoid $A\ C$.

Fig. 8.

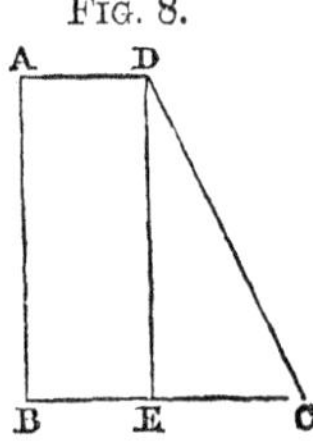

27. Projection Upward.—The space described by a body ascending for a given time is equal to the space described uniformly with the velocity of projection in that time, diminished by the space fallen through from rest in the same time.

Let $B\ C$ (Fig. 9) be the velocity of projection, and $A\ B$ the time in which a body would acquire that velocity in falling from rest. Then the triangle $A\ B\ C$ represents the space through which it would ascend before the velocity is lost. Let $B\ E$ be the given time of ascent; then the rectangle $B\ D$ is the space described in the time $B\ E$, with the velocity $B\ C$ continued uniformly, and $C\ D\ F$ (similar to $A\ B\ C$) the space fallen through in the same time. But the part $B\ E\ F\ C$ of the triangle $A\ B\ C$ is the space through which the body ascends in the time $B\ E$; and this is equal to the difference of the rectangle $B\ D$ and the triangle $C\ D\ F$.

Fig. 9.

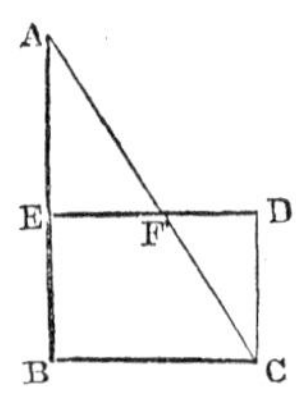

28. Formulæ for the Fall of Bodies.—The distance through which a body falls in a vacuum in one second of time varies on different parts of the earth. Between latitudes 40° and 50°, it is very nearly $16\frac{1}{12}$ feet, or 193 inches. Therefore (Art. 25), at the end of the second the body is moving with a velocity which, if gravity were to cease, would carry it over $32\frac{1}{6}$ feet per second. Let $g = 32\frac{1}{6}$ feet, the velocity acquired in one second of fall. Then $\frac{1}{2}\,g = 16\frac{1}{12}$, the distance of fall in the first second. Let s be the space described, and v the velocity acquired, in any other time t. Then, according to the laws of variation (Art. 23), we have:

(1.) $s : \frac{1}{2}\,g :: t^2 : 1^2$; ∴. . . . . $s = \frac{1}{2}\,g\,t^2.$

(2.) . . . . . . and . . . . $t = \sqrt{\frac{2\,s}{g}}.$

(3.) $s : \frac{1}{2} g :: v^2 : g^2$; $\therefore$ . . . . $s = \dfrac{v^2}{2g}$.

(4.) . . . . and . . . . $v = \sqrt{2gs}$.

(5.) $1 : t :: g : v$; $\therefore$ . . . . $t = \dfrac{v}{g}$.

(6.) . . . . . . and . . . . $v = g\,t$.

29. Space and Time Represented by Co-ordinates.— The relation of space and time in different kinds of motion may be well represented by the rectangular co-ordinates of certain lines. Thus, in *uniform* motion we have

$$s = v\,t,$$

in which v is constant. This may be regarded as the equation of a straight line passing through the origin, and making with the axis of abscissas an angle, whose tangent is v. Therefore, if any abscissa $A\,C$ (Fig. 10) represents the number of units of time occupied in the motion, the corresponding ordinate $C\,D$ will represent the space passed over.

Fig. 10.

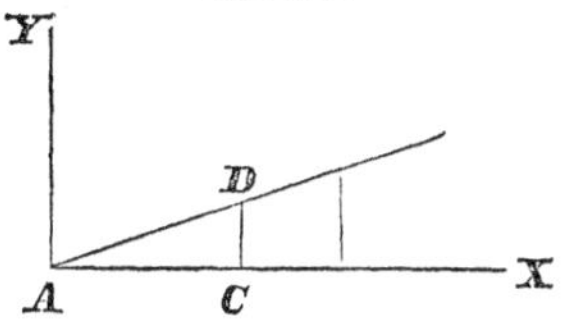

Again, for the *uniformly accelerated* motion of a falling body we have

$$s = \tfrac{1}{2}\,g\,t^2, \text{ or } t^2 = \frac{2}{g}\,s.$$

This is the equation of a parabola whose parameter is $\dfrac{2}{g}$. Therefore, if the parabola $A\,B$ (Fig. 11) be described, having $\dfrac{2}{g}$ for its parameter, and the time of falling is represented by any ordinate $C\,D$, the corresponding abscissa $A\,C$ will represent the space fallen through.

Fig. 11.

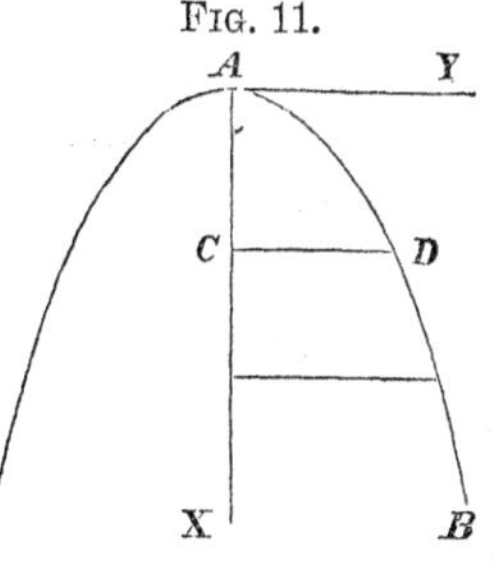

This is illustrated by Morin's apparatus, where a body falls parallel to the axis of a uniformly revolving cylinder, wrapped with paper, against which a pencil, attached to the falling body, gently presses. When the paper is unwound and developed upon a plane, the curve traced by the pencil is found to be a parabola.

30. Applications of Formulæ for the Fall of Bodies.—

1. A body falls 6 seconds; what *space* does it pass over, and what *velocity* does it acquire? *Ans.* $s = 579$ ft. $v = 193$ ft. per sec.

2. *How far* must a body fall to acquire a velocity of 50 feet per second, and *how long* will it be in falling?

Ans. $s = 38.86$ ft. $t = 1.55$ sec.

3. A body fell from the top of a tower 150 feet high; *how long* was it in falling, and what *velocity* did it have at the bottom?

Ans. $t = 3.054$ sec. $v = 98.237$ ft.

4. If a ball be thrown upward with a velocity of 100 feet per second, what *height* will it reach? *Ans.* 155.44 ft.

5. Suppose a body to fall during 3 seconds, and then to move uniformly during 2 seconds more, with the velocity acquired; what is the whole *distance* passed over?

The space fallen through is $16\frac{1}{12} \times 9 = 144\frac{3}{4}$ feet. The velocity acquired is $32\frac{1}{6} \times 3 = 96\frac{1}{2}$ feet. The space described uniformly is $96\frac{1}{2} \times 2 = 193$ feet. Therefore the whole space is $144\frac{3}{4} + 193 = 337\frac{3}{4}$ feet.

6. A ball fired perpendicularly upward was gone 10 seconds, when it returned to the same place; how *high* did it rise, and with what *velocity* was it projected? *Ans.* $s = 402\frac{1}{12}$ ft., $v = 160\frac{5}{6}$ ft.

31. Space in any Given Second or Seconds of Fall.—Since the spaces described in the successive units of time are as the odd numbers, and as $\frac{1}{2}g$ is described in the first second, therefore $3 \times \frac{1}{2}g$ is described in the second, $5 \times \frac{1}{2}g$ in the third, and generally $(2n - 1) \times \frac{1}{2}g$ in the nth second.

1. How far does a body move in the 14th second of its fall?

Ans. $434\frac{1}{4}$ ft.

2. A body had been falling 2 minutes; how far did it move in the last second? *Ans.* $3843\frac{11}{12}$ feet.

The space described in the last m seconds is found thus: The space in the whole time t, $= \frac{1}{2}gt^2$; and in the time $t - m$, the space $= \frac{1}{2}g(t - m)^2$. Subtracting the latter from the former, we find the space described in the last m seconds to be $\frac{1}{2}g(2mt - m^2)$. When $m = 1$, this becomes for the space in the last second $\frac{1}{2}g(2t - 1)$. This is the same form of expression as was found above, where n was a whole number of seconds. Therefore, the space described in any *one* second of the fall, whether the time from the beginning is an integral or a fractional number, is found by multiplying $\frac{1}{2}g$ by twice the number of seconds minus one.

3. What space was described in the last two seconds by a body which had fallen 300 feet? *Ans.* 213.58 feet.

4. A body had been falling $8\frac{1}{2}$ seconds; how far did it descend in the next second? *Ans.* $289\frac{1}{2}$ ft.

32. Calculation for Projection Upward or Downward.—A body projected downward describes tv feet by the force of projection, and $\frac{1}{2}gt^2$ feet by the force of gravity (Art. 26). A body projected upward describes $t\,v$ by the force of projection; but this is diminished by $\frac{1}{2}\,g\,t^2$, which gravity would cause it to describe in the same time (Art. 27). Therefore the formula for space described by a body projected downward is $t\,v + \frac{1}{2}\,g\,t^2$; by a body projected upward, the formula is $t\,v - \frac{1}{2}\,g\,t^2$.

1. A body is projected downward with a velocity of 30 feet in a second; *how far* will it fall in 4 seconds? *Ans.* $377\frac{1}{3}$ ft.

2. A body is projected upward with a velocity of 120 feet in a second; *how far* will it rise in 3 seconds? *Ans.* $215\frac{1}{4}$ ft.

3. Suppose at the same instant that a body begins to fall from rest from the point D (Fig. 12), another body is projected upward from B with a velocity which would carry it to A; it is required to find the point where they would meet.

Fig. 12.

A

D

C

B

Let C be the point where the bodies would meet; and let $A\,B = a$, $B\,D = b$, $D\,C = x$; then will $A\,D = a - b$, $A\,C = a - b + x$.

Now the time of descending through $D\,C = \left(\frac{2\,x}{g}\right)^{\frac{1}{2}}$; and the time of ascending through $B\,C$ (= time down $A\,B$ − time down $A\,C$) $= \left(\frac{2\,a}{g}\right)^{\frac{1}{2}} - \left(\frac{2\,(a - b + x)}{g}\right)^{\frac{1}{2}}$; but the time down $D\,C$ must be equal to the time up $B\,C$; hence we have $\left(\frac{2\,x}{g}\right)^{\frac{1}{2}} = \left(\frac{2\,a}{g}\right)^{\frac{1}{2}} - \left(\frac{2\,(a - b + x)}{g}\right)^{\frac{1}{2}}$, or $x^{\frac{1}{2}} = a^{\frac{1}{2}} - (a - b + x)^{\frac{1}{2}}$;

$\therefore (a - b + x)^{\frac{1}{2}} = a^{\frac{1}{2}} - x^{\frac{1}{2}}$, and $a - b + x = a + x - 2\,(a\,x)^{\frac{1}{2}}$;

$$\therefore 2\,(a\,x)^{\frac{1}{2}} = b, \text{ or } 4\,a\,x = b^2, \text{ and } x = \frac{b^2}{4\,a}.$$

4. Suppose a body to have fallen from A to B (Fig. 13), when another body begins to fall from rest at D; *how far* will the latter body fall before it is overtaken by the former?

Fig. 13.

A

B

D

C

Let C be the point where one body overtakes the other, and let $A\,B = a$, $B\,D = b$, $D\,C = x$; then $A\,C = a + b + x$. Now time down $D\,C = \left(\frac{2\,x}{g}\right)^{\frac{1}{2}}$, and time down $B\,C$ = time down $A\,C$ − time down $A\,B = \left(\frac{2\,(a + b + x)}{g}\right)^{\frac{1}{2}} - \left(\frac{2\,a}{g}\right)^{\frac{1}{2}}$; but at the moment when the lower body is overtaken, time down $D\,C$ = time down $B\,C$, or

$$\left(\frac{2\,x}{g}\right)^{\frac{1}{2}} = \left(\frac{2\,(a + b + x)}{g}\right)^{\frac{1}{2}} - \left(\frac{2\,a}{g}\right)^{\frac{1}{2}};$$

$$\therefore x = \frac{b^2}{4\,a}.$$

33. Questions on Falling Bodies.—

1. The momentum of a meteoric stone at the instant of

striking the earth was estimated at 18435 foot-pounds, and it had been falling 10 seconds; from what height did it fall, and what was its weight? *Ans.* $1608\frac{1}{3}$ ft.; 57.31 lbs.

2. An archer wishing to know the height of a tower, found that an arrow sent to the top of it occupied 8 seconds in going and returning; what was the height of the tower? *Ans.* $257\frac{1}{3}$ ft.

3. In what time would a man fall from a balloon three miles high, and what velocity would he acquire?

Ans. $t = 31.38$ sec.; $v = 1009.39$ ft.

4. A body having fallen for $3\frac{1}{2}$ seconds, was afterwards observed to move with the velocity which it had acquired for $2\frac{1}{2}$ seconds more; what was the whole space described by the body?

Ans. $478\frac{1}{2}$ ft., very nearly.

5. Through what space would the aeronaut (in Question 3) fall during the last second? *Ans.* 993.3 feet.

6. A body has fallen from the top of a tower 340 feet high; what was the space described by it in the last three seconds?

Ans. 298.957 ft.

7. Suppose a body be projected downward with a velocity of 18 feet in a second; how far will it descend in 15 seconds?

Ans. $3888\frac{3}{4}$ ft.

8. A body is projected upward with a velocity of 65 feet in a second; how far will it rise in two seconds? *Ans.* $65\frac{2}{3}$ ft.

9. With what velocity must a stone be projected into a well 450 feet deep, that it may arrive at the bottom in four seconds?

Ans. $48\frac{1}{6}$ ft. in a second.

10. The space described in the fourth second of fall was to the space described in the last second except four, as 1 : 3; what was the whole space described by the body? *Ans.* $3618\frac{3}{4}$ ft.

11. A staging is at the height of 84 feet above the earth. A ball thrown upward from the earth, after an absence of 7 seconds, fell on the staging; what was the velocity of projection?

Ans. 124.58 ft. per second.

12. A body is projected upward with a velocity of 483 feet in a second; in what time will it rise to a height of 1610 feet?

Ans. $t = 3.82$ sec., or 26.2 sec.

13. From a point $214\frac{2}{3}$ feet above the earth a body is projected upward with a velocity of 161 feet in a second; in what time will it reach the surface of the earth, and with what velocity will it strike? *Ans.* $t = 11.2$ sec., $v = 199$ ft.

14. Suppose a body to have fallen through 50 feet, when a second body begins to fall just 100 feet below it; how far will the latter body fall before it is overtaken by the former?

Ans. 50 ft.

15. A body is projected upward with a velocity of $64\frac{1}{3}$ feet in a

second; how far above the point of projection will it be at the end of 4 seconds? *Ans.* 0 ft.

16. A body is projected upward with a velocity of $128\frac{2}{3}$ feet in a second; where will it be at the end of 10 seconds?

Ans. $321\frac{2}{3}$ ft. below the point of projection.

[See Appendix for the discussion of the fall of bodies by the Calculus.]

34. Atwood's Machine.—Accurate observations on the direct fall of a body cannot be easily made, on account of its great velocity; and if they could be, the relations between time, space, and acquired velocity, would not be found to agree with those obtained by calculation, on account of the resistance of the air. Experiments on falling bodies are usually performed by an instrument known as Atwood's machine, represented in Fig. 14. From the base of the instrument, which is furnished with leveling screws, rises a substantial pillar, about seven feet high, supporting a small table upon the top.

Fig. 14.

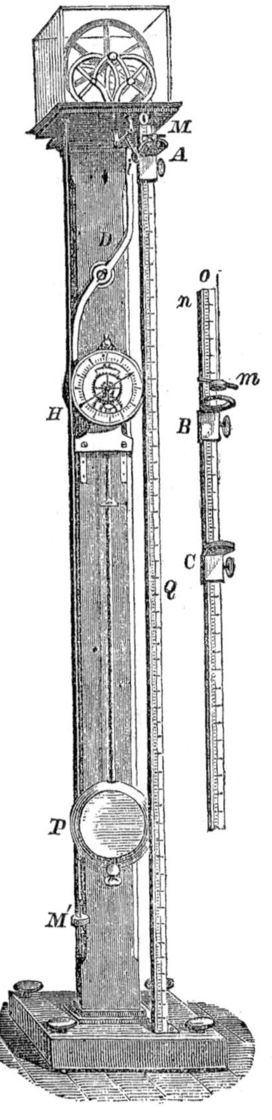

Above the table is a grooved wheel, delicately suspended on friction-wheels, and protected from dust by a glass case. Two equal poises, M and M', are attached to the ends of a fine cord, which passes over the groove of the wheel. As gravity exerts equal forces on M and M', they are in equilibrium. To set them in motion, a small bar m is placed on M, which will immediately begin to descend, and M' to rise. But this motion will be slower than in falling freely, because the force which gravity exerts on the bar must be communicated to the poises, and also to the revolving wheel over which the cord passes. By increasing the poises M, M', and diminishing the bar m, the motion may be made as slow as we please. H is a simple clock attached to the pillar for measuring seconds, and for dropping the poise M at the beginning of a vibration of the pendulum. Q is a scale of inches extending from the base to the table. The stage A may be clamped to any part of the scale, in order to stop the poise M in its descent, as represented at C. The ring B, which is large enough to allow the poise, but not the bar, to pass through it, is also clamped to the scale wherever the acceleration is to cease.

Let M be raised to the top, and held in place by a support, and then let the pendulum be set vibrating. When the index passes the zero point, the clock causes the support to drop away, and the poise descends. The pendulum shows how many seconds elapse before the bar is arrested by the ring, and how many more before the poise strikes the stage. From the top to the ring the motion is accelerated by the constant fraction of gravity acting on it; from the ring to the stage the poise moves uniformly with the acquired velocity. All the formulæ relating to the fall of a body can therefore be illustrated by these slow motions. Moreover, the resistance of the air is so much diminished when the motion is slow, that a good degree of correspondence is found to exist between the experiments and the results of calculation.

35. Living Force.—We have seen that when a body is projected upward in a vacuum, the height to which it will rise varies as the *square* of the velocity of projection. In the ascent, the body is constantly and uniformly opposed by gravity. If the motion of a given body were opposed by any other uniform obstruction, the distance it would proceed before coming to rest would also vary as the square of the velocity. This power to overcome a constant resistance varies, therefore, not as the momentum—that is, as the product of mass and velocity $q\ v$—but as the product of mass and *square* of velocity $q\ v^2$; and it is called the *vis viva* or *living force*, in distinction from *vis inertiæ* or *dead force*. If a ball weighing *one* pound move with the velocity of 2,000 feet, and another ball weighing *two* pounds move with the velocity of 1,000 feet, then the *momentum* ($q\ v$) of the first equals that of the second. But the *living force* ($q\ v^2$) of the first is twice as great as that of the second; for $1 \times \overline{2000}^2 : 2 \times \overline{1000}^2 :: 2 : 1$. The first body, in its ascent, will reach *four times* as great a height as the second; or, if the two balls be fired into a bank of earth, the first will penetrate four times as far as the second.

In practical mechanics, the living force is generally called the *working power;* and the *work* which it will perform is therefore measured by the *mass* multiplied by the *square* of the *velocity*. The working power of the steam in a locomotive is employed in maintaining a certain velocity in the train, in spite of grade, friction of rails and machinery, and resistance of the air. If the power of the steam were wholly cut off, the train would be uniformly retarded by the constant resistances, until, after running a certain distance, it would come to rest. But if the velocity of the train had been *twice* as great, it would have run *four* times as far before stopping; it would also have required *four* times as great a force to give the train this double velocity. Both of these facts, one relating to the effect, the other to the cause, show that the working power is to be estimated according to the square of the velocity. So the *working power* employed in moving any kind of machinery, which presents a constant resistance, varies as the square of the velocity imparted, and the *work* performed by the machinery is reckoned in the same way. To give a missile greater *velocity* is more advantageous than to increase its *mass*. A 40-pound ball, with 1400 feet velocity, is 7 times more efficient in penetrating the walls of forts and the hulks of ships than a 280-pound ball with 200 feet velocity, though the momentum is the same in each case.

36. Measure of Force.—In Art. 12 force is said to be measured by *momentum*, or $f \propto q\ v$; and in Art. 35 it is said to be measured by the *work* performed, or $f \propto q\ v^2$. But these statements are not to be considered as inconsistent with each other; for in the first case, force has reference to *inertia;* in the second case it has reference to *work*. When a force acts on a body that is free to move without obstruction (which is, however, only a supposable case), the effect is perpetual; the body will move on uniformly forever. If the force had been greater, the velocity would have been greater in the same ratio. But when resistances oppose (as is always true in practice), then the force is expended in overcoming them, and this is the *work* to be performed; and if the force ceases to operate, the motion will at length cease also; but, as has been shown, the space passed over, and therefore the work performed, will vary as the square of the velocity.

When force is employed to perform work, it is by some writers called *energy*, to distinguish it from force as used in producing momentum. Hence *energy* varies as the product of the *mass* and the *square of the velocity.*

CHAPTER III.

COMPOSITION AND RESOLUTION OF MOTION.

37. Motion by Two or More Forces.—Motion produced by a single force, either impulsive or continued, has been already considered. But motion is more generally caused by several forces acting in different directions.

When two or more forces act at once on a body, each force is called a *component*, and the joint effect is called the *resultant*. Forces may be represented by the *straight lines* along which they would move a body in a given time; the lines represent the forces in two particulars, the *directions* in which they act and their *relative magnitudes*. Whenever an arrow-head is placed on a line, it shows in which of the two directions along that line the force acts.

38. The Parallelogram of Forces.—This is the name given to the relation which exists between any two components and their resultant, and is stated as follows:

If two forces acting at once on a body are represented by the adjacent sides of a parallelogram, their resultant is expressed by the diagonal which passes through the intersection of those sides.

Suppose that a body situated at *A* (Fig. 15) receives an impulse which, acting alone, would carry it over *A B* in a given time, and another which would carry it over *A C* in the same length of time. If both impulses are given at the same instant, the body describes *A D* in the same time as *A B* by the first force, or *A C* by the second, and the motion in *A D* is uniform.

FIG. 15.

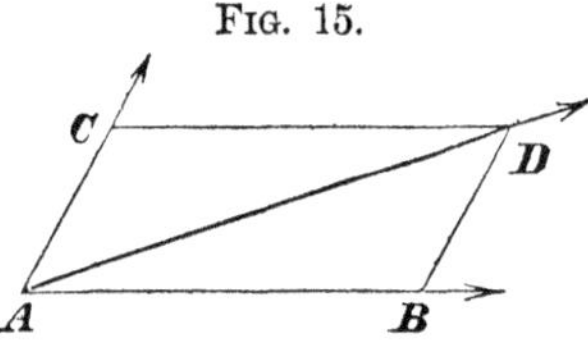

This is an instance of the coexistence of motions, stated in the second law of motion (Art 13). For the body, in passing directly from *A* to *D*, is making progress in the direction *A C* as rapidly as though the force *A B* did not exist; and at the same time it advances in the direction *A B* as fast as though that were the only force. When the body reaches *D*, it is as far from the line *A B* as if it had passed over *A C*; it is also as far from the line *A C* as if it had gone over *A B*. Thus it appears that both motions *A B* and *A C* fully coexist in the progress of the body along the diagonal *A D*. That the motion is uniform in the diagonal is evident from the law of inertia; for the body is not acted on after it leaves *A*.

It is evident that a single force might produce the same effect; that force would be represented, both in direction and magnitude, by the line *A D*. The force *A D* is said to be equivalent to the two forces *A B* and *A C*.

39. Velocities Represented.—The lines *A B* and *A C* are described by the components separately, and the line *A D* by their joint action, *in the same length of time*. Hence the *velocities* in those lines are as the lines themselves. In the parallelogram of forces, therefore, two adjacent sides and the diagonal between them represent—

1st. The *directions* of the components and resultant;
2d. Their relative *magnitudes;* and
3d. The relative *velocities* with which the lines are described.

40. The Triangle of Forces.—For purposes of calculation, it is more convenient to represent two components and their resultant by the sides of a triangle, than by the sides and diagonal of a parallelogram. In Fig. 15, *C D*, which is equal and parallel to *A B*, may represent in direction and magnitude the same force which *A B* represents. Therefore, the components are *A C* and *C D*, while their resultant is *A D*; and the angle *C* in the triangle is the supplement of *C A B*, the angle between the components. Care should be taken to construct the triangle so that the sides representing the components may be taken in succession in the directions of the forces, as, *A C*, *C D*; then *A D* correctly represents their resultant. But, although *A C* and *A B* represent the components, the third side, *C B*, of the triangle *A C B*, does not represent their resultant, since *A C* and *A B* cannot be taken successively in the direction of the forces. It is necessary to go back to *A* in order to trace the line *A B*. It should be observed, that though *C D* represents the *magnitude* and *direction* of the component, it is not in the *line* of its action, because both forces act through the same point *A*.

41. The Forces Represented Trigonometrically.—Since the sides of a triangle are proportional to the sines of the opposite angles, these sines may also represent two components and their resultant. Thus, the sine of *C A D* corresponds to the component *A B* (= *C D*); the sine of *C D A* (= *D A B*) corresponds to the component *A C*; and the sine of *C* (= sine of *C A B*) corresponds to the resultant *A D*. Each of the three forces is therefore represented by the sine of the angle between the other two.

42. Greatest and Least Values of the Resultant.—A change in the angle between the components alters the value of the resultant; as the angle increases from 0° to 180°, the resultant diminishes from the *sum* of the components to their *difference*. In Fig. 16, let *C A B* and *D A B* be two different angles between the same components *A C* (or *A D*) and *A B*. As *C A B* is less than *D A B*, its supplement *A B F* is greater than *A B E*, the supplement of *D A B*; therefore *A F* is greater than *A E*. When the angle *C A B* is dimin-

FIG. 16.

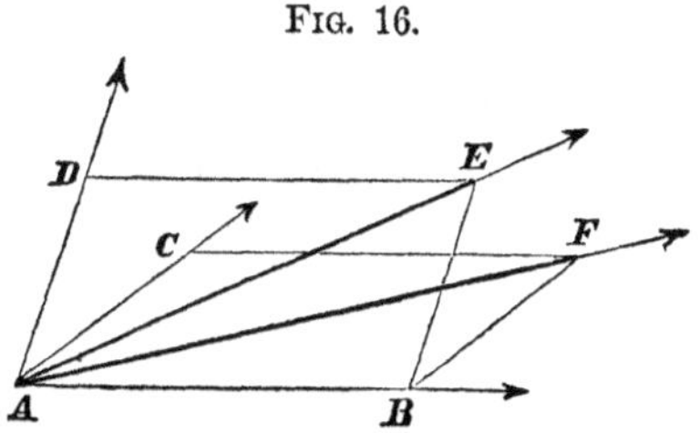

ished to 0°, the sides *A B*, *B F*, become one straight line, and *A F* equals their *sum;* when *D A B* is enlarged to 180°, *E* falls on *A B*, and *A E* equals the *difference* of *A B* and *A C*. Between the sum and difference of the components, the resultant may have all possible values.

43. The Polygon of Forces.—All the sides of a polygon except one may represent so many forces acting at the same time on a body, and the remaining side will represent their resultant. In Fig. 17, suppose *A B*, *A C*, and *A D*, to represent three forces acting together on a body at *A*. The resultant of *A B* and *A C* is represented by the diagonal *A E*; and the resultant of *A E* and *A D* by the diagonal *A F*. As *A F* is equivalent to *A E* and *A D*, and *A E* is equivalent to *A B* and *A C*, therefore *A F* is equivalent to the three, *A B*, *A C*, and *A D*. But if we substitute *B E* for *A C*, and *E F* for *A D*, then the three components are *A B*, *B E*, and *E F*, three sides of a polygon, and the resultant *A F* is the fourth side of the same polygon.

Fig. 17.

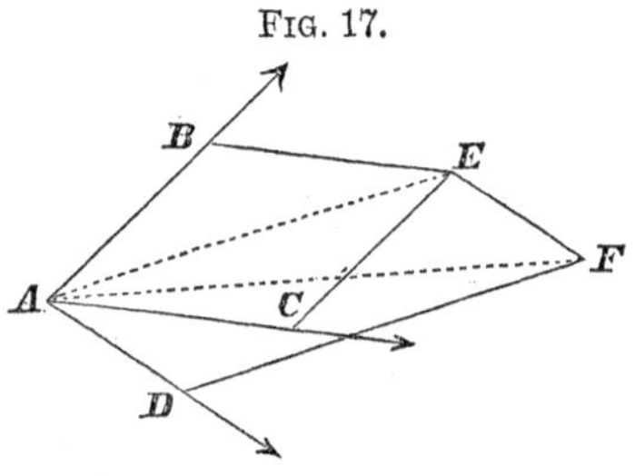

So, in Fig. 18, *A B*, *B C*, *C D*, *D E*, and *E F*, may represent the directions and relative magnitudes of five forces, which act simultaneously on a body at *A*. The resultant of *A B* and *B C* is *A C*; the resultant of *A C* and *C D* is *A D*; the resultant of *A D* and *D E* is *A E*; and the resultant of *A E* and *E F* is *A F*; which last is therefore the resultant of all the forces, *A B*, *B C*, *C D*, *D E*, and *E F*; the components being represented by five sides, and their resultant by the sixth side, of a polygon of six sides.

Fig. 18.

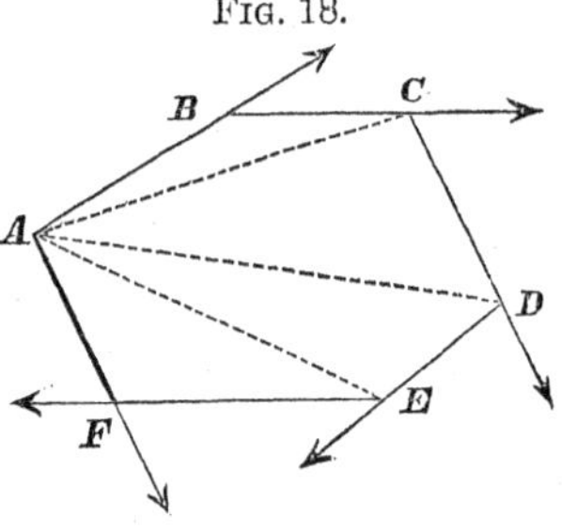

44. Curvilinear Motion.—Since, according to the first law of motion, a moving body proceeds in a straight line, if no force disturbs it, whenever we find a body describing a *curve*, it is certain that some force is continually deflecting it from a straight line. Besides the original impulse, therefore, which gave it motion in one direction, it is subject to the action of a continued force, which operates in another direction. A familiar example occurs in the path of a projectile. Suppose a body to be thrown from P (Fig. 19), with an impulse which would alone carry it to N, in the same time in which gravity alone would carry it to V. Complete the parallelogram $P\ Q$; then, as both motions coexist (2d law), the body at the end of the time will be found at Q. Let t be the time of describing $P\ N$ or $P\ V$; and let t' be the time of describing $P\ M$ by the impulse, or $P\ L$ by gravity. Then, at the end of the time t', the body will be at O. Now, as $P\ N$ is described uniformly, $P\ N : P\ M :: t : t'$; $\therefore P\ N^2 : P\ M^2 :: t^2 : t'^2$.

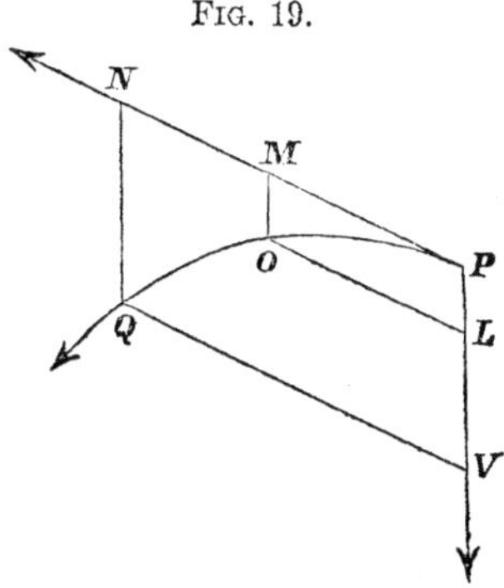

Fig. 19.

But (Art. 23), $P\ V : P\ L :: t^2 : t'^2$;
$\therefore P\ V : P\ L :: P\ N^2 : P\ M^2$; or $Q\ V^2 : O\ L^2$.

Hence, the curve is such that $P\ V \propto Q\ V^2$; that is, the abscissa varies as the square of the ordinate, which is a property of the parabola. $P\ O\ Q$ is therefore a parabola, one of whose diameters is $P\ V$, and the parameter to that diameter is $\frac{Q\ V^2}{P\ V}$

Owing to the resistance of the air, the curve deviates sensibly from a parabola, especially in swift motions.

45. Calculation of the Resultant of Two Impulsive Forces.—When two components and the angle between them are given, the resultant may be found both in direction and magnitude by trigonometry. The theorem required is that for solving a triangle, when two sides and the included angle are given; but the included angle is not that between the components, but its supplement (Art. 40). In Fig. 15, if $A\ B = 54$, and $A\ C = 22$, and $C\ A\ B = 75°$, then $A\ C\ D$ is the triangle for solution, in which $A\ C = 22$, $C\ D = 54$, and $A\ C\ D = 105°$. Performing the calculation, we find the resultant $A\ D = 63.363$, and the angle $D\ A\ B$, which it makes with the greater force, $= 19°\ 35'\ 43''$. This method will apply in all cases.

1. A foot-ball received two blows at the same instant, one directly east, at the rate of 71 feet per second, the other exactly northwest, at the rate of 48 feet per second; in what direction and with what velocity did it move? *Ans.* N. 47° 30′ 52″ E. Vel. = 50.253.

The process is of course abridged, if the forces act at a right angle with each other, as in the following example:

2. A balloon rises 1120 feet in one minute, and in the same time is borne by the wind 370 feet; what angle does its path make with the vertical, and what is its velocity per second?

Ans. 18° 16′ 53″; $v = 19.659$.

In the next example, one component and the angle which each component makes with the resultant, are given to find the resultant and the other component.

3. From an island in the Straits of Sunda, we sailed S. E. by S. (33° 45′) at the rate of 6 miles an hour; and being carried by a current, which was running toward the S. W. (making an angle with the meridian of 64° 12¼′), at the end of four hours we came to anchor on the coast of Java, and found the said island bearing due north; required *the length of the line* actually described by the ship, and *the velocity of the current?*

Ans. $s = 26.4$ miles.
$v = 3.7024$ miles per hour.

If the magnitudes and directions of any number of forces are given, the resultant of them all is obtained by a repetition of the same process as for two. In Fig. 18, first calculate $A\ C$, and the angle $A\ C\ B$, by means of $A\ B$, $B\ C$, and the angle B. Subtracting $A\ C\ B$ from $B\ C\ D$, we have the same data in the next triangle, to calculate $A\ D$, and thus proceed to the final resultant, $A\ F$.

As it is immaterial in what order the components are introduced into the calculation, it will diminish labor, to find first the resultant of any two *equal* components, or any two which make a *right angle* with each other; since it can be done by the solution of an isosceles, or a right-angled triangle.

Fig. 20.

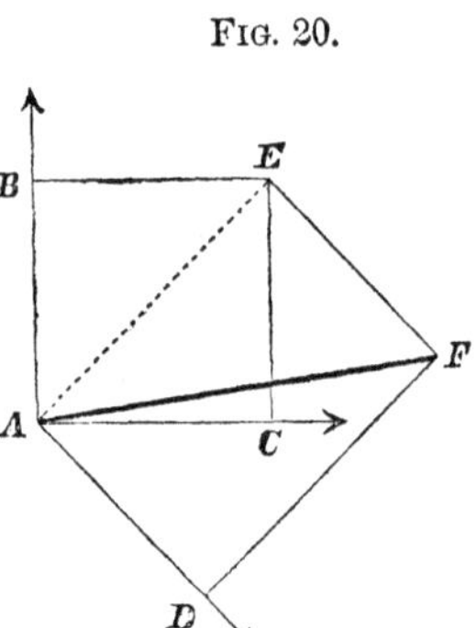

4. The particle A (Fig. 20) is urged by three equal forces $A\ B$, $A\ C$, and $A\ D$; the angle $B\ A\ C = 90°$, and $C\ A\ D = 45°$; what is the direction of the resultant, and how many times $A\ B$?

Ans. $B\ A\ F = 80°\ 16'$, and
$A\ F : A\ B :: \sqrt{3} : 1$.

5. Five sailors raise a weight by means of five separate ropes, in the same plane, connected with the main rope that is fastened to the weight in the manner represented in Fig. 22. B pulls at an angle with A of 20°; C with B, 19°; D with C, 21° 30′; and E with D, 25°. A, B, and C, pull with equal forces, and D and E with forces one-half greater; required the magnitude and direction of the resultant.

Ans. Its angle with A is 46° 33′ 10″. Its magnitude is 5.1957 times the force of A.

If the polygon $O\ A\ B\ C\ D\ E$ (Fig. 21) be constructed for the above case, $A\ C$ and $D\ F$ are easily calculated in the isosceles triangles $A\ B\ C$ and $D\ E\ F$, after which $A\ D$ and then $A\ F$ are to be obtained by the general theorem.

Fig. 21. Fig. 22.

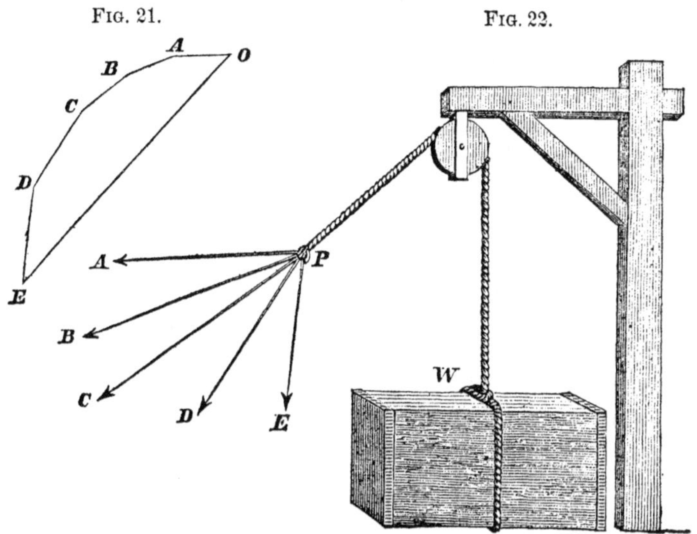

46. The Resultant and all Components, except one, being given, to Find that one Component.—If $A\ B$ (Fig. 23) is the resultant to be produced, and there already exists the force $A\ C$, a second force can be found, which acting jointly with $A\ C$, will produce the motion required. Join $C\ B$, and draw $A\ D$ equal and parallel to it, then $A\ D$ is the force required; for $A\ B$ is equivalent to $A\ C$ and $C\ B$. Therefore $C\ B$ has the magnitude and direction of the required force; $A\ D$ is the line in which it must act.

Fig. 23.

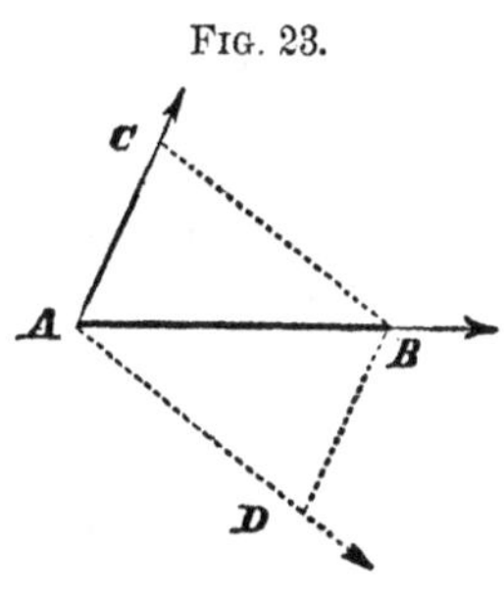

Again, suppose that *several* forces act on A, and it is required to find the force which, in conjunction with them all, shall produce the resultant $A\ B$. Let the several forces be combined into one resultant, and let $A\ C$ represent that resultant. Then $A\ D$ may be found as before.

The trigonometrical process for finding a component is essentially the same as for finding a resultant.

1. A ferry-boat crosses a river $\frac{3}{4}$ of a mile broad in 45 minutes, the current running all the way at the rate of 3 miles an hour; at what angle with the direct course must the boat head up the stream in order to move perpendicularly across? *Ans.* 71° 34'.

2. A sloop is bound from the mainland of Africa to an island bearing W. by N. (78° 45′) distant 76 miles, a current setting N. N. W. (22° 30′) 3 miles an hour; what is the *course* to arrive at the island in the shortest time, supposing the sloop to sail at the rate of 6 knots per hour; and what *time* will she take?

Ans. Course, S. 76° 41′ 4″ W. Time, 10 h. 40m. 7 sec.

3. The resultant of two forces is 10; one of them is 8, and the direction of the other is inclined to the resultant at an angle of 36°. Find the angle between the two forces.

Ans. 47° 17′ 5″ or 132° 42′ 55″.

4. A ball receives two impulses: one of which would carry it N. 27 feet per second; the other, E. 30° N. with the same velocity; what third impulse must be conjoined with them, to make the ball go E. with a velocity of 21 feet? *Ans.* S. 3° 22′ W. $v = 40.57$.

47. Resolution of Motion.—In the *composition* of motions or forces, the resultant of any given components is found; in the *resolution* of motion or force, the process is reversed; the resultant being given, the components are found, which are equivalent to that resultant.

If it be required to find what two components can produce the resultant $A\,B$ (Fig. 24), we have only to construct on $A\,B$, as a base, any triangle whatever, as $A\,B\,C$ or $A\,B\,D$ (Art. 40); then, if $A\,C$ is one component, the other is $A\,F$, equal and parallel to $C\,B$; or if $A\,D$ is one, the other is $A\,E$, equal and parallel to $D\,B$; and so for any triangle whatever on the base $A\,B$. The number of pairs is therefore infinite, whose resultant in each case is $A\,B$.

Fig. 24.

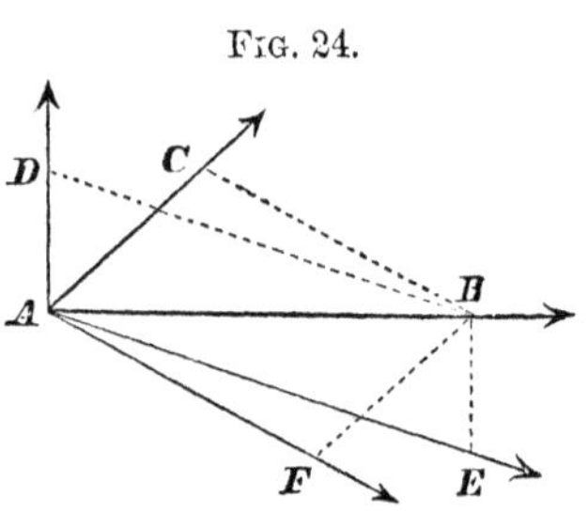

The *directions* of the components may be chosen at pleasure, provided the sum of the angles made with $A\,B$ is less than two right angles.

The *magnitude* and *direction* of one component may be fixed at pleasure.

The *magnitudes* of both components may be what we please, provided their difference is not greater, and their sum not less, than the given resultant.

These conditions are obvious from the properties of the triangle.

When a given force has been resolved into two others, each of those may again be resolved into two, each of those into two others still, and so on. Hence it appears that a given force may be resolved into any number of components whatever, with such limitations as to direction and magnitude as accord with the foregoing statements.

1. A motion of 153 toward the north is produced by the forces 100 and 125; how are they inclined to the meridian?

Ans. 54° 28′ and 40° 37′ 7″.

2. A resultant of 617 divides the angle between its components into 28° and 74°; what are the components?

Ans. 606.34 and 296.14.

48. Resolution into Pairs, with Certain Conditions.— In some cases, in which a condition is imposed, a simple construction will enable us to find the pairs of components which fulfill that condition.

1st. A given force is to be resolved into pairs which make a *given angle* with each other.

Let $A\ B$ (Fig. 25) be the given force. On $A\ B$ as a chord, construct the segment of a circle $A\ D\ B$, containing an angle equal to the supplement of the given angle. Then all the possible pairs of components fulfilling the condition will be found by drawing lines from A and B to points of the curve, as $A\ D$, $D\ B$, and $A\ C$, $C\ B$, &c.

FIG. 25.

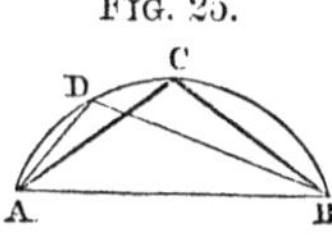

The segment must contain, not the given angle itself, but its supplement, because the given angle is at A, between $A\ D$ and a parallel to $D\ B$, or $C\ B$, &c.

If the given angle were a right angle, the segment to be constructed is a semicircle.

2d. To resolve a given force into two components, making a *given sum;* the sum must not be less than the given force (Art. 47).

Let $A\ B$ (Fig. 26) be the given force, and $M\ N$ the given sum. Having placed $A\ B$ on $M\ N$, so that $A\ N = B\ M$, construct a semi-ellipse on $M\ N$ as a transverse axis, with A and B for the foci. Lines drawn from A and B to any point of the curve will represent a pair of the required components; for, by a property of the ellipse, $A\ D + D\ B$, or $A\ C + C\ B = M\ N$.

FIG. 26.

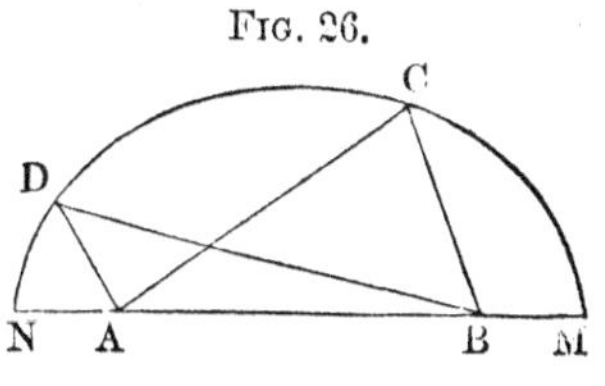

3d. To resolve a given force into two components, having a *given difference;* the difference must not be greater than the given force (Art. 47).

Let $A\ B$ (Fig. 27) be the given force, and $M\ N$ the given difference. Place $M\ N$ on $A\ B$ so that $A\ N = B\ M$, and construct the hyperbola $M C$, $N D$, having $M N$ for its transverse axis, with A and B for foci. Then $A D$, $D B$, or $A C$, $C B$, or any other lines from the foci to a point of the curve, will fulfill the condition required, because their difference equals the transverse axis.

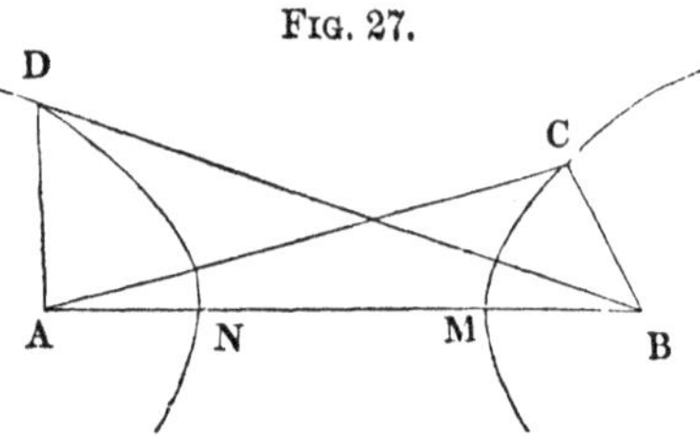

FIG. 27.

1. It is required to resolve the force 194 into pairs of components acting at an angle of 135° with each other; what is the radius of the circle whose segment is employed in the construction? Let $A B = 194$ (Fig. 28), and $A D B = 45°$; then $A C B = 90°$. Let $C H$ be perpendicular to $A B$; then $A B : A C :: A C : \frac{A B}{2}$, and

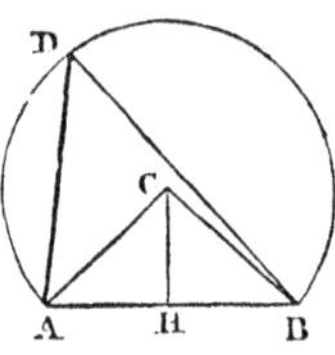

FIG. 28.

$$A C = \frac{A B}{\sqrt{2}} = 137.18.$$

2. To find the radius of the circle whose segment includes the components of the force a acting at *any* given angle with each other. Make $A B = a$, and let $A D B = A$, the supplement of the given angle. Then $A C B = 2 A$, and $A C H = A$; therefore,

$$\sin A : 1 :: \frac{a}{2} : A C = \frac{a}{2 \sin A}.$$

49. Resolution of a Force, to Find its Efficiency in a Given Direction.—By the resolution of a force into two others acting at right angles with each other, it is ascertained how much efficiency it exerts to produce motion in any given direction. For example, a weight W (Fig. 29), lying on a horizontal plane, and

FIG. 29.

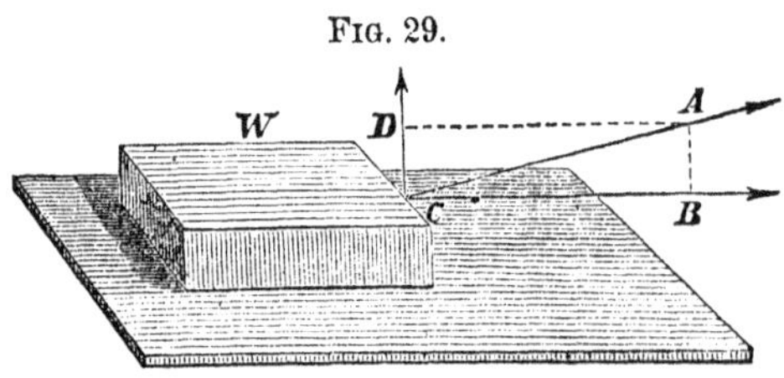

pulled by the oblique force $C A$, is prevented by gravity from moving in the line $C A$, and is compelled to remain on the plane. Resolve $C A$ into $C B$, *in* the plane, and $C D$ *perpendicular* to it; then the former represents the component which is efficient to cause motion along the plane; the latter has no influence to aid or hinder that motion; it simply diminishes pressure upon the plane. In like manner, if $A\ C$ is an oblique force, *pushing* the weight, its horizontal component, $B\ C$, is alone efficient to move it; the other, $A\ B$, merely increasing the pressure. In either case, the whole force is to that component which is efficient to move the body along the plane, *as radius to the cosine of inclination*. Also, the whole force is to that component which increases or diminishes pressure on the plane, *as radius to the sine of inclination*.

If only 88 per cent. of the strength of a horse is efficient in moving a boat along a canal, what angle does the rope make with the line of the tow-path? *Ans.* $28° \ 21' \ 27''$.

50. Resultant found by means of Rectangular Axes.—When several forces act in one plane upon a body, their resultant may be conveniently found by the use of right-angled triangles alone. Select at pleasure two lines at right angles to each other, both of them lying in the plane of the forces, and passing through the point at which the forces are applied. These lines are called *axes*. The following example illustrates their use:

Let $P\ A$, $P\ B$, $P\ C$, $P\ D$, $P\ E$ (Fig. 30) represent the forces in Question 5 (Art. 45). Let one axis, for convenience, be chosen in the direction $P\ A$, and let $P\ H$ be drawn at right angles to it for the other axis. These axes are supposed to be of indefinite length. Then proceed as in Art. 49, to resolve each force into two components on these axes. As $P\ A$ acts in the direction of one axis, it does not need to be resolved. To resolve $P\ B$, say

Fig. 30.

$$R : \cos 20° :: P\ B : P\ b, \text{ and}$$
$$R : \sin 20° :: P\ B : P\ b';$$

again,

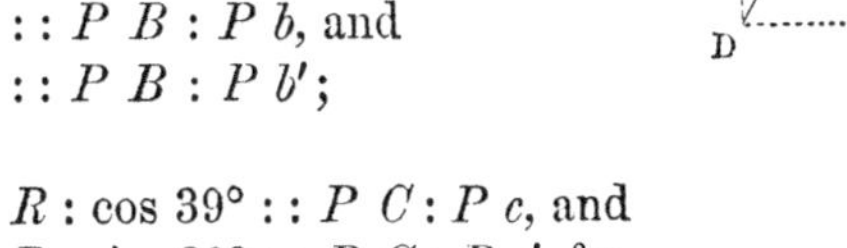

$$R : \cos 39° :: P\ C : P\ c, \text{ and}$$
$$R : \sin 39° :: P\ C : P\ c', \&c.$$

Suppose $P\ A$ produced so as to equal $P\ A + P\ b + P\ c + P\ d + P\ e = M$, and $P\ H$ produced so as to equal $P\ b' + P\ c' + P\ d' + P\ e' = N$. Now, as M acts in the line $P\ A$, and N at right angles to it, their resultant and the angle which it makes with $P\ A$ are found by the solution of another right-angled triangle. The resultant is 5.1957, and the angle is $46° \ 33' \ 10''$, as in Art. 45.

If any components of the resolved forces are opposite to $P\ A$ or $P\ H$, they are reckoned as negative quantities.

3

51. Analytical Expression for the Resultant.—Put $A\ C$ (Fig. 31) $= P$, $A\ B = P'$, $A\ D = R$, angle $C\ A\ B = a$; then by Trig.

$$\overline{A\ D}^2 = \overline{A\ C}^2 + \overline{C\ D}^2 - 2\ A\ C \times C\ D \cos A\ C\ D, \text{ or}$$

$$R^2 = P^2 + P'^2 + 2\ P\ P' \cos a; \text{ whence}$$

$$R = \sqrt{P^2 + P'^2 + 2\ P\ P' \cos a} \ldots (1). \text{ Hence}$$

The resultant of any two forces, acting at the same point, is equal to the square root of the sum of the squares of the two forces, plus twice the product of the forces into the cosine of the included angle.

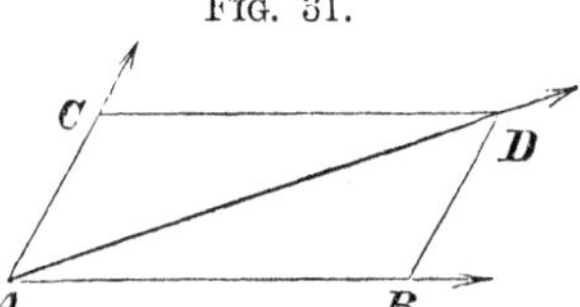

FIG. 31.

If $a = 0$, its cosine will be 1, and (1) becomes

$$R = P + P'.$$

If $a = 90°$, its cosine will be 0, and we shall have

$$R = \sqrt{P^2 + P'^2}.$$

If $a = 180°$, its cosine will be $-$ 1, and we shall have

$$R = P - P'.$$

1. Two forces, P and P', are equal in intensity to 24 and 30, respectively, and the angle between them is 105°; what is the intensity of their resultant? *Ans.* 33.21.

2. Two forces, P and P', whose intensities are, respectively, equal to 5 and 12, have a resultant whose intensity is 13; required the angle between them. *Ans.* 90°.

3. A boat is impelled by the current at the rate of 4 miles per hour, and by the wind at the rate of 7 miles per hour; what will be her rate per hour when the direction of the wind makes an angle of 45° with that of the current? *Ans.* 10.2 miles.

4. Two forces and their resultant are all equal; what is the value of the angle between the two forces? *Ans.* 120°.

52. Principle of Moments.—The *moment* of a force, with respect to a point, is the product of the force into the perpendicular let fall from the point to the line of direction of the force.

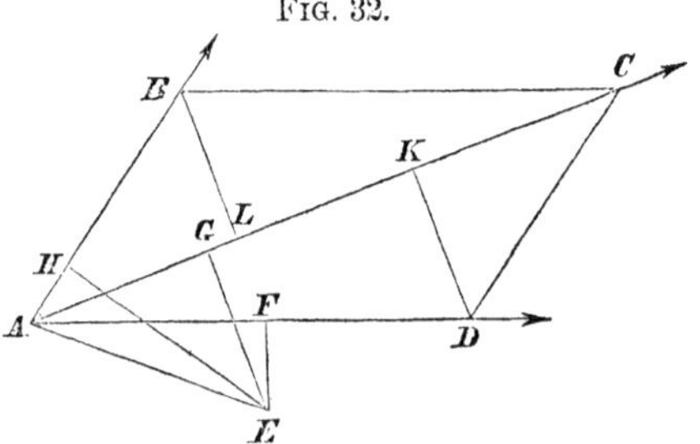

FIG. 32.

The fixed point is called the centre of *moments;* the perpendicular distance, the *lever-arm of the force;* and the *moment* measures the tendency of the force to produce rotation about the centre of moments. Denote the two forces $A\ D$, $A\ B$, and their resultant $A\ C$ (Fig. 32) by P, P', and R, respectively. From E, any point in the plane of the forces, let fall, upon the directions of the forces, the perpendiculars $E\ F$, $E\ H$, $E\ G$. Represent these perpendiculars by p, p', r. Draw $D\ K$ and $B\ L$ perpendicular to $A\ C$. Put $\alpha = C\ A\ D$, $\beta = C\ A\ B$, $\theta = C\ A\ E$. Then

$$R = A\ L + C\ L = P' \cos\beta + P \cos\alpha \ldots (1);$$

$$D\ K = P' \sin\beta,\ B\ L = P \sin\alpha;\ \therefore,\ \text{since } D\ K = B\ L,$$

$$P' \sin\beta = P \sin\alpha,\ \text{or},\ 0 = P' \sin\beta - P \sin\alpha \ldots (2).$$

Multiplying both members of (1) by $\sin\theta$, both members of (2) by $\cos\theta$, adding and reducing, we have

$$R \sin\theta = P' \sin(\theta + \beta) + P \sin(\theta - \alpha) \ldots (3).$$

$$\text{But } \sin\theta = \frac{r}{A\ E},\ \sin(\theta + \beta) = \frac{p'}{A\ E},\ \sin(\theta - \alpha) = \frac{p}{A\ E};$$

$\therefore$ (3) reduces to

$$R\ r = P'\ p' + P\ p \ldots (4).$$

If the point E falls within the angle $C\ A\ D$, $\sin(\theta - \alpha)$ becomes negative, and (3) becomes

$$R\ r = P'\ p' - P\ p \ldots (5).$$

Hence, *the moment of the resultant of two forces is equal to the algebraic sum of the moments of the forces taken separately.*

53. Forces Acting at Different Points. Parallel Forces.—We have thus far considered forces acting upon a single particle, or upon one point of a body. If, however, two forces P and P', in the same plane, act upon A and B, two different points of a rigid body, they may still have a resultant.

FIG. 33.

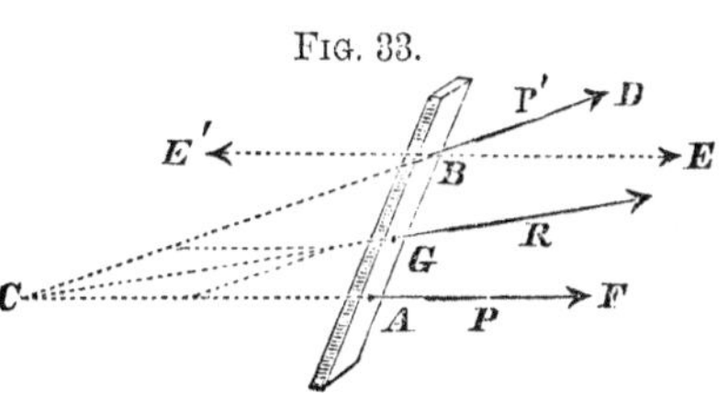

Let the lines of directions of the two forces $A\ F$ and $B\ D$ (Fig. 33) be produced to meet in C. The two forces may then be considered as acting at C, and thus compounded into a single force at that point, or at the point G of the body.

By (1) of the last article this resultant is

$$R = P' \cos\beta + P \cos\alpha \ldots (1).$$

When the forces become parallel, as $A\ F$ and $B\ E$, $\beta = 0$, and $\alpha = 0$, and (1) becomes

$$R = P' + P \ldots (2).$$

If the parallel forces act in opposite directions, as $A\ F$ and $B\ E'$, then $\alpha = 180°$, and $\beta = 0$, and (1) becomes

$$R = P' - P \ldots (3).\ \text{ Hence,}$$

The resultant of two parallel forces is in a direction parallel to them and equal to their algebraic sum.

54. Point of Application of the Resultant.—Let P and P' (Figs. 34, 35) be two parallel forces acting in the same or in

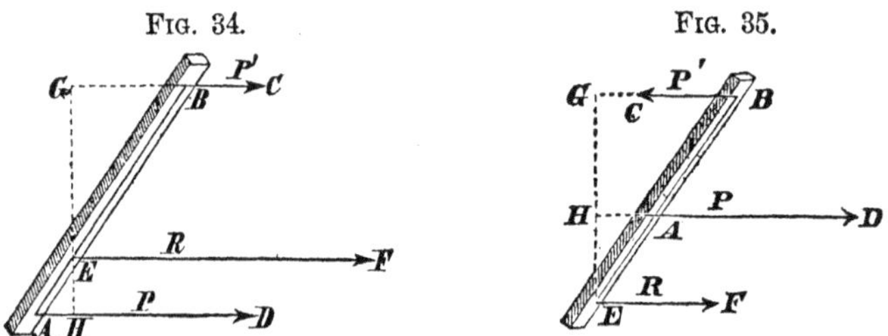

opposite directions, and let E be the point of application of the resultant. Assume this point as a centre of moments; then from (5) of Art. 52, since $r = 0$,

$P \times H\,E = P' \times G\,E$, or, in the form of a proportion,

$P' : P :: H\,E : G\,E$. But by similar triangles,

$H\,E : G\,E :: A\,E : E\,B$; $\therefore$

$P' : P :: A\,E : E\,B$.

That is, *the line of direction of the resultant of two parallel forces divides the line joining the points of application of the components, inversely as the components.*

By composition (Fig. 34) and division (Fig. 35) we obtain

$P' + P : P :: A\,B : E\,B$, and

$P - P' : P :: A\,B : E\,B$.

That is, *if a straight line be drawn to meet the lines of two parallel forces and their resultant, each of the three forces will be proportional to that part of the line contained between the other two.*

When the forces act in the same direction, we have

$E\,B = \frac{P \times A\,B}{P' + P}$, and when they act in opposite directions,

$E\,B = \frac{P \times A\,B}{P - P'}$.

If, in the last case, $P = P'$, then $E\,B$ will be infinite. The two forces in this case constitute what is called a *couple.* Their effect is to produce rotation about a point between them.

Any number of parallel forces may be reduced to a single force (or to a couple) by first finding the resultant of two forces, then the resultant of that and a third force, and so on to the last. And any single force may be resolved into two or any number of parallel forces by a method the reverse of this.

55. The Parallelopiped of Forces.—Hitherto forces have been considered as acting in the same plane. But if forces act in different planes, the solution of every case may be reduced to the following principle, called the *parallelopiped of forces*.

Any three forces acting in different planes upon a body may be represented by the adjacent edges of a parallelopiped, and their resultant by the diagonal which passes through the intersection of those edges.

Let *A C*, *A D*, and *A E* (Fig. 36), be three forces applied in different planes to the body at *A*. Construct the parallelopiped *C P*, having *A C*, *A D*, and *A E*, for its adjacent edges, and from *A* draw the diagonal *A B*. The section through the opposite edges *A C* and *P B* is a parallelogram, and therefore *A B* is the resultant of *A C* and *A P*, and *A P* is the resultant of *A D* and *A E*. Hence *A B* is the resultant of *A C*, *A D*, and *A E*.

FIG. 36.

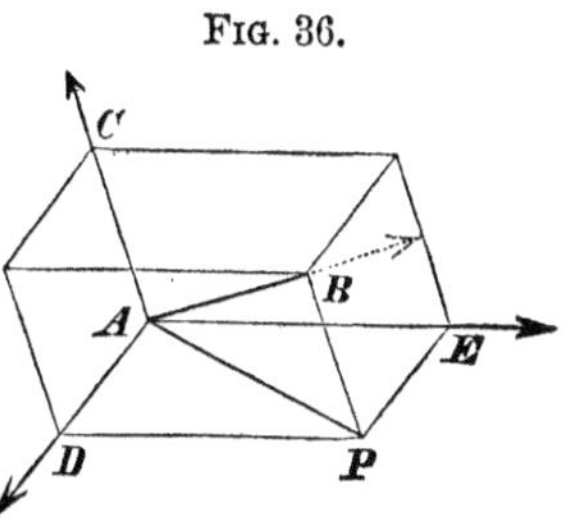

This process may obviously be reversed, and a given force may be resolved into three components in different planes along the edges of a parallelopiped, having such inclinations as we please.

56. Rectangular Axes. — The parallelopiped generally chosen is that whose sides are rectangles; the three adjacent edges of such a solid are called *rectangular axes*. All the forces which can possibly act on a body may be resolved into equivalent forces in the direction of three such axes. And since all forces which act in the direction of any one line may be reduced to a single force by taking their algebraic sum, therefore any number of forces acting through one point may be reduced to *three* in the direction of three axes chosen at pleasure.

Let *A X*, *A Y* (Fig. 37) be at right angles with each other,

FIG. 37.

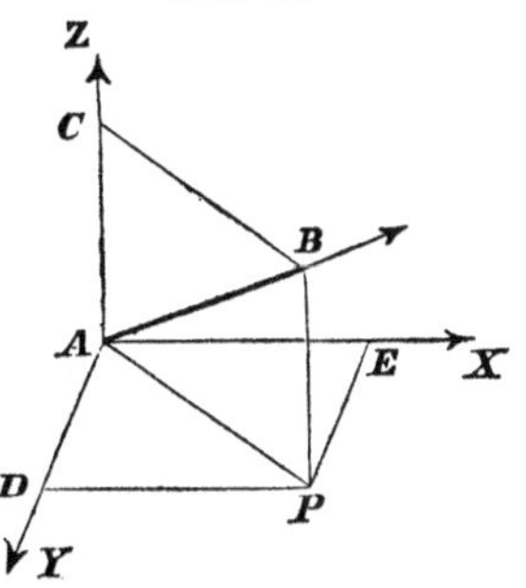

FIG. 38.

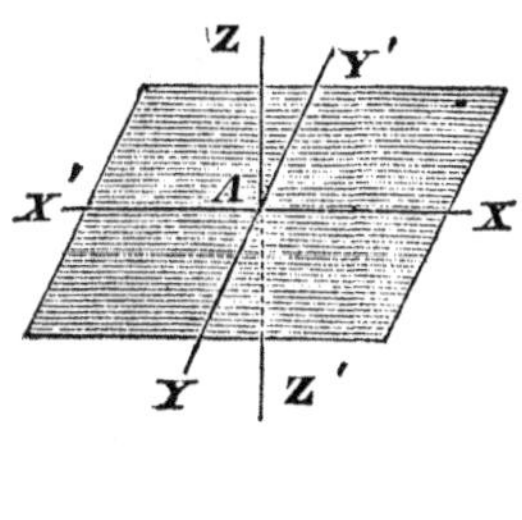

and *A Z* perpendicular to the plane of *A X* and *A Y*. Let *A B* represent a force acting on *A*. Resolve *A B* into *A C* on the axis

$A\,Z$, and $A\,P$ in the plane of $A\,X$, $A\,Y$; then resolve $A\,P$ into $A\,D$ and $A\,E$ on the other two axes. Therefore, $A\,C$, $A\,D$, and $A\,E$ are three rectangular forces, whose resultant is $A\,B$.

Let the axes $A\,X$, $A\,Y$, $A\,Z$, be produced indefinitely (Fig. 38) to X', Y', Z'; then their planes will divide the angular space about A into eight solid right angles, namely: A-$X\,Y\,Z$, A-$X\,Y'Z$, A-$X'\,Y'Z$, A-$X'Y\,Z$, above the plane of X and Y, and A-$X\,Y\,Z'$, A-$X\,Y'\,Z'$, A-$X'\,Y'Z'$, A-$X'\,Y\,Z'$ below it.

57. Geometrical Relation of Components and Resultant.—A force acting on the body A may be situated in any one of the eight angles, and its value may be expressed in terms of the squares of its three components. Let $A\,B$ (Fig. 39) be resolved as before into the rectangular components $A\,C$, $A\,D$, and $A\,E$. Then, by the right-angled triangles, we find

FIG. 39.

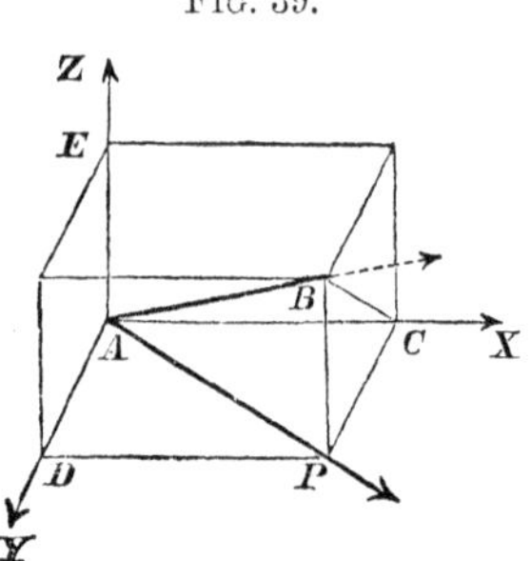

$A\,B^2 = B\,P^2 + A\,P^2 = A\,E^2 + A\,P^2$;

and

$A\,P^2 = A\,C^2 + C\,P^2 = A\,C^2 + A\,D^2$;

$$\therefore A\,B^2 = A\,C^2 + A\,D^2 + A\,E^2;$$

$$\text{and } A\,B = \sqrt{A\,C^2 + A\,D^2 + A\,E^2}.$$

If $A\,B$ is in the plane of X and Y, the component on the axis of Z becomes zero, and $A\,B = \sqrt{A\,C^2 + A\,D^2}$, and similarly for the other planes.

58. Trigonometrical Relation of Components and Resultant.—Let the angles which $A\,B$ makes with the axes of X, Y, Z, respectively, be a, β, γ; that is, $B\,A\,C = a$, $B\,A\,D = \beta$, $B\,A\,E = \gamma$. In the triangle $A\,B\,C$, right-angled at C, we have $A\,B : A\,C ::$ rad : cos a; therefore, making rad $= 1$,

$$A\,C = A\,B\,.\cos a.$$

In like manner, $A\,D = A\,B\,.\cos\beta$;

$$\text{and } A\,E = A\,B\,.\cos\gamma.$$

And since $A\,B$ is the resultant of the forces $A\,C$, $A\,D$, and $A\,E$, it is the resultant of $A\,B\,.\cos a$, $A\,B\,.\cos\beta$, $A\,B\,.\cos\gamma$. In general, the components of any force P, when resolved upon three rectangular axes, are $P\,.\cos a$, $P\,.\cos\beta$, $P\,.\cos\gamma$.

59. Any Number of Forces Reduced to Three on Three Rectangular Axes.—Suppose the body at A to be acted upon by a second force P', whose direction makes with the axes the angles a', β', γ'; then, as before, P' is the resultant of $P'.\cos a'$, $P'.\cos\beta'$, $P'.\cos\gamma'$; and a third force P'', in like manner, has for its components $P''.\cos\alpha''$, $P''.\cos\beta''$, $P''.\cos\gamma''$; and so of any number of forces.

Now, all the components on one axis may be reduced to one force by adding them together. Hence, the whole force in the axis of $X = P.\cos\alpha + P'.\cos\alpha' + P''.\cos\alpha'' + P'''.\cos\alpha''' +$ &c.; the whole in the axis of Y,

$$= P.\cos\beta + P'.\cos\beta' + P''.\cos\beta'' + P'''.\cos\beta''' + \&c.;$$

and that in the axis of Z,

$$= P.\cos\gamma + P'.\cos\gamma' + P''.\cos\gamma'' + P'''.\cos\gamma''' + \&c.$$

If any component acts in a direction opposite to others in the same axis, it is affected by a contrary sign, so that the force in the direction of any axis is the algebraic sum of all the individual forces in that axis.

If the sum of the components in *one* axis is reduced to zero by contrary signs, the effect of all the forces is limited to the plane of the other axes, and is to be obtained as in Art. 50, where two axes were employed. If the sum of the components on each of *two* axes is reduced to zero, then the whole force is exerted in the direction of the remaining axis, and is therefore perpendicular to the plane of the other two.

60. Equilibrium of Forces.—

1. *Two forces produce equilibrium when they are equal and act in opposite directions.*

It was shown (Art. 42) that two forces produce the least resultant when they act at an angle of 180° with each other, and that the resultant then equals the *difference* of the forces. If the forces are equal, their difference is zero, and the resultant vanishes; that is, the two forces produce equilibrium.

2. *Three forces produce equilibrium when they may be represented in direction and magnitude by the three sides of a triangle taken in order.*

For, when three forces are in equilibrium, one of them must be equal to, and opposite to, the resultant of both the others. But the forces $A\ C$ and $A\ B$ (Fig. 40) produce the resultant $A\ D$; therefore the equal and opposite force $D\ A$, since it is in equilibrium with $A\ D$, is also in equilibrium with $A\ C$ and $A\ B$, or $A\ C$ and $C\ D$. Hence the three forces $A\ C$, $C\ D$, and $D\ A$, taken in order around the figure, produce equilibrium.

Fig. 40.

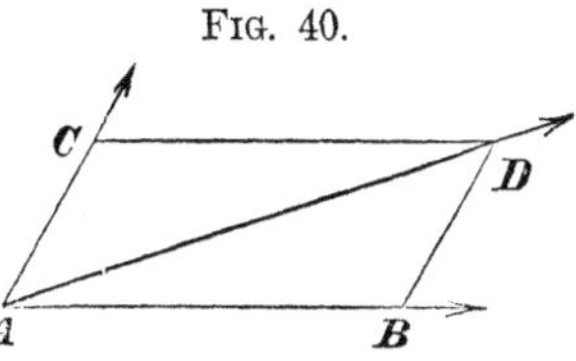

It is obvious that three forces in equilibrium must all be directed through one point, else each force could not be opposed to the resultant of the other two.

3. *More than three forces in one plane will produce equilibrium when they can be represented by the sides of a polygon taken in order.*

In Art. 43 it was shown that if several forces acting on a body, are represented by all the sides of a polygon except one, their resultant is represented by the remaining side. Thus, the resultant of the forces $A\,B$, $B\,C$, $C\,D$, and $D\,E$ (Fig. 41), is $A\,E$. Now, the force $E\,A$, equal and opposite to $A\,E$, since it would be in equilibrium with $A\,E$, is therefore in equilibrium with all the others. Hence the forces $A\,B$, $B\,C$, $C\,D$, $D\,E$, and $E\,A$, taken in order around the figure, are in equilibrium.

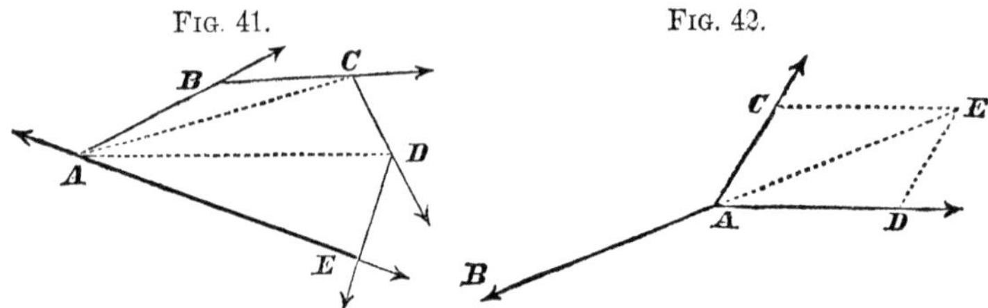

FIG. 41. FIG. 42.

61. Trigonometrical Representation of Three Forces in Equilibrium.—*When three forces are in equilibrium, each may be represented by the sine of the angle between the other two.*

Let $A\,E$ (Fig. 42) be the resultant of $A\,D$ and $A\,C$; then, if we apply a force $A\,B$ equal and opposite to $A\,E$, the forces $A\,D$, $A\,C$, and $A\,B$ will be in equilibrium. From the triangle $A\,E\,D$ we have the proportions

$$A\,D : A\,E : E\,D :: \sin A\,E\,D : \sin D : \sin E\,A\,D.$$

But $\sin A\,E\,D = \sin C\,A\,E = \sin B\,A\,C$; $\sin D = \sin C\,A\,D$; $\sin E\,A\,D = \sin B\,A\,D$; $A\,E = A\,B$; and $A\,C = E\,D$; $\therefore$ $A\,D : A\,B : A\,C :: \sin B\,A\,C : \sin C\,A\,D : \sin B\,A\,D$.

62. Equilibrium of Parallel Forces.—In order that a force may be in equilibrium with two parallel forces,

1. *It must be parallel to them.*
2. *It must be equal to their algebraic sum.*
3. *The distances of its line of action from the lines in which the two forces act, must be inversely as the forces.*

These three conditions belong to the *resultant* of two parallel forces, and therefore belong to that force which is in equilibrium with the resultant.

63. Equilibrium of Couples.—If two parallel forces are such as to constitute a couple, no *one* force can be in equilibrium with them. For the resultant of a couple has its point of application at an infinite distance (Art. 54). But a couple can be held in equilibrium by another couple; and the second couple may be either larger or smaller than the given couple, or it may be equal to it.

Let the couple P and P' (Fig. 43) act on a body at the points A and B; they tend to produce rotation about the middle point C. If another couple, Q and Q', equal to P and P', should be applied to produce equilibrium, one must directly oppose P, and the other P'. Then A and B, being each held at rest, all the forces are in equilibrium.

Fig. 43.

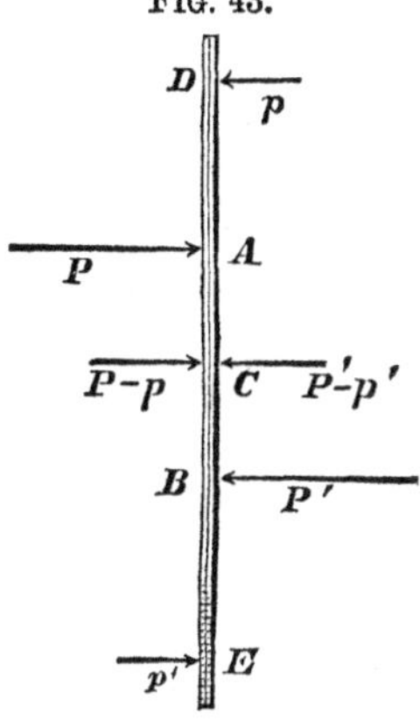

But if the second couple is less than P and P', they must act at distances from C, which are as much greater as the forces are less; or, if the second couple is greater than the first, they must act at distances which are as much less. Thus, the couple p and p', acting at D and E, tend to produce rotation about C in one direction, and P and P' in the opposite; and these tendencies are equal when $D\,C : A\,C :: P : p$. For, since the opposite forces, P and p, are inversely as their distances from C, their resultant is at C, and is equal to $P - p$ (Art. 53). For the same reason, the resultant of P' and p' is at C, and equal to $P' - p'$. But $P - p = P' - p'$, and they act in opposite directions. Hence C is at rest, and therefore all the forces are in equilibrium.

64. Equilibrium of Forces in Different Planes.—Since all the forces which can operate on a body may be reduced to three forces on rectangular axes, it is obvious that the whole system of forces cannot be in equilibrium till the sum of the components on each axis is reduced to zero. We must have, therefore, in Art. 59, as conditions of equilibrium, these three equations for the three axes, X, Y, and Z:

$$P\,.\cos a + P'\,.\cos a' + P''\,.\cos a'' +, \&c., = 0;$$
$$P\,.\cos \beta + P'\,.\cos \beta' + P''\,.\cos \beta'' +, \&c., = 0;$$
$$P\,.\cos \gamma + P'\,.\cos \gamma' + P''\,.\cos \gamma'' +, \&c., = 0.$$

65. Forces Resisted by a Smooth Surface.—Whenever any forces cause pressure upon a smooth surface, and are held in equilibrium by its resistance, the resultant of those forces must be at right angles to the surface. Suppose that $D A$ (Fig. 44) is either a single force or the resultant of two or more forces, and that it is held in equilibrium by the reaction of $A B$, a smooth surface. If $D A$ is not perpendicular to the surface, it can be resolved into two components, one perpendicular to the surface $A B$, the other parallel to it. The former, $D B$, is neutralized by the resistance of the surface; the latter, $B A$, is not resisted, and produces motion parallel to the surface, contrary to the supposition. Therefore $D A$, if held in equilibrium by the surface $A B$, must be perpendicular to it.

FIG. 44.

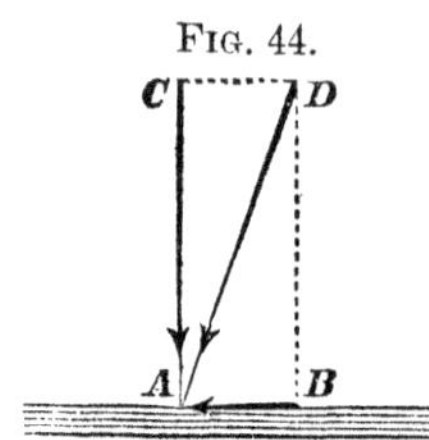

CHAPTER IV.

THE CENTRE OF GRAVITY.

66. The Centre of Gravity Defined.—In every body and in every system of bodies, there is a point so situated that all the parts acted on by the force of gravity balance each other about it in every position. That point is called the *centre of gravity*. The force of gravity acts in parallel lines on every particle of a body; the centre of gravity must therefore be the point through which the resultant of all these parallel forces is directed, in every position of the body. Hence, if the centre of gravity is supported, the body is supported. As to the support of the body, therefore, we may imagine all parts of it to be collected in its centre of gravity. When a system of bodies is considered, they are conceived to be united to each other by inflexible rods, which are without weight.

67. Centre of Gravity of Equal Bodies in a Straight Line.—The centre of gravity of two equal particles is in the middle point between them. Let A and B (Fig. 45), two equal particles, be joined by a straight line, and let $A\,a$ and $B\,b$ represent the forces of gravity. The resultant of these forces, since they are parallel and equal, will pass through the middle of $A B$ (Art. 54); G is therefore the centre of gravity. In like manner it is proved that the centre of gravity of two equal *bodies* is in the middle point between their respective centres of gravity.

FIG. 45.

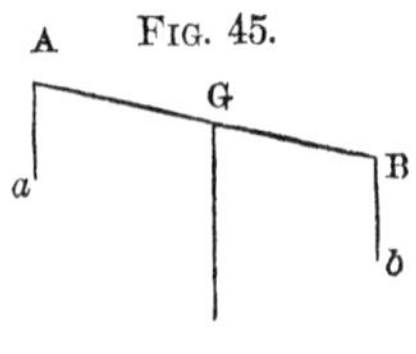

Any number of equal particles or bodies, arranged at equal distances on a straight line, have their common centre of gravity in the middle; since the above reasoning applies to each pair, taken at equal distances from the extremes. Hence, the centre of gravity of a material straight line (e.g., a fine straight wire) is in the middle point of its length.

68. Centre of Gravity of Regular Figures.—In the discussion of the centre of gravity in relation to *form*, *bodies* are considered uniformly dense, and *surfaces* are regarded as thin laminæ of matter.

In plane figures the centre of gravity coincides with the centre of magnitude, when they have such a degree of regularity that there are two diameters, each of which divides the figure into equal and symmetrical parts.

The circle, the parallelogram, the regular polygon, and the ellipse, are examples.

For instance, the regular hexagon (Fig. 46) is divided symmetrically by $A\,B$, and also by $C\,D$. Conceive the figure to be composed of material lines parallel to $A\,B$. Each of these has its centre of gravity in its middle point, that is, in $C\,D$, which bisects them all (Art. 67). Hence, the centre of gravity of the whole figure is in $C\,D$. For the same reason it is in $A\,B$. It is, therefore, at their intersection, which is also the centre of magnitude.

FIG. 46.

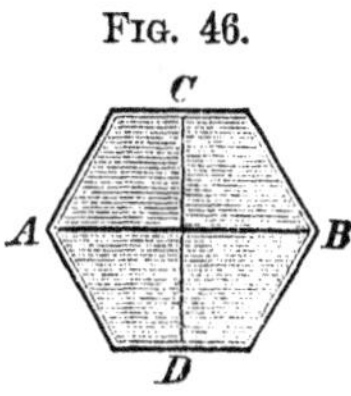

By a similar course of reasoning it is shown that in *solids* of uniform density, which are so far regular that they can be divided symmetrically by three different planes, the centres of gravity and magnitude coincide; e.g., the sphere, the parallelopiped, the cylinder, the regular prism, and the regular polyhedron.

69. Centre of Gravity between Two Unequal Bodies.—The centre of gravity of two unequal bodies is in a straight line joining their respective centres of gravity, and at the point which divides their distance in the inverse ratio of their weights. Let $A\,a$ and $B\,b$ (Fig. 47), passing through the centres of gravity of A and B, be proportional to their weights, and therefore represent the forces of gravity exerted upon them. By the laws of parallel forces, the resultant $G\,g = A\,a + B\,b$ (Art. 53), and $A\,a : B\,b :: B\,G : A\,G$. Therefore the centre of gravity must be at G, through which the resultant passes (Art. 66). This obviously includes the case of *equal* weights (Art 67).

FIG. 47.

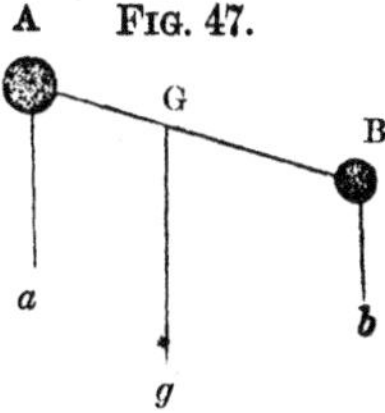

It appears from the foregoing that the whole pressure on a support at G is $A + B$, and that the system is kept in equilibrium by such support.

70. Equal Moments with Respect to the Centre of Gravity.—If A is put for the weight of A, and B for that of B, the above proportion becomes $A : B :: B\,G : A\,G$.

Let the proportion be changed to an equation, and we have $A \times A\,G = B \times B\,G$. Suppose now that $A\,B$ is an inflexible rod, without weight, and free to revolve about G. Since the bodies balance each other about that point, the equal products, $A \times A\,G$ and $B \times B\,G$, may be taken to represent the equal tendencies of the bodies to turn the system about G. The tendency of the body A, expressed by $A \times A\,G$, is called the *moment* of A, with reference to the point G; and, similarly, $B \times B\,G$ is the *moment* of B, with reference to the same point. Hence the proposition, that *the moments of two bodies with reference to their centre of gravity are equal.*

71. Centre of Gravity between Three or More Bodies.—The method of determining the centre of gravity of two bodies may be extended to any number.

Let A, B, C, D, &c. (Fig. 48), be the weights of the bodies, and let the centres of gravity of A and B be connected together by the inflexible line $A\,B$.

Fig. 48.

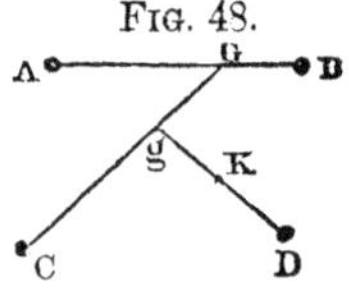

Divide $A\,B$ so that $A : B :: B\,G : A\,G$, or $A + B : B :: A\,B : A\,G$; then G is the centre of gravity of A and B. Join $C\,G$; and since $A + B$ may be considered as at the point G, divide $C\,G$ so that $A + B + C : C :: C\,G : G\,g$. In like manner, K, the centre of gravity of four bodies, is found by the proportion, $A + B + C + D : D :: D\,g : g\,K$. The same plan may be pursued for any number of bodies.

72. Centre of Gravity of a Triangle.—*The centre of gravity of a triangle is one-third of the distance from the middle of a side to the opposite angle.* Bisect $A\,C$ in D (Fig. 49), and $B\,C$ in E; join $A\,E$, $B\,D$, and $D\,E$. $B\,D$ bisects all lines across the triangle parallel to $A\,C$; therefore the centre of gravity of all those lines—that is, of the triangle—is in $B\,D$. For a like reason, it is in $A\,E$, and therefore at their intersection, G. Since $E\,C = \frac{1}{2}\,B\,C$, and $D\,C = \frac{1}{2}\,A\,C$, $\therefore E\,D = \frac{1}{2}\,A\,B$. But $E\,G\,D$ and $A\,G\,B$ are similar; $\therefore D\,G : B\,G :: D\,E : A\,B :: 1 : 2$; $\therefore D\,G = \frac{1}{2}\,B\,G = \frac{1}{3}\,B\,D$.

Fig. 49.

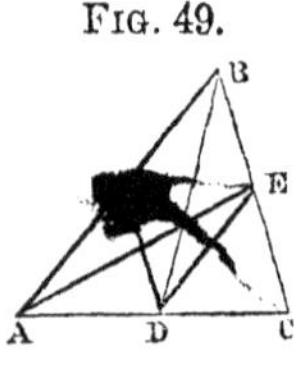

73. Centre of Gravity of an Irregular Polygon.—Divide the polygon into triangles by diagonals drawn through one of its angles, and then proceed according to the methods already given. Let $A\ C\ E$ (Fig. 50) be an irregular polygon, whose centre of gravity is to be found. Divide it into the triangles P, Q, R, S, by diagonals through A, and find their centres of gravity a, b, c, d (Art. 72). Join $a\ b$, and divide it so that $a\ b : a\ G :: P + Q : Q$; then G is the centre of gravity of the quadrilateral $P + Q$. Then join $G\ c$, and make $G\ c : G\ g :: P + Q + R : R$. By proceeding in this manner till all the triangles are used, the centre of gravity of the polygon is found at the last point of division.

FIG. 50.

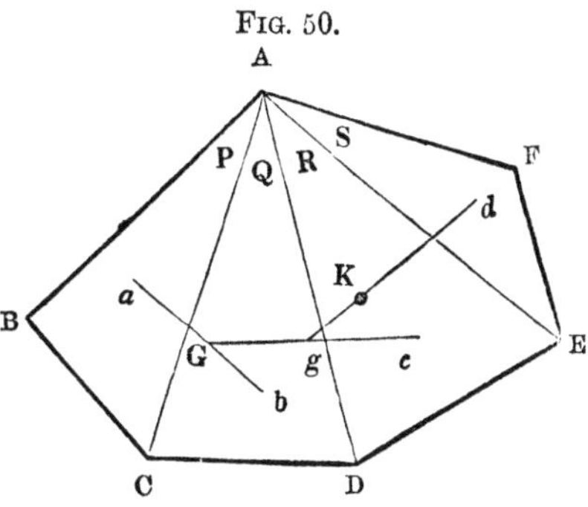

74. Centre of Gravity of the Perimeter of an Irregular Polygon.—Find the centre of gravity of each side, which is at its middle point, and then proceed as in Art. 71, the weight of each line being considered proportional to its length. Thus, let a, b, c, &c., be the centres of gravity of the sides, $A\ B$, $B\ C$, $C\ D$, &c. (Fig. 51); join $a\ b$, and divide it so that $a\ b : a\ G :: A\ B + B\ C : B\ C$; then G is the centre of gravity of $A\ B$ and $B\ C$. Next join $G\ c$, and make $G\ c : G\ g :: A\ B + B\ C + C\ D : C\ D$; then g is the centre of gravity of those three sides. Proceed in this manner till all the sides are used.

FIG. 51.

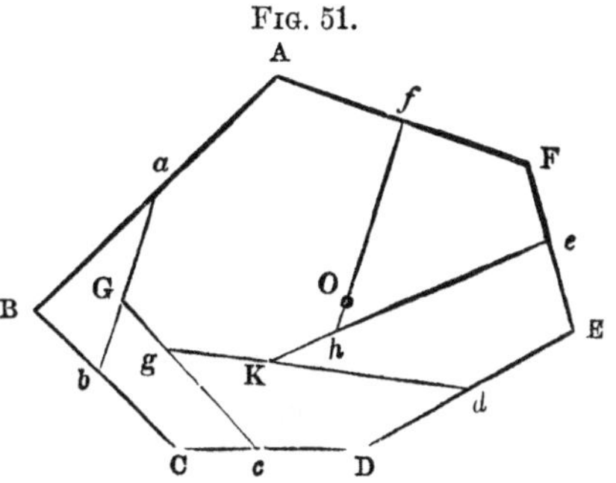

The perimeter of a polygon having the degree of regularity described in Art. 68, has its centre of gravity at the centre of the figure, as may be easily proved. If a polygon has a less degree of regularity than that, the centre of gravity both of its area and its perimeter may usually be found by methods more direct and simple than those given for polygons wholly irregular.

75. Centre of Gravity of a Pyramid.—*The centre of gravity of a triangular pyramid is in the line joining the vertex and the centre of gravity of the base, at one-fourth of the distance from the base to the vertex.*

Let G (Fig. 52) be the centre of gravity of the base $B\,D\,C$; and g that of the face $A\,B\,C$. The line $A\,G$ passes through the centre of gravity of every lamina parallel to $D\,B\,C$, on account of the similarity and similar position of all those laminæ; ∴ the centre of gravity of the pyramid is in $A\,G$. For a similar reason, it is in $D\,g$; and therefore at their intersection, O. Now $E\,G = \frac{1}{3}\,E\,D$, and $E\,g = \frac{1}{3}\,E\,A$; hence, by similar triangles, $g\,G = \frac{1}{3}\,A\,D$. But $G\,g\,O$ and $A\,O\,D$ are also similar; ∴ $G\,O = \frac{1}{3}\,A\,O = \frac{1}{4}A\,G$.

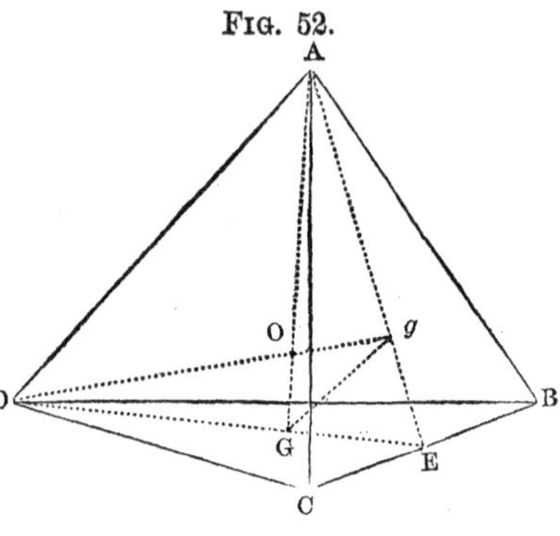

FIG. 52.

From this it is readily proved that the centre of gravity of *every* pyramid and cone is one-fourth of the distance from the centre of gravity of the base to the vertex.

76. Examples on the Centre of Gravity.—

1. A, B, and C (Fig. 53), weigh, respectively, 3, 2, and 1 pounds, $A\,B = 5$ ft., $B\,C = 4$ ft., and $C\,A = 2$ ft. Find the distance of their centre of gravity from C.

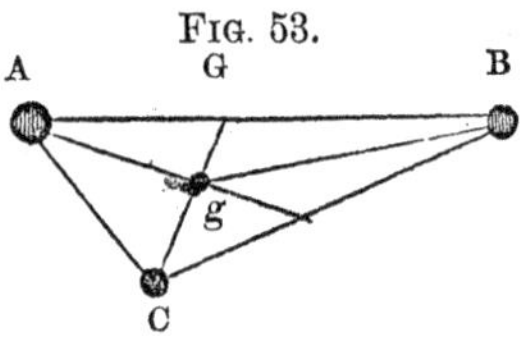

FIG. 53.

First, from the given sides of the triangle $A\,B\,C$, calculate the angles. A is found to be $49°\ 27\frac{1}{2}'$. Next find the place of G, the centre of gravity of A and B, by the proportion, $A + B : B :: A\,B : A\,G$; $A\,G$ is 2 feet, equal to $A\,C$. Calculate $C\,G$, the base of the isosceles triangle $A\,G\,C$. Its length is 1.673. Then find $C\,g$ by the proportion $C\,G : C\,g :: A + B + C : A + B$; therefore $C\,g = 1.394$.

2. $A = 5$ lbs., $B = 3$ lbs., and $C = 12$ lbs.; $A\,B = 8$ ft., $A\,C = 4$ ft., and the angle A is 90°; find the distance of the centre of gravity of A, B, and C, from C. *Ans.* 2 ft.

3. Three equal bodies are placed at the angles of any triangle whatever; show that the common centre of gravity of those bodies coincides with the centre of gravity of the triangle.

4. Find the centre of gravity of five equal heavy particles placed at five of the angular points of a regular hexagon.

Ans. It is one-fifth of the distance from the centre to the third particle.

5. A regular hexagon is bisected by a line joining two opposite angles; where is the centre of gravity of one-half?

Ans. Four-ninths of the distance from the centre to the middle of the second side.

6. A square is divided by its diagonals into four equal parts, one of which is removed; find the distance from the opposite side of the square to the centre of gravity of the remaining figure.

Ans. $\frac{7}{18}$ of the side of the square.

7. Two isosceles triangles are constructed on opposite sides of the same base, the altitude of the greater being h, and of the less, h'; where is the centre of gravity of the whole figure?

Ans. On the altitude of the greater triangle, at a distance from the common base equal to $\frac{1}{3}(h - h')$.

8. The base and the place of the centre of gravity of a triangle being given, required to construct the triangle.

9. Given the base and altitude of a triangle; required to construct the triangle, when its centre of gravity is perpendicularly over one end of the base.

10. On a cubical block stands a square pyramid, whose base, volume, and mass are respectively equal to those of the cube; where is the centre of gravity of the figure?

Ans. One-eighth of the height of the cube above its upper surface.

77. Centre of Gravity of Bodies in a Straight Line referred to a Point in that Line.—If several bodies are in a straight line, their common centre of gravity may be referred to a point in that line; and its distance from that point is obtained by *multiplying each weight into its own distance from the same point, and dividing the sum of the products by the sum of the weights.* Let A, B, C, and D, represent the weights of several bodies, whose centres of gravity are in the straight line $O\,D$ (Fig. 54). Required

Fig. 54.

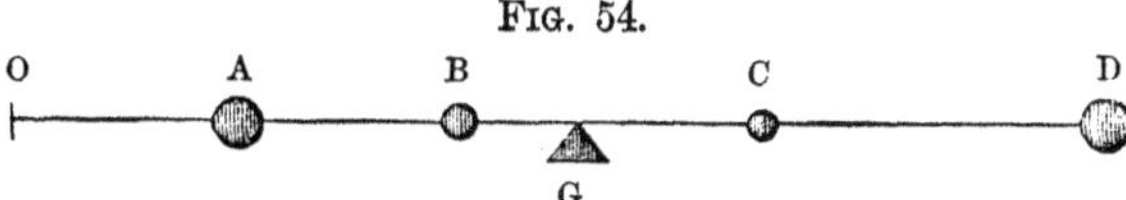

the distance of their common centre of gravity from any point O assumed in the same line. Let G be their common centre of gravity, then the moments of A and B must be equal to the opposing moments of C and D with reference to the point G (Art. 70). That is,

$$A \times A\,G + B \times B\,G = C \times C\,G + D \times D\,G; \text{ or,}$$

$$A \times (O\,G - O\,A) + B \times (O\,G - O\,B) = C \times (O\,C - O\,G) + D \times (O\,D - O\,G);$$

expanding, transposing the negative products, and factoring, we have

$$(A+B+C+D) \times O\,G = A \times O\,A + B \times O\,B + C \times O\,C + D \times O\,D.$$

Therefore,

$$O\,G = \frac{A \times O\,A + B \times O\,B + C \times O\,C + D \times O\,D}{A + B + C + D}$$

78. Centre of Gravity of a System referred to a Plane.—If the bodies are not in a straight line, they may be referred to a plane, which is assumed at pleasure. The distance of their common centre of gravity from that plane is expressed as before: *multiply each weight into its own distance from the plane, and divide the sum of the products by the sum of the bodies.*

Let p, p', p'' (Fig. 55), represent the weights of several bodies, whose centres of gravity are at those points respectively, and let $A\,C$ be the plane of reference. Join $p\,p'$, and let g be the common centre of gravity of p and p'; draw $p\,x$, $g\,k$, $p'\,x'$ at right angles to the plane $A\,C$, and consequently parallel to each other; join $x\,x'$, and since the points p, g, p', are in a straight line, the points x, k, x' will also be in a straight line, and therefore $x\,x'$ will pass through k. Join $g\,p''$, and let G be the common centre of gravity of p, p', p''; draw $G\,K$, $p''\,x''$, perpendicular to the plane; and through g draw $m\,n$ parallel to $x\,x'$ meeting $p\,x$ produced in n.

Fig. 55.

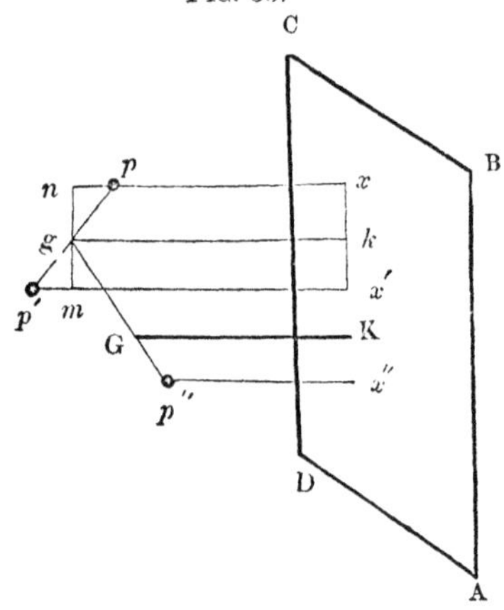

Now $p : p' :: p'\,g : p\,g ::$ (by sim. triangles) $p'\,m : p\,n$;

$$\therefore p \times p\,n = p' \times p'\,m, \text{ or } p \times (n\,x - p\,x) = p' \times (p'\,x' - m\,x');$$

but

$$n\,x = g\,k = m\,x', \therefore p \times (g\,k - p\,x) = p' \times (p'\,x' - g\,k),$$

and

$$(p + p') \times g\,k = p \times p\,x + p' \times p'\,x' \therefore g\,k = \frac{p \times p\,x + p' \times p'\,x'}{p + p'};$$

for the same reason, if $p + p'$ is placed at g, we have

$$G\,K = \frac{(p+p') \times g\,k + p'' \times p''\,x''}{(p + p') + p''} = \frac{p \times p\,x + p' \times p'\,x' + p'' \times p''\,x''}{p + p' + p''};$$

a formula which is applicable to any number of bodies.

Let the last equation be multiplied by the denominator of the fraction, and we have

$$(p + p' + p'' + \&c.)\; G\,K = p \times p\,x + p' \times p'\,x' + p'' \times p''\,x'' + \&c.;$$

that is, *the moment of any system of bodies with reference to a given plane, equals the sum of the moments of all the parts of the system with reference to the same plane.*

79. Centre of Gravity of a Trapezoid.—As an example of the foregoing principle, let it be proposed to find the centre of gravity of a trapezoid, considered as composed of two triangles. The centre of gravity of the trapezoid $A\,C$ (Fig. 56) is in $E\,F$, which bisects all the lines of the figure parallel to $B\,C$. Suppose G to be the centre of gravity of the trapezoid; through G draw $K\,M$ perpendicular to the bases. Let $K\,M = h$, $B\,C = B$, $A\,D = b$, and join $B\,D$.

FIG. 56.

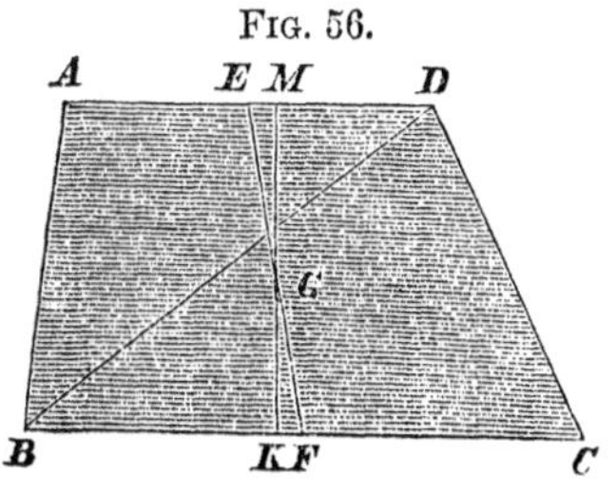

The moment of the trapezoid with reference to $B\,C$ is $(B + b)\frac{h}{2} \cdot G\,K$. The moment of the upper triangle is $\frac{b\,h}{2} \cdot \frac{2}{3}h$; the moment of the lower triangle is $\frac{B\,h}{2} \cdot \frac{h}{3}$; $\therefore (B + b)\frac{h}{2} \cdot G\,K = \frac{B\,h}{2} \cdot \frac{h}{3} + \frac{b\,h}{2} \cdot \frac{2}{3}h$; whence

$$G\,K = \frac{B + 2\,b}{B + b} \cdot \frac{h}{3}. \quad \text{But } G\,M = h - \frac{B + 2\,b}{B + b} \cdot \frac{h}{3} =$$

$$\frac{2\,B + b}{B + b} \cdot \frac{h}{3}; \therefore G\,M : G\,K :: 2\,B + b : B + 2\,b.$$

By similar triangles

$G\,M : G\,K :: E\,G : G\,F$; $\therefore E\,G : G\,F :: 2\,B + b : B + 2\,b$; or *the centre of gravity of a trapezoid is on the line which bisects the parallel bases, and divides it in the ratio of twice the longer plus the shorter to twice the shorter plus the longer.*

1. Four bodies, A, B, C, D, weighing, respectively, 2, 3, 6, and 8 pounds, are placed with their centres of gravity in a right line, at the distance of 3, 5, 7, and 9 feet from a given point; what is the distance of their common centre of gravity from that given point; and between which two of the bodies does it lie?

Ans. Between C and D; and its distance from the given point $7\frac{2}{19}$ feet.

2. There are five bodies, weighing, respectively, 1, 14, $21\frac{1}{2}$, 22, and $29\frac{1}{2}$ pounds; a plane is assumed passing through the last body, and the distances of the other four from the plane are, respectively, 21, 5, 6, and 10 feet; how far from the plane is the common centre of gravity of the five bodies? *Ans.* 5 feet.

[See Appendix for calculations of the place of the centre of gravity of curvilinear bodies.]

80. Centrobaric Mensuration.—The properties of the centre of gravity furnish a very simple method of measuring surfaces and solids of revolution. This method is comprehended in the two following propositions, known as the theorems of Guldinus:

1. *If any line revolve about a fixed axis, which is in the plane of that line, the* SURFACE *which it generates is equal to the product of the given line into the circumference described by its centre of gravity.*

Let any line, either straight or curved, revolve about a fixed axis which is in the plane of that line; and let f, f', f'', f''', etc., denote elementary portions of the line, d, d', d'', d''', &c., the distances of these portions, respectively, from the axis; then the surface generated by f, in one revolution, will be $2\pi\, d\, f$; hence the surface generated by the whole line will be

$$S = 2\pi\,(d f + d' f' + d'' f'' + d''' f''' + \text{\&c.}) \ldots (1).$$

Put L = the length of the revolving line, and G = the distance from the axis to the centre of gravity of the line; then (Art. 78)

$$G\, L = d f + d' f' + d'' f'' + d''' f''' + \text{\&c.} \ldots (2).$$

Combining (1) and (2), we have

$$S = 2\pi\, G\, L \ldots (3).$$

2. *If a plane surface, of any form whatever, revolve about a fixed axis which is in its own plane, the* VOLUME *generated is equal to the product of that surface into the circumference described by its centre of gravity.*

Let any plane surface revolve about an axis which is in the plane of that surface; and let f, f', f'', f''', &c., denote elementary portions of the surface, d, d', d'', d''', &c., the distances of these portions, respectively, from the axis; then the volume generated by f in one revolution will be $2\pi\, d\, f$; hence the volume generated by the whole surface will be

$$V = 2\pi\,(d f + d' f' + d'' f'' + d''' f''' + \text{\&c.}) \ldots (4).$$

Put A = the area of the revolving surface, and G = the distance from the axis to the centre of gravity of that surface; then (Art 78)

$$A\, G = d f + d' f' + d'' f'' + d''' f''' + \text{\&c.}, \ldots (5).$$

Substituting in (4), we have

$$V = 2\pi\, A\, G \ldots (6).$$

As an illustration of the first theorem, the straight line $C\,D$ (Fig. 57), revolving about the centre C, describes a circle whose surface is equal to $C\,D$ into the circumference of the circle described by its centre of gravity, E. This is evident also from the consideration that, since E is the centre of the line $C D$, the circumference described by it will be half the length of the circumference $A D B$; and the area of a circle is equal to the product of the radius into half the circumference.

FIG. 57.

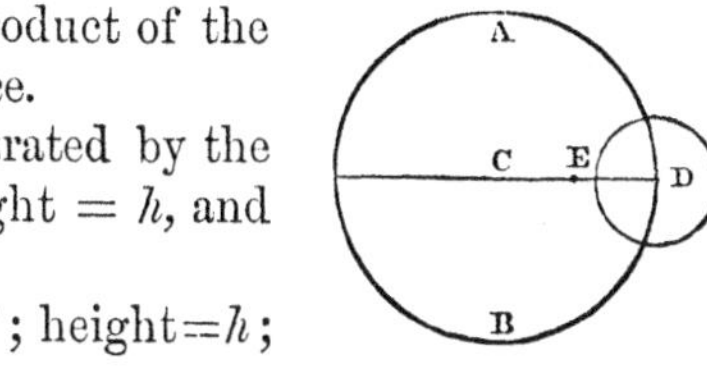

The second theorem is illustrated by the volume of a cylinder, whose height $= h$, and the radius of whose base $= r$.

Common method; base $= \pi r^2$; height $= h$; $\therefore$ vol. $= \pi r^2 h$.

Centrobaric method; revolving area $= r h$; circumference described by the centre of gravity $= \frac{1}{2} r \times 2 \pi$; $\therefore$ vol. $= r h \cdot \frac{1}{2} r \cdot 2 \pi = \pi r^2 h$.

81. Examples.—

1. Suppose the small circle (Fig. 57) to be placed with its plane perpendicular to the plane of the paper, and revolved about C, the point D describing the line $D B A$; required the content of the solid ring. If $C D = R$, and $E D = r$, then the area revolved $= \pi r^2$, and the circumference $D B A = 2 \pi R$; $\therefore$ the ring $= 2 \pi^2 R r^2$. It is equal to a cylinder whose base is the circle $E D$, and whose height equals the line $D B A$.

2. Find the convex surface of a cone; slant height $= s$; and rad. of base $= r$. The line revolved being s, and the distance from the axis to its centre of gravity, $\frac{1}{2} r$, the surface is $\pi r s$.

3. A square, whose side is one foot, is revolved about an axis which passes through one of its angles, and is parallel to a diagonal; required the volume of the figure thus formed.

Ans. $\pi \sqrt{2}$, or 4.4429 cubic ft.

4. Find the surface of a sphere whose radius is r. (The distance from the centre of a circle to the centre of gravity of its semi-circumference is $\frac{2 r}{\pi}$. Appendix, Art. 92.)

Ans. $\pi r \cdot \frac{2 r}{\pi} \cdot 2 \pi = 4 \pi r^2$.

5. Find the volume of a sphere whose radius is r. (Appendix, Art. 95.)

Ans. $\frac{4}{3} \pi r^3$.

82. Support of a Body.—A body cannot rest on a smooth plane, unless it is horizontal; for the pressure on a plane (Art. 65) cannot be balanced by the resistance of that plane, except when perpendicular to it; therefore, as the force of gravity is vertical, the resisting plane must be horizontal.

The *base of support* is that area on the horizontal plane which is comprehended by lines joining the extreme points of contact.

If there are *three* points of contact, the base is a triangle; if *four*, a quadrilateral, &c.

When the vertical through the centre of gravity (called *the line of direction*) falls within the base, the body is supported; if without, it is not supported. In the body *A* (Fig. 58) the force of gravity acts in the line *G F*, and there are lines of resistance on both sides of *G F*, as *G C* and *G E*, so that the body cannot turn on the edge of the base, without *rising* in an arc whose radius is *G C* or *G E*. But, in the body *B*, there is resistance only on one side; and therefore, if the force of gravity be resolved on *G C* and a perpendicular to it, the body is not prevented from moving in the direction of the latter, that is, in the arc whose radius is *G C*.

Fig. 58.

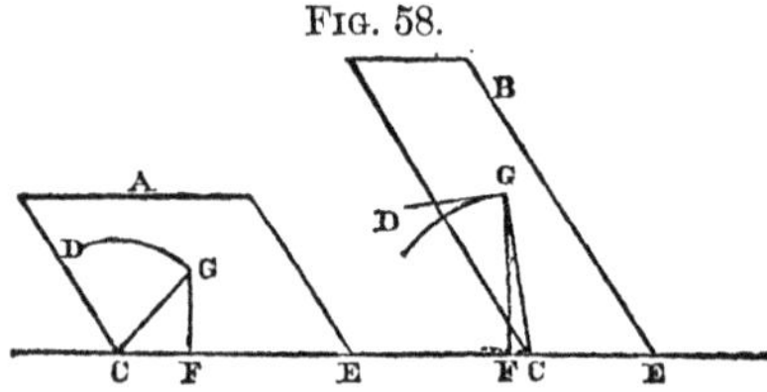

If the line of direction fall at the edge of the base, the least force will overturn it.

83. Different Kinds of Equilibrium.—If the base is reduced to a line or point, then, though there may be support, there is no *firmness* of support; the body will be moved by the least force. But it is affected very differently in different cases.

When it is moved from its position of support and left, it will in some cases return to it, pass by, and return again, and continue thus to vibrate till it settles in its place of support by friction and other resistances. This condition is called *stable equilibrium.*

In other cases, when moved from its position of support and left, it will depart further from it, and never recover that position again. This is called *unstable equilibrium.*

In other cases still, the body, when moved from its place of support and left, will remain, neither returning to it nor departing further from it. This is called *neutral equilibrium.*

Fig. 59.

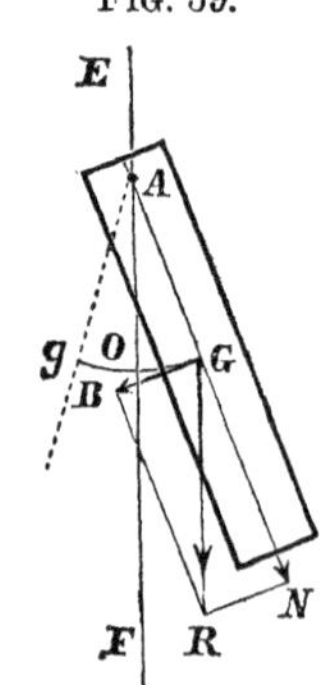

84. Stable Equilibrium.—Let the body (Fig. 59) be suspended on the pivot *A*. This is its base of support. While the centre of gravity is below *A*, the line of direction *E O F* passes through the base, and the body is supported. Let it be moved aside, and the centre of gravity be left at *G*. Let *G R* represent the force of gravity, and resolve it into *G N* on the line *A G*, and *N R*, or *G B*, perpendicular to *A G*. *G N* is resisted by the strength of *A*, and *G B* moves the centre of gravity in the arc whose radius is *A G*. Hence the body swings with accelerated motion till the centre of gravity reaches *O*, where the force *G B* becomes zero. But by its inertia, the body passes beyond that position, and ascends on the other side, till the retarding force of gravity stops it at *g*, as far from *O* as *G* is. It then descends again, and would never cease to oscillate were there no obstructions.

85. Unstable Equilibrium.—Next, let the body be turned on the pivot till the centre of gravity *G* is at *P*, above *A* (Fig. 60). Then, as well as when *G* is below *A*, the body is supported, because the line of direction *E P F* passes through the base *A*. But if turned and left in the slightest degree out of that position, it cannot recover it again, but will depart further and further from it. Let *G R* represent the force of gravity, and let it be resolved into *G N*, acting through *A*, and *G B* perpendicular to it. The former is resisted by *A*; the latter moves *G* away from *P*, the place of support. If the body is free to revolve about *A*, without falling from it, the centre of gravity will, by friction and other resistances, finally settle below *A*, as in the case of stable equilibrium.

Fig. 60.

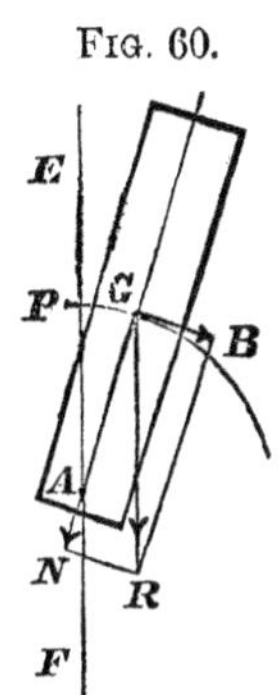

86. Neutral Equilibrium.—Once more, suppose the pivot supporting the body to be at *G*, the centre of gravity; then, in whatever situation the body is left, the line of direction passes through the base, and the body rests indifferently in any position.

These three kinds of equilibrium may be illustrated also by bodies resting by curved surfaces on a horizontal plane. Thus, if a cylinder is uniformly dense, it will always have a *neutral* equilibrium, remaining wherever it is placed. But if, on account of unequal density, its centre of gravity is not in the axis, then its equilibrium is *stable*, when the centre of gravity is below the axis, and *unstable* when above it.

In general, there is stable equilibrium when the centre of gravity, on being disturbed in either direction, begins to *rise;* unstable when, if disturbed either way, it begins to *descend;* and neutral when the disturbance neither raises nor lowers the centre of gravity.

Fig. 61.

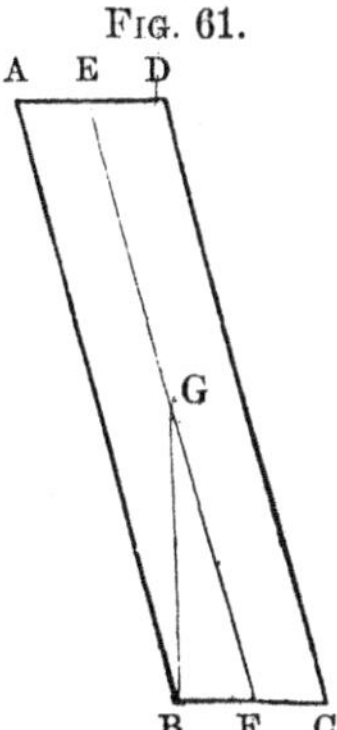

87. Questions on the Centre of Gravity.—

1. A frame 20 feet high, and 4 feet in diameter, is racked into an oblique form (Fig. 61),

till it is on the point of falling; what is its inclination to the horizon? *Ans.* 78° 27′ 47″.

2. A stone tower, of the same dimensions as the former, is inclined till it is about to fall, but preserves its rectangular form; what is its inclination? *Ans.* 78° 41′ 24″.

3. A cube of uniform density lies on an inclined plane, and is prevented by friction from sliding down; to what inclination must the plane be tipped, that the cube may just begin to roll down? *Ans.* 45°.

4. What must be the inclination of a plane, in order that a regular prism of any given number of sides may be at the limit between sliding and rolling down?

Ans. Equal to half the angle at the centre of the prism, subtended by one side.

5. A body weighing 83 lbs. is suspended, and drawn aside from the vertical 9°; what pressure is there on the point of support, and what force urges it down the arc?

Ans. Pressure on the support, 81.978 lbs.
Moving force, 12.984 lbs.

88. Motion of the Centre of Gravity of a System when one of the Bodies is Moved.—

When one body of a system is moved, the centre of gravity of the system moves in a similar path, and its velocity is to that of the moving body as the mass of that body is to the mass of the whole system.

If the system contains but two bodies, A and B (Fig. 62), suppose A to remain at rest, while B describes the straight lines $B\,C$, $C\,D$, &c., the centre of gravity G will in the same time describe the similar series, $G\,H$, $H\,J$, &c. When B is in the position B, and the centre of gravity at G, $A\,G : A\,B :: B : A + B$; when B is at C, $A\,H : A\,C :: B : A + B$; $\therefore A\,G : A\,B :: A\,H : A\,C$. Hence $G\,H$ is parallel to $B\,C$, and $G\,H : B\,C :: B : A + B$. In like manner, $H\,J : C\,D :: B : A + B$, &c. Thus, all the parts of one path are parallel to the corresponding parts of the other, and have a constant ratio to them. Therefore the paths are similar. As the corresponding parts are described in equal times, their lengths are as the velocities. But the lengths are as $B : A + B$; therefore the velocity of the common centre of gravity is to that of the moving body as the mass of the moving body is to the mass of both. The same reasoning is applicable when the body moves in a curve.

Fig. 62.

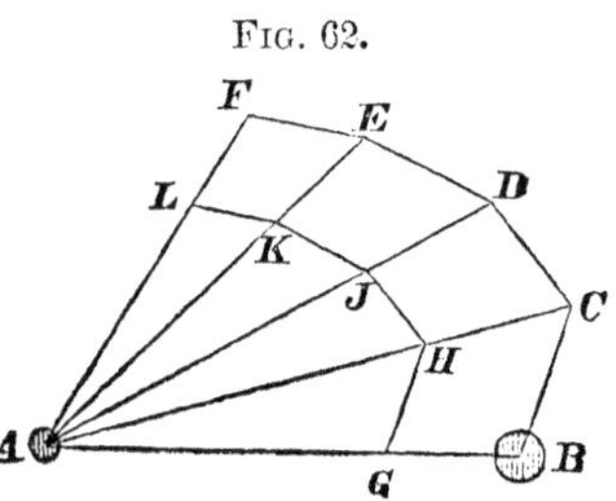

If the system contain any number of bodies, and the centre of gravity of the whole be at G, then the centre of gravity of all except B must be in the line $B\ G$ beyond G. Suppose it to be at A, and to remain at rest, while B moves; then it is proved in the same manner as before, that G, the centre of gravity of the whole system, moves in a path parallel to the path of B, and with a velocity which is to B's velocity as the mass of B to the mass of the entire system.

89. Motion of the Centre of Gravity of a System when Several of the Bodies are Moved.—

When any or all of the bodies of a system are moved, the centre of gravity moves in the same manner as if all the system were collected there, and acted on by the forces which act on the separate bodies.

Let A, B, C, &c. (Fig. 63), belong to a system containing any number of bodies, and let M be the mass of the system. Let A be moved over $A\ a$, B over $B\ b$, C over $C\ c$, &c. And first suppose the motions to be made in equal successive times. If the centre of gravity of the system is first at G, then that of all the bodies except A is in $A\ G$ produced, as at g. While A moves to a, G moves in a parallel line to H (Art. 88), and $G\ H : A\ a :: A : M$. In like manner, when B describes $B\ b$, the centre of gravity of the other bodies being at h, the centre of gravity of the system describes the parallel line, $H\ K$, and $H\ K : B\ b :: B : M$; and when C moves, $K\ L : C\ c :: C : M$, &c. Now, $A\ a$ and $G\ H$ represent the respective velocities of the body A, and the system M; therefore, if we convert the proportion $G\ H : A\ a :: A : M$ into an equation, we have $A \times A\ a = M \times G\ H$; that is, the momentum of the body A equals the momentum of the system M. It therefore requires the same force to move A over $A\ a$ as to move the system M over $G\ H$. The same is true of the other bodies. If then the several forces which move the bodies, limiting the number to three, for the present, were applied successively to the system collected at G, they would move it over $G\ H$, $H\ K$, $K\ L$. But if applied at once, they would move it over $G\ L$, the remaining side of the polygon. If, therefore, the forces, instead of acting successively on the bodies, were to move A over $A\ a$, B over $B\ b$, and C over $C\ c$, at the same time, the centre of gravity of the system would describe $G\ L$ in the same time. In the same way it may be proved, that whatever forces are applied to the

FIG. 63.

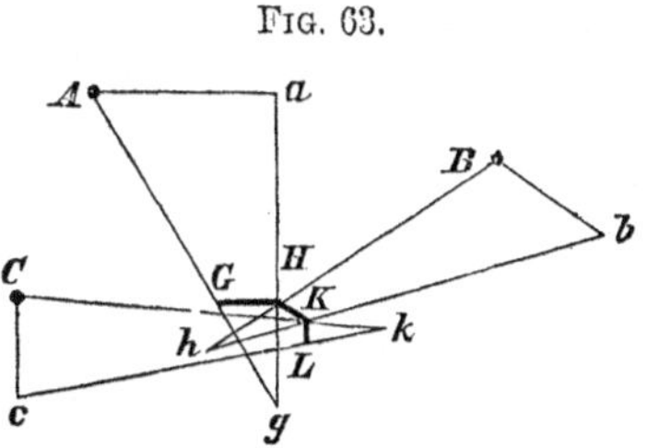

several bodies of a system, the centre of gravity of the system is moved in the same manner as a body equal to the whole system would be moved, if all the same forces were applied to it.

It is possible that the centre of gravity of a system should remain at rest, while all the bodies in it are in motion. For, suppose all the forces acting on the bodies to be such that they might be represented in direction and intensity by all the sides of a polygon, then, since a single body acted on by them would be in equilibrium, therefore the centre of gravity of the system would remain at rest, though the bodies composing it are in motion.

90. Mutual Action among the Bodies of a System.—The forces which have been supposed to act on the several bodies of a system are from without, and not forces which some of the bodies within the system exert on others. If the bodies of a system mutually attract or repel each other, such action cannot affect the centre of gravity of the whole system. For action and reaction are always opposite and equal. Whatever force one body exerts on any other to move it, that other exerts an equal force on the first, and the two actions produce equal and opposite effects on the centre of gravity between them. Therefore the centre of gravity of a system remains at rest, if the bodies which compose it are acted on only by their mutual attractions or repulsions.

91. Examples on the Motion of the Centre of Gravity.—

1. Two bodies, A and B, of given weights, start together from D (Fig. 64), and move uniformly with given velocities in the directions $D\,A$ and $D\,B$; required the direction and velocity of their centre of gravity.

Fig. 64.

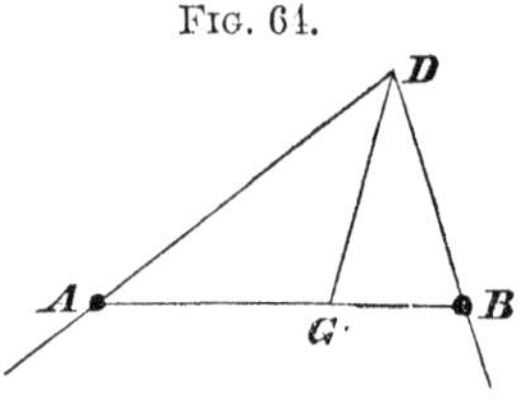

As the directions of $D\,A$ and $D\,B$ are given, we know the angle $A\,D\,B$; from the given velocities, we also know the lines $D\,A$ and $D\,B$, described in a certain time. Calculate the side $A\,B$, and the angles A and B. Find the place of the centre of gravity G between the bodies at A and B. Then, in the triangle $D\,B\,G$, $D\,B$, $B\,G$, and angle B are known, by which may be found the distance $D\,G$ passed over by the centre of gravity in the time, and $B\,D\,G$ the angle which its path makes with that of the body B.

2. The bodies A and B, of given weights, start together from D (Fig. 65), and move with equal velocities in opposite directions around the circumference of a circle, meeting again at D; what is the path of their centre of gravity?

Draw the diameter, $D\,E$, and join A and B, the points which the bodies have reached after any given time. As $D A$ and $D B$ are equal arcs, $A B$ is perpendicular to $D E$, and is bisected by it. Let G be the common centre of gravity of A and B, then

$$A : B :: B G : A G;$$
$$\therefore A + B : A - B :: B G + A G : B G - A G;$$
$$:: \quad A B \quad : \quad G M;$$
$$:: \quad A N \quad : \quad G N.$$

Therefore $A N$, the ordinate of the circle, is to $G N$, the corresponding ordinate of the figure described by the centre, always in the same constant ratio, of the sum of the bodies to their difference. But this is a property of the ellipse, that, when its axis is the diameter of a circle, the corresponding ordinates of the two figures are in a constant ratio. Hence the centre of gravity of A and B describes an ellipse, while they move, in the manner before stated, round the circle.

Fig. 65.

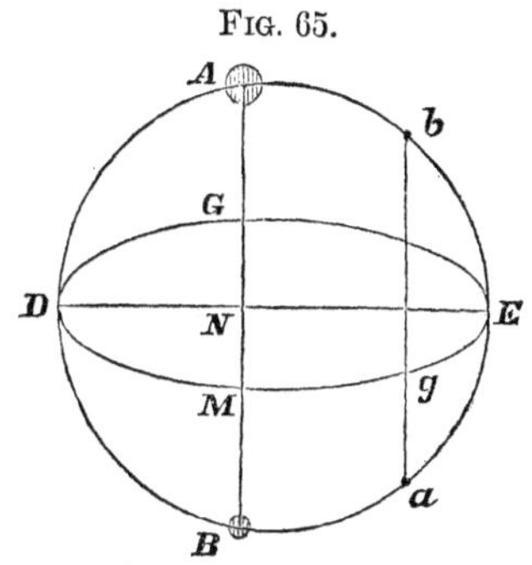

If the bodies approach equality, their difference grows less, and therefore the ellipse more eccentric, till, when the bodies are equal, the path of the centre of gravity is a straight line, as it evidently should be, in order to bisect the chords, $A B$, $b a$, &c.

3. Three bodies of given weight, A, B, C, in the same time and in the same order, describe with uniform velocity the three sides of the given triangle $A B C$ (Fig. 66); required the path of their centre of gravity.

Let G be their centre of gravity before they move. If they move successively, G describes $G K$, $K L$, $L M$, parallel to the sides of the triangle, and having to them respectively the same ratios as the corresponding moving bodies have to the sum of the bodies (Art. 89). Thus, three sides of the polygon are known; and the angle $K = B$, and $L = C$. These data are sufficient for calculating the fourth side, $G M$, which the centre of gravity describes, when the bodies move together.

Fig. 66.

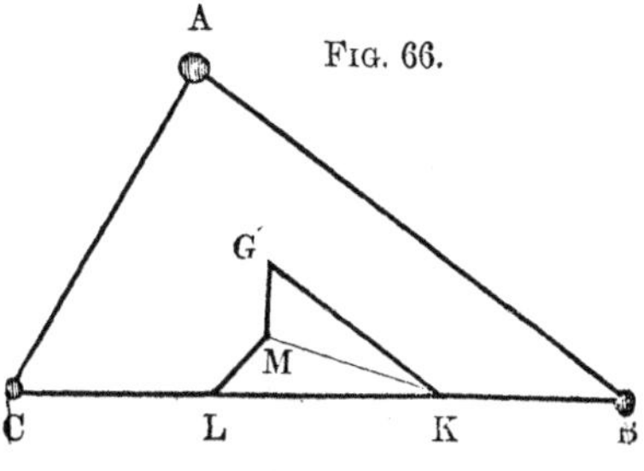

4. Show that when the three bodies in Example 3d are equal, the centre of gravity will remain at rest.

5. *A* (Fig. 67) weighs *one* pound; *B* weighs *two* pounds, and lies directly east of *A*; they move simultaneously, *A* northward, and *B* eastward, at the same uniform rate of 40 feet per second; required the direction and velocity of their centre of gravity.

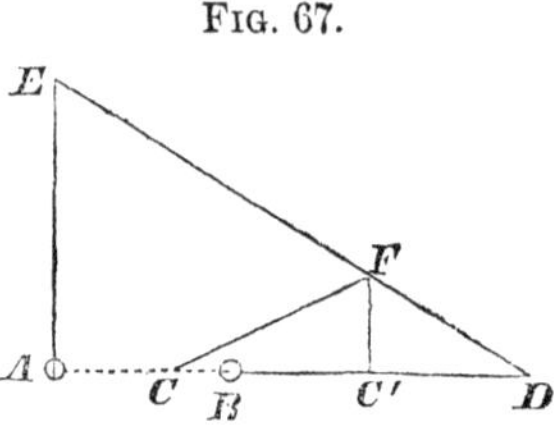

FIG. 67.

Ans. Velocity is 29.814 feet per second.
Direction is E. 26° 33′ 54″ N.

CHAPTER V.

THE COLLISION OF BODIES.

92. Elastic and Inelastic Bodies.—*Elastic* bodies are those which, when compressed, or in any way altered in form, tend to return to their original state. Those which show no such tendency are called *inelastic* or *non-elastic.* No substance is known which is entirely destitute of the property of elasticity: but some have it in so small a degree, that they are called inelastic, such as lead and clay. Elasticity is *perfect* when the *restoring* force, whether great or small, is equal to the *compressing* force. Air, and the gases generally, seem to be almost perfectly elastic; ivory, glass, and tempered steel, are imperfectly, though highly, elastic; and in different substances, the property exists in all conceivable degrees between the above-named limits.

93. Mode of Experimenting.—Experiments on collision may be made with balls of the same density suspended by long threads, so as to move in the line which joins their centres of gravity. If the arcs through which they swing are short compared with their radii, the balls, let fall from different heights, will reach the bottom sensibly at the same time, and will impinge with velocities which are very nearly proportional to the arcs. Thus *A* (Fig. 68), falling from 6,

FIG. 68

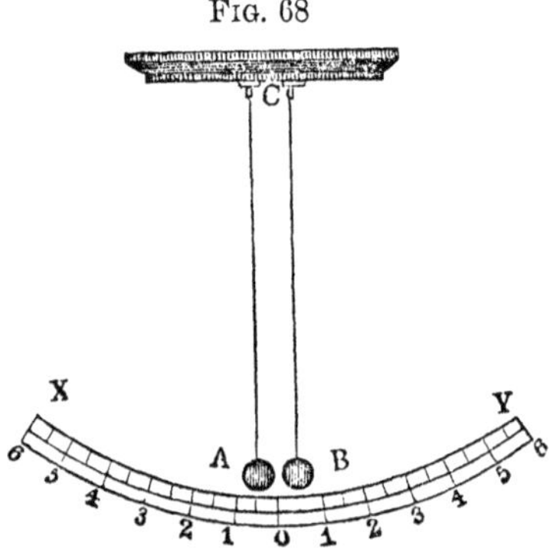

and B from 3, will come into collision at 0, with velocities which are as 2 : 1.

94. Collision of Inelastic Bodies.—Such bodies, after impact, move together as one mass.

The velocity of two inelastic bodies after collision is equal to the algebraic sum of their momenta, divided by the sum of the bodies.

Let A, B, represent the masses of the two bodies, and a, b, their respective velocities. Considering a as positive, if B moves in the opposite direction, its velocity must be called $-b$. Let v be the common velocity after impact.

1. *Same directions.*—The momentum of A is $A\,a$; that of B is $B\,b$; and the momentum of both after collision is $(A + B)\,v$. According to the third law of motion (Art. 13), whatever momentum A loses, B gains, so that the whole momentum is the same after collision as before; therefore

$$A\,a + B\,b = (A + B)\,v;\ \therefore v = \frac{A\,a + B\,b}{A + B}.$$

If B is at rest before impact, $b = 0$, and $v = \frac{A\,a}{A+B}$.

To find the *loss* or *gain* of velocity for either body, multiply the *other* body by the *difference* of velocities, and divide by the sum of the bodies. For, A's velocity before impact was a; after impact it is $\frac{A\,a + B\,b}{A + B}$; therefore the *loss* $= a - \frac{A\,a + B\,b}{A + B} = \frac{B\,(a - b)}{A + B}$. But B's *gain* is the velocity after impact diminished by the velocity before, i. e., $\frac{A\,a + B\,b}{A + B} - b = \frac{A\,(a - b)}{A + B}$.

When B is at rest, these expressions become

$$\frac{B\,a}{A + B} \text{ for } A\text{'s loss; } \frac{A\,a}{A + B} \text{ for } B\text{'s gain.}$$

2. *Opposite directions.*—Since b is negative, $v = \frac{A\,a - B\,b}{A + B}$.

To find *loss* or *gain* in this case, multiply the *other* body by the *sum* of the velocities, and divide by the sum of the bodies. For, *A's loss*

$$= a - \frac{A\,a - B\,b}{A + B} = \frac{B\,(a + b)}{A + B};$$

and *B's gain*

$$= \frac{A\,a - B\,b}{A + B} - (-\,b) = \frac{A\,a - B\,b}{A + B} + b = \frac{A\,(a + b)}{A + B}$$

In the case of opposite motions, the formula for v becomes zero, when $A\,a = B\,b$; but in that case, $A : B :: b : a$. Hence, if bodies which meet each other have velocities inversely as the quantities, they will *be at rest* after the collision.

95. Questions on Inelastic Bodies.—

1. A, weighing 3 oz., and moving 10 feet per second, overtakes B, weighing 2 oz., and moving 3 feet per second; what is the common velocity after impact? *Ans.* $7\frac{1}{5}$ feet per second.

2. A weight of 7 oz., moving 11 feet per second, strikes upon another at rest weighing 15 oz.; required the velocity after impact? *Ans.* $3\frac{1}{2}$ feet per second.

3. A weighs 4 and B 2 pounds; they meet in opposite directions, A with a velocity of 9, and B with one of 5 feet per second; what is the common velocity after impact?

Ans. $4\frac{1}{3}$ feet per second.

4. $A = 7$ pounds, $B = 4$ pounds; they move in the same direction, with velocities of 9 and 2 feet per second; required the velocity lost by A and gained by B? *Ans.* A $2\frac{6}{11}$, B $4\frac{5}{11}$.

5. A body moving 7 feet per second, meets another moving 3 feet per second, and thus loses half its momentum; what are the relative masses of the two bodies?

Ans. $A : B :: 13 : 7$.

6. A weighs 6 pounds and B 5; B is moving 7 feet per second, in the same direction as A; by collision B's velocity is doubled; what was A's velocity before impact?

Ans. $19\frac{5}{6}$ feet per second.

96. Collision of Elastic Bodies.—Elastic bodies after collision do not move together, but each has its own velocity. These velocities are found by *doubling* the *loss* and *gain* of inelastic bodies. When the elastic body A impinges on B, it loses velocity while it is becoming compressed, and again, while recovering its form, it loses as much more, because the restoring force is equal to the compressing force. For a like reason, B gains as much velocity while recovering its form as it gained while being compressed by the action of A. Hence, doubling the expressions for loss and gain given in Art. 94, and applying them to the original velocities, we find the velocity of each body after collision, on the supposition of perfect elasticity.

When the directions are the same,

$$\text{the velocity of } A = a - \frac{2\,B\,(a - b)}{A + B};$$

$$\text{that of } B = b + \frac{2\,A\,(a - b)}{A + B}.$$

When the directions are opposite,

$$\text{the velocity of } A = a - \frac{2\,B\,(a + b)}{A + B};$$

$$\text{that of } B = -\,b + \frac{2\,A\,(a + b)}{A + B}.$$

Reducing these expressions, we have for the velocities of elastic bodies after collision the following formulæ:

(1.) Same direction. Velocity of $A = \frac{(A - B)\,a + 2\,B\,b}{A + B}$.

(2.) Same direction. Velocity of $B = \frac{(B - A)\,b + 2\,A\,a}{A + B}$.

(3.) Opposite directions. Velocity of $A = \frac{(A - B)\,a - 2\,B\,b}{A + B}$.

(4.) Opposite directions. Velocity of $B = \frac{(A - B)\,b + 2\,A\,a}{A + B}$.

97. Equal Elastic Bodies.—After the impact of *equal* elastic bodies, *each takes the original velocity of the other.* When $A = B$, formula (1) is reduced to b; and formula (2) to a; that is, A has B's former velocity, and B has A's. The same is true if they move in opposite directions. For, when $A = B$, formula (3) becomes $-\,b$, which was B's original velocity, and formula (4) becomes a, which was A's. Therefore, in the *opposite* motions of equal elastic bodies, collision causes *each* to rebound, since $+\,a$ is exchanged for $-\,b$, and $-\,b$ for $+\,a$.

If we reduce these four formulæ for the case in which $A = B$, and B is at rest, we find the same interchange of conditions; for formula (1) becomes 0, and formula (2) becomes a; so formula (3) becomes 0, and formula (4) becomes a.

98. Unequal Elastic Bodies.—

1. If a *greater* body impinge on a *less one at rest*, the impinging body goes forward, but slower than before, and the other precedes it with a greater velocity than the impinging body first had. For, formula (1) becomes $\frac{(A - B)\,a}{A + B}$, which is positive, but less than a; therefore it advances, though slower than before. But formula (2) becomes $\frac{2\,A\,a}{A + B}$, which is greater than a; hence, B goes on faster than A did before collision.

2. If a *less* body impinge on a *greater one at rest*, it rebounds, and the other goes forward, but with a less velocity than that which the impinging body first had. For, $\frac{(A - B)\,a}{A + B}$ is negative, and $\frac{2\,A\,a}{A + B}$ is less than a.

3. If two elastic bodies, having *equal* velocities, meet each other, and one of them is brought to *rest*, its mass is three times as great as that of the other. For, as the velocities are equal, by

formula (3), $\frac{(A - B)\,a - 2\,B\,a}{A + B} = 0$; $\therefore (A - B)\,a - 2\,B\,a = 0$; $\therefore A = 3\,B$.

99. Series of Elastic Bodies.—

1. *Equal bodies.*—Let a row of equal elastic bodies, $A, B, C \ldots F$ (Fig. 69) be suspended in contact; then (Art. 97), if A be drawn back and left to fall against B, it will rest after impact, and B will tend to move on with A's velocity; after the impact of B on C, B will remain, and C tend to move with the same velocity; and so the motion will be transmitted through the series, and F will move away, while all the others remain at rest.

Fig. 69.

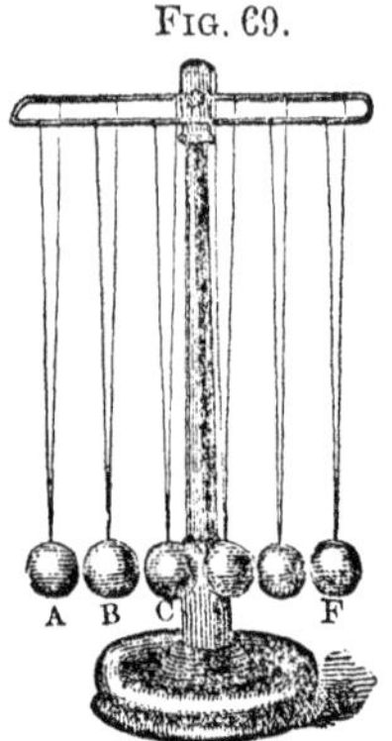

2. *Decreasing series.*—If the bodies decrease, as A, B, C, &c. (Fig. 70), and A be drawn back to A', and allowed to fall against B, then (Art. 98) A still moves forward, while B receives a greater velocity than A had, C still greater, &c. The last of the series, therefore, moves with the greatest velocity, and each one with a greater velocity than that which impinged on it.

Fig. 70.

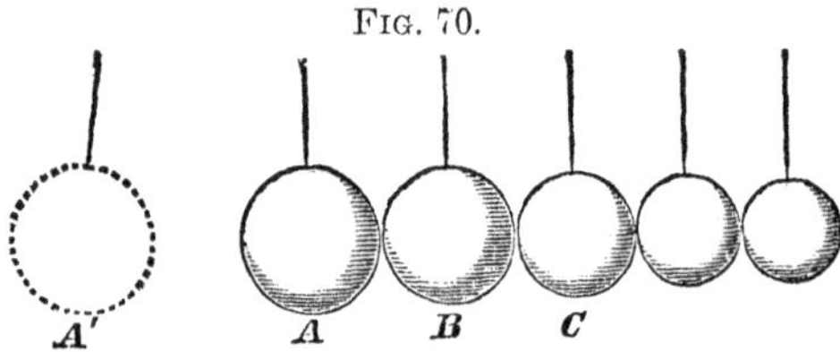

3. *Increasing series.*—If the bodies increase, as A, B, C, &c. (Fig. 71), then, when A falls from A' against B, it imparts to B a

Fig. 71.

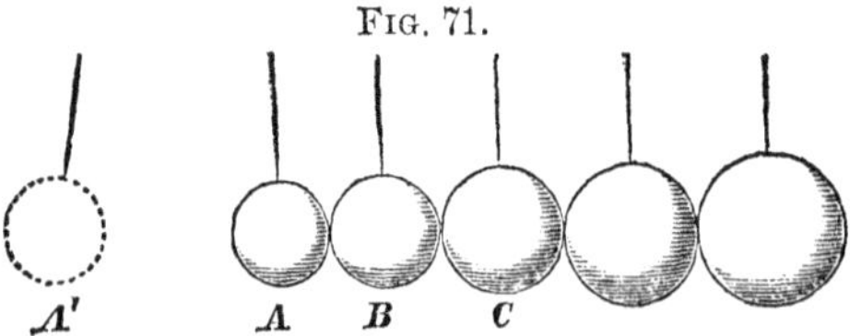

less velocity than it had itself, and rebounds (Art 98); in like manner B rebounds from C, and so on; while the last of the series goes forward with less velocity than any previous one would have had if it had been the last.

If the bodies in Fig. 70 are in geometrical progression, the velocity of the first to that of the last is as $1 : \left(\frac{2}{1+r}\right)^{n-1}$

Let the series be $A, Ar, Ar^2 \dots . Ar^{n-1}$.

By Art. 98, when A impinges on B at rest the velocity communicated to B is $\frac{2\,A\,a}{A+B} = \frac{2\,A\,a}{A+A\,r} = \frac{2\,a}{1+r} = b$.

Again, the velocity imparted to C is

$$\frac{2\,B\,b}{B+C} = \frac{2\,A\,r}{A\,r + A\,r^2} \times \frac{2\,a}{1+r} = \frac{2^2\,a}{(1+r)^2}.$$

Hence the successive velocities are $a, \frac{2\,a}{1+r}, \frac{2^2\,a}{(1+r)^2}$, &c., from which it appears that any term in the series is found by multiplying the original velocity by 2, raised to a power one less than the number of terms and divided by $1 + r$ raised to the same power. Consequently, the last term is $\frac{2^{n-1}\,a}{(1+r)^{n-1}}$. Hence, vel. of the first : vel. of the last :: $a : \frac{2^{n-1}\,a}{(1+r)^{n-1}} :: 1 : \left(\frac{2}{1+r}\right)^{n-1}$

100. Questions on Elastic Bodies.—

1. A, weighing 10 lbs. and moving 8 feet per second, impinges on B, weighing 6 lbs. and moving in the same direction, 5 feet per second; what are the velocities of A and B after impact?
Ans. A's $= 5\frac{3}{4}$, B's $= 8\frac{3}{4}$.

2. $A : B :: 4 : 3$; directions the same; velocities $5 : 4$; what is the ratio of their velocities after impact? *Ans.* 29 : 36.

3. A, weighing 4 lbs., velocity 6, meets B, weighing 8 lbs., velocity 4; required their respective directions and velocities after collision? *Ans.* A is reflected back with a velocity of $7\frac{1}{3}$, and B with a velocity of $2\frac{2}{3}$.

4. A and B move in opposite directions; A equals $4\,B$, and $b = 2\,a$; how do the bodies move after collision?
Ans. A returns with $\frac{1}{5}$, B with $1\frac{3}{5}$ its original velocity.

5. There are ten bodies whose masses increase geometrically by the constant ratio 3, and the first impinges on the second with the velocity of 5 feet per second; required the motion of the last body? *Ans.* The last body would move with the velocity of $\frac{5}{512}$ feet per second.

101. Living Force lost in the Collision of Inelastic Bodies.—The amount of living force (Art. 35) before collision is $A\,a^2 + B\,b^2$; and after collision it is $(A+B) \times \frac{(A\,a + B\,b)^2}{(A+B)^2} = \frac{(A\,a + B\,b)^2}{A+B}$. Subtract the latter from the former, and call the remainder d. Then $d = A\,a^2 + B\,b^2 - \frac{(A\,a + B\,b)^2}{A + B}$. Expanding and uniting terms, $d = \frac{A\,B\,(a - b)^2}{A + B}$. This value of d is positive, because $(a - b)^2$ is necessarily positive, as well as A and B. Therefore there is always a loss of living force in the collision of inelastic bodies.

102. Living Force Preserved in the Collision of Elastic Bodies.—The living force of A before collision is $A\,a^2$; after collision, it is $A \times \frac{\{(A - B)\,a + 2\,B\,b\}^2}{(A + B)^2}$. Subtracting the latter from the former, the loss (supposing there is loss) is

$$\frac{(A+B)^2\,A\,a^2 - (A-B)^2\,A\,a^2 - 4\,(A-B)\,A\,B\,a\,b - 4\,A\,B^2\,b^2}{(A + B)^2} \quad . . (1.)$$

The living force of B before collision is $B\,b^2$; after collision, it is $B \times \frac{\{(B - A)\,b + 2\,A\,a\}^2}{(A + B)^2}$; and the expression for loss is

$$\frac{(A+B)^2\,B\,b^2 - (B-A)^2\,B\,b^2 - 4\,(B-A)\,A\,B\,a\,b - 4\,A^2\,B\,a^2}{(A+B)^2} \quad . . . (2.)$$

Therefore the total loss of living force is the sum of the expressions (1.) and (2.).

Reducing the two first terms in each fraction to one, the fractions become

$$\frac{4\,A^2\,B\,a^2 - 4\,(A - B)\,A\,B\,a\,b - 4\,A\,B^2\,b^2}{(A + B)^2} \quad . . . . (3.)$$

and

$$\frac{4\,A\,B^2\,b^2 - 4\,(B - A)\,A\,B\,a\,b - 4\,A^2\,B\,a^2}{(A + B)^2} \quad . . . . (4.)$$

If the fractions (3) and (4) be added, it is evident that the numerators cancel each other, and therefore the sum of the fractions is zero. Hence, there is no loss of living force in the collision of elastic bodies.

103. Impact on an Immovable Plane.—If an inelastic body strikes a plane perpendicularly, its motion is simply *destroyed;* in strictness, however, it imparts an infinitely small velocity to the body called immovable. If it strikes obliquely, and the plane is smooth, it slides along the plane with a diminished velocity. Let $A\,L$ (Fig. 72) represent the motion of the body before impact on the plane $P\,N$, and resolve it into $A\,C$, perpendicular, and $C\,L$, parallel to the plane. Then $A\,C$, as before, is destroyed, but $C\,L$ is not affected; hence the former velocity is to its velocity on the plane, as $A\,L : C\,L$ :: radius : cosine of the inclination.

Fig. 72.

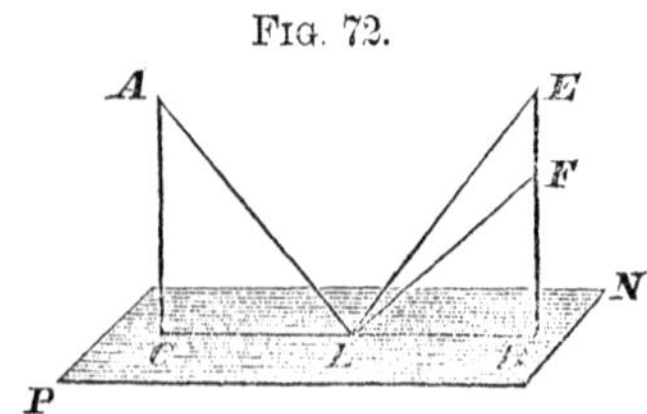

If a perfectly elastic body impinges perpendicularly upon a plane, then, after its motion is destroyed, the force by which it resumes its form causes an equal motion in the opposite direction; that is, the body rebounds in its own path as swiftly as it struck. But if the impact is oblique, the body rebounds at an equal angle on the opposite side of the perpendicular. For, resolve $A\ L$, as before, into $A\ C$, $C\ L$; the latter continues uniformly; but, instead of the component $A\ C$, there is an equal motion in the opposite direction. Therefore, if $L\ D$ is made equal to $C\ L$, and $D\ E$ equal to $A\ C$, the resultant of $L\ D$ and $D\ E$ is $L\ E$, which is equal to $A\ L$, and has the same inclination to the plane. Hence, the angles of incidence and reflection are equal, and on opposite sides of the perpendicular to the surface at the place of impact.

104. Imperfect Elasticity.—The formulæ for the velocity of bodies after collision, and the statements of the preceding article, are correct only on the supposition that bodies are, on the one hand, entirely destitute of elasticity, or on the other perfectly elastic. As no solid bodies are known, which are strictly of either class, these deductions are found to be only near approximations to the results of experiment. In all practical cases of the impact of movable bodies, the loss and gain of velocity are *greater* than if they were inelastic, and *less* than if perfectly elastic. And in cases of impact on a plane, there is always *some* velocity of rebound, but less than the previous velocity; and therefore, if the collision is oblique, the body has less velocity, and makes a smaller angle with the plane than before. For, making $D\ F$ less than $A\ C$, the resultant $L\ F$ is less than $A\ L$, and the angle $D\ L\ F$ is smaller than $D\ L\ E$, or $A\ L\ C$.

CHAPTER VI.

SIMPLE MACHINES.

105. Classification of Machines.—In the preceding chapters, the motion of bodies has been supposed to arise from the immediate action of one or more forces. But a force may produce effects *indirectly*, by means of something which is interposed for the purpose of changing the mode of action. These intervening bodies are called, in general, *machines;* though the names, *tools, instruments, engines*, &c., are used to designate particular classes of them. The elements of machinery are called *simple machines.* The following list embraces those in most common use:

1. The lever.
2. The wheel and axle.

3. The pulley.
4. The rope machine.
5. The inclined plane.
6. The wedge.
7. The screw.
8. The knee-joint.

In respect to principle, these eight, and all others, may be reduced to three.

1. The law of *equal moments*, applicable in those cases in which the machine turns on a pivot or axis, as in the lever and the wheel and axle.

2. The principle of *transmitted tension*, to be applied wherever the force is exerted through a flexible cord, as in the pulley or rope machine.

3. The principle of *oblique action*, applicable to all the other machines, the force being employed to balance or overcome one component only of the resistance.

The force which ordinarily puts a machine in motion is called the *power;* the force which resists the power, and is balanced or overcome by it, is called the *weight*.

A *compound* machine is one in which two or more simple machines are so connected that the weight of the first constitutes the power of the second, the weight of the second the power of the third, &c.

I. The Lever.

106. The Three Orders of Straight Lever.—The lever is a bar of any form, free to turn on a fixed point, which is called the *fulcrum*. In the *first* order of lever, the *fulcrum* is between the power and weight (Fig. 73); in the *second*, the *weight* is between the power and fulcrum (Fig. 74); in the *third*, the *power* is between the weight and fulcrum (Fig. 75).

Fig. 73.

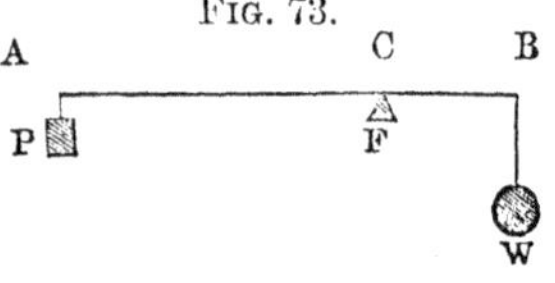

Fig. 74 Fig. 75.

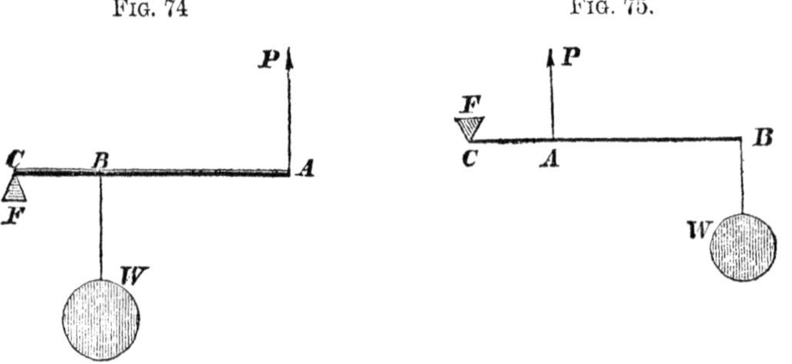

If P and W, in either of these figures, represent forces acting in vertical lines, then the circumstances of equilibrium are determined by the laws of *parallel forces* (Art. 54). In Fig. 73, if P and W are in equilibrium, their resultant will be so situated at C, that $P : W :: B\ C : A\ C$; and the fulcrum must be at that point, and be able to sustain a pressure equal to $P + W$. In Fig. 74, P and the reaction of F at C are two upward forces, whose resultant is counterbalanced by W; then W is represented by the whole line, and P by the part $B\ C$; $\therefore P : W :: B\ C : A\ C$, as before. The pressure on F equals $W - P$. In Fig. 75, W and the reaction of F are downward forces, whose resultant is at A, in equilibrium with P. Here P is represented by the whole line $B\ C$, and W by the part $A\ C$; $\therefore P : W :: B\ C : A\ C$. The upward pressure against F is equal to $P - W$.

Hence, in each order of the straight lever, when the forces act in parallel lines,

The power and weight are inversely as the lengths of the arms on which they act.

107. Equal Moments in Relation to the Fulcrum.—Changing the proportion into an equation, we find for each order of the lever, $P \times A\ C = W \times B\ C$; that is,

The power and weight have equal moments in relation to the fulcrum.

The *moment* of either force is the measure of its efficiency to turn the lever; for, since the lever is in equilibrium, the efficiency of the power to turn it in one direction must equal the efficiency of the weight to turn it in the opposite direction. We may therefore use $P \times A\ C$ to represent the former, and $W \times B\ C$, the latter.

If *several* forces, as in Fig. 76, are in equilibrium, some tending

FIG. 76.

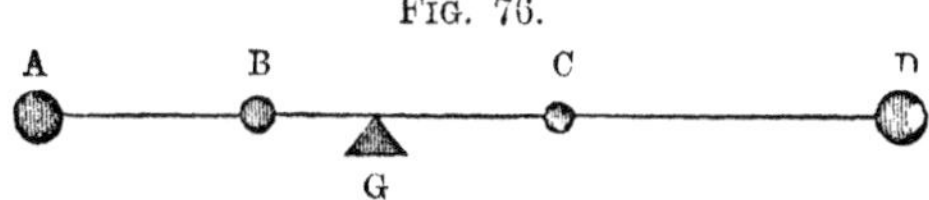

to turn the bar in one direction, and others in the opposite, then A and B must have the same efficiency to produce one motion as C and D have to produce the opposite; that is, $A \times A\ G + B \times B\ G = C \times C\ G + D \times D\ G$; or,

The *sum of the moments* of A and B equals the *sum of the moments* of C and D.

In order to allow for the influence of the weight of the lever itself, consider it to be collected at its centre of gravity, and add its moment to that of the power or weight, according as it aids the one or the other. In Fig. 73, let the weight of the lever $= w$, and the distance of its centre from C on the side of $P = m$; then $P \times A\ C + m\ w = W \times B\ C$. In the 2d and 3d orders, the moment of the lever necessarily aids the weight; and hence, in each case, $P \times A\ C = W \times B\ C + m\ w$.

If a weight hangs on a bar between two supports, as in Fig. 77, it may be regarded as a lever of the 2d order, the reaction of either support being considered as a power. Let F denote the reaction at A, and F' at C; then by the theorems of parallel forces, we have the pressures at A and C inversely as their distances from B, and $W = F + F'$.

FIG. 77.

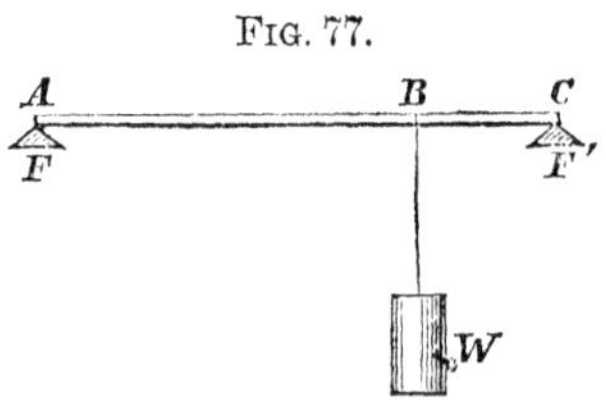

108. The Acting Distance.—In the three orders, as above described, the equilibrium is not destroyed by inclining the lever to any angle whatever with the horizon, provided the centre of motion C is at the centre of gravity of the bar, and not *above* or *below* it, and provided the directions of the forces remain vertical. For, by the principle of parallel forces, *any* straight line intersecting the lines of the forces is divided by the line of the resultant into parts which are inversely as the forces; therefore (Fig. 78) $b\ C : a\ C :: P : W$. Hence, the resultant of P and W remains at C, in every position of the lever. By similar triangles, $b\ C : a\ C :: C\ N : C\ M$; $\therefore P : W :: C\ N : C\ M$; $\therefore P \times C\ M = W \times C\ N$. The lines $C\ M$ and $C\ N$, which are drawn from the fulcrum perpendicular to the lines in which the forces act, are called the *acting distances* of the power and weight, respectively. And as they may be employed in levers of irregular form, the moments of power and weight are usually measured by the products, $P \times C\ M$ and $W \times C\ N$; therefore, *the power multiplied by its acting distance equals the weight multiplied by its acting distance;* or, more briefly, the moment of the power equals the moment of the weight, as in Art. 107. In Figs. 73, 74, and 75, the acting distances are in each case identical with the arms of the lever.

FIG. 78.

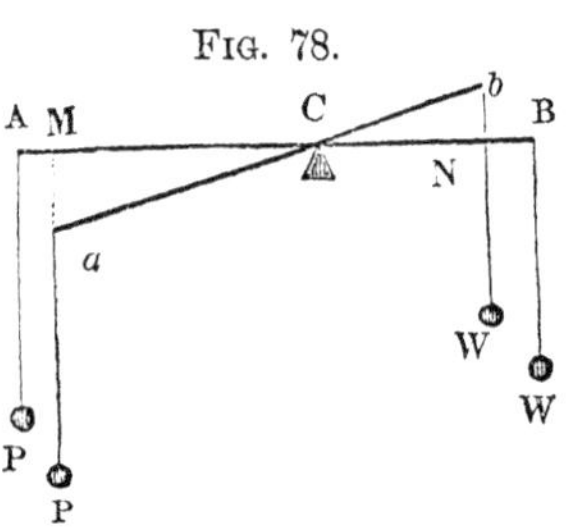

109. Lever not Straight, and Forces not Parallel.—Let $A\ C\ B$ (Fig. 79) be a lever of any form, and let it be in equilibrium by the forces P and P', acting in any oblique directions in the same plane. Produce $P\ A$ and $P'\ B$ till they meet in D; then, if the fulcrum is at C, the resultant must be in the direction

$D\,C$; otherwise the reaction of the fulcrum cannot keep the system in equilibrium (Art. 60, 2). Therefore (Art. 61) $P : P' :: \sin B\,D\,C : \sin A\,D\,C$.

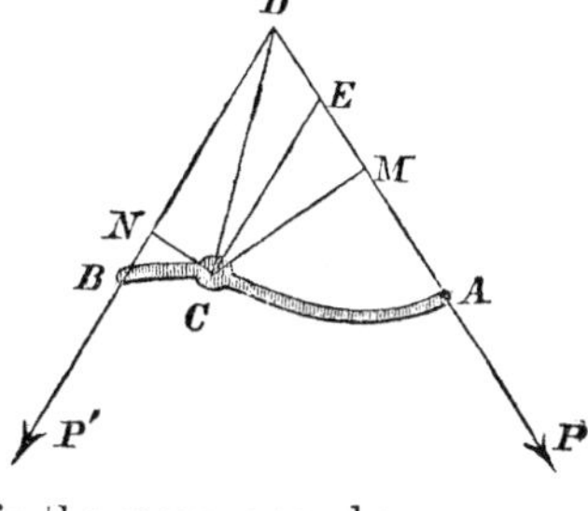

Fig. 79.

Draw $C\,M$ perpendicular to $A\,D$, and $C\,N$ to $B\,D$, and they are the sines of $A\,D\,C$ and $B\,D\,C$, to the same radius $D\,C$.

$\therefore P : P' :: C\,N : C\,M$; and $P \times C\,M = P' \times C\,N$.

The lines $C\,M$ and $C\,N$ are the acting distances of P and P'; therefore the law of the lever in all cases is the same, namely:

The moment of the power equals the moment of the weight.

When the forces act obliquely, the pressure on the fulcrum is less than the sum of the forces; for, if $C\,E$ is parallel to $B\,D$, then $D\,E$, $E\,C$, and $C\,D$, represent the three forces which are in equilibrium. But $C\,D$ is less than the sum of $D\,E$ and $E\,C$.

110. The Compound Lever.—When a lever acts on a second, that on a third, &c., the machine is called a *compound lever*. The law of equilibrium is—

The power is to the weight as the product of the acting distances on the side of the weight is to the product of the acting distances on the side of the power.

Let the force exerted by $A\,B$ on $B\,D$ (Fig. 80) be called x, and

Fig. 80.

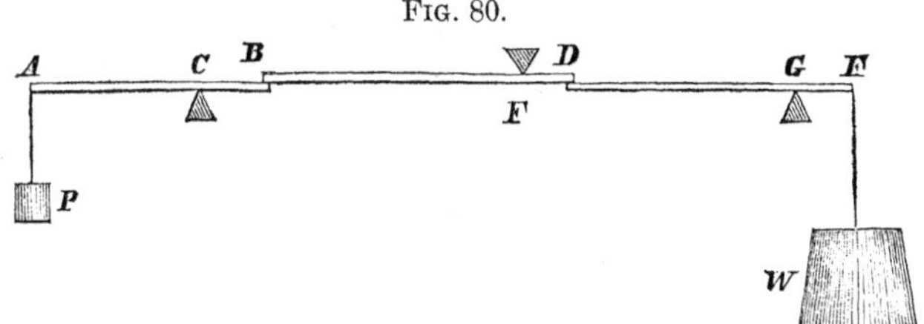

that of $B\,D$ on $D\,E$ be called y; then

$$P : x :: B\,C : A\,C;$$
$$\text{and } x : y :: D\,F : B\,F;$$
$$\text{and } y : W :: E\,G : D\,G.$$

Compounding these proportions, and dividing the first couplet by the common factors, we have

$$P : W :: B\,C \times D\,F \times E\,G : A\,C \times B\,F \times D\,G.$$

If the levers were of irregular forms, the acting distances might not be identical with the arms, as they are in the figure.

111. The Balance.—This is a common and valuable instrument for weighing. It is a straight lever with equal arms, having scale-pans, either suspended at the ends, or standing upon them, one to contain the poises, and the other the substance to be weighed. For scientific purposes, particularly for chemical analysis, great care is bestowed on the construction of the balance.

The arms of the balance, measured from the fulcrum to the points of suspension, must be precisely equal.

The knife-edges forming the fulcrum, and the points of suspension, are made of hardened steel, and arranged exactly in a straight line.

The centre of gravity of the beam is *below* the fulcrum, so that there may be a stable equilibrium; and yet below it by an exceedingly small distance, in order that the balance may be very sensitive.

To preserve the edge of the fulcrum from injury, the beam is raised by supports called *Y's*, when not in use.

A long index at right angles to the beam, points to zero on a scale when the beam is horizontal.

To protect the instrument from dust and moisture at all times, and from air-currents while weighing, the balance is in a glass case, whose front can be raised or lowered at pleasure.

Fig. 81.

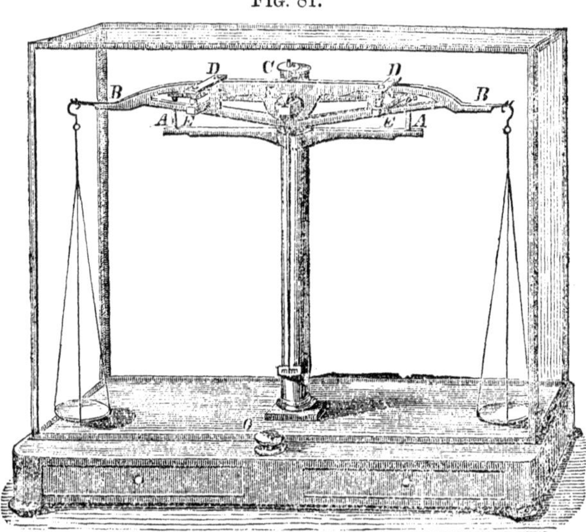

A balance for chemical analysis is shown in Fig. 81. By turning the knob *O*, the beam can be raised on the *Y's A A* from the surface on which the fulcrum *K* rests. The screw *C* raises and lowers the fulcrum in relation to the centre of gravity of the beam, in order to increase or diminish the sensitiveness of the instrument. In the most carefully made balances, the index will make a perceptible change, by adding to the scale *one millionth* of the poise.

For commercial purposes, it is convenient to have the scale-pans above the beam. This is done by the use of additional bars, which with the beam form parallelograms, whose upright sides are rods, projecting upward and supporting the scales. Such contrivances necessarily increase friction; but balances so constructed are sufficiently sensitive for ordinary weighing.

112. The Steelyard.—This is a weighing instrument, having a graduated arm, along which a poise may be moved, in order to balance various weights on the short arm. While the moment of the article weighed is changed by increasing or diminishing its quantity, that of the poise is changed by altering its acting distance. Since $P \times A\ C = W \times B\ C$ (Fig. 82), and P is constant,

FIG. 82.

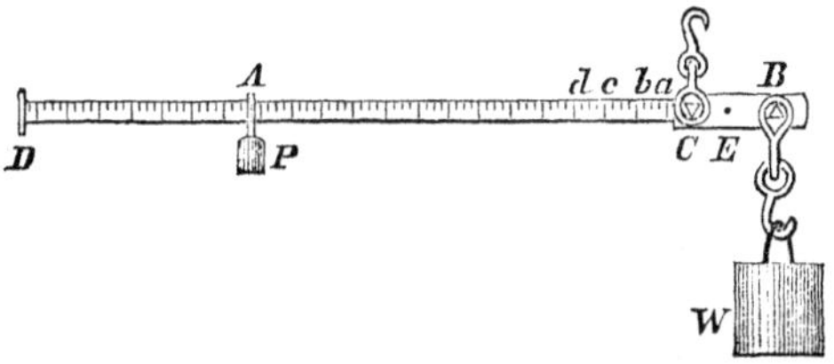

and also the distance $B\ C$ constant, $A\ C \propto W$; hence, if W is successively 1 lb., 2 lbs., 3 lbs., &c., the distances of the notches, a, b, c, &c., are as 1, 2, 3, &c.; in other words, the bar $C\ D$ is divided into equal parts. In this case, the graduation begins from the fulcrum C as the zero point.

But suppose, what is often true, that the centre of gravity of the steelyard is on the long arm, and that P placed at E would balance it; then the moment of the instrument itself is on the side $C\ D$, and equals $P \times C\ E$. Hence, the equation becomes

$$P \times A\ C + P \times C\ E = W \times B\ C; \text{ or}$$
$$P \times A\ E = W \times B\ C.$$

$\therefore W \propto A\ E$; and the graduation must be considered as commencing at E for the zero point. Such a steelyard cannot weigh below a certain limit, corresponding to the first notch a.

To find the length of the divisions on the bar, divide $A\ E$, the distance of the poise from the zero point, by W, the number of units balanced by P, when at that distance.

The steelyard often has *two* fulcrums, one for less and the other for greater weights.

113. Platform Scales.—This name is given to machines arranged for weighing heavy and bulky articles of merchandise. The largest, for cattle, loaded wagons, &c., are constructed with the platform at the surface of the ground. In order that the platform may stand firmly beneath its load, it rests by four feet on as many levers of the second order, whose arms have equal ratios. *A F*, *B F*, *C G*, *D G* (Fig. 83), are four such levers, resting on the

Fig. 83.

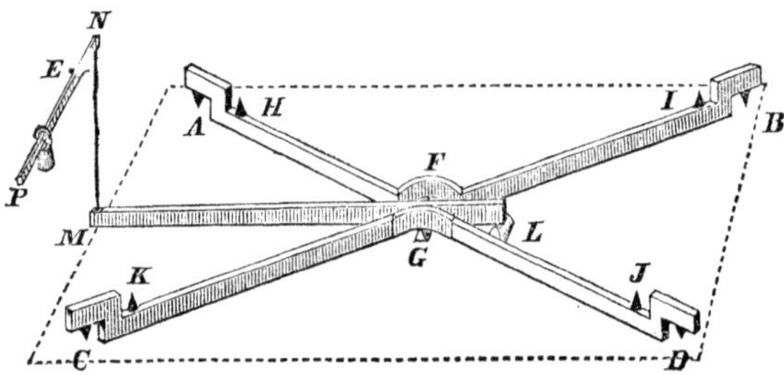

fulcrums, *A*, *B*, *C*, *D*, while the other ends meet on the knife-edge, *F G*, of another lever, *L M*. This fifth lever has its fulcrum at *L*, and its outer extremity is attached by a vertical rod, *M N*, to a steelyard, whose fulcrum is *E*, and poise *P*. The five levers are arranged in a square cavity just below the surface of the ground. The dotted line shows the outline of the cavity. On the bearing points of the four levers, *H*, *I*, *J*, *K*, rest the feet of the platform (not represented), which is firmly built of plank, and just fits into the top of the cavity without touching the sides. The machine is a compound lever of three parts; for the four levers act as one at *F G*, and are used to give steadiness to the platform which rests upon them.

A construction quite similar to the above is made of portable size, and used in all mercantile establishments for weighing heavy goods.

114. Questions on the Lever.—1. *A B* (Fig. 84) is a uniform bar, 2 feet long, and weighs 4 oz.; where must the fulcrum be put, that the bar may be balanced by *P*, weighing 5 lbs.?

Ans. $\frac{4}{7}$ of an inch from *A*.

Fig. 84.

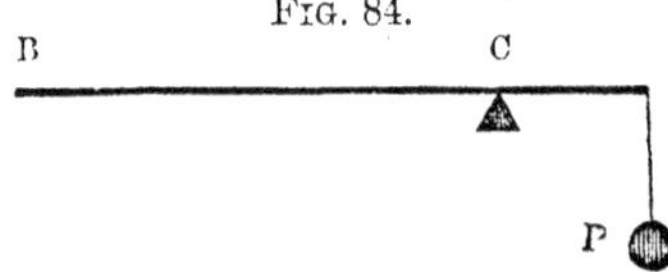

2. A lever of the second order is 25 feet long; at what distance from the fulcrum must a weight

of 125 pounds be placed, so that it may be supported by a power able to sustain 60 pounds, acting at the extremity of the lever.

Ans. 12 feet.

3. *A* and *B* are of the same height, and sustain upon their shoulders a weight of 150 pounds, placed on a pole $9\frac{1}{3}$ feet long; the weight is placed $6\frac{2}{3}$ feet from *A*; what is the weight sustained by each person?

Ans. *A* sustains $42\frac{6}{7}$ lbs., and *B* sustains $107\frac{1}{7}$ lbs.

4. The longer arm of a steelyard is 2 feet 2 inches in length, and the shorter $2\frac{2}{3}$ inches; and its apparatus of hooks, &c., is so contrived that a weight of 2 pounds, placed upon the longer arm, at the distance of 10 inches from the centre of motion, will balance 8 pounds placed at the extremity of the shorter arm; the movable weight (of 2 pounds) cannot conveniently be placed nearer to the fulcrum than $\frac{2}{3}$ of an inch; what must be the graduation of the steelyard that it may weigh ounces, and what will be the greatest and least weights that can be ascertained by it?

Ans. The graduation is to 12ths of an inch; and it will weigh from 1 to 20 pounds.

5. A *bent lever*, *A C B* (Fig. 85), has the arm $A\ C = 3$ feet, $C\ B = 8$ feet, $P = 5$ lbs., and the angle $A\ C\ B = 140°$; what weight, *W*, must be attached at *B*, in order to keep *A C* horizontal?

FIG. 85.

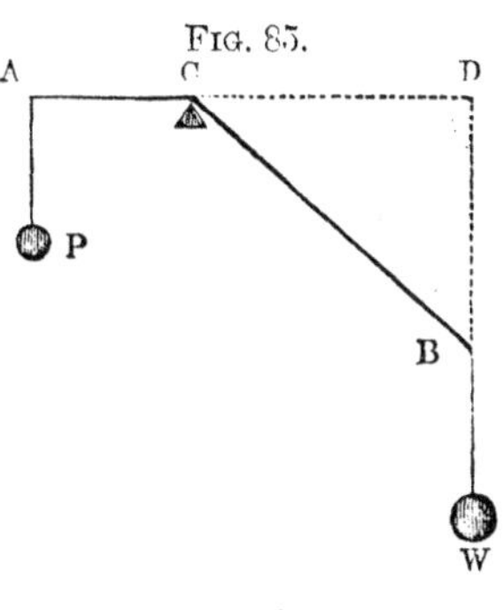

Ans. 2.4476 lbs.

6. A cylindrical straight lever is 14 feet long, and weighs 6 lbs. 5 oz.; its longer arm is 9, and its shorter 5 feet; at the extremity of its shorter arm a weight of 15 lbs. 2 oz. is suspended; what weight must be placed at the extremity of the longer arm to keep it in equilibrium?

Ans. 7 lbs.

7. A uniform bar, 12 feet long, weighs 7 lbs.; a weight of 10 lbs. hangs on one end, and 2 feet from it is applied an upward force of 25 lbs.; where must the fulcrum be put to produce equilibrium?

Ans. 1 foot from the 10 lbs.

8. The lengths of the arms of a balance are *a* and *b*. When *p* ounces are hung on *a*, they balance a certain body; but it requires *q* ounces to balance the same body, when placed in the other scale. What is the true weight of the body? According to the first weighing, $a\,p = b\,x$; according to the second, $b\,q = a\,x$. $\therefore a\,b\,p\,q = a\,b\,x^2$, and $x = \sqrt{p\,q}$. Hence, the true weight is a geometrical mean between the apparent weights.

9. On one arm of a false balance a body weighs 11 lbs.; on the other, 17 lbs. 3 oz.; what is the true weight?

Ans. 13 lbs. 12 oz.

10. Four weights of 1, 3, 5, 7 lbs., respectively, are suspended from points of a straight lever, eight inches apart; how far from the point of suspension of the first weight must the fulcrum be placed, that the weights may be in equilibrium?

Ans. 17 inches.

11. Two weights keep a horizontal lever at rest, the pressure on the fulcrum being 10 lbs., the difference of the weights 4 lbs., and the difference of the lever arms 9 inches; what are the weights and their lever arms?

Ans. Weights, 7 lbs. and 3 lbs.; arms, $6\frac{3}{4}$ in. and $15\frac{3}{4}$ in.

II. The Wheel and Axle.

115. Description and Law of the Machine.—The wheel and axle consists of a cylinder and a wheel, firmly united, and free to revolve on a common axis. The power acts at the circumference of the wheel in the direction of a tangent, and the weight in the same manner, at the circumference of the cylinder or axle; so that the acting distances are the radii at the two points of contact. As the system revolves, the radii successively take the place of acting distances, without altering at all the relation of the forces to each other. The wheel and axle is therefore a kind of endless lever.

Let W (Fig. 86) be the weight suspended from the axle, tending to revolve it on the line $L\,M$; and P, the power acting on the wheel, tending to revolve the system in the opposite direction. It is plain that the acting distances are the radius of the axle, and $A\,C$ the radius of the wheel. In case of equilibrium, the moment of W equals the moment of P. Calling the radius of the axle r, and the radius of the wheel R, then $W \times r = P \times R$; or

Fig. 86.

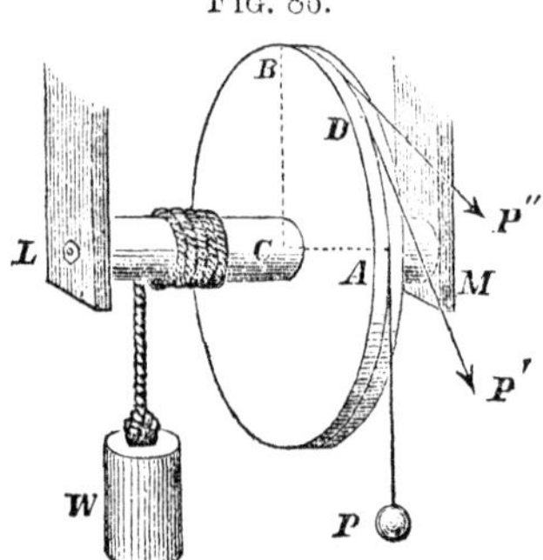

$$P : W :: r : R.$$

If, instead of the weight P, suspended on the wheel, the rope be drawn by any force in the direction P' or P'', it is still tangent to the circumference, and therefore its acting distance, $C\,D$ or $C\,B$, the same as before. In general, the law of equilibrium for this machine is,

The power is to the weight as the radius of the axle to the radius of the wheel.

If the rope on the wheel, being fastened at A (Fig. 87) is drawn by the side of the wheel, as $A\ P'$, the acting distance of the power is diminished from $C\ A$ to $C\ E$, and therefore its efficiency is diminished in the same ratio. Were the rope drawn away from the wheel, as $A\ P''$, making an equal angle on the other side of $A\ P$, the same effect is produced, the acting distance now becoming $C\ F$.

Fig. 87.

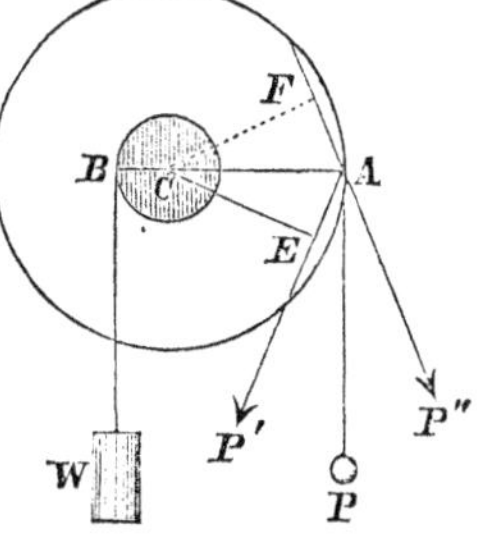

The radius of the wheel and the radius of the axle should each be reckoned from the axis of rotation to the *centre of the rope;* that is, half of the thickness of the rope should be added to the radius of the circle on which it is coiled. Calling t the half thickness of the rope on the axle, and t' that of the rope on the wheel, the proportion for equilibrium is $P : W :: r + t : R + t'$.

116. The Compound Wheel and Axle.—When a train of wheels, like that in Fig. 88, is put in motion, those which *communicate* motion by the circumference are called *driving wheels*, as A and C; those which *receive* motion by the circumference are called *driven wheels*. And the law of equilibrium is,

Fig. 88.

The power is to the weight as the product of the radii of the driving wheels to the product of the radii of the driven wheels.

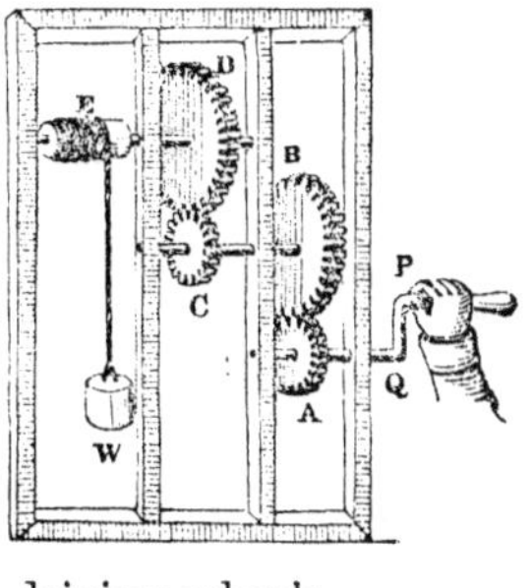

The crank $P\ Q$ is to be reckoned among driven wheels; the axle E among driving wheels.

Let the radius of B be called R; of D, R'; of A, r; of C, r'; of E, r''. Call the force exerted by A on B, x; that of C on D, y. Then

$$P : x :: r : P\ Q;$$
$$x : y :: r' : R;$$
$$y : W :: r'' : R';$$
$$\therefore P : W :: r \times r' \times r'' : P\ Q \times R \times R'.$$

If the driving wheels are equal to each other, and also the driven wheels, and the number of each is n, then

$$P : W :: r^n . R^n.$$

117. Direction and Rate of Revolution.—When two wheels are geared together by teeth, they necessarily revolve in contrary directions. Hence, in a train of wheels, the alternate axles revolve the same way.

The circumferences of two wheels which are in gear move with the same velocity; hence the number of revolutions will be reciprocally as the radii of the wheels.

Since teeth which gear together are of the same size, the relative *number of teeth* is a measure of the relative circumferences, and therefore of the relative radii of the wheels. If the wheel A (Fig. 88) has 20 teeth, and B has 40, and again if C has 15, and D 45, then for every revolution of B, A revolves twice, and for every revolution of D, C revolves three times. Therefore, six turns of the crank are necessary to give one revolution to the axle E.

By cutting the teeth of wheels on a conical instead of a cylindrical surface, the axles may be placed at any angle with each other, as represented in Fig. 89.

FIG. 89.

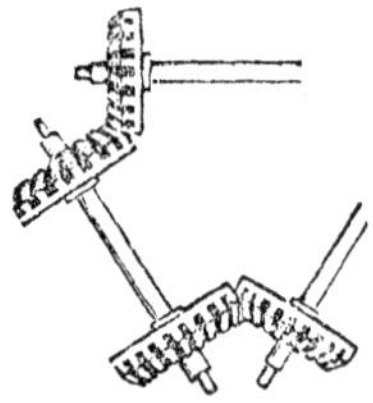

Whether axles are parallel or not, *bands* instead of teeth may be used for transmitting rotary motion. But as bands are liable to slip more or less, they cannot be employed in cases requiring exact relations of velocity.

118. Questions on the Wheel and Axle.—

1. A power of 12 lbs. balances a weight of 100 lbs. by a wheel and axle; the radius of the axle is 6 inches; what is the *diameter* of the wheel?
Ans. 8 ft. 4 in.

2. $W = 500$ lbs.; $R = 4$ ft.; $r = 8$ in.; the weight hangs by a rope 1 inch thick, but the power acts at the circumference of the wheel without a rope; what power will sustain the weight?
Ans. 88.54 lbs.

3. $R = 1$ ft.; $r = 2$ in.; the well-stone weighs 256 lbs.; the bucket, empty, weighs 18 lbs.; the bucket, filled, weighs 65 lbs.; what force must a person apply to the bucket-rope, in each case, for equilibrium? *Ans.* 1st, down, $24\frac{2}{3}$ lbs.; 2d, up, $22\frac{1}{3}$ lbs.

4. In Fig. 88, A and C have each 15 teeth, B and D each 40 teeth; the radius of the axle E is 4 inches; the rope on it 1 inch in diameter; and the radius of the crank $P\ Q$ is 18 inches; what is the ratio of power to weight in equilibrium? *Ans.* $1 : 28\frac{4}{9}$.

III. The Pulley.

119. The Pulley Described.—The pulley consists of one or more wheels or rollers, with a rope passing over the edge in which a groove is sunk to keep the rope in place. The axis of the roller is in a *block*, which is sometimes fixed, and sometimes rises and falls with the weight; and the pulley is accordingly called a *fixed pulley* or a *movable pulley*. The principle which explains the relation of power and weight in every form of pulley, is this:

Whatever strain or tension is applied to one end of a cord, is transmitted through its whole length, if it does not branch, however much its direction is changed.

In the pulley, the sustaining portions of the rope are assumed to be parallel to each other.

Fig. 90. Fig. 91.

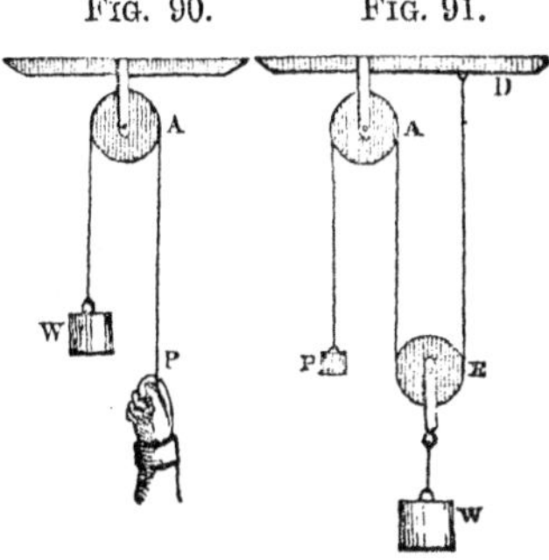

120. The Fixed Pulley.—In the fixed pulley, A (Fig. 90), the force P produces a tension in the string, which is transmitted through its whole length, and which can be balanced only when W equals P. Hence, in the fixed pulley, *the power and weight are equal.* This machine is useful for changing the *direction* in which the force is applied to the weight; and if the power only acts in the plane of the groove of the wheel, it is immaterial what is its direction, horizontal, vertical, or oblique.

Fig. 92.

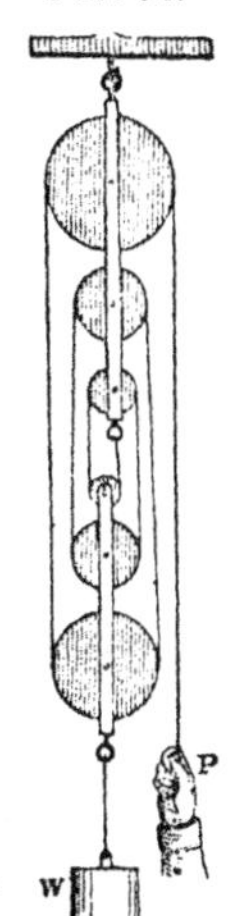

121. The Movable Pulley.—In Fig. 91, the tension produced by P, is transmitted from A down to the wheel E, and thence up to D; therefore W is sustained by *two* portions of the rope, each of which exerts a force equal to P.

$$\therefore W = 2P; \text{ or } P : W :: 1 : 2.$$

The same reasoning applies, where the rope passes between the upper and lower blocks any number of times, as in Fig. 92. The force causes a tension in the rope, which is transmitted to every portion of it. If n is the number of portions which sustain the lower block, then W is upheld by $n\,P$; and if there is equilibrium, $P : W :: 1 : n$. In the figure, the weight equals *six times* the power. The law of equilibrium, therefore, for the movable pulley with one rope, is this,

The power is to the weight as one to the number of

the sustaining portions of the rope; or, as one to twice the number of movable pulleys.

122. The Compound Pulley.—Wherever a system of pulleys has separate ropes, the machine is to be regarded as compound, and its efficiency is calculated accordingly. Figures 93 and 94 are examples. In Fig. 93, call the weight sustained by F, x, and that sustained by D, y. Then (Art. 121),

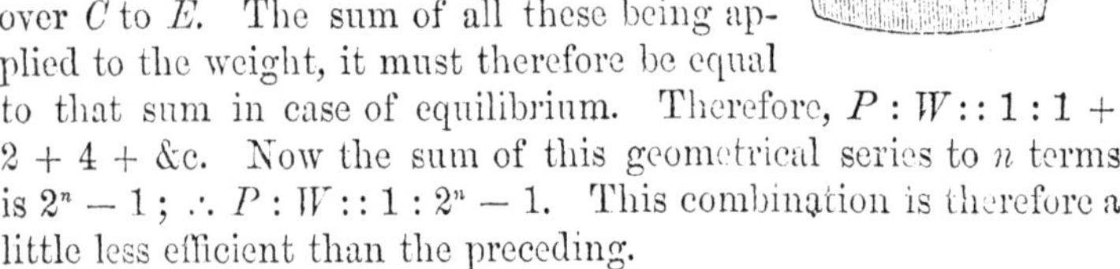

FIG. 93. FIG. 94.

$$P : x :: 1 : 2;$$
$$x : y :: 1 : 2;$$
$$y : W :: 1 : 2.$$
$$\therefore P : W :: 1 : 2^3 :: 1 : 8.$$

And if n is the number of ropes,

$$P : W :: 1 : 2^n.$$

In Fig. 94, the tension P is transmitted over A directly to the weight at G; the wheel A is loaded, therefore, with 2 P, and a tension of 2 P comes upon the second rope, which is transmitted over B to the weight at F. In like manner, a tension of 4 P is transmitted over C to E. The sum of all these being applied to the weight, it must therefore be equal to that sum in case of equilibrium. Therefore, $P : W :: 1 : 1 + 2 + 4 +$ &c. Now the sum of this geometrical series to n terms is $2^n - 1$; $\therefore P : W :: 1 : 2^n - 1$. This combination is therefore a little less efficient than the preceding.

Since the several ropes have different tensions, the weight cannot be balanced upon them, unless those of greatest tension are nearest the line of direction of the body. For example, if the rope F is directed toward the centre of gravity of the weight, the rope G should be attached four times as far from it as the rope E, in order to prevent the weight from tipping.

The pulley owes its efficiency as a machine to the fact, that the tension produced by the power is applied *repeatedly* to the weight. The only use of the wheels is to diminish friction. Were it not for friction, the rope might pass round fixed pins in the blocks, and the ratio of power to weight would still be in every case the same as has been shown.

IV. The Rope Machine.

123. Definition and Law of this Machine.—

The rope machine is one in which the power and weight are in equilibrium by the tension of one or more ropes.

According to this definition the pulley is included. It is that particular form of the rope machine in which the sustaining parts of the ropes are parallel; and it is treated as a separate machine, because its theory is very simple, and because it is used far more extensively than any other forms.

Fig. 95.

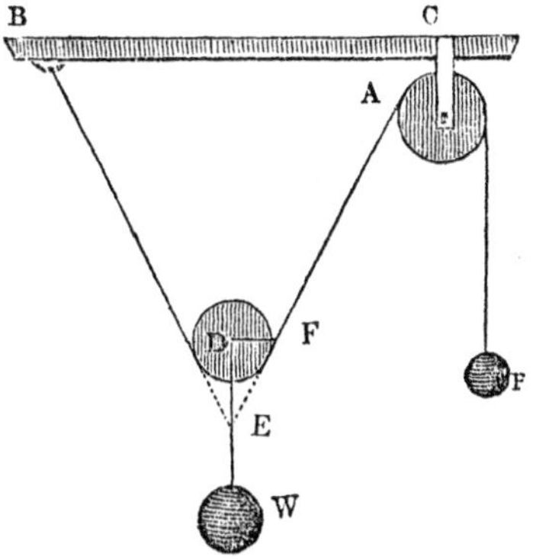

If the two portions of rope which sustain the weight are inclined, as in Fig. 95, then W is no longer equal to the sum of their tensions, as it is in the pulley, but is always less than that, according to the following law:

The power is to the weight as radius is to twice the cosine of half the angle between the parts of the rope.

Put $A\ E\ B = 2\ a$; then $F\ E\ D = a$, and since $\sin B\ E\ W = \sin B\ E\ D = \sin a$, we shall have (Art. 61) $P : W :: \sin a : \sin 2\ a$; but $\sin a : \sin 2\ a :: R : 2 \cos a$; $\therefore\ P : W :: R : 2 \cos a$.

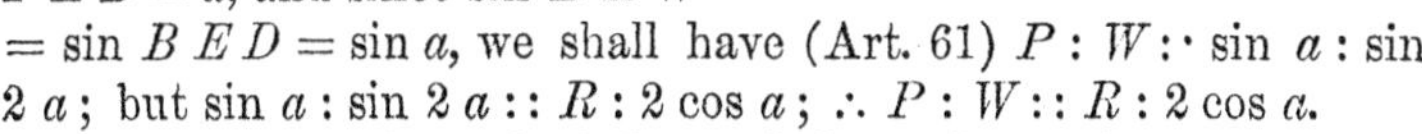

If in Fig. 96, the end of the cord, instead of being attached to the beam, is carried over another fixed pulley, and a weight equal to P is hung upon it, the equilibrium will be preserved, because all parts of the rope have a tension equal to P; therefore, as before,

$$P : W :: R : 2 \cos a;\ \text{or}\ 2\ P : W :: R : \cos a :: B\ C : C\ D.$$

124. Change in the Ratio of Power and Weight.—If P is given, all the possible values of W are included between $W = 0$, and $W = 2\ P$.

When the rope is straight from A to B, so that $C\ D = 0$, then, by the above proportion, $W = 0$. As W is increased from zero, the point C descends; and when $D\ C = \frac{1}{2}\ B\ C$, then, by the proportion, $W = P$. In that case $D\ C\ B = 60°$, and the angles, $A\ C\ B$, $A\ C\ W$, and $B\ C\ W$, are equal (each being $120°$), as they should be, because each of the equal forces, P, P, and W, is as the sine of the angle between the directions of the other two.

Fig. 96.

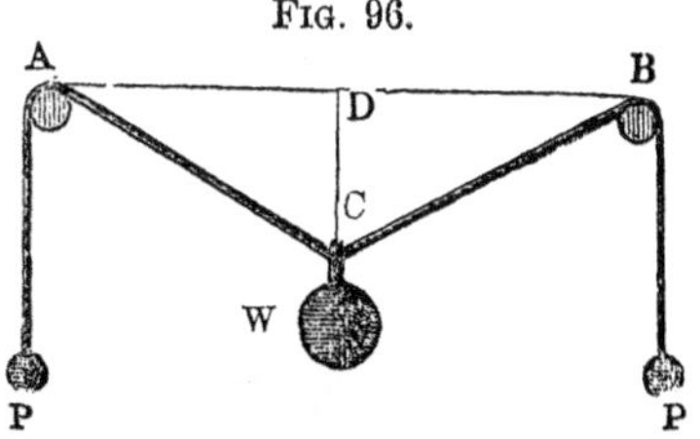

But when W has increased to $2P$, it descends to an infinite distance; for then, by the proportion, $CD = BC$, that is, the side of a right-angled triangle is equal to the hypothenuse. Thus, the extreme values of W are 0 and $2P$.

It appears from the foregoing, that a perfectly *flexible rope having weight* cannot be drawn into a straight horizontal line, by any force however great; for C cannot coincide with D, except when $W = 0$.

125. The Branching Rope.—When C, where the weight is suspended, is a *fixed* point of the rope, we have a branching rope, and the principle of transmitted tension does not apply beyond the point of division.

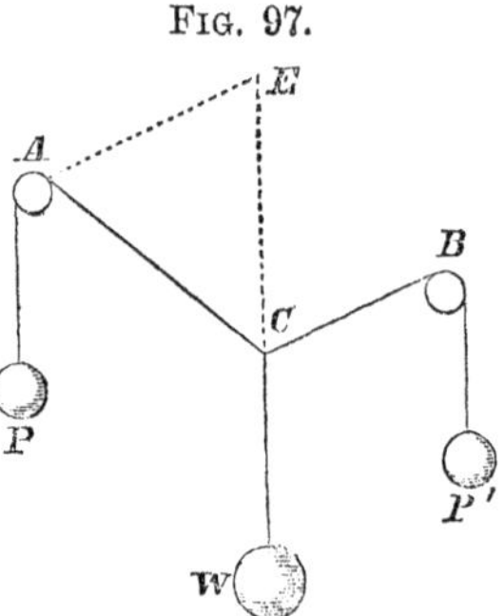

FIG. 97.

Let P, P' and W (Fig. 97), be given, and C a fixed point of the rope. Produce WC, and let AE, drawn parallel to CB, intersect it in E. The sides of ACE are proportional to the given forces; therefore its angles can be found, and the inclinations of AC and BC to the vertical CW are known.

126. The Funicular Polygon.—If several weights are attached at fixed points along the cord ACB (Fig. 98), the combination is called the funicular polygon; and the fact that there are opposite and equal tensions in any portion of the cord will enable us to transfer all the forces to one point.

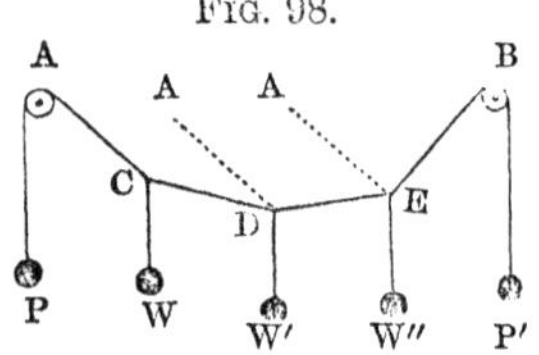

FIG. 98.

Let the tension of $CD = T$; and that of $DE = T'$. C is kept at rest by P, T, and W; hence T is equal to the resultant of P and W. But the same T (in the opposite direction) equals and balances the resultant of T' and W'. Suppose, now, CD to vanish, by removing C to D; draw AD parallel to AC; let P act in the line DA, and W in the line DW'. D will now be in equilibrium, as before, because there has merely been made a substitution of P and W in their original directions for T, their equivalent. Now consider D to be acted on by three forces, T', P, and $W + W'$; $\therefore T' =$ the resultant of P and $W + W'$, and the two latter can be transferred, as before, to E, AE being parallel to AC or AD. E is therefore kept at rest by the three forces, P, P', and $W + W' + W''$. We can now use the triangle of forces, as in the preceding article, to determine the directions of BE and AE, or its parallel, AC, and hence, of the parts, CD and DE.

If the weights are all equal, and their number $= n$, the three forces at E are P, P', and $n W$.

An example of this kind occurs in the suspension bridge, whose weight is distributed at equal distances along the supporting chains. And an extreme case is that of a heavy rope or chain suspended loosely over pulleys, as in Fig. 99. Equal weights are suspended at an infinite number of points, and therefore the funicular polygon becomes a curve, and is called the *catenary* curve. Its directions at the extremities, A and B, and the law of the curve, may be determined by the principles given above.

Fig. 99.

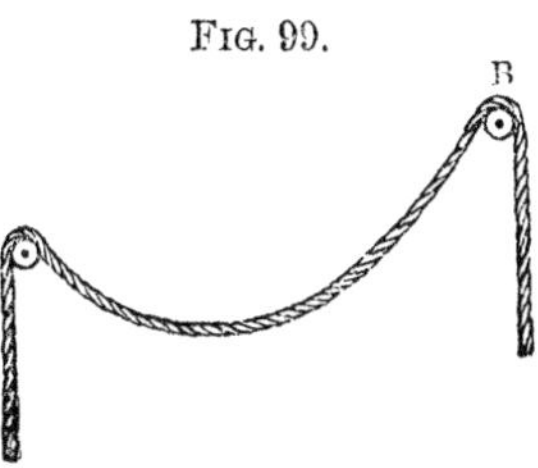

V. The Inclined Plane.

127. Relation of Power, Weight, and Pressure on the Plane.—The mechanical efficiency of the inclined plane is explained on the principle of *oblique action;* that is, it enables us to apply the power to balance or overcome only *one component* of the weight, instead of the whole. Let the weight of the body G, lying on the inclined plane $A\ C$ (Fig. 100), be represented by W; and resolve it into F parallel, and N perpendicular to the plane. N represents the perpendicular pressure, and is equal to the reaction of the plane; F is the force by which the body tends to move down the plane.

Let $a =$ the angle C, the inclination of the plane; therefore $W\ G\ N = a$. Then $F = W \cdot \sin a$; and $N = W \cdot \cos a$.

Fig. 100. Fig. 101.

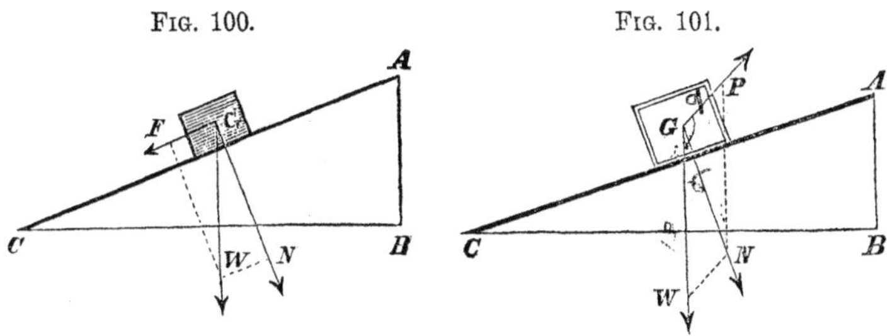

Now suppose a force P is applied at G (Fig. 101), which keeps the body at rest. Then the resultant of W and P must be N, which is resisted by the plane; therefore,

$$P : W :: \sin G\ N\ P, \text{ or } \sin a : \sin P\ G\ N.$$

When the power acts parallel to the plane, $P\ G\ N = 90°$, and we have $P : W :: \sin a : \sin 90° :: A\ B : A\ C$. Hence, when the power acts in a line parallel to the inclined plane, which is the most common direction,

The power is to the weight as the height to the length of the inclined plane.

When the power acts in a line parallel to the base of the inclined plane, $P\ G\ N = 90° - a$, and we have $P : W :: \sin a : \cos a :: A\ B : B\ C$. Hence, when the power acts in a line parallel to the base of the inclined plane,

The power is to the weight as the height is to the base of the inclined plane.

128. Power most Efficient when Acting Parallel to the Plane.—From the proportion

$$P : W :: \sin a : \sin P\ G\ N, \text{ we derive}$$

$$W = \frac{P \cdot \sin P\ G\ N}{\sin a}.$$

Now as P and $\sin a$ are given, W varies as $\sin P\ G\ N$, which is the greatest possible when $P\ G\ N = 90°$; that is, when the power acts in a line parallel to the plane.

Whether the angle $P\ G\ N$ diminishes or increases from 90°, its sine diminishes, and becomes zero, when $P\ G\ N = 0°$, or 180°. Therefore $W = 0$, or no weight can be sustained, when the power acts in the line $G\ N$, perpendicular to the plane, either toward the plane or from it.

129. Expression for Perpendicular Pressure.—From the triangle $P\ G\ N$ we obtain

$$N : W :: \sin G\ P\ N : \sin P\ G\ N,$$

$$\text{or } N : W :: \sin P\ G\ W : \sin P\ G\ N;$$

$$\therefore N = W \frac{\sin P\ G\ W}{\sin P\ G\ N}.$$

If the power acts in a line parallel to the inclined plane, $P\ G\ W = 90° + a$, $P\ G\ N = 90°$, and $N = W \dfrac{\sin (90° + a)}{\sin 90°} = W \cos a$.

If the power acts in a line parallel to the base of the inclined plane, $P\ G\ W = 90°$, $P\ G\ N = 90° - a$, and $N = W \dfrac{1}{\cos a} = W \sec a$.

If the power acts in a line perpendicular to the inclined plane, $P\ G\ W = a$, $P\ G\ N = 0°$, and $N = W \dfrac{\sin a}{0} = \infty$.

130. Equilibrium between Two Inclined Planes.—If a body rests, as represented in Fig. 102, between two inclined planes, the three forces which retain it are its weight, and the resistances of the planes. Draw $H F$ and $L F$ perpendicular to the planes through the points of contact, and $G F$ vertically through the centre of gravity of the body. Since the body is in equilibrium, these three lines will pass through the same point (Art. 60, 2). Let that point be F, and draw $G P$ parallel to $L F$, and $M K$ parallel to the horizon. $G P F$ is similar to $K C M$. Therefore (since Pressure on $A C$: Pr. on $D C :: P G : F P$),

Fig. 102.

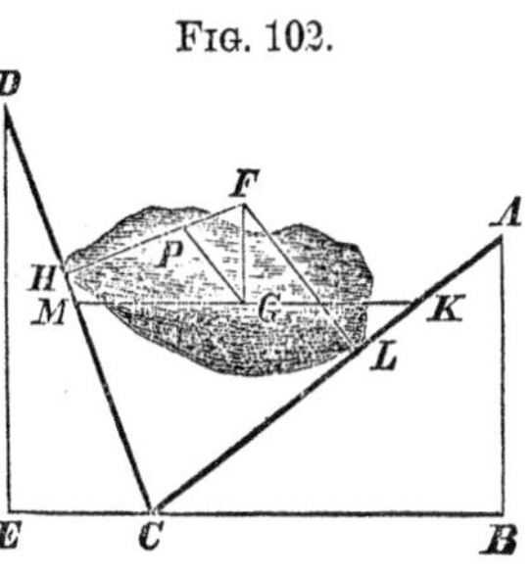

$$\text{Pressure on } A\,C : \text{Pr. on } D\,C :: K\,C : M\,C,$$
$$:: \sin M : \sin K,$$
$$:: \sin D\,C\,E : \sin A\,C\,B.$$

That is, when a body rests between two planes, it exerts pressures on them which are inversely as the sines of their inclinations to the horizon.

If, therefore, one of the planes is horizontal, none of the pressure can be exerted on any other plane. It is friction alone which renders it possible for a body on a horizontal surface to lean against a vertical wall.

131. Bodies Balanced on Two Planes by a Cord passing over the Ridge.—Let P and W balance each other on the planes $A D$ and $A C$ (Fig. 103), which have the common height $A B$, by means of a cord passing over the fixed pulley A. The tension of the cord is the common power which prevents each body from descending; and as the cord is parallel to each plane, we have (calling the tension t),

Fig. 103.

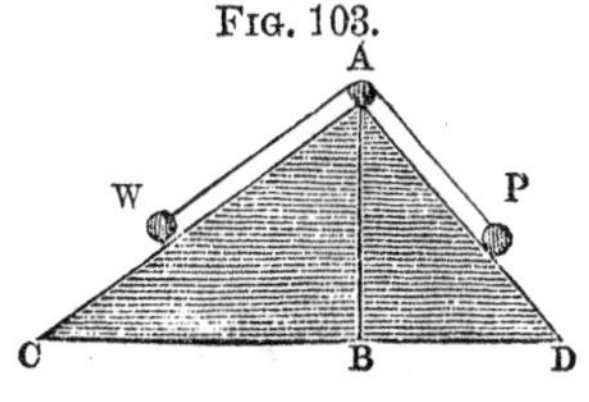

$$t : P :: A\,B : A\,D;$$
$$\text{and } t : W :: A\,B : A\,C;$$
$$\therefore P : W :: A\,D : A\,C;$$

that is, the weights, in case of equilibrium, are directly as the lengths of the planes.

132. Questions on the Inclined Plane.—

1. If a horse is able to raise a weight of 440 lbs. perpendicularly, what weight can he raise on a railway having a slope of five degrees? *Ans.* 5048.5 lbs.

2. The grade of a railroad is 20 feet in a mile; what power must be exerted to sustain any given weight upon it?

Ans. 1 lb. for every 264 lbs.

3. What force is requisite to hold a body on an inclined plane, by pressing perpendicularly against the plane?

Ans. An infinite force.

4. A certain power was able to sustain 500 tons on a plane of $7\frac{1}{2}°$; but on another plane, it could sustain only 400 tons; what was the inclination of the latter? *Ans.* $9° 23' 25''$.

5. Equilibrium on an inclined plane is produced when the power, weight, and perpendicular pressure are, respectively, 9, 13, and 6 lbs.; what is the inclination of the plane, and what angle does the power make with the plane?

Ans. $a = 37° 21' 26''$. Inclination of power to plane $= 28° 46' 54''$.

6. A power of 10 lbs., acting parallel to the plane, supports a certain weight; but it requires a power of 12 lbs. parallel to the base to support it. What is the weight of the body, and what is the inclination of the plane?

Ans. $W = 18.09$ lbs. $a = 33° 33' 25''$

7. To support a weight of 500 lbs. upon an inclined plane of 50° inclination to the horizon, a force is applied whose direction makes an angle of 75° with the horizon. What is the magnitude of this force, and the pressure of the weight against the plane?

Ans. $P = 422.6$ lbs. $N = 142.8$ lbs.

VI. The Screw.

133. Reducible to the Inclined Plane.—The screw is a cylinder having a spiral ridge or thread around it, which cuts at a constant oblique angle all the lines of the surface parallel to the axis of the cylinder. A hollow cylinder, called a *nut,* having a similar spiral within it, is fitted to move freely upon the thread of the solid cylinder. In Fig. 104, let the base $A B$ of the inclined

Fig. 104.

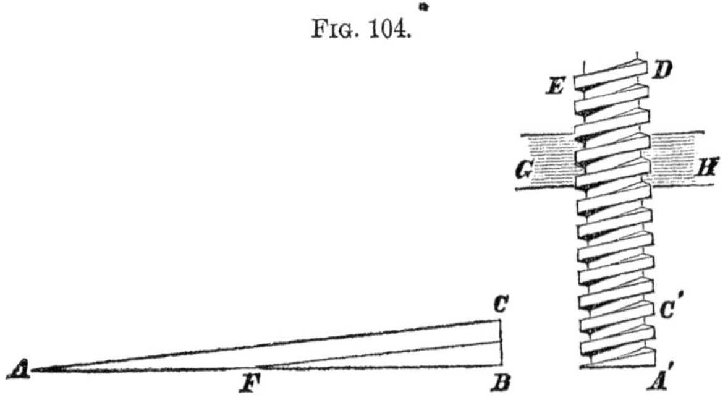

plane $A\ C$ be equal to twice the circumference of the cylinder $A'\ E$; then let the plane be wrapped about the cylinder, bringing the points A, F, and B, to the point A'; then will $A\ C$ describe two revolutions of the thread from A' to C'. Therefore the mechanical relations of the screw are the same as of the inclined plane.

If a weight be laid on the thread of the screw, and a force be applied to it horizontally in the direction of a tangent to the cylinder, the case is exactly analogous to that of a body moved on an inclined plane by a force parallel to the base. Let r be the radius of the cylinder, then $2\ \pi\ r$ is the circumference; also let d be the distance between the threads, (that is, from any point of one revolution to the corresponding point of the next,) measured parallel to the axis of the cylinder; then $2\ \pi\ r$ is the base of an inclined plane, and d its height. Therefore (Art. 127),

$$P : W :: d : 2\ \pi\ r; \text{ or,}$$

The power is to the weight as the distance between the threads measured parallel to the axis, is to the circumference of the screw.

If instead of moving the weight on the thread of the screw, the force is employed to turn the screw itself, while the weight is free to move in a vertical direction, the law is the same. Thus, whether the screw $A'\ E$ is allowed to rise and fall in the fixed nut $G\ H$, or whether the nut rises and falls on the thread of the screw, while the latter is revolved, without moving longitudinally, in each case, $P : W :: d : 2\ \pi\ r$.

134. The Screw and Lever Combined.—The screw is so generally combined with the lever in practical mechanics, that it is important to present the law of the compound machine. Let $A\ F$ (Fig. 105) be the section of a screw, and suppose $B\ C$, a lever of the second order, to be applied to turn it. The fulcrum is at C, the power acts at B, and the effect produced by the lever is at A, the surface of the cylinder. Call that effect x, and let $d =$ the distance between the threads; then,

FIG. 105.

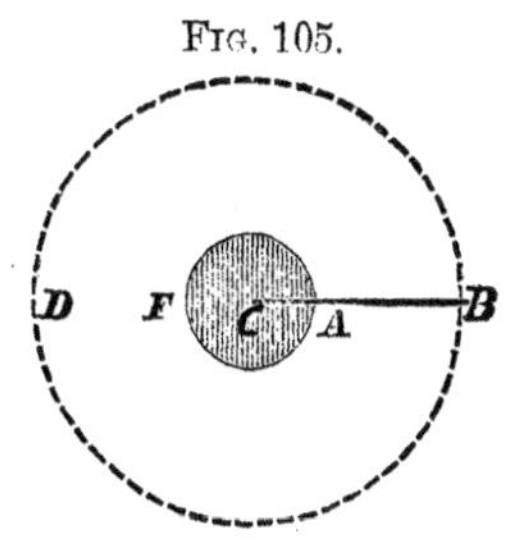

$$P : x :: A\ C : B\ C,$$

$$\text{and } x : W :: d : 2\ \pi\ A\ C;$$

compounding and reducing, we have

$$P : W :: d : 2\ \pi\ B\ C; \text{ that is,}$$

The power is to the weight as the distance between the threads, measured parallel to the axis, to the circumference described by the power.

The law as thus stated is applicable to the screw when used with the lever or without it.

135. The Endless Screw.—The screw is so called, when its thread moves between the teeth of a wheel, thus causing it to revolve. It is much used for diminishing very greatly the velocity of the weight.

Fig. 106.

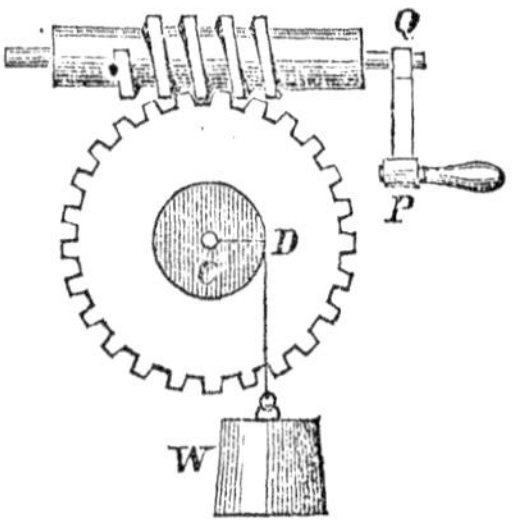

Let $P\ Q$ (Fig. 106) be the radius of the crank to which the power is applied; d, the distance between the threads; R, the radius of the wheel; r, the radius of the axle; and call the force exerted by the thread upon the teeth, x. Then,

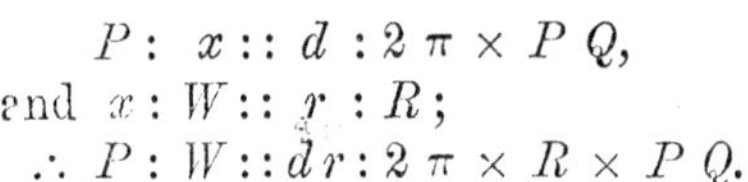

$$P : x :: d : 2\pi \times P\ Q,$$
and $x : W :: r : R$;
$$\therefore P : W :: d\,r : 2\pi \times R \times P\ Q.$$

If, for example, $P\ Q = 30$ inches, $d = 1$ in., $R = 18$ in.; $r + t = 2$ in.; then W moves with 1696 times less velocity than P.

136. The Right and Left Hand Screw.—The common form of screw is called the *right-hand* screw, and may be described thus; *if the thread in its progress along the length of the cylinder, passes from the left over to the right, it is called a right-hand screw.* Hence, a person in driving a screw forward turns it from his left *over* (not *under*) to his right, and in drawing it back he reverses this movement. Fig. 104 represents a right-hand screw.

The *left-hand* screw is one whose thread is coiled in the opposite direction,—that is, it advances by passing from right over to left. This kind is used only when there is special reason for it. For example, the screws which are cut upon the left-hand ends of carriage axles are left-hand screws; otherwise there would be danger that the friction of the hub against the nut might turn the nut off from the axle. Also, when two pipes for conveying gas or steam are to be drawn together by a nut, one must have a right-hand, and the other a left-hand screw.

137. Questions on the Screw.—

1. The distance between the threads of a screw is one inch, the bar is two feet long from the axis, and the power is 30 lbs.; what is the weight or pressure? *Ans.* 4523.89 lbs.

2. The bar is three feet long, reckoned from the axis, $P = 60$ lbs., $W = 2240$ lbs.; what is the distance between the threads? *Ans.* 6.058 inches.

3. A compound machine consists of a crank, an endless screw, a wheel and axle, a pulley, and an inclined plane. The radius of the crank is 18 inches; the distance between the threads of the screw, one inch; the radius of the wheel on which the screw acts, two feet; the radius of the axle, 6 inches; the pulley block has two movable pulleys with one rope; and the inclination of the plane to the horizon is 30°. What weight on the plane will be balanced by a power of 100 lbs. applied to the crank? *Ans.* 361911.168 lbs.

VII. The Wedge.

138. Definition of the Wedge, and the Mode of Using.—The usual form of the wedge is a triangular prism, two of whose sides meet at a very acute angle. This machine is used to raise a weight by being driven as an inclined plane underneath it, or to separate the parts of a body by being driven between them. When it is used by itself, and does not form part of a compound machine, force is usually applied by a blow, which produces an intense pressure for a short time, sufficient to overcome a great resistance.

139. Law of Equilibrium.—Whatever be the direction of the blow or force, we may suppose it to be resolved into two components, one perpendicular to the back of the wedge, and the other parallel to it. The latter produces no effect. The same is true of the resistances; we need to consider only those components of them which are perpendicular to the sides of the wedge.

Let $M\ N\ O$ (Fig. 107) represent a section of the wedge perpendicular to its faces; then $P\ A$, $Q\ A$, and $R\ A$, drawn perpendicular to the faces severally, show the directions of the forces which hold the wedge in equilibrium. Taking $A\ B$ to represent the power, draw $B\ C$ parallel to $R\ A$, and we have the triangle $A\ B\ C$, whose sides represent these forces. But $A\ B\ C$ is similar to $M\ N\ O$, as their sides are respectively perpendicular to each other. Hence, calling the forces P, Q, and R, respectively,

Fig. 107.

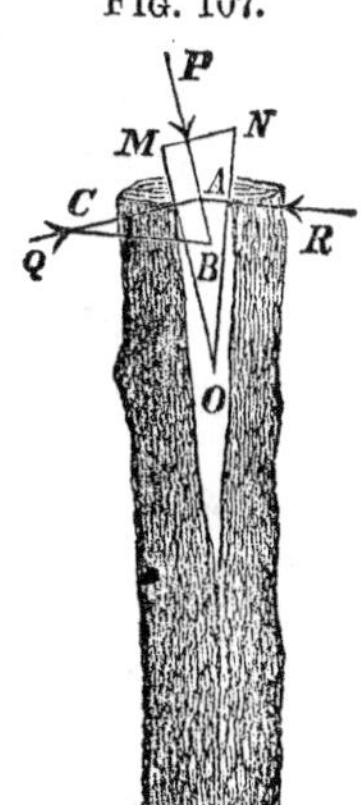

$$P : Q :: M\ N : M\ O;$$
$$\text{and } P : R :: M\ N : N\ O;$$

that is, there is equilibrium in a wedge, when

The power is to the resistances as the back of the wedge to the sides on which the resistances respectively act.

If the triangle is isosceles, the two resistances are equal, as the proportions show; and P is to either resistance, R, as the breadth of the back to the length of the side.

If the resisting surfaces touch the sides of the wedge only in one point each, then $Q\ A$ and $R\ A$, drawn through the points of contact, must meet $A\ P$ in the same point (Art. 60, 2); otherwise the wedge will roll, till one face rests against the resisting body in two or more points.

The efficiency of the wedge is usually very much increased by combining its own action with that of the lever, since the point where it acts generally lies at a distance from the point where the effect is to be produced. Thus, in splitting a log of wood, the resistance to be overcome is the cohesion of the fibers; and this force is exerted at a distance from the wedge, while the fulcrum is a little further forward in the solid wood.

VIII. The Knee-Joint.

140. Description and Law of Equilibrium.—The knee-joint consists of two bars, usually equal, hinged together at one end, while the others are at liberty to separate in a straight line. The power is applied at the hinge, tending to thrust the bars into a straight line; the weight is the force which opposes the separation.

Fig. 108.

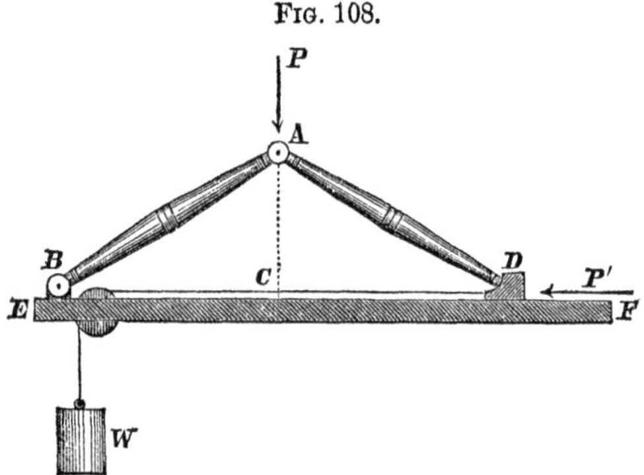

Suppose that $A\ B$ and $A\ D$ (Fig. 108) are equal bars, hinged together at A; and that the bar $A\ B$ is free only to revolve about the axis B, while the end D of the other bar can move parallel to the base $E\ F$. If P urges A toward the base, it tends to move D further from the fixed point B. The force P', which opposes that motion, is represented in the figure by the weight W. The law of equilibrium is,

The power is to the weight as twice radius to the tangent of half the angle between the bars.

The point A is held in equilibrium by three forces, the power P, the resistance along $B\ A$, and that along $D\ A$. As $A\ B\ D$ is isosceles, and $A\ C$ is perpendicular to $B\ D$, the angles $P\ A\ B$ and $P\ A\ D$ are equal; therefore the resistances $B\ A$ and $D\ A$ are equal (Art. 61). Let $a =$ the angle $B\ A\ C = \frac{1}{2}\ B\ A\ D$; and let $R =$ the resistance in the line $D\ A$. Then, since $\sin P\ A\ B = \sin a$, we have

$$P : R :: \sin 2\,a : \sin a.$$

But W is equal only to that component of R which is parallel to $B\ D$. Therefore, resolving R, we have (Art. 49),

$$R : W :: \text{rad} : \cos A\ D\ C;\ \text{or}$$
$$:: \text{rad} : \sin a.$$

Compounding, we find,

$$P : W :: \text{rad} \,.\, \sin 2\,a : \sin^2 a.$$

$$\text{But } \sin 2\,a = \frac{2 \sin a \,.\, \cos a}{\text{rad}};\ \text{therefore}$$

$$P : W :: \frac{2\ \text{rad} \,.\, \sin a \,.\, \cos a}{\text{rad}} : \sin^2 a;$$

$$:: 2\ \text{rad} : \frac{\text{rad} \,.\, \sin a}{\cos a} = \tan a;$$

or the power is to the weight as twice radius to the tangent of the half angle between the bars.

$$\text{Since } A\ C : C\ D :: \text{rad} : \tan a,$$
$$\therefore P : W :: 2\ A\ C : C\ D;\ \text{or},$$

The power is to the weight as twice the height of the joint to half the distance between the ends of the bars.

141. Ratio of Power and Weight Variable.—It is obvious that the ratio between power and weight is different for different positions of the bars. As A is raised higher, $B\ A\ D$ diminishes; and when $B\ A\ D = 0$, then $a = 0$, and $\tan a = 0$;

$$\therefore P : W :: 2\ \text{rad} : 0,$$

and the power has no efficiency. But as A approaches the base $E\ F$, a approaches 90°; therefore $\tan a$ increases, and the power is more efficient. When A, B, and D are in a straight line, $a = 90°$, and $\tan a$ is infinite, $\therefore P : W :: 2\ \text{rad} : \infty$. Hence, the efficiency of the power is infinitely great. The indefinite increase of efficiency in the power, which occurs during a single movement, renders this machine one of the most useful for many purposes, as printing and coining.

Questions on the knee-joint.—

1. A power of 50 lbs. is exerted on the joint A (Fig. 108); compare the weight which will balance it, when $B\ A\ D$ is 90°, and when it is 160°. *Ans.* 25 lbs. and 141.78 lbs.

2. When the angle between the bars is 110°, a certain power

just overcomes a weight of 65 lbs.; what must be the angle, in order that the weight overcome may be five times as great?

Ans. 164° 3′ 22″.

PRINCIPLE OF VIRTUAL VELOCITIES.

142. The Point of Application Moving in the Line of the Force.—In examining the simple machines, we have in each instance simply inquired for the relative *magnitude* of the forces, called the power and the weight, when in equilibrium. There is another important particular to be noticed, namely, the relative *velocity* of the power and weight, when they begin to move. It can be shown, in every case, that *the velocities, when reckoned in the direction in which the forces act, are inversely as the forces.*

Some examples are first given in which the point of application moves in the line in which the force acts.

In the *straight lever* (Fig. 109), which is in equilibrium by the weights P and W, suppose a slight motion to exist; then the velocity of each will be as the arc described in the same time; but the arcs are similar, since they subtend equal angles. Therefore, if V = velocity of P, and v = velocity of W,

FIG. 109.

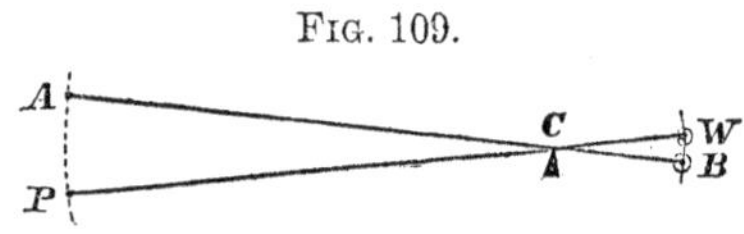

$$V : v :: A\,P : B\,W :: A\,C : B\,C;$$

but it has been shown (Art. 106) that

$$P : W :: B\,C : A\,C;$$
$$\therefore V : v :: W : P;$$

that is, the velocity of the power is to the velocity of the weight as the weight to the power. Hence, $P \times$ its velocity $= W \times$ its velocity; that is, the momentum of the power equals the momentum of the weight.

In the *wheel and axle*, let R and r be the radii, and suppose the machine to be revolved; then while P descends a distance equal to the circumference of the wheel $= 2\,\pi\,R$, the weight ascends a distance equal to the circumference of the axle $= 2\,\pi\,r$. Therefore,

$$V : v :: 2\,\pi\,R : 2\,\pi\,r :: R : r;$$

but (Art. 115),

$$P : W :: r : R;$$
$$\therefore V : v :: W : P;$$

or, the velocities are inversely as the weights; and $P \times V = W \times v$, the momentum of the power equals the momentum of the weight.

In the *fixed pulley* the velocities are obviously equal; and we have before seen that the power and weight are equal; therefore the proportion holds true, $V : v :: W : P$; and the momenta are equal.

In the *movable pulley*, if n is the number of sustaining parts of the cord, when W rises any distance $= x$, each portion of cord is shortened by the distance x, and all these n portions pass over to P, which therefore descends a distance $= n\,x$.

Hence, $V : v :: n\,x : x :: n : 1$;

but (Art. 121), $P : W :: 1 : n$;

$\therefore\ V : v :: W : P$;

as in all the preceding cases.

In the screw (Fig. 104), while the power describes the circumference $= 2\,\pi \times B\,C$, the weight moves only the distance $= d$; therefore,

$$V : v :: 2\,\pi \times B\,C : d;$$

but (Art. 133), $P : W :: d : 2\,\pi \times B\,C$;

$\therefore\ V : v :: W : P$;

therefore the momentum of the power equals the momentum of the weight, as before.

143. The Point of Application Moving in a Different Line from that in which the Force Acts.—The cases thus far noticed are the most obvious ones, because the points of application of power and weight *actually move* in the directions in which their force is exerted. But the principle we are considering is that of *virtual velocities*. If the force is exerted in one line, and the motion of the point of application is in a different line, then its *virtual velocity* is merely that component which lies in the former line. The case of the inclined plane will illustrate the principle.

First, let P (Fig. 110) act parallel to the plane, and suppose the body to be moved either up or down the plane a distance equal to $G\,d$. That is the velocity of the *power*. But in the direction of the *weight* (force of gravity), the body moves only the distance $b\,d$. Therefore the velocity of the power is to the velocity of the weight (each being reckoned in the line of its action) as $G\,d$ to $b\,d$.

Fig. 110.

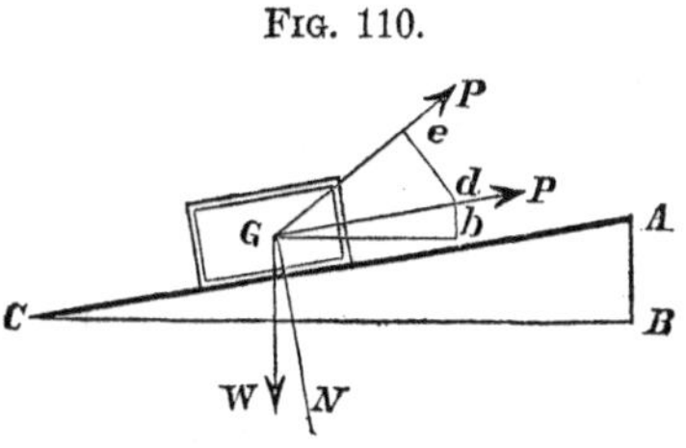

By similar triangles, $G\,d : b\,d :: A\,C : A\,B$;

or $V : v :: A\,C : A\,B$.

But (Art. 127), $P : W :: A\,B : A\,C$;

$\therefore\ V : v :: W : P$.

Again, let the power act in any oblique direction, as $G\ e$. If the body moves over $G\ d$, draw $d\ e$ perpendicular to $G\ e$; then $G\ e$ is the distance passed over in the direction of the power, and $b\ d$ in the direction of the weight. $G\ d$ being taken as radius, $G\ e$ is $\cos d\ G\ e = \cos (P\ G\ N - 90°) = \sin P\ G\ N$; and $b\ d = \sin a$. Therefore, the virtual velocity of the power is to the virtual velocity of the weight as $\sin P\ G\ N$ to $\sin a$;

$$\text{or } V : v :: \sin P\ G\ N : \sin a.$$

But (Art. 128), $$P : W :: \sin a : \sin P\ G\ N;$$

$$\therefore V : v :: W : P.$$

We learn from the foregoing principle, that a machine does not enable us to obtain any *greater* effect than the power could produce without its aid, but only to produce an effect in a *different form*. A given power, for instance, may move a much *greater quantity* of matter by the aid of a machine, but it will move it as much more slowly. On the other hand, a power, by means of a machine, may produce a far *greater velocity* than would be possible without such aid; but the quantity moved, or the intensity of the force exerted, would be proportionally less. By machines, therefore, we do not increase the effects of a power, but only modify them.

FRICTION IN MACHINERY.

144. The Power and Weight not the only Forces in a Machine.—For each machine a certain proportion has been given, which ensures equilibrium. And it is implied that if either the power or the weight be altered, the equilibrium will be destroyed. But practically this is not true; the power or weight may be considerably changed, or possibly one of them may be entirely removed, and the machine still remain at rest. The obstruction which prevents motion in such cases, and which always exists in a greater or less degree, arises from *friction;* and friction is caused by roughness in the surfaces which rub against each other. The minute elevations of one surface fall in between those of the other, and directly interfere with the motion of either, while they remain in contact. Polishing diminishes the friction, but can never remove it, for it never removes all roughness.

As friction always tends to prevent motion, and never to produce it, it is called a *passive* force. It assists the power, when the weight is to be kept at rest, but opposes it, when the weight is to be moved. There are other passive forces to be considered in the study of science, but no other has so much influence in the operations of machinery as friction.

145. Modes of Experimenting.—When one surface slides on another, the friction which exists is called the *sliding* friction ;

but when a wheel rolls along a surface, the friction is called *rolling* friction. The sliding friction occurs much more in machines than the rolling friction.

Experiments for ascertaining the laws of friction may be performed by placing on a table a block of three different dimensions, and measuring its friction under different circumstances by weights acting on the block by means of a cord and pulley, as represented in Fig. 111. This was the method by which Coulomb first ascertained the laws of friction.

FIG. 111.

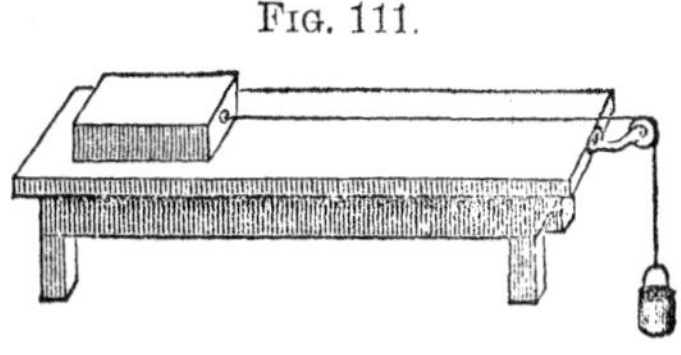

Another mode is to place the block on an inclined plane, whose angle can be varied, and then find the relative friction in different cases, by the largest inclination at which it will prevent the block from sliding. For, when *W* on the inclined plane *A B* (Fig. 112), is on the point of sliding down, friction is the power, which acting parallel to the plane, is in equilibrium with the weight. In such cases, the power is to the weight as the height to the length.

FIG. 112.

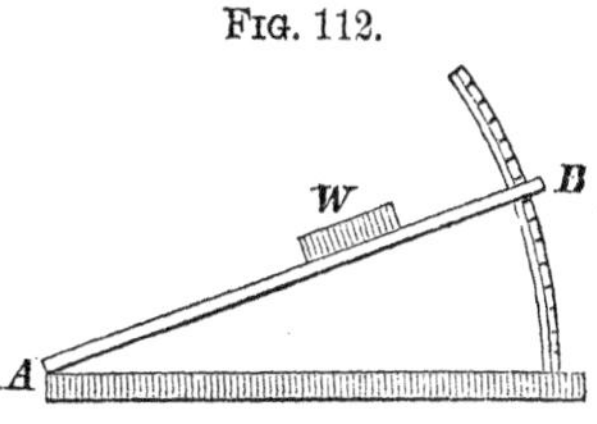

The *coefficient* of friction is the fraction whose numerator is the force required to overcome the friction, and its denominator the weight of the body.

146. Laws of Sliding Friction.—The laws of sliding friction on which experimenters are generally agreed are the following:

1. *Friction varies as the pressure.*—If weights are put upon the block, it is found that a double weight requires a double force to move it, a triple weight a triple force, &c.

2. *It is the same, however great or small the surface on which the body rests.*—If the block be drawn, first on its broadest side, then on the others in succession, the force required to overcome friction is found in each case to be the same. Extremes of size are, however, to be excepted. If the loaded block were to rest on three or four very small surfaces, the obstruction might be greatly increased by the indentations thus occasioned in the surface beneath them.

3. *Friction is a uniformly retarding force.*—That is, it destroys equal amounts of motion in equal times, whatever may be the velocity, like gravity on an ascending body.

4. *Friction at the first moment of contact is less than after contact has continued for a time.*—And the time during which friction increases, varies in different materials. The friction of wood on wood reaches its maximum in three or four minutes; of metal on metal, in a second or two; of metal on wood, it increases for several days.

5. *Friction is less between substances of different kinds than between those of the same kind.*—Hence, in watches, steel pivots are made to revolve in sockets of brass or of jewels, rather than of steel.

147. Friction of Axes.—In machinery, the most common case of friction is that of an axis revolving in a hollow cylinder, or the reverse, a hollow cylinder revolving on an axis. These are cases of sliding friction, in which the power that overcomes the friction, usually acts at the circumference of a wheel, and therefore at a mechanical advantage. Thus, the friction on an axis, whose coefficient is as high as 20 per cent., requires a power of only *two* per cent. to overcome it, provided the power acts at the circumference of a wheel whose diameter is ten times that of the axis.

148. Rolling Friction.—This form of friction is very much less than the sliding, since the projecting points of the surfaces do not directly encounter each other, but those of the rolling wheel are lifted up from among those of the other surface, as the wheel advances.

By the use of the apparatus described in Art. 145, the laws of the rolling are found to be the same as those of the sliding friction. But on account of the manner in which this form of friction is overcome, there is this additional law:

The force required to roll the wheel varies inversely as the diameter.

For the power, acting at the centre of the wheel to turn it on its lowest point as a momentary fulcrum, has the advantage of greater acting distance as the diameter increases.

It is the rolling friction which gives value to *friction wheels*, as they are called. When it is desirable that a wheel should revolve with the least possible friction, each end of its axis is made to rest in the angle between two other wheels placed side by side, as shown in Fig. 113. The wheel is obstructed only by the rolling friction on the surfaces of the four wheels, and the retarding effect of the sliding friction at the pivots of the latter is greatly reduced on the principle of the wheel and axle.

The sliding friction is diminished by lubricating the surface, the rolling friction is not.

Fig. 113.

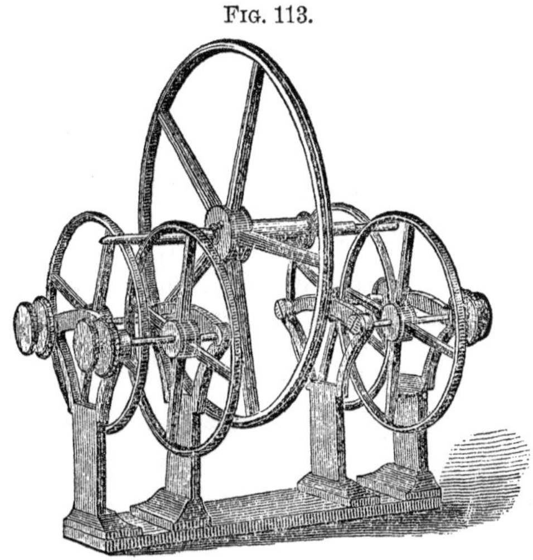

149. Advantages of Friction.—Friction in machinery is generally regarded as an evil, since more power is on this account required to do the work for which the machine is made. But it is easy to see, that in general friction is of incalculable value, or rather, that nothing could be accomplished without it. Objects stand firmly in their places by friction; and the heavier they are, the more firmly they stand, because friction increases with the pressure. All fastening by nails, bolts, and screws, is due to friction. The fibers of cotton, wool or silk, when intertwined with each other, form strong threads or cords, only because of the power of friction. Without friction, it would be impossible to walk or even to stand, or to hold anything by grasping it with the hand.

CHAPTER VII.

MOTION ON INCLINED PLANES.—THE PENDULUM.

150. The Force which Moves a Body Down an Inclined Plane.—It was shown (Art. 127) that when the power acts in a line parallel to the inclined plane, $P : W :: A\,B : A\,C$. If, therefore, P ceases to act, the body descends the plane only with a force equal to P.

Let g (the velocity acquired in a second in falling freely) $=$ the force of gravity, $f =$ the force acting down the plane, $h =$ the height, $l =$ the length; then by substitution,

$$f : g :: h : l, \text{ and}$$
$$f = \frac{h}{l} g.$$

Therefore, the force which moves a body down an inclined plane is equal to that fraction of gravity which is expressed by the height divided by the length. This is evidently a constant force on any given plane, and produces uniformly accelerated motion. Therefore the motion on an inclined plane does not differ from that of free fall in kind, but only in degree. Hence the formulæ for time, space, and velocity on an inclined plane are like those relating to free fall, if the value of f be substituted for g.

151. Formulæ for the Inclined Plane.—The formulæ for free fall (Art. 28) are here repeated, and against them the corresponding formulæ for descent on an inclined plane.

	Free fall.	Descent on an inclined plane.
1.	$s = \frac{1}{2} g t^2$	$s = \frac{g h t^2}{2 l}$.
2.	$t = \sqrt{\frac{2 s}{g}}$	$t = \sqrt{\frac{2 l s}{g h}}$.
3.	$s = \frac{v^2}{2g}$	$s = \frac{l v^2}{2 g h}$.
4.	$v = \sqrt{2 g s}$	$v = \sqrt{\frac{2 g h s}{l}}$.
5.	$t = \frac{v}{g}$.	$t = \frac{l v}{g h}$.
6.	$v = g t$	$v = \frac{g h t}{l}$.

By formula 1, $s \propto t^2$, and by formula 3, $s \propto v^2$. It follows that in equal successive times the spaces of descent are as the odd numbers, 1, 3, 5, &c., and of ascent as these numbers inverted; also, that with the acquired velocity continued uniformly, a body moves twice as far as it must descend to acquire that velocity. If a body be projected up an inclined plane, it will ascend as far as it must descend in order to acquire the velocity of projection. The distance passed over in the time t by a body projected with the velocity v, down or up an inclined plane, equals $t v \pm \frac{g h t^2}{2 l}$. These statements are proved as in the case of free fall, Chapter II.

152. Formulæ for the whole Length of a Plane.—

1. *The velocity acquired in descending a plane is the same as that acquired in falling down its height.*

For now $s = l$; hence (formula 4), $v = \left(\frac{2 g h s}{l}\right)^{\frac{1}{2}} = (2 g h)^{\frac{1}{2}}$, which is the formula for free fall through h, the height of the plane.

On different planes, therefore, $v \propto h^{\frac{1}{2}}$.

2. *The time of descending a plane is to the time of falling down its height as the length to the height.*

For (formula 2) $t = \left(\frac{2\,l\,s}{g\,h}\right)^{\frac{1}{2}} = l\left(\frac{2}{g\,h}\right)^{\frac{1}{2}}$ But the time of fall down the height is $\left(\frac{2\,h}{g}\right)^{\frac{1}{2}}$. Therefore,

$$t \text{ down plane} : t \text{ down height} :: l\left(\frac{2}{g\,h}\right)^{\frac{1}{2}} : \left(\frac{2\,h}{g}\right)^{\frac{1}{2}};$$

$$:: l\left(\frac{2}{g}\right)^{\frac{1}{2}} : h\left(\frac{2}{g}\right)^{\frac{1}{2}};$$

$$:: l : h.$$

On different planes, $t \propto \frac{l}{\sqrt{h}}$.

It follows that if several planes have the same height, the velocities acquired in descending them are equal, and the times of descent are as the lengths of the planes. For, let $A\,C$, $A\,D$, $A\,E$, (Fig. 114) have the same height $A\,B$; then, since $v \propto h^{\frac{1}{2}}$, and h is the same for all, v is the same. And since $t \propto \frac{l}{\sqrt{h}}$, and h is the same for all the planes, $t \propto l$.

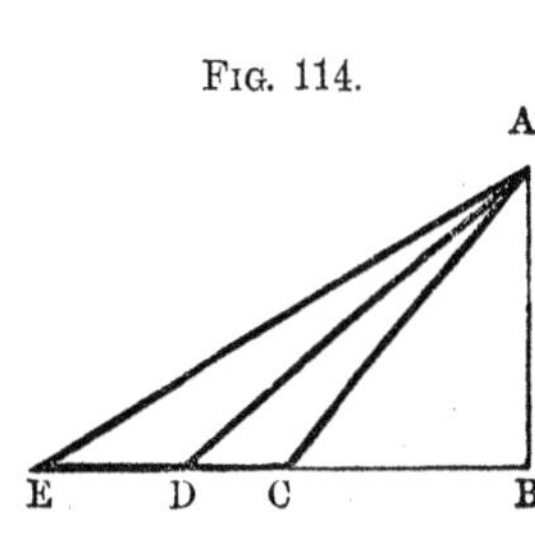

Fig. 114.

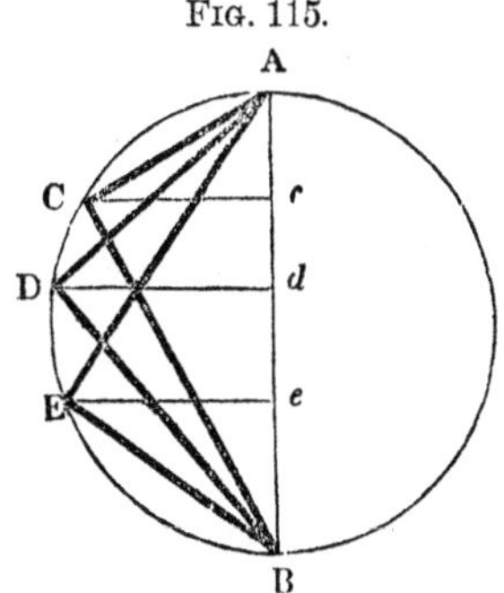

Fig. 115.

153. Descent on the Chords of a Circle.—In descending the chords of a circle which terminate at the ends of the vertical diameter, *the acquired velocities are as the lengths, and the times of descent are equal to each other and to the time of falling through the diameter.*

For (Art. 152) the velocity acquired on $A\,C$ (Fig. 115) =

$(2\,g\,.\,A\,c)^{\frac{1}{2}} = \left(2\,g\,.\,\frac{A\,C^2}{A\,B}\right)^{\frac{1}{2}} = A\,C\left(\frac{2\,g}{A\,B}\right)^{\frac{1}{2}}$, which, since $\left(\frac{2g}{A\,B}\right)^{\frac{1}{2}}$ is constant, varies as $A\,C$, the length.

Again (Art. 152), the time down $A\,C = \left(\frac{2\,A\,C^2}{g\,.\,A\,c}\right)^{\frac{1}{2}} = \left(\frac{2\,A\,B\,.\,A\,c}{g\,.\,A\,c}\right)^{\frac{1}{2}} = \left(\frac{2\,A\,B}{g}\right)^{\frac{1}{2}}$, which is equal to the time of falling freely through $A\,B$, the diameter.

154. Velocity Acquired on a Series of Planes.—If no velocity be lost in passing from one plane to another, the velocity acquired in descending a series of planes is equal to that acquired in falling through their perpendicular height. For, in Fig. 116, the velocity at B is the same, whether the body comes down $A\,B$ or $E\,B$, as they are of the same height, $F\,b$. If, therefore, the body enters on $B\,C$ with the acquired velocity, then it is immaterial whether the descent is on $A\,B$ and $B\,C$ or on $E\,C$; in either case, the velocity at C is equal to that acquired in falling $F\,c$. In like manner, if the body can change from $B\,C$ to $C\,D$ without loss of velocity, then the velocity at D is the same, whether acquired on $A\,B$, $B\,C$, and $C\,D$, or on $F\,D$, which is the same as down $F\,G$.

Fig. 116.

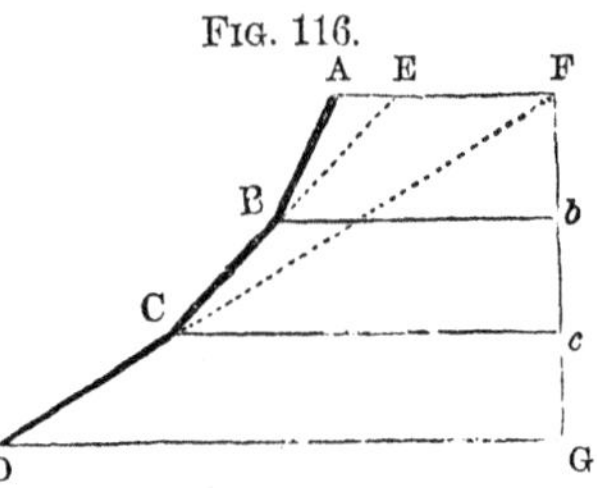

155. The Loss in Passing from one Plane to Another.—The condition named in the foregoing article is not fulfilled. A body *does* lose velocity in passing from one plane to another. And the *loss* is to the *whole previous velocity* as the *versed sine of the angle* between the planes to *radius*.

Let $B\,F$ (Fig. 117) represent the velocity which the body has at B. Resolve it into $B\,D$ on the second plane, and $D\,F$ perpendicular to it. $B\,D$ is the initial velocity on $B\,C$; and, if $B\,I = B\,F$, $D\,I$ is the loss. But $D\,I$ is the versed sine of the angle $F\,B\,D$, to the radius $B\,F$; and $\therefore$ the loss is to the velocity at B as $D\,I : B\,F$: : ver. sin B : rad.

Fig. 117.

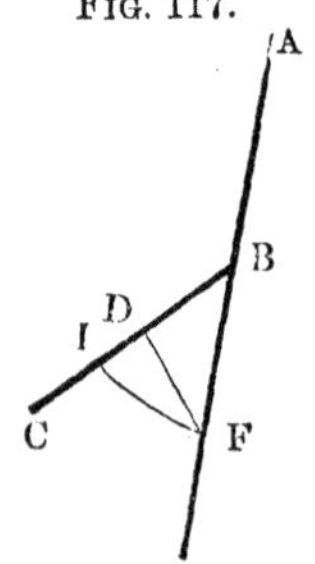

156. No Loss on a Curve.—Suppose now the number of planes in a system to be infinite; then it becomes a curve (Fig. 118). As the angle between two successive elements of the curve is infinitely small, its chord is also infinitely small; but its versed sine is *infinitely smaller still,* i. e., an infinitesimal of the *second* order; for diam. : chord :: chord : ver. sin. Therefore, although the sum of all the infinitely small angles is a finite angle 180° — $A\ G\ D$, yet, as the loss of velocity at each point is an infinitesimal of the *second* order, the *entire loss* (which is the sum of the losses at all the points of the curve) is an infinitesimal of the *first* order.

FIG. 118.

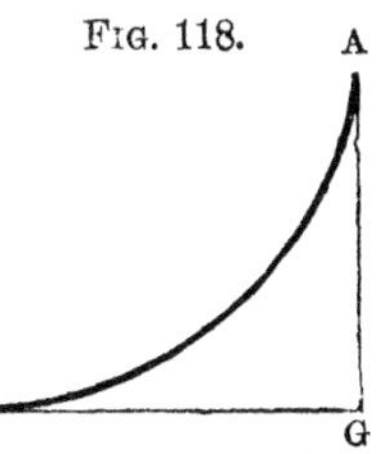

Hence, a body loses no velocity on a curve, and therefore acquires at the bottom the same velocity as in falling freely through its height.

It appears, therefore, that whether a body descends *vertically,* or on an *inclined plane,* or on a *curve* of any kind, the *acquired velocity is the same,* if the height is the same.

157. Times of Descending Similar Systems of Planes and Similar Curves.—If planes are equally inclined to the horizon, the *times* of describing them are as *the square roots of their lengths.* For, if the height and base of each plane be drawn, similar triangles are formed, and $h : l$ is a constant ratio for the several planes. By Art. 152, $t \propto \frac{l}{\sqrt{h}} \propto \frac{l}{\sqrt{l}} \propto \sqrt{l}$; that is, the time varies as the square root of the length.

If two *systems* of planes are similar, i. e., if the corresponding parts are proportional and equally inclined to the horizon, it is still true that *the times of descending them are as the square roots of their lengths.*

Let $A\ B\ C\ D$ and $a\ b\ c\ d$ (Fig. 119) be similar, and let $A\ F$ and

FIG. 119.

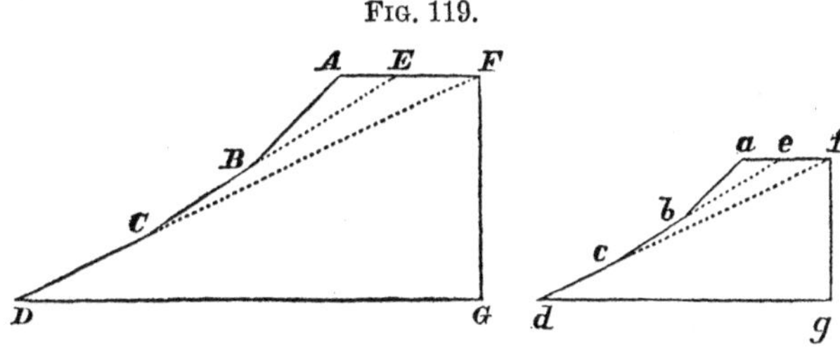

$a\,f$ be drawn horizontally, and the lower planes produced to meet them, then it is readily proved that all the homologous lines of the figures are proportional, and their square roots also proportional. Then (reading $t, A\ B$, *time down* $A\ B$, &c.),

we have

$$t, A\,B : t, a\,b :: \sqrt{A\,B} : \sqrt{a\,b};$$

$$t, E\,B : t, e\,b :: \sqrt{E\,B} : \sqrt{e\,b} :: \sqrt{A\,B} : \sqrt{a\,b};$$

and $t, EC : t, ec :: \sqrt{EC} : \sqrt{ec} :: \sqrt{AB} : \sqrt{ab}$;

∴ (by subtraction) $t, BC : t, bc :: \sqrt{AB} : \sqrt{ab}$.

In like manner, $t, CD : t, cd :: \sqrt{AB} : \sqrt{ab}$.

∴ (by addition)

$$t, (AB + BC + CD) : t, (ab + bc + cd) :: \sqrt{AB} : \sqrt{ab}$$
$$:: \sqrt{(AB + BC + CD)} : \sqrt{(ab + bc + cd)}.$$

Though there is a loss of velocity in passing from one plane to another, the proposition is still true; because, the angles being equal, the losses are proportional to the acquired velocities; and therefore the initial velocities on the next planes are still in the same ratio as before the losses; hence the ratio of times is not changed.

The reasoning is applicable when the number of planes in each system is infinitely increased, so that they become *curves*, similar, and similarly inclined to the horizon. Suppose these curves to be *circular* arcs; then, as they are similar, they are proportional to their radii. Hence, the times of descending similar circular arcs are as the square roots of the radii of those arcs.

158. Questions on the Motions of Bodies on Inclined Planes.—

1. How long will it take a body to descend 100 feet on a plane whose length is 150 feet, and whose height is 60 feet?

Ans. 3.9 sec.

2. There is an inclined railroad track, $2\frac{1}{2}$ miles long, whose inclination is 1 in 35. What velocity will a car acquire, in running the whole length of the road by its own weight?

Ans. 106.2 miles per hour.

3. A body weighing 5 lbs. descends vertically, and draws a weight of 6 lbs. up a plane whose inclination is 45°. How far will the first body descend in 10 seconds? *Ans.* 110.74 feet.

4. A body descending vertically drew another body of half the weight up an inclined plane. When the bodies had described a space c the cord broke, and the smaller body continued its motion through an additional space c before it began to descend. What is the inclination of the plane? *Ans.* 30°.

159. The Pendulum.—A pendulum is a weight attached by an inflexible rod to a horizontal axis of suspension, so as to be free to vibrate by the force of gravity. If it is drawn aside from its position of rest, it descends, and by the momentum acquired, rises on the opposite side to the same height, when gravity again causes its descent as before. If unobstructed, its vibrations would never cease.

A *single vibration* is the motion from the highest point on one side to the highest point on the other side. The motion from the highest point on one side to the same point again is called a *double vibration.*

The *axis of the pendulum* is a line drawn through its centre of gravity perpendicular to the horizontal axis about which the pendulum vibrates.

The *centre of oscillation* of a pendulum is that point of its axis at which, if the entire mass were collected, its time of vibration would be unchanged.

The *length* of a pendulum is that part of its axis which is included between the axis of suspension and the centre of oscillation.

All the particles of a pendulum may be conceived to be collected in points lying in the axis. Those which are *above* the centre of oscillation tend to vibrate quicker (Art. 157), and therefore accelerate it; those which are *below* tend to vibrate slower, and therefore retard it. But, according to the definition of the centre of oscillation, these accelerations and retardations exactly balance each other at that point.

160. Calculation of the Length of a Pendulum.—Let $C\,q$ (Fig. 120) be the axis of a pendulum in which all its weight is collected, C the point of suspension, G the centre of gravity, O the centre of oscillation, a, b, &c., particles above O, which accelerate it, p, q, &c., particles below O, which retard it. $C\,O = l$, is the length of the pendulum required. Denote the masses concentrated in $a, b \ldots p, q$, by $m, m' \ldots m''\, m'''$, and their distances from C by $r, r' \ldots r'', r'''$; and denote the distance from C to G by k. Denote the angular velocity by θ; then the velocity of m will be $r\,\theta$ and its momentum will be $m\,r\,\theta$.

Fig. 120.

If m had been placed at O, the moving force would have been $m\,l\,\theta$. The difference $m\,(l - r)\,\theta$, is that portion of the force which accelerates the motion of the system.

The moment of this force with respect to C is $m\,(l - r)\,r\,\theta$.

In like manner the moment of m' is $m'\,(l - r')\,r'\,\theta$, and so on for all the particles between C and O.

The moments of the forces tending to retard the system applied at the points p, q, &c., are

$$m''\,(r'' - l)\,r''\,\theta,\; m'''\,(r''' - l)\,r'''\,\theta, \text{ \&c.}$$

But since these forces are to balance each other, we have

$$m\,(l - r)\,r\,\theta + m'\,(l - r')\,r'\,\theta + \text{\&c.} = m''\,(r'' - l)\,r''\,\theta + m'''\,(r''' - l)\,r'''\,\theta + \text{\&c.};$$

whence $$l = \frac{m\,r^2 + m'\,r'^2 + m''\,r''^2 + \&c.}{m\,r + m'\,r' + m''\,r'' + \&c.}$$

Or $l = \frac{S\,(m\,r^2)}{S\,(m\,r)}$, where S denotes the sum of all the terms similar to that which follows it.

The numerator of this expression is called the *moment of inertia* of the body with respect to the axis of suspension, and the denominator is called the *moment of the mass,* with respect to the axis of suspension.

By the principle of moments (Art. 77) $m\,r + m'\,r' + \&c.$, or $S\,(m\,r) = M\,k$, where M denotes the entire mass of the pendulum; hence, by substitution, $l = \frac{S\,(m\,r^2)}{M\,k}$.

That is, *the distance from the axis of suspension to the centre of oscillation is found by dividing the moment of inertia, with respect to that axis, by the moment of the mass with respect to the same axis.*

161. The Point of Suspension and the Centre of Oscillation Interchangeable.—Let the pendulum now be suspended from an axis passing through O, and denote by l' the distance from O to the new centre of oscillation. The distances of $a, b \ldots . p, q$, from O, will be $l - r$, $l - r'$, &c., and the distance $G\,O$ will be $l - k$.

Hence, from the principle just established, we have

$$l' = \frac{S\,[m\,(l - r)^2]}{M\,(l - k)} = \frac{S\,(m\,l^2 - 2\,m\,r\,l + m\,r^2)}{M\,(l - k)}$$
$$= \frac{S\,(m\,l^2) - 2\,S\,(m\,r\,l) + S\,(m\,r^2)}{M\,(l - k)}$$

But $S\,(m\,r^2) = M\,k\,l$; and since l is constant,

$$l' = \frac{M\,l^2 - 2\,l\,S\,(m\,r) + M\,k\,l}{M\,(l - k)} = \frac{M\,l^2 - 2\,M\,k\,l + M\,k\,l}{M\,(l - k)}$$
$$= \frac{M\,(l - k)\,l}{M\,(l - k)} = l.$$

This last equation shows that the centre of oscillation and the point of suspension are interchangeable; that is, if the pendulum were suspended from O, it would vibrate in the same time as when suspended from C.

162. The Cycloid.—One of the methods of investigating the theory of the pendulum is by means of the properties of the cycloid. This curve is described by a point situated on the circumference of a circle, as it rolls on a straight line.

Let the circle $A\;H\;B$ (Fig. 121) make one revolution upon the line $C A X$, equal to its circumference; the curve line $C D B X$, traced out by that point of the circle which was in contact with C when the circle began to roll, is called a cycloid. If $C X$ be bisected in A, and $A B$ be drawn at right angles to it, it is evident, from the manner in which the curve is generated, that it will have similar branches on both sides of $A B$, and that its vertex B will be so placed as to make the *axis* $A B$ equal to the diameter of the generating circle. The properties of the cycloid, as applied to the vibration of the pendulum, are the following.

FIG. 121.

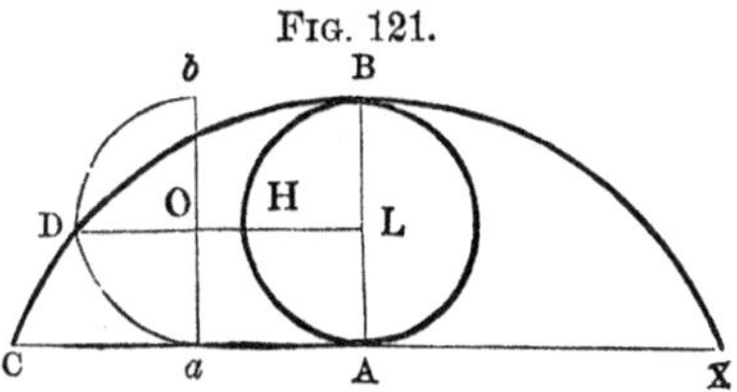

163. The Cycloidal Ordinate $D H$ *equals the circular arc* $B H$.—For, let $b D a$ (Fig. 121) be the position of the circle when the generating point is at D; draw the diameter $b a$ parallel to $B A$, and from D draw $D H L$ parallel to $C A$; then the arc $D a$ = arc $H A$, ∴ the sines $D O$, $H L$, are equal; hence $D H = O L$; but from the mode in which the cycloid is generated, $C a$ = arc $D a$, and $C A$ = semi-circumference $B H A$; hence $D H = O L = a A = C A - C a$ = semi-circumference $B H A$ – arc $H A$ = arc $B H$.

164. A Tangent to the Cycloid *at any point*, E (*Fig.* 122), *is parallel to the corresponding chord* $B K$ *of the generating circle.*—Draw $D H L$ infinitely near to $E K M$; join $B K$, and produce it to k. The elementary triangle $H K k$ is similar to the triangle $K R B$ formed by the tangents ($K R$, $B R$) to the circle at the points K, B, and is consequently isosceles; ∴ $K H = H k$. Now (Art. 163), arc $B K H = D H$; from which equation subtract the previous one, and arc $B K = D k$. But arc $B K = E K$; ∴ $E K = D k$. Hence, since $E K$ and $D k$ are equal and parallel, $E D$ and $K k$ must also be equal and parallel; and as the tangent at the point E may be considered as coinciding with $E D$, it must therefore be parallel to the chord $B K$.

FIG. 122.

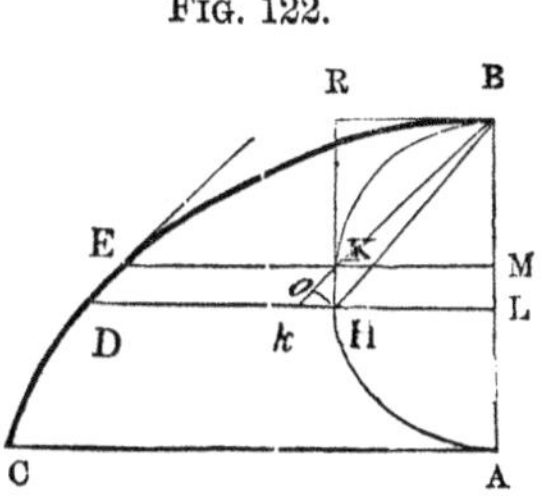

Hence the ends of the cycloid meet the base at right angles; for the tangent at C is parallel to $B A$, the axis.

165. The Cycloidal Arc $B E$ *is equal to twice the corresponding chord* $B K$ *of the generating circle.*—Draw $H o$ perpendicular to $K k$; and since the triangle $K H k$ is isosceles, $H o$ bisects the base $K\,k$, $\therefore K\,k$ or $E\,D = 2\,K\,o$; and since $H\,o$ may be considered as a small circular arc described with radius $B\,H$, $K\,o = B\,o - B\,K = B\,H - B\,K$; hence $E\,D$ and $K\,o$ are corresponding increments of the cycloidal arc $B\,E$ and the chord $B\,K$; and as the arc and chord begin together from the point B, and every increment of the former is twice the corresponding increment of the latter, the arc $B\,E$ must be equal to twice the chord $B\,K$; consequently, the whole arc $B\,C$ = twice the diameter $A\,B$; and the length of the whole curve $C\,B\,X$ (Fig. 121) $= 4\,A\,B$. And as $C\,A\,X = \pi\,.\,A\,B$, therefore the whole cycloid : its base : : 4 : π.

166. Descent by Gravity on a Cycloid.—$E\,F\,M$ (Fig. 123) is a circle whose diameter $E\,M$ is perpendicular to the horizon, and $B\,G\,M$ is the corresponding semicycloid.

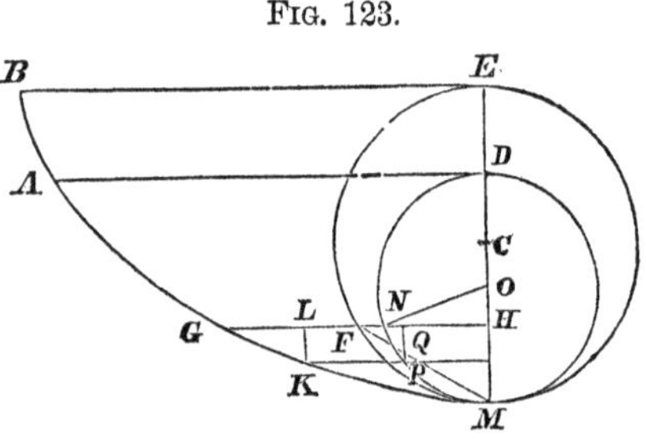

Fig. 123.

Let the body begin to descend from any point A.

Draw $A\,D$ parallel to $B\,E$, and upon $M\,D$ as a diameter describe the circle $D\,N\,P\,M$, with its centre at O.

Put $h = D\,M$, $r = C\,E$, $x = D\,H$; then the time of describing the elementary arc $G\,K$ will be $\frac{G\,K}{v} = \frac{G\,K}{\sqrt{2\,g\,x}}$.

By similar triangles $\frac{G\,K}{K\,L} = \frac{F\,M}{M\,H} = \frac{\sqrt{M\,H\,.\,M\,E}}{M\,H} = \sqrt{\frac{M\,E}{M\,H}}$; (1) and $\frac{N\,P}{P\,Q} = \frac{O\,N}{N\,H} = \frac{O\,N}{\sqrt{M\,H\,.\,D\,H}}$. (2) Now $K\,L = P\,Q$; hence, dividing (1) by (2),

$$\frac{G\,K}{N\,P} = \frac{\sqrt{M\,E\,.\,D\,H}}{O\,N} = \frac{\sqrt{2\,r\,x}}{\frac{1}{2}h} = \frac{2\sqrt{2\,r\,x}}{h};\text{ whence}$$

$$G\,K = \frac{2}{h}\sqrt{2\,r\,x}\,.\,N\,P;$$

and the time of descending $G\,K$ is

$$\frac{\frac{2}{h}\sqrt{2\,r\,x}\,.\,N\,P}{\sqrt{2\,g\,x}} = \frac{2}{h}\sqrt{\frac{r}{g}}\,.\,N\,P.$$

In like manner, the time of describing any other elementary arc will be found to be $\frac{2}{h}\sqrt{\frac{r}{g}}$ times the corresponding arc on the circumference $D\,N\,P\,M$; hence the time of describing the cycloidal arc $A\,M$ will be $\frac{2}{h}\sqrt{\frac{r}{g}} \times \text{arc } D\,N\,P\,M = \frac{2}{h}\sqrt{\frac{r}{g}} \cdot \frac{\pi h}{2} = \pi\sqrt{\frac{r}{g}}$.

This expression for the time down $A\,M$ being independent of h, is very remarkable, for it proves that

The time of descent on a cycloid to the lowest point is always the same, from whatever point in the curve the body begins to descend.

The time of falling through $E\,M$ is $2\sqrt{\frac{r}{g}}$; $\therefore$ time down $A\,M$: time down $E\,M$ :: $\pi\sqrt{\frac{r}{g}} : 2\sqrt{\frac{r}{g}}$ :: $\pi : 2$ :: semi-circumference : diameter.

167. The Involute of a Semicycloid.—The *involute* of any curve is another curve described by the extremity of a tangent as it unwinds from the former, which is called the *evolute*. If, for example, a tangent of a circle unwinds from it, the circumference of the circle is the *evolute*, and the spiral described by the end of the tangent is the *involute* of the circle. The involutes of most curves are different from their evolutes; but in the case of the semicycloid, the involute and evolute are of the same form and size.

FIG. 124.

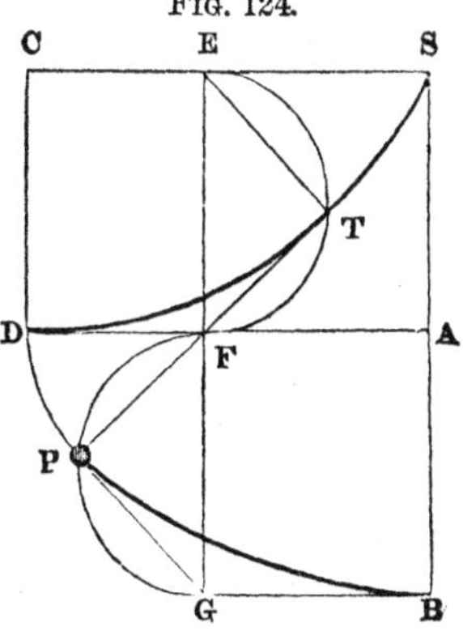

Take any line $S\,C$ (Fig. 124), and draw $S\,A$ at right angles to it; make $S\,C : S\,A$ :: semi-circumference of a circle : its diameter; and complete the parallelogram $S\,C\,D\,A$. Produce $S\,A$ to B, making $A\,B = S\,A$; upon $S\,C$, $A\,D$, describe two semicycloids $S\,D$, $D\,B$, the vertex of the former of which is at D, and the latter at B; then if the tangent unwinds, beginning at D, until the point of contact reaches S, its extremity will always be found in the semicycloid $D\,B$. For, through any point F on $A\,D$, draw $E\,F\,G$ perpendicular to $S\,C$, and through B draw $B\,G$ parallel to $S\,C$; then $E\,G = S\,B$; on $E\,F$, $F\,G$, describe the semicircles $E\,T\,F$, $F\,P\,G$, and draw the chords $T\,F$, $F\,P$, the former of which (Art. 164) is a tangent to the cycloid $S\,D$ at T. Now $S\,E = \text{arc } E\,T$, and $S\,C = E\,T\,F$; $\therefore C\,E\ (= D\,F) = \text{arc } T\,F$; but $D\,F = F\,P$; $\therefore$ arcs $F\,T$, $F\,P$ are equal, and also the angles subtended, $T\,E\,F$, $F\,G\,P$. Therefore, as T and P are right angles, $E F T = P F G$, and $P F T$ is a straight line; moreover, $T P = 2 T F =$ (Art. 165) the cycloidal arc $T D$. Therefore, $T P$ is a tangent unwound from D, and P is its extremity; and P having been assumed anywhere on the semicycloid $D B$, it follows that $D P B$ is the involute of $S T D$.

168. The Cycloidal Pendulum.—A pendulum may be made to vibrate in a cycloid by attaching the weight P (Fig. 125) to a flexible cord, whose point of suspension is at S, where two semicycloids meet. The cord and the semi-cycloid should be of the same length, and then (Art. 167) the weight P will, at each vibration, describe arcs of the cycloid $D B E$, as involutes of $S D$ and $S E$. Hence, the conclusion arrived at in Art. 166 applies to this motion; namely, *the time down $P B$ from any point P : time down $A B$: : semi-circumference : diameter ;* ∴ doubling the antecedents, *the time of a single vibration : time of falling half the length of the pendulum* : : π : 1.

FIG. 125.

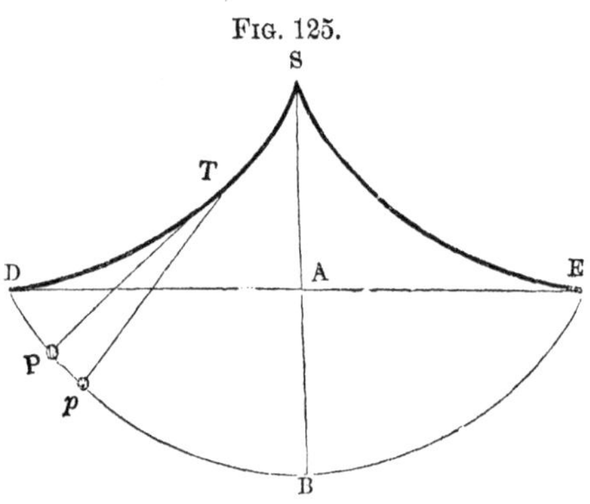

169. Application to the Circular Pendulum.—Since the proportion at the close of the foregoing article is always true, from whatever point the descent commences, therefore

All the vibrations of a cycloidal pendulum are performed in equal times, however large or small the extent of swing.

This is not true of any other curve. But it is evident that a very short arc of a cycloid at the lowest point B is coincident with the arc of a circle whose centre is S. Hence, if a pendulum vibrate through *very short arcs*, the conclusions are practically true, that in *circular* pendulums also unequal arcs are described in equal times, and that the time of a vibration is to the time of falling through half the length of the pendulum as π is to 1. For this reason, the pendulum of an astronomical clock is so connected with the machinery by its scapement, as to vibrate in small arcs.

170. Relation of Time, Length, and Force of Gravity.—Let $l =$ the length of a pendulum, that is, the distance from the point of suspension to the centre of oscillation. Then the time of falling half its length $= \left(\frac{2s}{g}\right)^{\frac{1}{2}} = \left(\frac{l}{g}\right)^{\frac{1}{2}}$. Hence, putting $t =$ time of a single vibration,

$$t : \left(\frac{l}{g}\right)^{\frac{1}{2}} :: \pi : 1; \text{ or } t = \left(\frac{\pi^2 l}{g}\right)^{\frac{1}{2}}; \text{ and } l = \frac{g t^2}{\pi^2}.$$

Therefore, the length of a pendulum being known, the time of one vibration is found; and on the other hand, if the time of a vibration is known, the length of the pendulum is obtained from it.

From the same formulæ, we find that $t \propto \sqrt{l}$, or

The time in which a pendulum makes a vibration varies as the square root of the length.

As $t \propto \sqrt{l}$, $\therefore l \propto t^2$; hence, if the length of a seconds pendulum equals l, then a pendulum which vibrates *once in two seconds* equals $4\,l$, and one which beats *half seconds* $= \frac{1}{4}\,l$, &c.

Again, by observing the *length* of a pendulum which vibrates in a given *time*, the *force of gravity*, g, may be found. For, as $l = \frac{g t^2}{\pi^2}$, $\therefore g = \frac{\pi^2 l}{t^2}$. And if g varies, as it does in different latitudes and at different altitudes, then $l = \frac{g t^2}{\pi^2} \propto g\, t^2$; and if the time is constant (as, for example, *one second*), then $l \propto g$. Hence,

The length of a pendulum for beating seconds varies as the force of gravity.

Also, $t \propto \left(\frac{l}{g}\right)^{\frac{1}{2}}$; that is, the time of a vibration varies directly as the square root of the length, and inversely as the square root of the force of gravity.

Since the *number*, n, of vibrations in a given time varies inversely as the *time* of one vibration, therefore $n \propto \left(\frac{g}{l}\right)^{\frac{1}{2}}$, and $g \propto l\, n^2$. Hence, if the time and the length of a pendulum are given,

The force of gravity varies as the square of the number of vibrations.

1. What is the length of a pendulum to beat seconds, at the place where a body falls $16\frac{1}{12}$ ft. in the first second?

Ans. 39.11 inches, nearly.

2. If 39.11 inches is taken as the length of the seconds pendulum, how long must a pendulum be to beat 10 times in a minute?

Ans. $117\frac{1}{3}$ feet.

3. In London, the length of a seconds pendulum is 39.1386

inches; what velocity is acquired by a body falling one second in that place? *Ans.* 32.19 feet.

171. The Compensation Pendulum.—This name is given to a pendulum which is so constructed that its length does not vary by changes of temperature. As all substances expand by heat, and contract by cold, therefore a pendulum will vibrate more slowly in warm than in cold weather. This difficulty is overcome in several ways, but always by employing two substances whose rates of expansion and contraction are unequal. One of the most common is the *grid-iron pendulum*, represented in Fig. 126. It consists of alternate rods of steel and brass, connected by cross-pieces at top and bottom. The rate of longitudinal expansion and contraction of brass to that of steel is about as 100 to 61; so that *two* lengths of brass will increase and diminish more than *three* equal lengths of steel. Therefore, while there are three expansions of steel downward, two upward expansions of brass can be made to neutralize them. In the figure the dark rods represent steel, the white ones brass. Suppose the temperature to rise, the two outer steel rods (acting as one) let down the cross-bar *d*; the two brass rods standing on *d* raise the bar *b*; the steel rods suspended from *b* let down the bar *e*, on which the inner brass rods stand, and raise the short bar *c*; and finally, the centre steel rod, passing freely through *d* and *e*, lets down the disk of the pendulum. These lengths (counting each pair as a single rod) are adjusted so as to be in the ratio of 100 for the steel to 61 for the brass; in which case the upward expansions just equal those which are downward, and therefore the centre of oscillation remains at the same distance from the point of suspension.

Fig. 126.

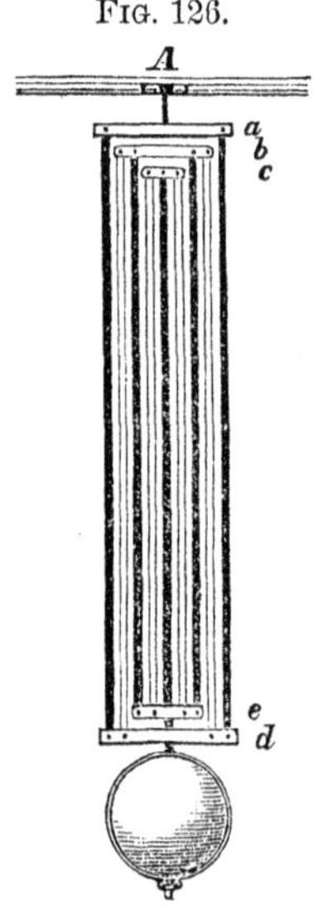

If the temperature falls, the two contractions of brass are equal to the three of steel, so that the pendulum is not shortened by cold.

The *mercurial pendulum* consists of a steel rod terminating at the bottom with a rectangular frame in which is a tall narrow jar containing mercury, which is the weight of the pendulum. It requires only 6.31 inches of mercury to neutralize the expansions and contractions of 42 inches of steel.

CHAPTER VIII.

PROJECTILES. CENTRAL FORCES.

172. Path of a Projectile.—It has been shown already (Art. 44), that a body projected in any direction not coincident with the vertical, describes a parabola. In swift motions, however, the path of a projectile differs widely from a parabola; and the laws of atmospheric resistance must be employed to obtain corrections for the conclusions deduced in this chapter.

173. Formulæ Investigated.—In order to investigate the *general formula*, let A (Fig. 127) be the point of projection, $A\ B$ the plane over which the body is projected, passing through A. $A\ B$ also denotes the *range* or distance to which the body is thrown. Let $A\ C$ be drawn parallel, and $B\ C\ D$ perpendicular to the horizon. Put $a = C\ A\ D$, the angle of elevation; $b = C\ A\ B$, the angle of elevation or depression of the plane of the range; $v =$ the velocity of projection; $t =$ the time of flight; $r =$ the range; and $g = 32\frac{1}{6}$ feet, the velocity imparted by gravity in one second.

FIG. 127.

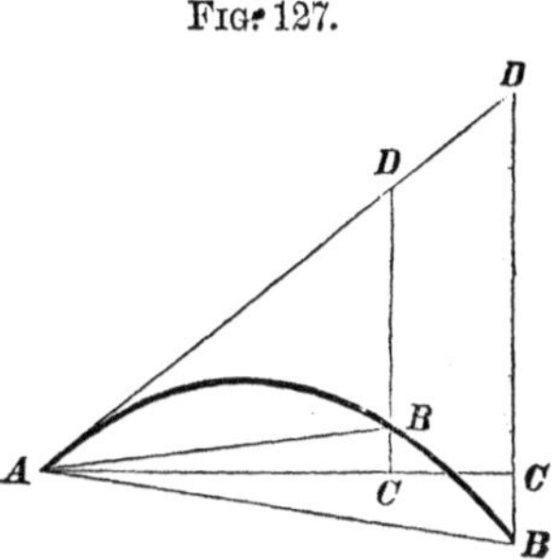

Then, by the laws of uniform motion, at the end of the time t, if gravity did not act, the body would be found in the point D, while, by the laws of falling bodies, it would in the same time pass through the perpendicular $D\ B$; consequently,

$$A\ D = t\ v;\ \text{and}\ D\ B = \tfrac{1}{2}\ g\ t^2.$$

In the right-angled triangles $A\ B\ C$ and $A\ D\ C$, the angle B is the complement of b, and the angle D is the complement of a; and, since the sides are as the sines of the opposite angles,

$$\cos b : \sin (a \pm b) :: t\ v : t\ v\ \frac{\sin (a \pm b)}{\cos b} = \tfrac{1}{2}\ g\ t^2.$$

Plus is used when the plane $A\ B$ descends; *minus*, when it ascends.

Or,
$$\frac{g\ t}{2\ v} = \frac{\sin (a \pm b)}{\cos b}. \quad . \quad . \quad . \quad . \quad . \quad . \quad . \quad . \quad (1)$$

Again, $\cos a : \sin (a \pm b) :: r : \dfrac{r \sin (a \pm b)}{\cos a} = \tfrac{1}{2}\ g\ t^2.$

Or, $$\frac{g\,t^2}{2\,r} = \frac{\sin\,(a \pm b)}{\cos\,a}. \quad . \quad . \quad . \quad . \quad . \quad . \quad . \quad . \qquad (2)$$

Eliminating t from (1) and (2), we have

$$\frac{r}{v^2} = \frac{2\,\sin\,(a \pm b)\,\cos\,a}{g\,\cos^2\,b}. \quad . \quad . \quad . \quad . \quad . \quad . \qquad (3)$$

From these three equations, all the relations between the time, velocity, range, and angle of elevation, are readily determined; so that any two of these four quantities being given, the other two may be found. Thus,

By equation (1) $$v = \frac{g\,t\,\cos\,b}{2\,\sin\,(a \pm b)}.$$

By equation (2) $$r = \frac{g\,t^2\,\cos\,a}{2\,\sin\,(a \pm b)}.$$

The *range* and *elevation* being given, to find the *time* and *velocity*.

By equation (2) $$t = \left(\frac{2\,r\,\sin\,(a \pm b)}{g\,\cos\,a)}\right)^{\frac{1}{2}}.$$

By equation (3) $$v = \left(\frac{r\,g\,\cos^2\,b}{2\,\sin\,(a \pm b)\,\cos\,a}\right)^{\frac{1}{2}}.$$

The *velocity* and *elevation* being given, to find the *time* and *range*.

By equation (1) $$t = \frac{2\,v\,\sin\,(a \pm b)}{g\,\cos\,b}.$$

By equation (3) $$r = \frac{2\,v^2\,\sin\,(a \pm b)\,\cos\,a}{g\,\cos^2\,b}$$

If any two of the above quantities be given to find the *angle of elevation*, then (b being known) in order to find the value of a we substitute in formulæ (1), (2), and (3), for $\sin\,(a \pm b)$, its value, viz., $\sin\,a\,\cos\,b \pm \sin\,b\,\cos\,a$, and, in reducing, put tan for $\frac{\sin}{\cos}$.

Formula (1) becomes $\sin\,a \pm \tan\,b\,\cos\,a = \frac{g\,t}{2\,v}$, whence by eliminating $\cos\,a$, the value of $\sin\,a$ can be found. The resulting equation being a quadratic, there will be, in general, two values of $\sin\,a$; that is, two angles of elevation for the same value of v and t.

Formula (2) becomes $\tan\,a\,\cos\,b \pm \sin\,b = \frac{g\,t^2}{2\,r}$, whence

$$\tan\,a = \frac{g\,t^2}{2\,r\,\cos\,b} \mp \tan\,b.$$

Formula (3) becomes $\sin\,a\,\cos\,a \pm \tan\,b\,\cos^2\,a = \frac{g\,r\,\cos\,b}{2\,v^2}$.

Put $c = \frac{g\,r \cos b}{2\,v^2}$, and $x = \sin a$, then $(1 - x^2)^{\frac{1}{2}} = \cos a$; and $x\,(1 - x^2)^{\frac{1}{2}} \pm \tan b\,(1 - x^2) = c$, from which x or $\sin a$ may be found.

174. Different Angles of Elevation for the Same Range.—As this last equation is a biquadratic, it will give four values of x; the two positive values indicate that there are two different angles of elevation corresponding to the same values of v and r. When these two values are *equal*, then, as shown below, $a = \frac{1}{2}\,(90° \mp b)$, in which case the range (r) is a maximum; and there is the same range for any two angles equally above and below that which gives the maximum. For, since $r = 2\,v^2\,\frac{\sin(a \pm b)\cos a}{g \cos^2 b}$, if v and the angle b are given, the range will vary as $\sin(a \pm b)\cos a$. But

$$\begin{aligned}\sin(a \pm b)\cos a &= \sin a \cos a \cos b \pm \sin b \cos^2 a \\ &= \tfrac{1}{2}\sin 2\,a \cos b \pm \sin b\,(\tfrac{1}{2} + \tfrac{1}{2}\cos 2\,a) \\ &= \tfrac{1}{2}\sin 2\,a \cos b \pm \tfrac{1}{2}\cos 2\,a \sin b \pm \tfrac{1}{2}\sin b \\ &= \tfrac{1}{2}\sin(2\,a \pm b) \pm \tfrac{1}{2}\sin b;\end{aligned}$$

and since the second part of this expression is constant, the range will be a maximum when $\sin(2\,a \pm b)$ is a maximum; that is, when $2\,a \pm b = 90°$.

$$\therefore a = \tfrac{1}{2}\,(90° \mp b).$$

Therefore the range will be a maximum when the angle of elevation is equal to $\frac{1}{2}\,(90° \mp b)$.

When $a = \frac{1}{2}\,(90° \mp b) + c$, $\quad r = v^2\,\frac{\sin(90° + 2\,c) \pm \sin b}{g \cos^2 b}$,

and when $a = \frac{1}{2}\,(90° \mp b) - c$, $\quad r = v^2\,\frac{\sin(90° - 2\,c) \pm \sin b}{g \cos^2 b}$.

But $\sin(90° + 2\,c) = \sin(90° - 2\,c)$, since the sines of supplementary arcs are equal; hence all angles of elevation, equally above and below that which gives the maximum, have equal ranges. Thus, a cannon ball fired at an angle of 60° above a horizontal plane, would reach the plane at the same distance from the point of projection as if fired at an angle of 30°. When the data of the problem give or require a greater value for $\sin(2\,a \pm b)$ than 1, the sine of 90°, the problem, under the proposed conditions, is impossible.

That the two values of $\sin a$ are equal when the range is a maximum, may be shown as follows:

Let x and y be two *varying supplementary* arcs, and let $a = \frac{1}{2}\,(x \mp b)$; then $r = v^2\,\frac{\sin x \pm \sin b}{g \cos^2 b}$.

Again, let $a = \frac{1}{2}(y \mp b)$; then $r = v^2 \dfrac{\sin y \pm \sin b}{g \cos^2 b}$.

Now, although these two values of a may be different, yet the ranges corresponding to them will be equal, because $\sin x = \sin y$. Suppose x to *increase* from 0 to 90°; then y will *decrease* from 180° to 90°, and the two values of a will become equal, each being $\frac{1}{2}(90° \mp b)$. But, as has been shown, this value of a gives a maximum for r.

175. The Greatest Height of a Projectile.—To find the greatest height to which the projectile will ascend, it must be considered that a body projected perpendicularly upward, will rise to the same height from which it must have fallen to acquire the velocity of projection (Art. 24). Since v represents the whole velocity of projection in an oblique direction, and since a is the angle of elevation, therefore $v \sin a$ is that component of the velocity which acts directly upward. And the space described vertically by this component of the velocity is [(3) Art. 28] $s = \dfrac{v^2}{2g}$. Hence, substituting h for s, and $v \sin a$ for v, we have

$$h = \frac{v^2 \sin^2 a}{2g}. \quad . \quad . \quad . \quad . \quad . \quad . \quad . \quad . \quad . \quad . \quad (4)$$

If, therefore, the angle of elevation and the velocity of projection are given, the greatest height is found as above. Or, if the angle of elevation, and that of the plane, be given, along with the *range* (r) or the *time* (t), then let v be found first, as in Art. 173; after which h may be obtained from equation (4).

If the velocity of projection, and the greatest height to which the projectile rises, were given, equation (4) will determine the angle of elevation. For since $h = \dfrac{v^2 \sin^2 a}{2g}$, $\therefore \sin^2 a = \dfrac{2gh}{v^2}$, and $\sin a = \dfrac{\sqrt{2gh}}{v}$.

176. Particular Formulæ for a Horizontal Plane.—The preceding equations become much more simple when the projection is above a *horizontal* plane; for then $b = 0$; therefore $\sin b = 0$, and $\cos b = 1$; hence, from equations (1), (2), and (3), we have

$$t = \frac{2v \sin a}{g} \propto v \sin a \quad . \quad . \quad . \quad . \quad . \quad . \quad (1'),$$

$$\tan a = \frac{gt^2}{2r} \quad . \quad . \quad . \quad . \quad . \quad . \quad . \quad . \quad . \quad . \quad . \quad (2'),$$

$$r = \frac{v^2 \sin 2a}{g} \propto v^2 \sin 2a \quad . \quad . \quad . \quad . \quad (3').$$

On a horizontal plane, therefore, we have the following theorems:

I. *The* TIME OF FLIGHT *varies as the velocity of projection multiplied by the sine of the angle of elevation.*

II. *The* RANGE *varies as the square of the velocity of projection, multiplied by the sine of twice the angle of elevation.*

Moreover, since the sine of twice 45° equals the sine of 90°, which equals radius, hence, by Theorem II,

III. *The* RANGE IS GREATEST *when the angle of elevation is* 45°, *and is the same at elevations equally above and below* 45°.

IV. *The* TIME OF FLIGHT IS GREATEST *when the body is thrown perpendicularly upward.*

177. The Equation of the Path of a Projectile.—Suppose the body is projected from A (Fig. 128) in the direction $A\,T$ with a velocity v, and let $A\,X$, horizontal, and $A\,Y$, vertical, be rectangular axes.

Fig. 128.

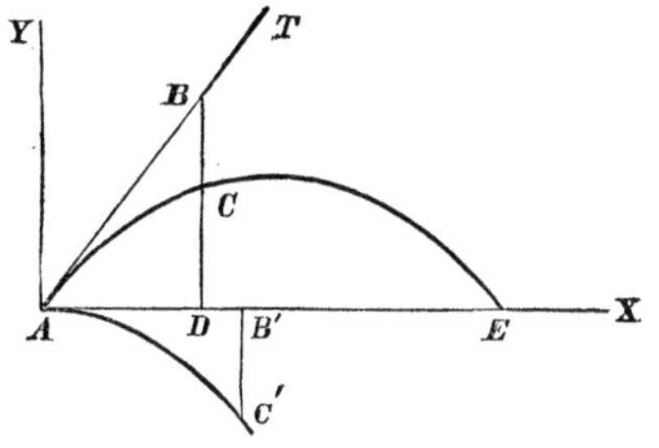

The components of v along the axes are, $v \cos a$ for $A\,X$, and $v \sin a$ for $A\,Y$. At the end of the time t, suppose the body to be at C. Denote the co-ordinates of the point C by x and y; then $x = t\,v \cos a$, and $y = B\,D - B\,C = t\,v \sin a - \frac{1}{2} g\,t^2$.

Eliminating t, we have

$$y = x \tan a - \frac{g\,x^2}{2\,v^2 \cos^2 a},$$

an equation expressing the relation between x and y for any value of t whatever, and consequently the equation of the path.

To find the range $A\,E$, make $y = 0$; then $x = 0$, or $x = \frac{2\,v^2 \sin a \cos a}{g}$.

The first value of x corresponds to the point A; the second is the *range* $A\,E$.

To find the time of flight, make $x = r$ in the equation $x = t\,v \cos a$, and we have $t = \frac{r}{v \cos a}$.

If $a = 0$, $y = -\frac{g\,x^2}{2\,v^2}$, the equation of the path when the body is thrown horizontally. Since y is negative for all values of x, every point of the path, except A, lies below a horizontal line drawn through the point of projection. *A C′* represents the path when the body is projected in the line *A X*.

If *positive* ordinates are estimated from *A X* downward, the equation may be written $y = \frac{g x^2}{2 v^2}$.

178. To Find the Range on an Oblique Plane.—Let b be the inclination of the plane to the horizon; then $y = \pm x \tan b$ is the equation of the line in which the oblique plane intersects the plane of the projectile's path. Combining this with the equation $y = x \tan a - \frac{g x^2}{2 v^2 \cos^2 a}$, we have $\pm x \tan b = x \tan a - \frac{g x^2}{2 v^2 \cos^2 a}$; whence $x = 0$, and $x = \frac{2 v^2 \cos^2 a}{g} (\tan a \mp \tan b) = \frac{2 v^2 \cos a \sin (a \mp b)}{g \cos b}$, and hence the range will be $r = \frac{x}{\cos b} = \frac{2 v^2 \cos a \sin (a \mp b)}{g \cos^2 b}$.

179. Questions on Projectiles.—

1. A gun was fired at an elevation of 50°, and the shot struck the ground at the distance of 4898 feet; with what velocity did it leave the gun, and how long was it in the air?

Ans. Velocity, 400 feet per second.
Time, 19.05 seconds.

2. Range 4898 feet, time of flight 16 seconds; required the angle of elevation and the velocity of projection?

Ans. $a = 40° 3'$, $v = 400$ feet per sec.

3. Range 2898 feet, velocity of projection 389.1 feet, what were the elevation and time of flight?

Ans. $a = 19°$ or $71°$, $t = 7.87$ or 22.86 sec.

4. Elevation 40°, range 4898; required the range when the elevation is $29\frac{1}{2}°$ *Ans.* 4263.

5. Elevation 40° 3′, time of flight 16 seconds; required the range and velocity of projection? *Ans.* $r = 4898$, $v = 400$ ft.

6. Velocity 510 feet per sec., time of flight 15 seconds, to find the elevation and range. *Ans.* $a = 28° 14'$, $r = 6740$.

7. On a slope ascending uniformly above a horizontal plane at an angle of 10° 20′, a ball was fired at an angle of elevation above the horizon of 34°, and with a velocity of 401 feet per second; what was the range on the slope when the gun was directed up the hill, and what when directed downward?

Ans. 3438 and 5985 feet.

8. What will be the time of flight for any given range, the angle of elevation being 45°?

Ans. $t = \sqrt{\frac{2 r}{g}}.$

9. Having given the angle of elevation, to determine the velocity, so that the projectile may pass through a given point.

Ans. $v = \frac{x'}{\cos. a} \sqrt{\frac{g}{2(x' \tan a - y')}}$, where x' and y' are the co-ordinates of the given point.

10. Find the angle of elevation and velocity of projection of a shell, so that it may pass through two points, the co-ordinates of the first being $x' = 1700$ ft., $y' = 10$ ft., and of the second, $x'' = 1800$ ft., $y'' = 10$ ft. *Ans.* $a = 39' 19''$, $v = 2218.3$ ft.

Central Forces.

180. Central Forces Described.—Motion in a curve is always the effect of two forces; one an impulse, which alone would cause uniform motion in a straight line; the other a continued force, which urges the body toward some point out of the original line of motion. The first is called the *projectile* force, the second the *centripetal* force.

The *centripetal* force may be resolved into two components; one in the direction of the tangent, the other perpendicular to it. The *tangential* component will accelerate or retard the motion in the curve according as it acts *with* the projectile force, or in *opposition* to it. When the body moves in the circumference of a circle, the *tangential* component of the centripetal force is 0, and hence the motion is uniform.

If the centripetal force should cease to act at any instant, the body, by its inertia, would immediately begin to move in a straight line tangent to the curve at the point where the body was when the force ceased to act.

Since the body, by its inertia, *tends* to move in a tangent, there is a continued *outward* pressure directed *from* the centre of curvature; this is called the *centrifugal* force. In circular motion it is equal to the *centripetal* force, and directly opposed to it.

181. Expressions for the Centrifugal Force in Circular Motion.—

1. Let r = the radius of the circle, v = the velocity of the body, c = the distance through which the centrifugal force causes the body to move in one second, and let $A B$ (Fig. 129) be the arc described in the infinitely small time t; then $A B = v t$, and, by a method similar to that employed in the discussion of the force of gravity, it may be shown that $B D = c t^2$.

Fig. 129.

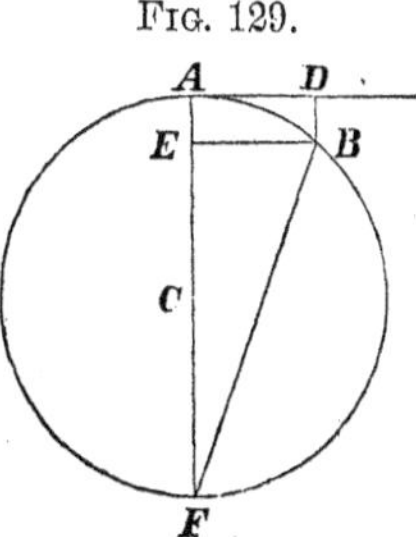

But $A\,B$, being a very small arc, may be considered as equal to its chord, which is a mean proportional between $A\,E$ and the diameter $2\,r$. Hence $c\,t^2 = \frac{v^2\,t^2}{2\,r}$, or

$$c = \frac{v^2}{2\,r} \quad . \quad . \quad . \quad . \quad . \quad . \quad . \quad . \quad . \quad . \quad (1)$$

If this be doubled, then (Art. 25) $\frac{v^2}{r}$ is the velocity which the centrifugal force is capable of generating in one second, and this is sometimes taken as the measure of the centrifugal force.

From (1) it follows, that *in equal circles the centrifugal force varies as the square of the velocity.*

2. The value of c may be expressed in a different form. Let t' = the time of a complete revolution; then $2\,\pi\,r = v\,t'$; whence $v = \frac{2\,\pi\,r}{t'}$. This substituted in (1) gives

$$c = \frac{2\,\pi^2\,r}{t'^2} \quad . \quad . \quad . \quad . \quad . \quad . \quad . \quad . \quad . \quad . \quad (2)$$

Hence, *The centrifugal force varies directly as the radius of the circle, and inversely as the square of the time of revolution.*

3. Let w = the weight of the revolving body, and c' = the centrifugal force expressed in pounds; then

$$w : c' :: \tfrac{1}{2}\,g : \frac{v^2}{2\,r}; \text{ whence } c' = \frac{w\,v^2}{r\,g} \quad . \quad . \quad (3)$$

Let n = the number of revolutions per second; then

$v = 2\,\pi\,r\,n$, and (3) becomes

$$c' = \frac{4\,\pi^2}{g} \,.\, w \,.\, r \,.\, n^2 \quad . \quad . \quad . \quad . \quad . \quad . \quad . \quad . \quad (4)$$

182. Two Bodies Revolving about their Centre of Gravity.—Let A and B (Fig. 130) be two bodies connected by a rod, and let them be made to revolve about the centre of gravity C; then by (4) the centrifugal force of A will be $\frac{4\,\pi^2}{g} \,.\, A \,.\, A\,C \,.\, n^2$, and of B, $\frac{4\,\pi^2}{g} \,.\, B \,.\, B\,C \,.\, n^2$.

FIG. 130.

B C A

But C being the centre of gravity of the two bodies, $A \,.\, A\,C = B \,.\, B\,C$; $\therefore$ the centrifugal force of A equals that of B. Hence, *If two bodies revolve in the same time about an axis passing through their centre of gravity, there will be no strain upon that axis.*

183. Centrifugal Force on the Earth's Surface.—As the earth revolves upon its axis, all free particles upon it are influenced by the centrifugal force. Let $N\,S$ (Fig. 131) be the axis, and A a particle describing a circumference with the radius $A\ O$. Put $r = C\ Q$, $r' = A\ O$, $l =$ the angle $A\ C\ Q$, the latitude, $c =$ the centrifugal force at the equator, $c' =$ the centrifugal force at A, $v =$ velocity of Q, and v' = velocity of A; then

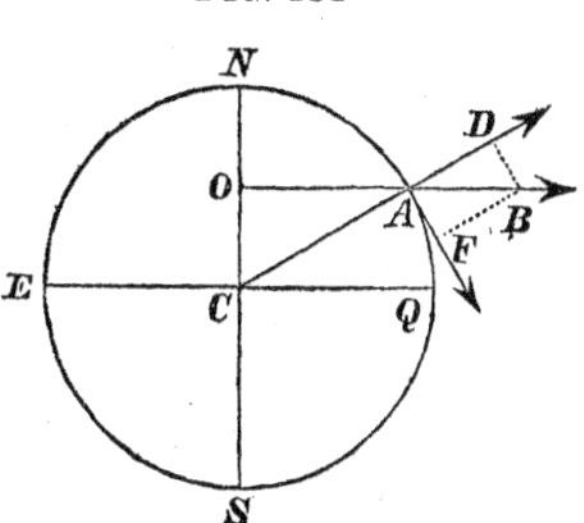

FIG. 131

$$c = \frac{v^2}{2\ r}, \text{ and } c' = \frac{v'^2}{2\ r'}.$$

But $v : v' :: r : r'$; whence $v' = \frac{v\ r'}{r}$. Again, from the triangle $A\ C\ O$ we have $r' = r \cos l$; hence $v' = v \cos l$, and $c' = \frac{v^2 \cos^2 l}{2\ r \cos l} = \frac{v^2 \cos l}{2\ r}$. Comparing this value of c' with that of c, we have

$$c' = c \cos l.$$

That is, *the centrifugal force at any point on the earth's surface is equal to the centrifugal force at the equator, multiplied by the cosine of the latitude of the place.*

Let $A\ B$ represent the centrifugal force at A, and resolve it into $A\ D$ on $C\ A$ produced, and $A\ F$, tangent to the meridian $N\ Q\ S$; then, since the angle $D\ A\ B = A\ C\ Q = l$, we have

$$A\ D = A\ B \cos l = c \cos l \,.\, \cos l = c \cos^2 l.$$

That is, *that component of the centrifugal force at any point, which opposes the force of gravity, is equal to the centrifugal force at the equator, multiplied by the square of the cosine of the latitude of the place.*

In like manner we find $A\ F = A\ B \sin l = c \cos l \sin l = \frac{c \sin 2\ l}{2}$. From this equation we see that the tangential component is 0 at the equator, increases till $l = 45°$, where it is a maximum; then goes on diminishing till $l = 90°$, when it again becomes 0.

The effect of $A\ D$ is to diminish the weight of the particle, while the effect of $A\ F$ is to urge it toward the equator.

184. Examples on Central Forces.—

1. A ball weighing 10 lbs. is whirled around in a circumference of 10 feet radius, with a velocity of 30 feet per second. What is the tension upon the cord which restrains the ball?

Ans. 28 lbs. nearly.

2. With what velocity must a body revolve in a circumference of 5 feet radius, in order that the centrifugal force may equal the weight of the body? *Ans.* $v = 12.7$ ft.

3. A ball weighing 2 lbs. is whirled round by a sling 3 feet long, making 4 revolutions per second. What is its centrifugal force?
Ans. 117.84 lbs.

4. A weight of 5 lbs. is attached to the end of a cord 3 feet long just capable of sustaining a weight of 100 lbs. How many revolutions per second must the body make in order that the cord may be upon the point of breaking? *Ans.* $n = 2.3$ nearly.

5. A railway carriage, weighing 7 tons, moving at the rate of 30 miles per hour, describes an arc whose radius is 400 yards. What is the outward pressure upon the track? *Ans.* 786 + lbs.

6. A hemisphere has its base fixed in a horizontal position, and a body, under the influence of gravity, moves down the convex side of it from the highest point. How far from the base will the body be when it leaves the surface of the hemisphere?
Ans. $\frac{2}{3}$ *r.*

185. Composition of two Rotary Motions.—*When a body is rotating on an axis, and a force is applied which alone would cause it to rotate on some other axis, the body will commence rotation on an axis lying between them, and the velocities of rotation on the three axes are such, that each may be represented by the sine of the angle between the other two.*

Suppose that the body $H K$ (Fig. 132) is rotating on $A B$, and that a force is applied to make it rotate on $C D$. Let these axes intersect within the body, and call the point of intersection, G. Imagine a perpendicular to the plane of the axes to be drawn through G, and let P be a particle of the body in this perpendicular. Suppose the particle P, in an infinitely small time t, to pass over $P a$ by the first rotation, and $P c$ by the second. Then, since the particle will describe the diagonal $P e$ in the time t, this line must indicate the direction and velocity of the resultant rotation. Therefore, if $E F$ be drawn through G, perpendicular to the plane $G P e$, $E F$ is the axis on which the body revolves in consequence of the two rotations given to it. Since $P G$ is perpendicular to the plane $A G C$, and also to the line $E F$, therefore $E F$ is in that plane; that is, the new axis of rotation is in the plane of the other two axes. The angles $A G E$ and $E G C$, are respectively equal to the angles $a P e$ and $e P c$, the inclinations of the planes of rotation. But the lines, $P a$, $P c$, $P e$, represent the velocities in those directions respectively; and (Art. 41) $P a : P c : P e :: \sin c P e : \sin a P e : \sin a P c$; there-

Fig. 132.

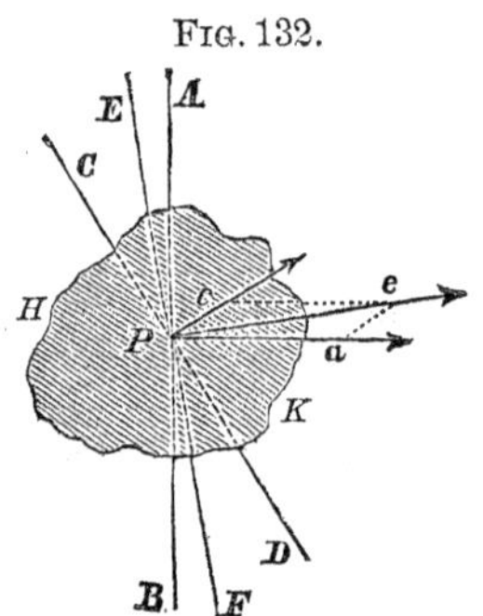

fore $P a : P c : P e :: \sin C G E : \sin A G E : \sin A G C$; or, the velocities on the three axes, (namely, the axes of the component rotations, and of the resultant rotation,) are such, that each may be represented by the sine of the angle between the other two axes.

186. The Gyroscope.—The *gyroscope* affords an illustration of the composition of two rotations imparted to a body. As usually constructed, it consists of a heavy wheel *G H* (Fig. 133), accurately balanced on the axis *a b*, which runs with as little friction as possible upon pivots in a metallic ring. In the direction of the axis, there is a projection *B* from the ring, having a socket sunk into it on the under side, so that it may rest on the pointed standard, *S*, without danger of slipping off.

Fig. 133.

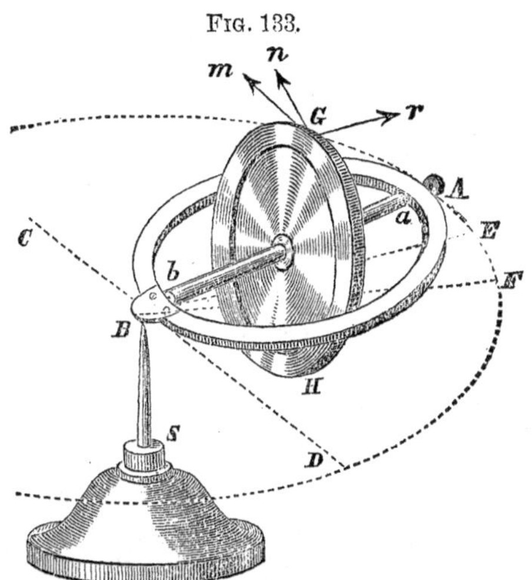

The wheel is made to rotate swiftly by drawing off a cord wound upon *a b*, and then the socket in *B* is placed on the standard, and the whole left to itself. Immediately, instead of falling, the ring and wheel commence a slow revolution in a horizontal plane around the standard, the point *A* following the circumference *A E F*, in a direction contrary to the motion of the top of the wheel.

This revolution is explained by applying the principle of composition of rotations given in the preceding article. The particles of the wheel are rotating about the horizontal axis *a b* by the force imparted by the string. The force of gravity tends to make it fall, that is, to revolve in a vertical circle around the axis *C D* at right angles to *a b*. Hence, in a moment after dropping the ring, the system will be found revolving on an axis which lies in the direction *E B*, between *A B* and *C D*, the other two axes. *Now*, gravity bears it down around a new axis perpendicular to *E B*. Therefore, as before, it changes to still another axis *F B*, and thus continues to go round in a horizontal circle.

The only way possible for it to rotate on an axis in a new position, is to turn its present axis of rotation into that position.

Hence, the whole instrument turns about, in order that its axis may take these successive positions.

The change of axis is seen also by observing the resultant of the motions of the particles at the top and bottom of the wheel. For example, G is moving swiftly in the direction m by the rotation around $a\ b$; by gravity it tends to move slowly in the line r, tangent to a vertical circle about the centre B. The resultant is in the line n, tangent to the wheel when its axis $a\ b$ has taken the new position $E\ B$.

The centre of gravity of the ring and wheel tends to remain at rest, while the resultant of the two rotations carries around it all other parts, standard included, in horizontal circles. But the standard by its inertia and friction resists this effort, and the reaction causes the ring and wheel to go around the standard.

CHAPTER IX.

STRENGTH OF MATERIALS.

187. Longitudinal Strength.—*Strength* is the power to resist fracture; *stress*, the power to produce fracture. When a force is applied to a bar or rod, to pull it asunder, its strength, called in this case *longitudinal strength, is proportional to the area of its cross-section.* Each line of particles in the direction of the length of the bar has its separate strength, and the whole strength therefore depends on their number, that is, on the area of the cross-section. The *form* of the cross-section is immaterial.

188. Lateral Strength, Support at Each End.—When a beam rests horizontally, supported at both ends, and pressed by a weight at the centre, its *strength at that point varies as the area of the cross-section, multiplied by the depth of the centre of gravity.*

Let $A\ B\ C\ D$ (Fig. 134) represent a longitudinal section of a

FIG. 134.

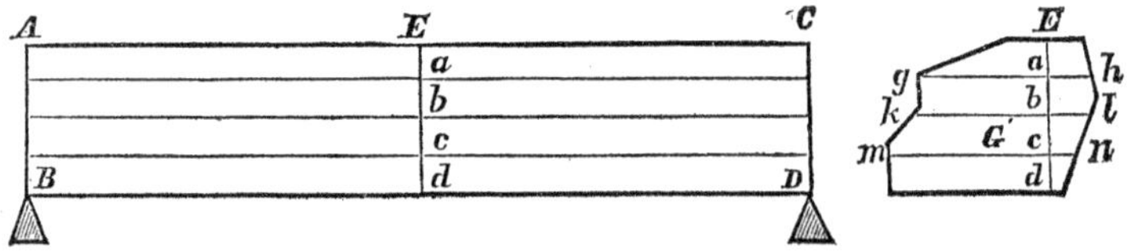

prismatic beam and $E\ k\ d\ l$ a section of any form whatever, at right angles to the axis of the beam. Let G be the centre of gravity of the cross-section, and $g\ h$, $k\ l$, $m\ n$, &c., the width of horizontal laminæ of the beam. Suppose this beam to be supported at its two ends, and that on the middle of it at E there is placed a weight, W. From E draw $E\,d$ perpendicular to the horizon, and cutting the several laminæ in the points a, b, c, d, &c.

The pressure of the weight W tends to produce a fracture in the beam, beginning at d, and passing through the laminæ in succession until it arrives at E.

The tendency of the beam to resist fracture depends partly upon the *cohesion* of its corresponding particles, and partly upon the *distance* from E at which the force of cohesion acts. E may then be considered as the centre of motion of a lever, at the extremities of whose arms $E\,d$, $E\,c$, $E\,b$, $E\,a$, &c., this force is applied. Let s = the cohesive force of one line of particles, then $s \times g\,h$ will represent the strength of the lamina whose width is $g\,h$, and $s \times g\,h \times E\,a$ will represent the power of this lamina to resist fracture; hence the whole power of the beam to resist fracture will be

$$s\,(g\,h \times E\,a + k\,l \times E\,b + m\,n \times E\,c + \text{\&c.})$$

Put A = the area of the cross-section and G = the depth of the centre of gravity below E; then (Art. 78)

$$G = \frac{g\,h \times E\,a + k\,l \times E\,b + m\,n \times E\,c + \text{\&c.}}{A}$$

or $A\,.\,G = g\,h \times E\,a + k\,l \times E\,b + m\,n \times E\,c + \text{\&c.}$, and hence the strength of the beam is equal to $s\,.\,A\,.\,G \propto A\,.\,G$.

189. Special Cases.—This proposition is general, and applies to a number of distinct cases. In *cylindrical* and *square* beams, since the area of the section varies as the square of its depth, and the distance of the centre of gravity from the point E varies as the depth, their strength is as the *cube* of the depth. In beams whose cross-section is a rectangle (Fig. 135), the strength varies as the *breadth* and *square of the depth;* for here the area being as the product of the two sides, and the distance of the centre of gravity from E being equal to half the vertical side, and therefore proportioned to that side, the proposition is, that the strength varies as the breadth × depth × depth, or as the breadth into the square of the depth. Hence, the same beam with its narrow side upward, is as much stronger than with its broad side upward, as the depth exceeds the breadth.

Fig. 135.

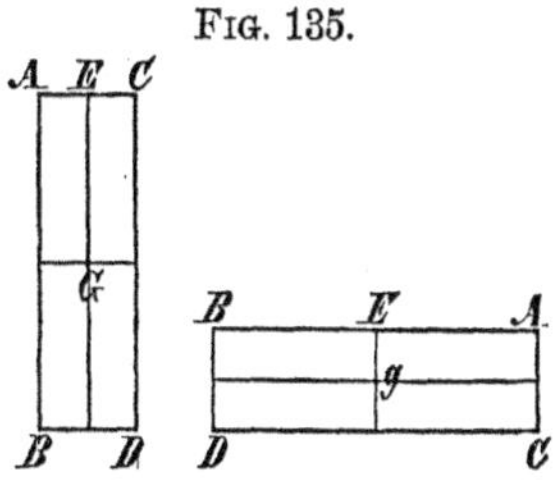

For the area being the same in both cases, the strengths are proportioned to $E\,G$ and $E\,g$, or as $A\,B$ to $A\,C$. Thus if a joist be 10 inches broad and $2\frac{1}{2}$ thick, it will bear four times as much weight when laid on its edge as when laid on its side. Hence the modern mode of flooring with thin but deep pieces of timber.

Again, a *triangular* beam is *twice* as strong when resting on a *side*, as when resting on an *edge*. For, the area being the same in both cases, the strength varies as $E\,G$ and $E\,g$ (Fig. 136), which are as 2 to 1 (Art. 72). These principles apply not only to beams, but to bars, and similar forms of every sort of matter.

Fig. 136.

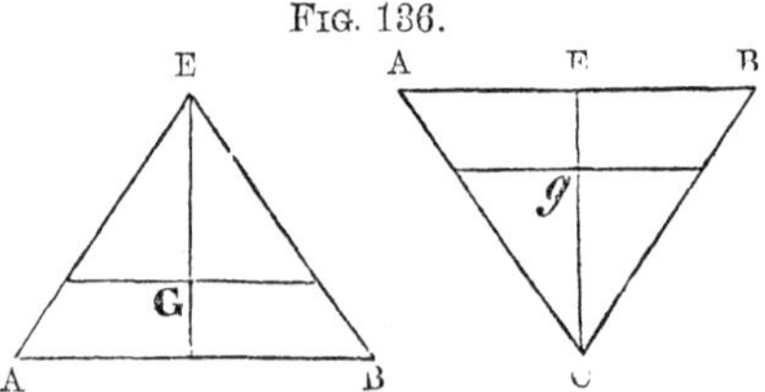

190. Stress from the Weight of the Beam.—The *stress* arising from the weight of a beam varies as the *product of the length and weight*. Let L = length, and W = weight of the whole beam, w = weight of the portion $A\,C$ (Fig. 137).

Fig. 137.

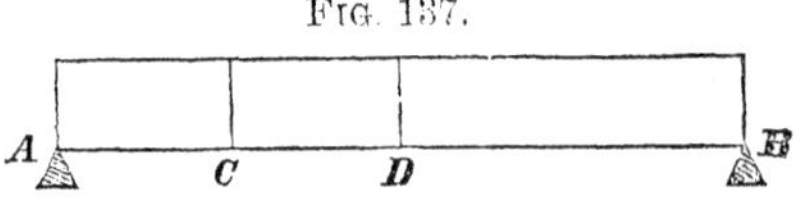

The pressure on each prop is $\frac{1}{2}\,W$; this, therefore is the force acting at A which tends to fracture the beam at C with an energy expressed by $\frac{1}{2}\,W \times A\,C$.

Now suppose the portion $C\,B$ to be held firmly in solid masonry, and a force $= \frac{1}{2}\,W$ to act upward at A, and another $= w$ at the middle of $A\,C$ to act downward, the tendency to produce fracture will be the same as before, and hence the stress at C will be $\frac{1}{2}\,W \times A\,C - w \times \frac{1}{2}\,A\,C = \frac{1}{2}\,(W - w)\,A\,C$. But $A\,B : A\,C :: W : w$; whence $w = \dfrac{A\,C}{A\,B}\,W$; so that the expression for the stress will be

$$\tfrac{1}{2}\left(W - \frac{A\,C}{A\,B}\,W\right)A\,C = \tfrac{1}{2}\,W\left(\frac{A\,B - A\,C}{A\,B}\right)A\,C = \tfrac{1}{2}\,W\frac{A\,C\,.\,B\,C}{A\,B}.$$

$A\,B$ being constant, the stress at any point C varies as the rectangle of the two lines $A\,C$ and $B\,C$, and is greatest when $A\,C = B\,C$, or when C is at the middle of the beam, where the stress is

$$\tfrac{1}{2}\,W\frac{A\,D^2}{A\,B} = \tfrac{1}{2}\,W\frac{\frac{1}{4}\,L^2}{L} = \tfrac{1}{8}\,W\,.\,L \propto W\,.\,L.$$

If, therefore, we use the term *relative strength* to denote the *ratio* of the *strength* to the *stress*, and represent this ratio by S, we shall have

$$S \propto \frac{A\,.\,G}{L\,.\,W}.$$

If beams are *similar*, then the above ratio is *inversely as any one of their three dimensions*. Being similar, their length, breadth, and thickness are proportional. Let D represent any one of these three dimensions. Then,

$$\text{Since } A \propto D^2 \text{ and } G \propto D,$$
$$A \,.\, G \propto D^3; \text{ also, } L \propto D, \text{ and } W \propto D^3;$$
$$\therefore S \propto \frac{D^3}{D^4} \propto \frac{1}{D}.$$

Hence the *relative strength* of large structures is less than that of smaller similar ones. If a model is *three* feet long, and the structure 75 feet long, then the structure is 25 times weaker relatively than the model.

191. Additional Weight at the Centre.—If a weight W' is *uniformly* distributed through a beam whose length is L, the stress arising from the weight is (Art. 190) $\frac{1}{8} L \,.\, W'$. If the same weight is placed at the *middle* of the beam, the stress is $\frac{1}{2} L \,.\, \frac{1}{2} W' = \frac{1}{4} L\ W'$; hence, if a weight is placed at the middle point, the stress is twice as great as when distributed uniformly through the beam.

If W is the weight of the beam, and a weight W' is placed at the middle, then, from what has just been shown, the relative strength will be

$$S = \frac{A \,.\, G}{\frac{1}{4} L \left(\frac{1}{2} W + W'\right)} \propto \frac{A \,.\, G}{L \left(\frac{1}{2} W + W'\right)}.$$

If the beam is small compared with the weight laid upon it, then

$$S \propto \frac{A \,.\, G}{L \,.\, W'}.$$

In order that the foregoing general formulæ may be applied to practice, so as to find the actual strength of bars or beams, it is necessary to have some standard of strength ascertained by experiment, which may be employed as the unit of comparison. For example, it is found by experiment that a stick of *oak*, *one foot long* and *one inch square*, is able, when supported at both ends, to sustain a weight of 600 pounds; and that a bar of *iron* of the same dimensions would sustain, in the same circumstances, 2190 pounds. The oak weighs half a pound, and the iron three pounds. With these data applied to the foregoing formulæ, we may solve such problems as the following:

1. What weight can be sustained at the middle point of a prismatic beam of oak, whose length is 6 feet, and its end 4 inches square?

If the weight of the bar is left out of account,

$$\frac{A \cdot G}{L \cdot W'} = \frac{1^2 \cdot \frac{1}{2}}{1 \cdot 600}, \text{ for one case; and } \frac{4^2 \cdot 2}{6 \cdot W'}, \text{ for the other.}$$

But these expressions are to be equal at the moment of rupture of both beams, since at that moment they have the same *relative* strength;

$$\therefore \frac{1}{1200} = \frac{16}{3\,W'}; \therefore W' = 6400 \text{ lbs.} \quad \textit{Ans.}$$

If the weight of the beams be considered, then

$$\frac{1^2 \cdot \frac{1}{2}}{1 \cdot (\frac{1}{4} + 600)} = \frac{4^2 \cdot 2}{6\,(24 + W')}; \therefore W' = 6378\tfrac{2}{3} \text{ lbs.}$$

In this example, the weight of the large beam is known to be 48 lbs., from the given dimensions and weight of the small one.

2. What must be the depth of a beam in the form of a rectangular prism, whose breadth is 2 inches and length 8 feet, to support a weight of 6400 pounds, its own weight not being taken into consideration?

$$\frac{1^2 \cdot \frac{1}{2}}{1 \cdot 600} = \frac{2 \cdot D \cdot \frac{1}{2} D}{8 \cdot 6400}; \therefore D = 6{,}53 \text{ in.} \quad \textit{Ans.}$$

3. What weight can be supported at the middle point of a bar of iron 10 feet long, and the side of whose square end is 3 inches, its own weight not being taken into consideration?

Ans. 5913 pounds.

192. Additional Weight, at Any Point.—The *stress* produced by a weight, at any point, is as *the product of the two distances from the ends.* In Fig. 138, according to the theorems for parallel forces, pressure at $A = \frac{W \times B\,C}{A\,B}$, and pressure at $B = \frac{W \times A\,C}{A\,B}$. But the *reaction* of either point of support is equal to the pressure on that point; and this force acts at C with a *leverage* $A\ C$ on one side, and $B\ C$ on the other, so that the stress at $C = \frac{W \times B\,C}{A\,B} \times A\,C$, or $\frac{W \times A C}{A\,B} \times B\,C$, either of which expressions = stress at C, and $\propto A\,C \times B\,C$. And since this rectangle is greatest when $A\,C = C\,B$, and diminishes as these lines become more and more unequal in length, so the tendency of a horizontal bar to break is greatest in the middle, and decreases toward the points of support.

FIG. 138.

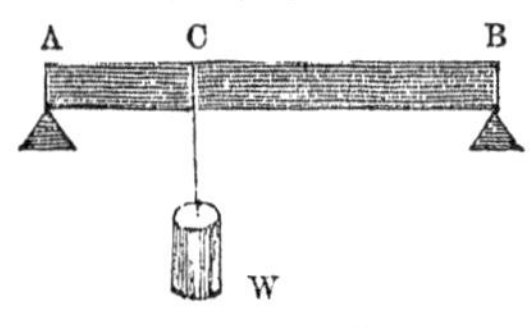

193. Form for Equal Strength.—Hence a beam, in order to be equally strong throughout, must be thickest in the middle;

and if the sides of such a beam are parallel planes, the figure of the beam must be *elliptical.*

For let the curve $A\ P\ D\ M$ (Fig. 139), whose axis is $A\ D$, represent a longitudinal section, and let $a =$ the thickness or breadth of the beam; then a section of the beam perpendicular to the axis at any point C will be a rectangle, whose breadth is a, depth $P\ M$, and the depth of its centre of gravity $\frac{1}{2}\ P\ M$. Hence, the tendency of the beam to resist fracture at any point C is as $a \times P\ M^2$; but the *stress* at C is as $A\ C \times C\ D$; therefore

FIG. 139.

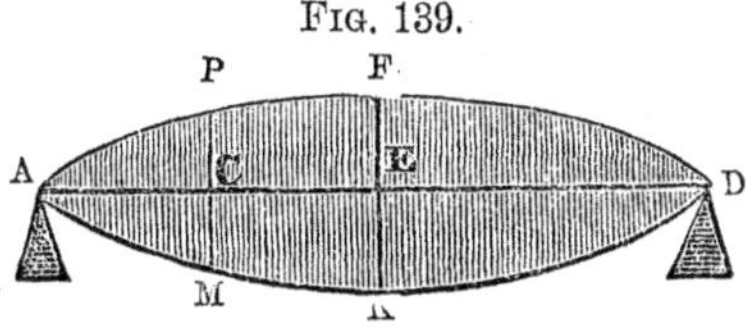

$$\text{the relative strength at } C \propto \frac{a \times P\ M^2}{A\ C \times C\ D} \propto \frac{P\ M^2}{A\ C \times C\ D};$$

hence, if $P\ M^2 \propto A\ C \times C\ D$, the strength will be the same at every point: but in this case the curve $A\ P\ D\ M$ is an ellipse, whose major and minor axes are $A\ D$ and $F\ K$.

Therefore, in the use of horizontal rectangular timbers in building, much of the timber is useless, though it would only be a waste of labor to remove the redundant parts. But iron beams are often made of curved forms, which combine strength with lightness and economy of material. The convex form has sometimes been adopted in the iron bars for railroad tracks, as shown in Fig. 140.

FIG. 140.

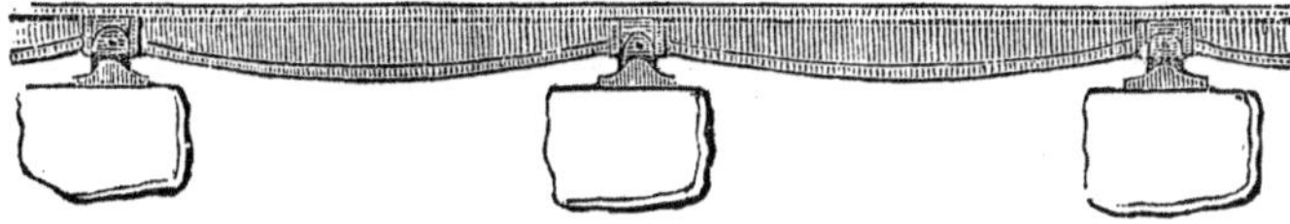

194. Lateral Strength, Support at One End.—When a prismatic beam is secured firmly in a wall at one end, the same general statements hold true as in relation to beams supported at both ends.

1. The strength varies as *the area of the cross-section multiplied by the height of the centre of gravity.*

Let $A\ B\ E\ F$, $a\ b\ e\ f$ (Fig. 141), represent the longitudinal sections of two prismatic beams fixed horizontally into the wall $H\ K\ L\ M$; then the tendency of these beams to *resist* fracture at the ends $E\ F$, $e\ f$,

FIG. 141.

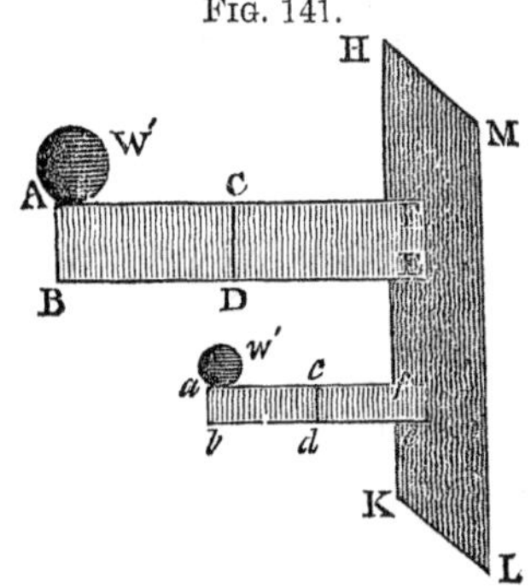

where they are inserted into the wall, will be measured *by the area of the cross-section into the height of its centre of gravity;* for in this case the fracture will begin at the upper points F, f, and end at the lower points E, e; that is, strength $\propto A \,.\, G$.

But the tendency to *produce* fracture will be the weight of the beams, acting at the distance of their centres of gravity from the ends $E\,F$, $e\,f$. Hence,

$$S \propto \frac{A \,.\, G}{L \,.\, W}.$$

2. If an additional weight lies on the end of the beam, then

$$S \propto \frac{A \,.\, G}{L\,(\frac{1}{2}\,W + W')}.$$

3. If beams are similar, $S \propto \frac{1}{D}$.

If any other cross-section be taken, as $C\,D$, and W represent the weight from it to the end $A\,B$, then, by the same course of reasoning as before, the relative strength at $C\,D\,(= S) \propto \frac{A \,.\, G}{A\,C \,.\, W}$.

If W' represent a weight at the end, the strength to support W', at any section $C\,D$, is as $\frac{A \,.\, G}{A\,C \,.\, W'}$; and if W' is constant, $S \propto \frac{A \,.\, G}{A\,C}$.

195. Forms of Equal Strength at Every Section.—

1. A beam supported at one end, and having the form of a wedge, whose triangular sides are parallel to the horizon, has equal strength at every section, for supporting a weight at the extremity. Let the wedge, in Fig. 142, have the uniform depth d; then $G = \frac{1}{2}\,d$; and the area of the section at C is $E\,D \,.\, d$; $\therefore A \,.\, G = \frac{1}{2}\,E\,D \,.\, d^2 \propto E\,D \propto A\,C$; hence the relative strength is as $\frac{A\,C}{A\,C}$; that is, it is constant.

Fig. 142.

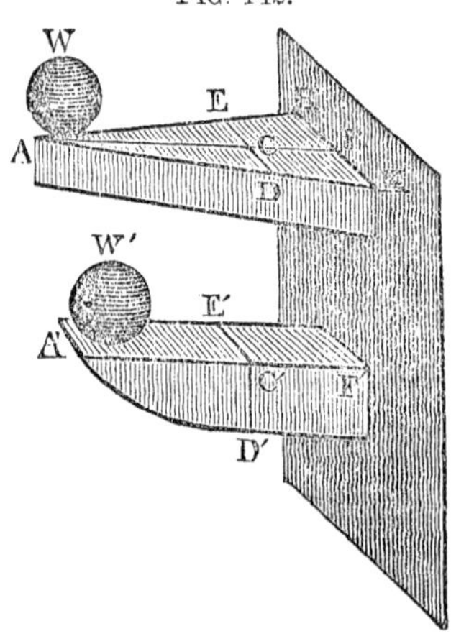

2. A beam, whose vertical sides are parallel, and whose longitudinal section, parallel to the sides, is a semi-parabola, has equal strength at every section. In Fig. 142, let d = the uniform thickness; then $A = d \,.\, C'\,D'$, $G = \frac{1}{2}\,C'\,D'$; $\therefore A \,.\, G$

$= \frac{1}{2} d \,.\, C' D'^2 \propto C' D'^2$. Hence, $S \propto \frac{C' D'^2}{A' C'}$; but $A' C' \propto C' D'^2$; $\therefore S \propto \frac{C' D'^2}{C' D'^2}$, which is constant.

196. Prismatic Beams Breaking by their own Weight.— Suppose beams of prismatic or cylindrical form to have *similar* cross-sections, one dimension of which is D, and to be supported at both ends, or only at one end; then $S \propto \frac{A \,.\, G}{L\left(\frac{1}{2} W + W'\right)} \propto \frac{D^3}{L\left(\frac{1}{2} W + W'\right)}$. This variation, thrown into the form of a full proportion, becomes $S : s :: \frac{D^3}{L\left(\frac{1}{2} W + W'\right)} : \frac{d^3}{l\left(\frac{1}{2} w + w'\right)}$. For example, let the beams be cylinders, whose lengths are L, l; their diameters, D, d; weights, W, w; while W', w', are additional weights laid on their middle points, or the unsupported ends. Then the above proportion gives their relative strength. Now let the second beam have no weight laid upon it; that is, let $w' = 0$; then $S : s :: \frac{D^3}{L\left(\frac{1}{2} W + W'\right)} : \frac{d^3}{\frac{1}{2} l \,.\, w}$. If $d = D$, $S : s :: \frac{1}{L\left(\frac{1}{2} W + W'\right)} : \frac{1}{\frac{1}{2} l \,.\, w}$. But since $w : W :: l : L$, $\therefore w = \frac{W \,.\, l}{L}$; hence $S : s :: \frac{1}{L\left(\frac{1}{2} W + W'\right)} : \frac{2 L}{W \,.\, l^2}$; which expresses their relative strength, when the diameters are equal, and the second beam is not loaded. If their lengths and weights become such as to cause the beams to break, then, since $S = s$, $\therefore \frac{1}{L\left(\frac{1}{2} W + W'\right)} = \frac{2 L}{W \,.\, l^2}$; $\therefore l^2 = \frac{L^2 \left(W + 2 W'\right)}{W}$, and $l = L\left(\frac{W + 2 W'}{W}\right)^{\frac{1}{2}}$; which gives the length of a beam that breaks by its own weight.

Let the prismatic or cylindrical beams be *similar* to each other; then $D^3 : d^3 :: L^3 : l^3$; $\therefore S : s :: \frac{L^2}{\frac{1}{2} W + W'} : \frac{l^2}{\frac{1}{2} w}$. But the weights of similar solids of the same density are as the cubes of their homologous dimensions; $\therefore w : W :: l^3 : L^3$; $\therefore w = \frac{W \,.\, l^3}{L^3}$; hence, by substitution, $S : s :: \frac{L^2}{\frac{1}{2} W + W'} : \frac{2 L^3}{W \,.\, l}$. And when the beams break, since $S = s$, therefore $\frac{L^2}{\frac{1}{2} W + W'} = \frac{2 L^3}{W \,.\, l}$; $\therefore \frac{1}{2} W + W' = \frac{W \,.\, l}{2 L}$, or $l = \frac{L\left(W + 2 W'\right)}{W}$. If, therefore, a cylindrical beam whose length is L breaks with the given weight W' placed upon it, a similar cylindrical beam whose length is $\frac{L(W+2W')}{W}$ will break with its own weight.

The same reasoning is applicable to a beam of any prismatic form, whatever the shape of the cross-section.

197. Comparison of Beams Supported at One End and at Both Ends.—If a horizontal beam be supported at *both ends*, the stress produced by its own weight, W, is measured by $\frac{1}{8} L \times W$ (Art. 190).

If the beam be supported at *one end* only, the stress is measured by the whole weight applied at the centre of gravity, and consequently the stress $= \frac{1}{2} L \times W$.

Therefore a beam supported at both ends has four times the relative strength of the same beam supported only at one end. And if a certain beam supported at one end breaks by its own weight, a beam of the same dimensions twice as long will break by its own weight when resting on two supports.

If, however, instead of the weight of the beam itself, this is left out of the account, and a weight W' be added, then the stress on the beam when supported at *one end* will be measured by $L \times W'$; while, in the case of the beam supported at *both ends*, the weight being at the middle point of the beam, the stress is measured as before, by $\frac{1}{4} L \times W$ (Art. 191). Therefore, a weight placed at the end of a beam supported only at one end produces four times the stress as the same weight placed at the middle of the beam when supported at both ends.

1. What must be the length of a beam of oak one inch square, supported at both ends, which is just capable of bearing its own weight?

By Art. 191, a beam of oak 1 foot long and 1 inch square, weighing $\frac{1}{2}$ pound, just supports 600 pounds. And by Art. 196, the expression $l = L\left(\frac{W+2W'}{W}\right)^{\frac{1}{2}}$ denotes that when a beam whose length is L breaks when W' is placed upon it, l is the length of a beam that will break with its own weight; consequently, since here $L = 1$, $W = \frac{1}{2}$, and $W' = 600$, $l = \left(\frac{\frac{1}{2}+1200}{\frac{1}{2}}\right)^{\frac{1}{2}} = (2401)^{\frac{1}{2}}$ $= 49$ feet.

2. Two beams are of equal length and weight, the first being a square prism whose section is 4 inches square, the second a rectangular prism, 8 by 2 inches; how much stronger is the second

beam than the first, and how much stronger when laid on the narrow than on the broad side? (Art. 189.)

Ans. The second beam is *twice* as strong as the first, and *four* times as strong when laid on the narrow as on the broad side.

198. Structures Relatively Weaker as they are Larger.—The foregoing articles explain the observed fact that the relative strength of every kind of structure becomes less as its size is increased. For, the absolute strength increases as the *square* of one of the dimensions, while the weight increases as the *cube* of the same. A *model*, therefore, has far greater relative strength than the building copied from it; and in respect to every kind of structure, there are limits of magnitude which cannot be exceeded.

The same fact is observed in the animal and vegetable world. Relatively to size, insects are very much stronger than large animals, and shrubs stronger than trees.

199. Strength of Solid and Hollow Cylinders.—If a solid and a hollow cylinder, of equal length, have the same quantity of matter, so that the area of their cross-sections shall be equal, then their strength will be in the ratio of the distances of their centres of gravity from the upper surfaces. But the centres of gravity being at the centres of the cross-sections, it follows that the strength of the solid cylinder will be less than that of the hollow cylinder in the ratio of the diameter of the former to that of the latter.

It appears, therefore, that the strength of a *tube* is always greater than the strength of the same quantity of matter made into a solid rod of the same length; and leaving out of view the diminished rigidity, there would seem to be no limit to the strength which might be given to such a cylinder by increasing its diameter.

Many illustrations are found in nature, such as the bones of animals, the quills of birds' feathers, the straw of grain, and the tubular stalks of some larger plants.

An interesting application of the principle has been made in modern times in the construction of iron tubular bridges.

---

200. MISCELLANEOUS PROBLEMS IN MECHANICS.

1. Two forces, F and F', acting in the diagonals of a parallelogram, keep it at rest in such a position that one of its edges is horizontal; show that $F \sec a' = F' \sec a = W \operatorname{cosec} (a + a')$, where W is the weight of the parallelogram, and a and a' the angles between the diagonals and the horizontal side.

2. Four parallel forces act at the angles of a plane quadrilateral, and are inversely proportional to the segments of its diagonals nearest to them; show that the point of application of their resultant lies at the intersection of the diagonals.

3. Find the centre of gravity of four equal heavy particles placed at the four angular points of a triangular pyramid.

4. Five pieces of a uniform chain are hung at equidistant points along a rigid rod without weight, and their lower ends are in a straight line passing through one end of the rod; find the centre of gravity of the system.

5. A right square pyramid, whose height is 8 feet, and the edge of its base 1 foot, is tipped on one edge till it is on the point of falling; what angle does its axis make with the horizon?

6. If three uniform rods be rigidly united so as to form half of a regular hexagon, prove that if suspended from one of the angles, one of the rods will be horizontal.

7. A cone of uniform density, whose slant height is 15 inches, is suspended by the edge of its base, when its axis is found to incline 12° to the horizon; required the other dimensions of the cone.

8. If $A B C$ be an isosceles triangle, having a right angle at C, and if D and E be the middle points of $A C$ and $A B$, respectively, prove that a perpendicular from E upon $B D$ will pass through the centre of gravity of the triangle $B D C$.

9. An oblique cylinder, inclining $62\frac{1}{2}°$ to the horizon, having slant height $= 14$ inches, and the diameter of the base $= 6\frac{1}{2}$ inches, has a ball of the same material hung upon its edge, which just upsets it; required the diameter of the ball.

10. A body, the lower surface of which is spherical, rests upon a horizontal plane; find in what case the equilibrium is stable, and in what case unstable.

11. A smooth circular ring rests on two pins projecting from a wall, and the pins are not in the same horizontal plane; find the pressure on each pin.

12. A given isosceles triangle is inscribed in a circle; find the centre of gravity of the remaining area of the circle.

13. A homogeneous hemisphere rests with its convex surface on a horizontal plane; at what points of the circumference of the plane base of the hemisphere must three weights of 10, 15, and 20 lbs. be suspended, in order that its position be not changed?

14. Two smooth cylinders of equal radii just fit in between two parallel vertical walls, and rest on a smooth horizontal plane, without pressing against the walls; if a third equal cylinder be placed on the top of them, find the resulting pressure against either wall.

15. A cylinder, suspended by a point on the side, inclines 40° to the horizon; the point is moved 3 feet lengthwise on the side, and then the cylinder inclines 24° to the horizon, with the other end down; find the point of suspension, that the cylinder may hang horizontal.

16. A flat semicircular board, with its plane vertical, and curved edge upward, rests on a smooth horizontal plane, and is pressed at two given points of its circumference by two heavy rods, which slide freely in vertical guides; find the ratio of the weights of the rods, that the board may be in equilibrium.

17. The radii of a wheel and axle are 12 inches and 3 inches; the power is 30 lbs., the weight 100 lbs.: as the power in this case preponderates, required how many degrees from the bottom of the wheel the end of the rope is when the forces are in equilibrium.

18. A frustum is cut from a right cone by a plane bisecting the axis, and parallel to the base; show that it will rest with its slant side on a horizontal plane, if the height of the cone have to the diameter of its base a greater ratio than $\sqrt{7}$ to $\sqrt{17}$.

19. Explain the action of an oar, when used in rowing, and determine the effect produced, having given the distances from the hands to the side of the boat, and from the side of the boat to the point where the oar may be considered as acting on the water.

20. A uniform wheel, free to revolve on its axis, has the weights, 21 lbs. and 13 lbs., attached to the circumference, 100° apart; how far from the bottom will the weights be, respectively, when the system is in equilibrium?

21. Two equal rods, without weight, are connected at their middle points by a pin, which allows free motion in a vertical plane; they stand upon a horizontal plane, and their upper extremities are connected by a thread, which carries a weight. Show that the weight will rest half way between the pin and the horizontal line joining the upper ends of the rods.

22. A uniform heavy rod, of given length, is to be supported in a given position, with its upper end resting against a smooth vertical wall, by a string fastened to its lower end; find the point in the wall to which the string must be attached.

23. A light cord, with one end attached to a fixed point, passes over a pulley in the same horizontal plane with the fixed point, and supports a weight hanging freely at its other end. A heavy ring being put upon the cord in different places between the fixed point and the pulley, it is required to show that, if the weight of the ring be small compared with the other weight, the positions of the ring, when in equilibrium, will be approximately in the arc of a circle.

24. If particles of unequal weight be placed in the angular

points of a triangular pyramid, and G be their common centre of gravity, G', G'', G''', &c., be the common centres of gravity for every possible arrangement of the particles; show that the centre of gravity of equal particles, placed at G, G', G'', &c., is the centre of gravity of the pyramid.

25. Two equal circular disks with smooth edges, placed on their flat sides in the corner between two smooth vertical planes, inclined to each other at a given angle, touch each other in the line bisecting that angle; find the radius of the least disk that may be pressed between them without causing them to separate.

26. A ladder of uniform weight throughout, 36 feet long, weighs 72 lbs., and leans against a vertical wall, making an angle of 66° 40′ with the horizon; a man, weighing 130 lbs., ascends 30 feet on the ladder; required the amount of pressure against the wall.

27. Where is the centre of gravity of the area included between two circles tangent to each other internally?

PART II.

HYDROSTATICS.

CHAPTER I.

LIQUIDS AT REST.

201. Liquids Distinguished from Solids and Gases.—A *fluid* is a substance whose particles are moved among each other by a very slight force. In solid bodies the particles are held by the force of cohesion in fixed relations to each other; hence such bodies retain their form in spite of gravity or other small forces exerted upon them. If a solid be reduced to the finest powder, still each grain of the powder is a solid body, and its atoms are held together in a determinate shape. A pulverized solid, if piled up, will settle by the force of gravity to a certain inclination, according to the smallness and smoothness of its particles, while a liquid will not rest till its surface is horizontal.

Fluids are of two kinds, liquids and gases. In a *liquid*, there is a perceptible cohesion among its particles; but in a *gas*, the particles mutually repel each other. These fluids are also distinguished by the fact that liquids cannot be compressed except in a very slight degree, while the gases are very compressible. A force of 15 pounds on a square inch, applied to a mass of water, will compress it only about .000046 of its volume, as is shown by an instrument devised by Oersted. But the same force applied to a quantity of air of the usual density at the earth's surface will reduce it to one-half of its former volume.

202. Transmitted Pressure.—It is an observed property of fluids that a force which is applied to one part is transmitted undivided to all parts. For instance, if a piston *A* (Fig. 143) is pressed upon the water in the vessel *A D C* with a force of *one pound*, every other piston of the same size, as *B, C, D*, or *E*, receives a pressure of *one pound* in addition to the previous pressure of the water itself. Hence the whole amount of bursting pressure exerted within the vessel by the weight

Fig. 143.

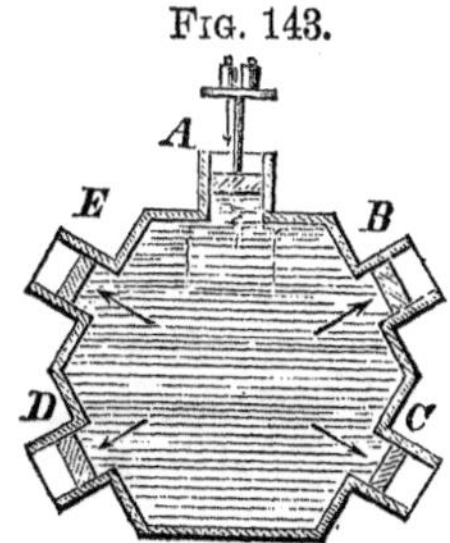

upon A equals as many pounds as there are portions of surface equal to the area of A. And if the pressure is increased till the vessel bursts, the fracture is as likely to occur in some other part as in that toward which the force is directed.

203. The Hydraulic Press.—An important application of the principle of transmitted pressure occurs in Bramah's hydraulic press, represented in Fig. 144. The walls of the cylinder and res-

FIG. 144.

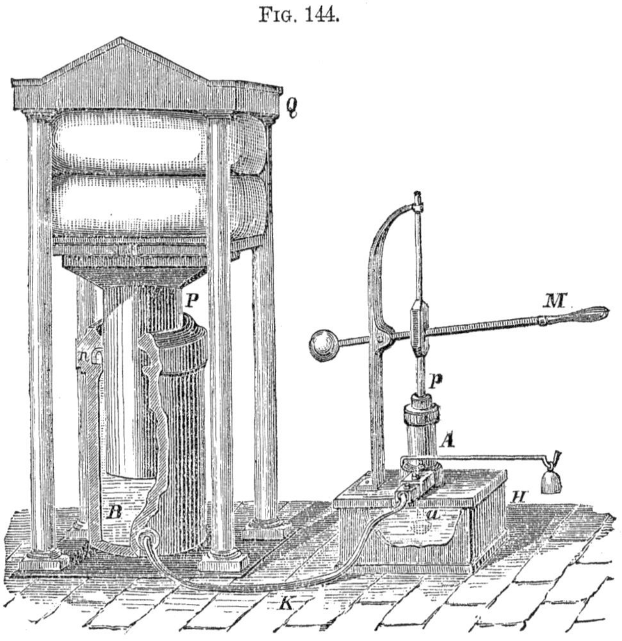

ervoir are partly removed, to show the interior. A is a small forcing pump, worked by the lever M, by which water is raised in the pipe a from the reservoir H, and driven through the tube K into the cylinder B, where it presses up the piston P, and the iron plate on the top of it, against the substance above. At each downward stroke of the small piston p, a quantity of water is transferred to the cylinder B, and presses up the large piston with a force as many times greater than that exerted on the small one as the under surface of P is greater than that of p (Art. 202). If the diameter of p is *one* inch, and that of P is *ten* inches, then any pressure on p exerts a pressure 100 times as great on P. The lever M gives an additional advantage. If the distances from the fulcrum to the rod p and to the hand are as 1 : 5, this ratio compounded with the other, 1 : 100, gives the ratio of power at M to the pressure at Q as 1 : 500; so that a power of 100 lbs. exerts a pressure of 50000 lbs.

This machine has the special advantage of working with a small amount of friction. It is used for pressing paper and books, packing cotton, hay, &c.; also for testing the strength of cables and steam-boilers. It has been sometimes employed to raise great weights, as, for instance, the tubular bridge over the Menai straits; the two portions, after being constructed at the water level, were raised more than 100 feet to the top of the piers, by two hydraulic presses. The weight of each length lifted at once was more than 1800 tons.

The relation of power to weight in the hydraulic press is in accordance with the principle of virtual velocities (Art. 142). For, while a given quantity of water is transferred from the smaller to the larger cylinder, the velocity of the large piston is as much less than that of the small one as its area is greater. But we have seen that the pressures are directly as the areas. Therefore, in this as in other machines, the intensities of the forces are inversely as their virtual velocities.

204. Equilibrium of a Fluid.—In order that a fluid may be at rest,

1. *The pressures at any one point must be equal in all directions.*

2. *The surface must be perpendicular to the resultant of the forces which act upon it.*

Both of these conditions result from the mobility of the particles. It is obvious that the first must be true, since, if any particle were pressed more in one direction than another, it would move in the direction of the greater force, and therefore not be at rest, as supposed.

In order to show the truth of the second condition, let $m\ p$ (Fig. 145) represent the resultant of the forces which act on the fluid. Then, if the surface is not perpendicular to $m\ p$, that force may be resolved into $m\ q$ perpendicular to the surface, and $m\ f$ parallel to it. The latter, $m\ f$, not being opposed, the particles move in that direction.

Fig. 145.

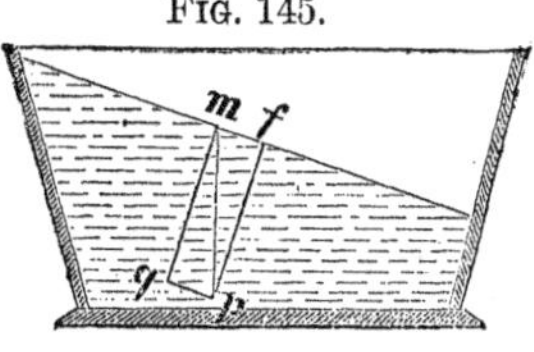

As gravity is the principal force which acts on all the particles, the surface of a fluid at rest is ordinarily *level*, that is, perpendicular to a vertical or plumb line. If the surface is of small extent, it is sensibly a plane, though it is really curved, because the vertical lines, to which it is perpendicular, converge toward the centre of the earth.

205. The Curvature of a Liquid Surface.—The earth being 7912 miles in diameter, a distance of 100 feet on its surface subtends an angle of about one second at the centre, and therefore the levels of two places 100 feet apart are inclined one second to each other.

The amount of depression for moderate distances is found by the formula, $d = \frac{2}{3} L^2$, in which d is the depression in feet, and L the length of arc in miles. Let $B\,E$ (Fig. 146) be a small arc of a great circle on the earth; then $C\,E$ is the depression. As $B\,E$ is small, its chord may be considered equal to the arc, and $B\,G$ equal to the depression. But $B\,G : B\,E :: B\,E : B\,A$; that is, $d : L :: L : 7912$; or $d = \frac{L^2}{7912}$. In order to express d in feet, while the other lines are in miles, we have

FIG. 146.

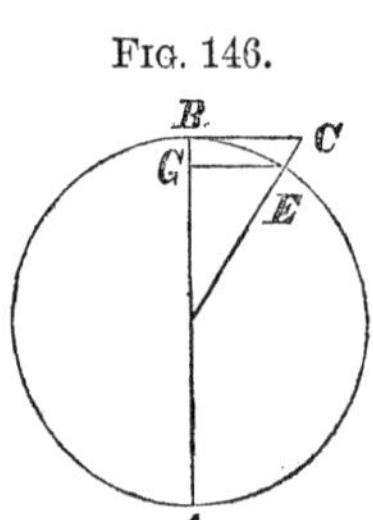

$$d = \frac{L^2 \times 5280^2}{7912 \times 5280} = \frac{L^2 \times 5280}{7912} = \frac{2}{3} L^2, \text{ very nearly.}$$

This gives, for one mile, $d = 8$ inches; for two miles, $d = 2$ ft. 8 in.; and for 100 miles, $d = 6667$ ft., &c. If a canal is 100 miles long, each end is more than a mile below the tangent to the surface of the water at the other end.

206. The Spirit Level.—Since the surface of a liquid at rest is level, any straight line which is placed parallel to such a surface is also level. Leveling instruments are constructed on this principle. The most accurate kind is the one called the *spirit* level. Its most essential part is a glass tube, $A\,B$ (Fig. 147), nearly filled with alcohol (because water would be liable to freeze), and hermetically sealed. The tube having a little convexity upward from end to end, though so slight as not to be visible, the bubble of air moves to the highest part, and changes its place by the least inclination of the tube. The tube is so connected with a straight bar of wood or metal, as $D\,C$ (Fig. 148), or for nicer purposes, with a telescope, that the bubble is at the middle M when the bar or the axis of the telescope is exactly level. The tube usually has graduation lines upon it for adjusting the bubble accurately to the middle.

FIG. 147.

B A

FIG. 148.

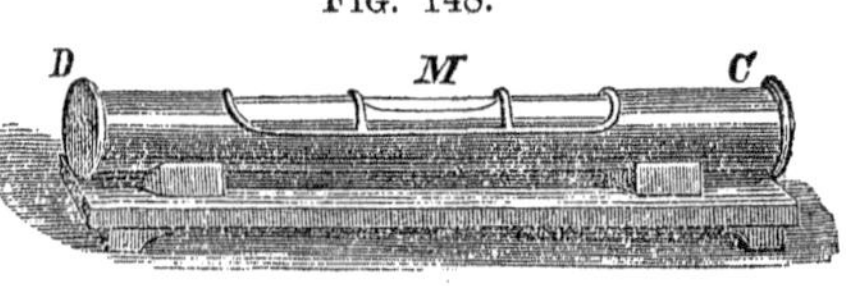

207. Pressure as Depth.—From the principle of equal transmission of force in a fluid, it follows that, if a liquid is uniformly dense, its pressure on a given area varies as the perpendicular depth, whatever the form or size of the reservoir. Let the vessel *A B C D* (Fig. 149), having the form of a right prism, be filled with water, and imagine the water to be divided by horizontal planes into strata of equal thickness. If the density is everywhere the same, the weights of these strata are equal. But the pressure on each stratum is the sum of the weights of all the strata above it. Therefore, in this case, the pressure varies as the depth.

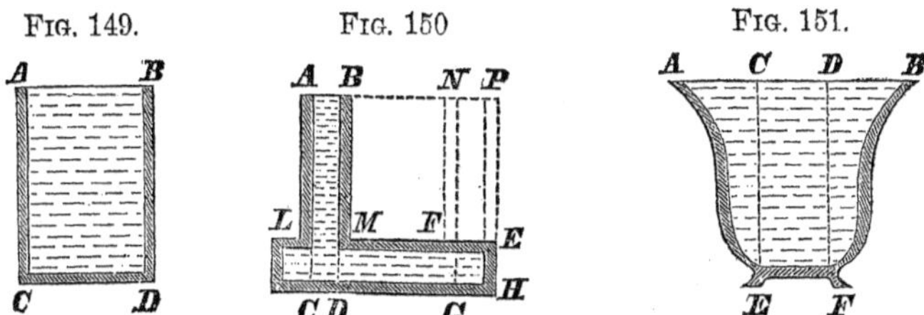

FIG. 149. FIG. 150 FIG. 151.

But let the reservoir *A B E H* (Fig. 150) contain water which is not directly beneath the highest part. The pressures in the column *A B C D* are transmitted laterally to *E H*, however far distant; so that the surface of each horizontal stratum sustains equal pressures in all parts, whether directly beneath *A B* or not. Hence, if *G H* is equal to *C D*, the downward pressure on *G H* is equal to the weight of *A B C D*; so, also, the upward pressure on *E F* is equal to the weight of *A B L M*, and would just sustain the column of water *E F N P*.

Again, if the base is smaller than the top, as in the vessel *A B E F* (Fig. 151), then the pressure on *E F* equals only the weight of the column *C D E F*. The water in the surrounding space *A C E*, *B D F*, simply serves as a vertical wall to balance the lateral pressures of the central column.

If the surface pressed upon is oblique or vertical, then the points of it are at unequal depths; in this case, the depth of the area is understood to be the *average* depth of all its parts; that is, the depth of its centre of gravity.

If the fluid were compressible, the lower strata would be more dense than the upper ones, and therefore the pressure would increase at a faster rate than the depth.

208. Amount of Pressure in Water.—One cubic foot of water weighs 1000 ounces, or 62.5 pounds. Therefore, the pressure on *one square foot*, at the depth of *one foot*, is 62.5 pounds. From this, as the *unit* of hydrostatic pressure, it is easy to determine the pressures on all surfaces, at all depths; for it is obvious that, when the depth is the same, the pressure varies as the surface pressed upon; and it has been shown that, on a given surface, the pressure varies as the depth of its centre of gravity; it therefore varies as the product of the two. Let p = pressure; a = area pressed upon; and d = the depth of its centre of gravity; then $p = a\,d \times 62.5$.

Depth.	Pounds per sq. ft.	Depth.	Pounds per sq. ft.
1 ft.	62.5	100 ft.	6,250
10.	625	1 mile.	330,000
16	1000	5 miles.	1,650,000

From the above table it may be inferred that the pressure on a square foot in the deepest parts of the ocean must be not far from two millions of pounds; for the depth in some places is more than five miles, and sea-water weighs 64.37 pounds, instead of 62.5 pounds. A brass vessel full of air, containing only a pint, and whose walls were one inch thick, has been known to be crushed in by this great pressure, when sunk to the bottom of the ocean.

Owing to the increase of pressure with depth, there is great difficulty in confining a high column of water by artificial structures. The strength of banks, dams, flood-gates, and aqueduct pipes, must increase in the same ratio as the perpendicular depth from the surface of the water, without regard to its horizontal extent.

Fig. 152.

209. Column of Water whose Weight Equals the Pressure.—A convenient mode of conceiving readily of the amount of pressure on an area, in any given circumstances, is this: consider the area pressed upon to form the horizontal base of a hollow prism; let the height of the prism equal the average depth of the area; and then suppose it filled with water. The weight of this column of water is equal to the pressure. For the contents of the prism (whose base = a, and its height = d), = $a\,d$; and the weight of the same = $a\,d \times 62.5$ lbs.; which is the same expression as was obtained above for the pressure.

On the bottom of a cubical vessel full of water, the pressure equals the *weight* of the water; on each side of the same the pressure is *one-half* the weight of the water; hence, on all the five sides the pressure is *three times* the weight of the water; and if the top were closed, on which the pressure is zero, the pressure on the six sides is the same, three times the weight of the water.

210. Illustrations of Hydrostatic Pressure.—A vessel may be formed so that both its base and height shall be great, but its cubical contents small; in which case, a great pressure is produced by a small quantity of water. The hydrostatic bellows is an example. In Fig. 152, the weight which can be sustained on the lid *D I* by the column *A D* is equal to that of a prism or cylinder of water, whose base is *D I*, and its height *D A*. It is immaterial how shallow is the stratum of water on the base, or how slender the tube *A D*, if greater than a capillary size.

In like manner, a cask, after being filled, may be burst by an additional pint of water; for, by screwing a long and slender pipe into the top of the cask, and filling it with water, the pressure is easily made greater than the strength of the cask can bear.

211. The Same Level in Connected Vessels.—In tubes or reservoirs which communicate with each other, water will rest only when its surface is at the same level in them all. If water is poured into *D* (Fig. 153), it will rise in the vertical tube *B*, so as to stand at the same level as in *D*. For, the pressure toward the right on any cross-section *E* of the horizontal pipe *m n* equals the product of its area by its depth below *D*. So the pressure on the same section towards the left equals the product of its area by its depth below *B*. But these pressures are equal, since the liquid is at rest. Therefore *E* is at equal depths below *B* and *D*; in other words, *B* and *D* are on the same level. The same reasoning applies to the irregular tubes *A* and *C*, and to any others, of whatever form or size.

Fig 153.

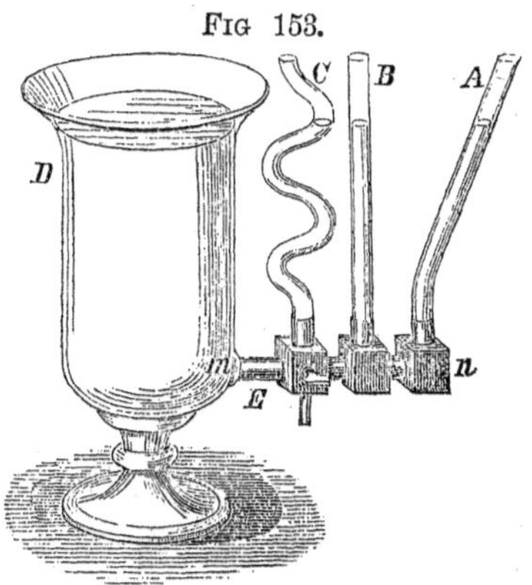

Water conveyed in aqueducts, or running in natural channels in the earth, will rise just as high as the source, but no higher.

Artesian wells illustrate the same tendency of water to rise to its level in the different branches of a tube. When a deep boring is made in the earth, it may strike a layer or channel of water which descends from elevated land, sometimes very distant. The pressure causes it to rise in the tube, and often throws it many feet above the surface. Fig. 154 shows an artesian well, through which is discharged the water that descends in the porous stratum *K K*, confined between the strata of clay *A B* and *C D*.

FIG. 154.

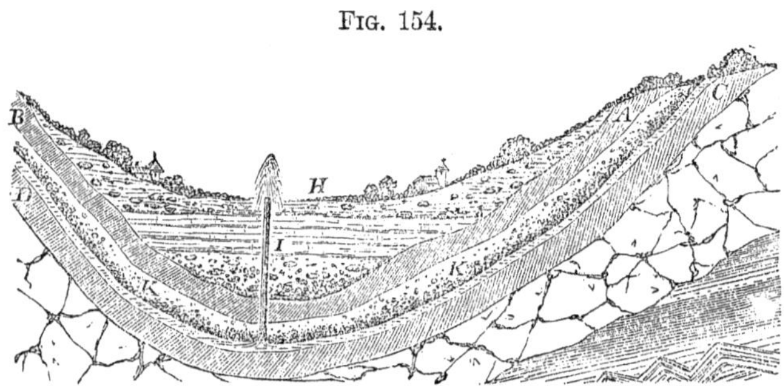

212. Centre of Pressure.—The centre of pressure of any surface immersed in water is that point through which passes the resultant of all the pressures on the surface. It is the point, therefore, at which a single force must be applied in order to counterbalance all the pressures exerted on the surface. If the surface be a plane, and horizontal, the centre of pressure coincides with the centre of gravity, because the pressures are equal on every part of it, just as the force of gravity is. But if the plane surface makes an angle with the horizon, the centre of pressure is lower than the centre of gravity, since the pressure increases with the depth. For example, if the vertical side of a vessel full of water is rectangular, the centre is *one-third* of the distance from the middle of the base to the middle of the upper side. If triangular, with its base horizontal, the centre of pressure is *one-fourth* of the distance from the middle of the base to the vertex. If triangular, with the top horizontal, the centre of pressure is *half* way up on the bisecting line.

[See Appendix for calculations of the place of the centre of pressure.]

213. The Loss of Weight in Water.—When a body is immersed in water, it suffers a pressure on every side, which is proportional to the depth. Opposite components of lateral pressures, being exerted on surfaces at the same depth, balance each other; but this cannot be true of the vertical pressures, since the top and bottom of the body are at unequal depths. The upward pressure on the bottom exceeds the downward pressure on the top;

and this excess constitutes the *buoyant power* of a fluid, which causes a loss of weight.

A body immersed in water loses weight equal to the weight of water displaced.

For before the body was immersed, the water occupying the same space was exactly supported, being pressed upward more than downward by a force equal to its own weight. The weight of the *body*, therefore, is diminished by this same difference of pressures, that is, by the weight of the displaced water.

On the supposition of the complete incompressibility of water, this loss is the same at all depths, because the weight of displaced water is the same. As water, however, is slightly compressible, its buoyant power must increase a little at great depths. Calling the compression .000046 for one atmosphere (= 34 feet of water), the bulk of water at the depth of a mile is reduced by about $\frac{1}{140}$, and its specific gravity increased in the same ratio; so that, *possibly*, a body might sink near the surface, and float at great depths in the ocean. But this is not *probable* in any case, since the same compressing force may reduce the volume of the solid as much as that of the water. And, furthermore, the increase of density by increased depth is so slow, that even if solids were incompressible, most of those which sink at all would not find their floating place within the greatest depths of the ocean. For example, a stone twice as heavy as water must sink 100 miles before it could float.

214. Equilibrium of Floating Bodies.—If the body which is immersed has the same density as water, it simply loses its whole weight, and remains wherever it is placed. But if it is less dense than water, the excess of upward pressure is more than sufficient to support it; it is, therefore, raised to the surface, and comes to a state of equilibrium after partly emerging. In order that a floating body may have a stable equilibrium, the three following conditions must be fulfilled:

1. *It displaces an amount of water whose weight is equal to its own.*

2. *The centre of gravity of the body is in the same vertical line with that of the displaced water.*

3. *The metacenter is higher than the centre of gravity of the body.*

The reason for the *first* condition is obvious; for both the body and the water displaced by it are sustained by the same upward pressures, and therefore must be of equal weight.

That the *second* is true, is proved as follows: Let *C* (Fig. 155, 1) be the centre of gravity of the displaced water, while that of the body is at *G*. Now the fluid, previous to its removal, was sustained by an upward force equal to its own weight, acting through its centre of gravity *C*; and the same upward force now acts upon

FIG. 155.

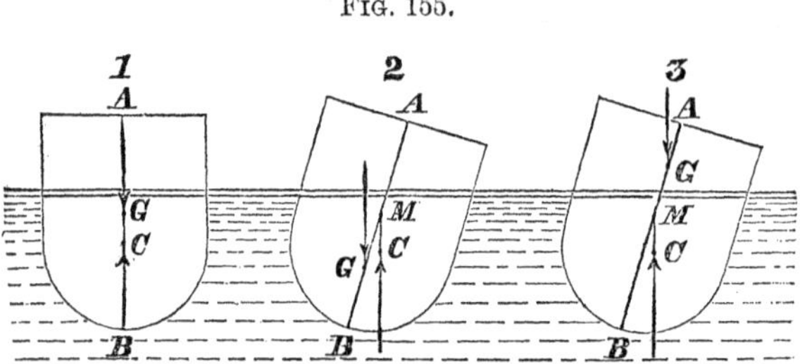

the floating body through the same point. But the body is urged downward by gravity in the direction of the vertical line *A G B*. Were these two forces exactly opposite and equal, they would keep the body at rest; but this is the case only when the points *C* and *G* are in the same vertical line: in every other position of these points, the two parallel forces tend to turn the body round on a point between them.

215. The Metacenter.—To understand the *third* condition, the metacenter must be defined. When a floating body is slightly inclined from its state of equilibrium, as in Fig. 155, 2 and 3, and a vertical is drawn through the new centre of gravity *C* of the displaced water, this vertical must intersect the former vertical *A B*; the intersection, *M*, is called the *metacenter*. When the centre of gravity of the body *G* is *lower* than the metacenter, as in Fig. 155, 2, the parallel forces, downward through *G* and upward through *C*, revolve the body back to its position of equilibrium, which is then called a stable equilibrium. But if the centre of gravity of the body is higher than the metacenter, as in Fig. 155, 3, the rotation is in the opposite direction, and the body is upset, the equilibrium being unstable. Once more, if the centre of gravity of the body is *at* the metacenter, the body rests indifferently in any position, as, for example, a sphere of uniform density. The equilibrium in this case is called neutral.

If only the first condition is fulfilled, there is *no* equilibrium; if only the first and second, the equilibrium is *unstable;* if all the three, the equilibrium is *stable.*

In accordance with the third condition, it is necessary to place the heaviest parts of a ship's cargo in the bottom of the vessel, and sometimes, if the cargo consists of light materials, to fill the bottom with stone or iron, called *ballast,* lest the masts and rigging should raise the centre of gravity too high for stability. On the same principle, those articles which are prepared for-life-preservers, in case of shipwreck, should be attached to the upper part of the body, that the head may be kept above water. The danger arising from several persons standing up in a small boat is quite apparent; for the centre of gravity is elevated, and liable to become higher than the metacenter, thus producing an unstable equilibrium.

216. Floating in a Small Quantity of Water.—As pressure on a given surface depends solely on the depth, and not at all on the extent or quantity of water, it follows that a body will float as freely in a space slightly larger than itself as on the open water of a lake. For instance, a ship may be floated by a few hogsheads of water in a dock whose form is adapted to it. In such a case, it cannot be literally true that the displaced water weighs as much as the vessel, when *all* the water in the dock may not weigh a hundredth part as much. The expression "displaced water" means the amount which would fill the place occupied by the immersed portion of the body. An experiment illustrative of the above is, to float a tumbler within another by means of a spoonful of water between.

217. Floating of Heavy Substances.—A body of the most dense material may float, if it has such a form given it as to exclude the water from the upper side, till the required amount is displaced. Ships are built of iron, and laden with substances of greater specific gravity than water, and yet ride safely on the ocean. A block of any heavy material, as lead, may be sustained by the upward pressure beneath it, provided the water is excluded from the upper side by a tube fitted to it by a water-tight joint.

218. Specific Gravity.—The weight of a body compared with the weight of the same volume of the standard, is called its *specific gravity.*

Distilled water, at about 39° F., the temperature of its greatest density, is the standard for all solids and liquids, and common air, at 32°, for gases. Therefore the specific gravity of a solid or a liquid body, is the ratio of its weight to the weight of an equal volume of water; and the specific gravity of an aeriform body is the ratio of its weight to the weight of an equal volume of air. Hence, to find the specific gravity of a solid or liquid, divide its weight by the weight of the same volume of water; but in the case of a gas, divide by the weight of the same volume of air.

219. Methods of Finding Specific Gravity.—

1. For a solid heavier than water, *divide its weight by its loss of weight in water.*

The reason for this rule is obvious. The weight which a submerged body loses (Art. 213) is equal to the weight of the displaced water, which has, of course, the same volume as the body; therefore, dividing by the loss is the same as dividing by the weight of the same volume of water.

2. For a solid lighter than water, *divide its weight by its weight added to the loss it occasions to a heavier body previously balanced in water.*

For, if the light body be attached to a body heavy enough to sink it, it loses all its own weight, and causes loss to the other which was previously balanced. And the whole loss equals the weight of water displaced by the light body. Hence, as before, we in fact divide the weight of the body by the weight of the same volume of water.

3. For a liquid, *find the loss which a body sustains weighed in the liquid and then in water, and divide the first loss by the second.*

For the first loss equals the weight of the displaced liquid, and the second that of the displaced water; and the volume in each case is the same, namely, that of the body weighed in them.

But the specific gravity of a liquid may be more directly obtained by measuring equal volumes of it and of water in a flask, and finding the weight of each. Then the weight of the liquid divided by that of the water is the specific gravity required.

220. The Hydrometer, or Areometer.—In commerce and the arts, the specific gravities of substances are obtained in a more direct and sufficiently accurate way, by instruments constructed for the purpose. The general name for such instruments is the *hydrometer*, or *areometer*. But other names are given to such as are limited to particular uses; as, for example, the *alcoömeter* for alcohol, and the *lactometer* for milk. The hydrometer, represented in Fig. 156, consists of a hollow ball, with a graduated stem. Below the ball is a bulb containing mercury, which gives the instrument a stable equilibrium when in an upright position. Since it will descend until it has displaced a quantity of the fluid equal in weight to itself, it will of course sink to a greater depth if the fluid is lighter. From the depths to which it sinks, therefore, as indicated by the graduated stem, the corresponding specific gravities are estimated.

FIG. 156.

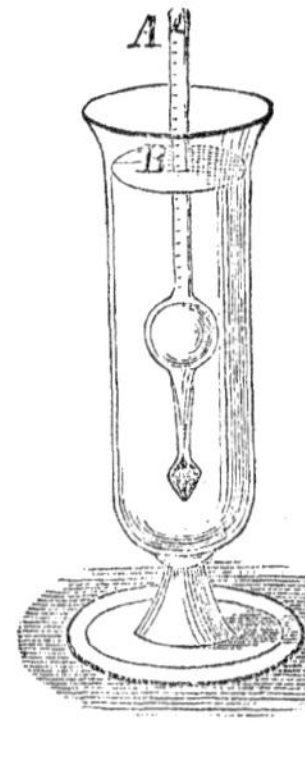

A

B

Nicholson's hydrometer (Fig. 157) is the most useful of this class of instruments, since it may be applied to finding the specific gravities of solid as well as liquid bodies. In addition to the hollow ball of the common hydrometer, it is furnished at the top with a pan A for receiving weights, and a cavity beneath for holding the substance under trial. The instrument is so adjusted that when 1000 grains are placed in the pan, the instrument sinks in distilled water at the temperature of $39\frac{1}{2}°$ F. to a fixed mark, 0, on the stem. Calling the weight of the instrument W, the weight of displaced water is $W + 1000$.

Fig. 157.

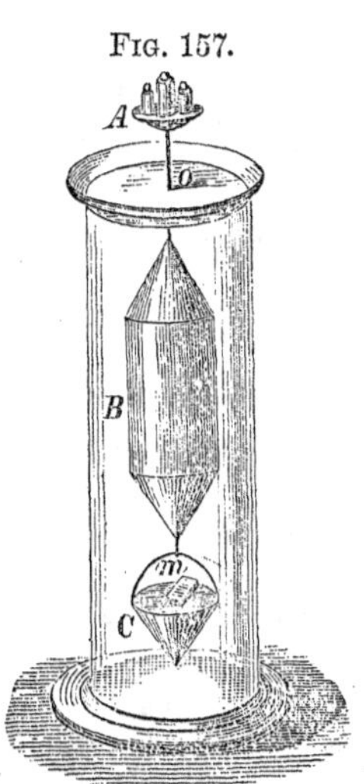

To find the specific gravity of a *liquid*, place in the pan such a weight w as will just bring the mark to the surface. Then the weight of the liquid displaced is $W + w$. But its volume is equal to that of the displaced water. Therefore its specific gravity is $\frac{W + w}{W + 1000}$.

To find the specific gravity of a *solid*, place in the pan a fragment of it weighing less than 1000 grains, and add the weight w required to sink the mark to the water-level. Then the weight of the substance in air is $1000 - w$. Remove the substance to the cavity at the bottom of the instrument, and add to the weight in the pan a sufficient number of grains w' to sink the mark to the surface. Then w' is the *loss* of weight in water; therefore, $\frac{1000 - w}{w'}$ is the specific gravity of the substance.

221. Specific Gravity of Liquids by Means of Heights.—The specific gravity of two liquids may be compared by their relative heights when in equilibrium. Let the tubes m and n (Fig. 158) communicate with each other, and be furnished with a scale of heights above the zero line $B\ C$. Suppose the column of water $A\ B$ to be in equilibrium with the column of mercury $C\ D$. Put h = the height of the water, h' = the height of the mercury, and s = the specific gravity of the latter; then, since pressure varies as the product of height and density, and the pressures in this case are equal, we have $h \times 1 = h' \times s$; whence $s = \frac{h}{h'}$; that is, *the specific*

Fig. 158.

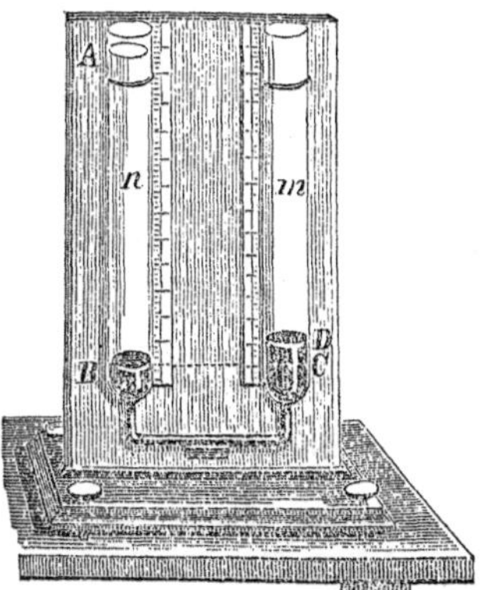

gravity is found by dividing the height of the water by the height of the liquid. Also, $h : h' :: s : 1$; that is, *the heights of two columns in equilibrium are inversely as their specific gravities.*

The heavier liquid should be poured in first, till it stands somewhat above *B C*, the zero mark of the scale; and then the lighter should be poured into one branch, till it presses the other down to the zero line. The heights of both are reckoned upward from *B C*, since the heavy liquid below *B C* balances itself.

222. Table of Specific Gravities.—An accurate knowledge of the specific gravities of bodies is important for many purposes of science and art, and they have therefore been determined with the greatest possible precision. The heaviest of all known substances is *platinum*, whose specific gravity, when compressed by rolling, is 22, water being 1; and the lightest is *hydrogen*, whose specific gravity is = .073, common air being 1. Now, as water is about 800 times as heavy as air, it is (800 ÷ .073 =) 10,959 times as heavy as hydrogen. Therefore platinum is about (10,959 × 22=) 241,000 times as heavy as hydrogen. Between these limits, 1 and 241,000, there is a wide range for the specific gravities of all other substances. As a class, the common metals are the heaviest bodies; next to these come the metallic ores; then the precious gems; minerals in general, animal and vegetable substances, as shown in the following table;

Metals (pure), not including the bases of the alkalies and earths, from . . . 5—22

Platinum	22.0	Copper	8.90
Gold	19.25	Steel	7.84
Mercury	13.58	Iron	7.78
Lead	11.35	Tin	7.29
Silver	10.47	Zinc	7.00

Metallic ores, lighter than the pure metals, but usually above	4.00
Precious gems, as the ruby, sapphire, and diamond	3—4
Minerals, comprehending most stony bodies	2—3
Liquids, from ether highly rectified to sulphuric acid highly concentrated	¾—2
Acids in general, heavier than water.	
Oils in general, lighter; but the oils of cloves and cinnamon are heavier than water; the greater part lie between .9 and 1	.9—1
Milk	1.032
Alcohol (perfectly pure)	.797
" of commerce	.835
Proof spirit	.923
Wines; the specific gravity of the lighter wines, as Champagne and Burgundy, is a little less, and of the heavier wines, as Malaga, a little greater than that of water.	
Woods, cork being the lightest, and lignum vitæ the heaviest	.24—1.34

223. Floating.—The human body, when the lungs are filled with air, is lighter than water, and but for the difficulty of keeping the lungs constantly inflated, it would naturally float. With a moderate degree of skill, therefore, swimming becomes a very easy process, especially in salt water. When, however, a man plunges, as divers sometimes do, to a great depth, the air in the lungs becomes compressed, and the body does not rise except by muscular effort. The bodies of drowned persons rise and float after a few days, in consequence of the inflation occasioned by putrefaction.

As rocks are generally not much more than twice as heavy as water, nearly half their weight is sustained while they are under water; hence, their weight seems to be greatly increased as soon as they are raised above the surface. It is in part owing to their diminished weight that large masses of rock are transported with great facility by a torrent. While bathing, a person's limbs feel as if they had nearly lost their weight, and when he leaves the water, they seem unusually heavy.

224. To Find the Magnitude of an Irregular Body.—It would be a long and difficult operation to find the exact contents of an irregular mineral by direct measurement. But it might be found with facility and accuracy by weighing it in air and then finding its loss of weight in water. The loss is the weight of a mass of water having the same volume. Now, as 1000 ounces of water measure 1728 cubic inches, a direct proportion will show what is the volume of the displaced water; that is, of the mineral itself.

225. Cohesion and Adhesion.—What distinguishes a liquid from a solid is not its want of cohesion so much as the mobility of its particles. It is proved in many ways that the particles of a liquid strongly attract each other. It is owing to this that water so readily forms itself into drops. The same property is still more observable in mercury, which, when minutely divided, will roll over surfaces in spherical forms. When a disk of almost any substance is laid upon water, and then raised gently, it lifts a column of water after it by adhesion, till at length the edge of the fluid begins to divide, and the column is detached, not in all parts at once, but by a successive rupturing of the lateral surface. It is proved that the whole attraction of the liquid would be far too great to be overcome by the force applied to pull off the disk, were it not that it is encountered by little and little, at the edges of the column. But it is the cohesion of the water which is overcome in this experiment; for the upper lamina still adheres to the disk. By a pair of scales we find that it requires the same force to draw off disks of a given size, whatever the materials may be, provided they are *wet* when detached. This is what might be expected, since in each case we break the attraction between two laminæ of water. But if we use disks which are not wet by the liquid, it is not generally true that those of different material will be removed by the same force; indicating that some substances adhere to a given liquid more strongly than others.

These molecular attractions extend to an exceedingly small distance, as is proved by many facts. A lamina of water adheres as strongly to the thinnest disk that can be used as to a thick one; so, also, the upper lamina coheres with equal force to the next below it, whether the layer be deep or shallow.

226. Capillary Action.—This name is given to the molecular forces, adhesion and cohesion, when they produce disturbing effects on the surface of a liquid, elevating it above or depressing it below the general level. These effects are called *capillary*, because most strikingly exhibited in very fine (*hair-sized*) tubes.

The liquid will be elevated in a concave curve, or depressed in a convex curve, by the side of the solid, according as the attraction of the liquid molecules for each other is less or greater than twice the attraction between the liquid and the solid.

Case 1st. Let *H K* (Fig. 159, 1) and *L M* be a section of the vertical side of a solid, and of the general level of the liquid. The

Fig. 159.

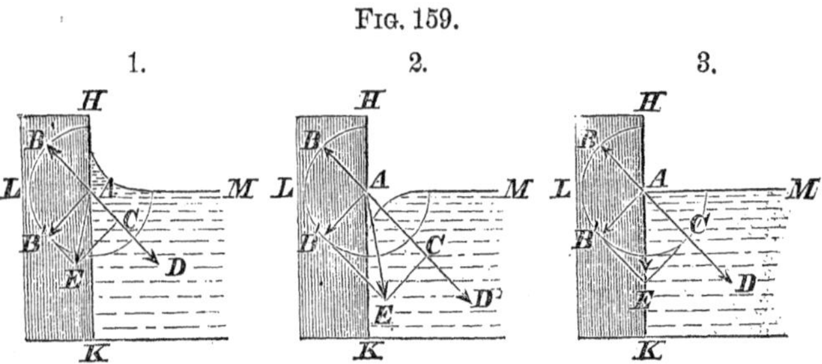

particle *A*, where these lines meet, is attracted (so far as this section is concerned) by all the particles of an insensibly small quadrant of the liquid, the resultant of which attractions is in the line *A D*, 45° below *A M*. It is also attracted by all the particles in two quadrants of the solid, and the resultants are in the directions *A B*, 45° above, and *A B'*, 45° below *L M*.

Now suppose the force *A D* to be *less than twice A B* or *A B'*. Cut off *C D* = *A B*; then *A B*, being opposite and equal to *C D*, is in equilibrium with it. The remainder *A C*, being less than *A B'*, their resultant *A E* will be directed toward the solid; and therefore the surface of the liquid, since it must be perpendicular to the resultant of forces acting on it (Art. 204), takes the direction represented; that is, concave upward.

Case 2d. Let *A D* (Fig. 159, 2), the attraction of *A* toward the liquid particles, be *more than twice A B*, the attraction toward a quadrant of the solid. Then, making *C D* equal to *A B*, these two resultants balance as before; and as *A C* is greater than *A B'*, the angle between *A C* and the resultant *A E* is less than 45°, and *A* is drawn away from the solid. Therefore the surface, being perpendicular to the resultant of the molecular forces acting on it, is convex upward.

Case 3d. If *A D* (Fig. 159, 3) be exactly twice *A B*, then *C D* balances *A B*, and the resultant of *A C* and *A B'* is *A E* in a vertical direction; therefore the surface at *A* is level, being neither elevated nor depressed.

Case 1st occurs whenever a liquid readily *wets* a solid, if brought in contact with it, as, for example, water and clean glass. Case 2d occurs when a solid *cannot be wet* by a liquid, as glass and mercury. Case 3d. is rare, and occurs at the limit between the other two; water and steel afford as good an example as any.

227. Capillary Tubes.—In fine tubes these molecular forces affect the entire columns as well as their edges. If the material of the tube can be wet by a liquid, it will raise a column of that liquid above the level, at the same time making the top of the column concave. If it is not capable of being wet, the liquid is depressed, and the top of the column is convex. The first case is illustrated by glass and water; the second by glass and mercury.

The materials being given, the distance by which the liquid is elevated or depressed varies inversely as the diameter. Therefore the product of the two is constant.

The amount of elevation and depression depends on the strength of the molecular forces, rather than on the specific gravity of the liquids. Alcohol, though lighter than water, is raised only half as high in a glass tube.

If the upper part of a tube is capillary, while the lower part is large, a liquid is sustained (after being raised by suction) at the same height as if the whole were capillary. But it is found that the large mass in the lower part is upheld by atmospheric pressure after the capillary part has been closed by the molecular attraction.

228. Parallel and Inclined Plates.—Between parallel plates a liquid rises or falls half as far as in a tube of the same diameter. This is because the sustaining force acts only on two sides of each filament, while in a tube it acts on all sides. Therefore, as in tubes the height varies inversely as the diameter, so in plates the height varies inversely as the distance between them.

If the plates are inclined to each other, having their edge of meeting perpendicular to the horizon, the surface of a liquid rising between them assumes the form of a *hyperbola*, whose branches approach the vertical edge, and the water-level, as the asymptotes of the curve. This results from the law already stated, that the height varies inversely as the distance between the plates. Let the edge of meeting, *A H* (Fig. 160), be the axis of ordinates, and the line in which the level surface of the water intersects the glass, *A P*, the axis of abscissas. Let *B C*, *D E*, be any ordinates, and *A B*, *A D*, their abscissas, and *B L*, *D K*, the distances between the plates. By the law of capillarity, the heights *B C*, *D E*, are inversely as *B L*, *D K*. But, by the similar triangles, *A B L*, *A D K*, *B L*, *D K*, are as *A B*, *A D*; therefore, *B C*, *D E*, are inversely as *A B*, *A D*; and this is a property of the hyperbola with reference to the centre and asymptotes, that the ordinates are inversely as the abscissas.

FIG. 160.

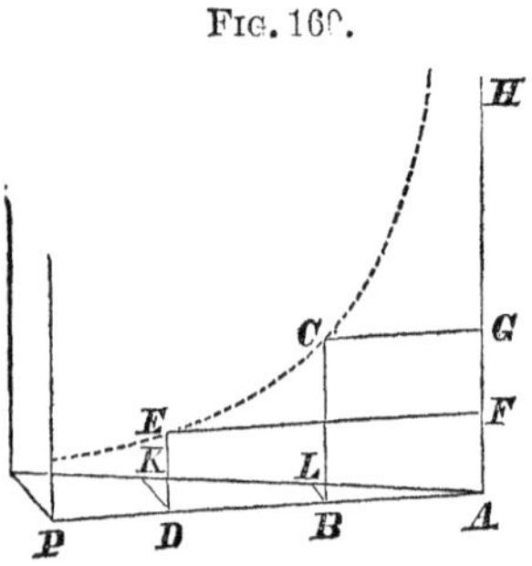

229. Effects of Capillarity on Floating Bodies.—Some cases of apparent attractions and repulsions between floating bodies are caused by the forms which the liquid assumes on the sides of the bodies. If two balls raise the water about them, and are so near to each other that the concave surfaces between them meet in one, they immediately approach each other till they touch; and then, if either be moved, the other will follow it. The water, which is raised and hangs suspended between them, draws them together.

Again, if each ball depresses the water around it, they will also move to each other, and be held together, so soon as they are near enough for the convex surfaces to meet. In this case, they are not pulled, but pushed together by the hydrostatic pressure of the higher water on the outside.

Once more, if one ball raises the water, and the other depresses it, and they are brought so near each other that the curves meet, they immediately move apart, as if repelled. For now the equilibrium is destroyed in a way just the reverse of the preceding cases. The water between the balls is too high for that which depresses, and too low for that which raises the water, so that the former is pushed away, and the latter is drawn away.

The first case, which is by far the most common, explains the fact often observed, that floating fragments are liable to be gathered into clusters; for most substances are capable of being wet, and therefore they raise the water about them.

230. Illustrations of Capillary Action.—It is by capillary action that a part of the water which falls on the earth is kept near its surface, instead of sinking to the lowest depths of the soil. This force aids the ascent of sap in the pores of plants. It lifts the oil between the fibres of the lamp-wick to the place of combustion. Cloth rapidly imbibes moisture by its numerous capillary spaces, so that it can be used for wiping things dry. If paper is not *sized*, it also imbibes moisture quickly, and can be used as *blotting-paper;* but when its pores are filled with sizing, to fit it for writing, it absorbs moisture only in a slight degree, and the ink which is applied to it must dry by evaporation.

The great strength of the capillary force is shown in the effects produced by the swelling of wood and other substances when kept wet. Dry wooden wedges, driven into a groove cut around a cylinder of stone, and then occasionally wet, will at length cause it to break asunder. As the pores between the fibres of a rope run around it in spiral lines, the swelling of a rope caused by keeping it wet will contract its length with immense force.

231. Questions in Hydrostatics.—

1. The diameters of the two cylinders of a hydraulic press are *one inch* and *one foot*, respectively; before the piston descends, the column of water in the small cylinder is *two feet* higher than the bottom of the large piston. Suppose that by a screw a force of 500 lbs. is applied to the small piston; what is the whole force exerted on the large piston at the beginning of the stroke?

Ans. 72098.17 lbs.

2. A junk bottle, whose lateral surface contained 50 square inches, being let down into the sea 3000 feet, what pressure do the sides of the bottle sustain, no allowance being made for the increased specific gravity of the sea-water?

Ans. 65104.166 lbs.

3. A Greenland whale sometimes has a surface of 3600 square feet; what pressure would he bear at the depth of 800 fathoms?

Ans. 1080,000,000 lbs., or more than 482,142 tons.

4. A mill-dam, running perpendicularly across a river, slopes at an angle of 25 degrees with the horizon. The average depth of the stream is 12 feet, and its breadth 500 yards; required the amount of pressure on the dam?

Ans. 15,971,906 lbs., or 7130 + tons.

5. A mineral weighs 960 grains in air, and 739 grains in water; what is its specific gravity? *Ans.* 4.344.

6. What are the respective weights of two equal cubical masses of gold and cork, each measuring 2 feet on its linear edge?

Ans. The gold weighs 9625 lbs. = 4.297 tons; the cork weighs 120 lbs.

7. A mass of granite contains 5949 cubic feet. The specific gravity of a fragment of it is found to be 2.6; what does the mass weigh? *Ans.* 431.568 tons.

8. An island of ice rises 30 feet out of water, and its upper surface is a circular plane, containing $\frac{3}{4}$ths of an acre. On the supposition that the mass is cylindrical, required its weight, and depth below the water, the specific gravity of sea-water being 1.0263, and that of ice .92.

Ans. Weight, 242,900 tons; depth, 259.64 feet.

9. Wishing to ascertain the exact number of cubic inches in a very irregular fragment of stone, I ascertained its loss of weight in water to be 5.346 ounces; required its volume.

Ans. 9.238 cubic inches.

10. Hiero, king of Syracuse, ordered his jeweller to make him a crown of gold containing 63 ounces. The artist attempted a fraud by substituting a certain portion of silver; which being suspected, the king appointed Archimedes to examine it. Archimedes, putting it into water, found it displaced 8.2245 cubic inches of the fluid; and having found that the inch of gold weighs 10.36 ounces, and that of silver 5.85 ounces, he discovered what part of the king's gold had been purloined; it is required to repeat the process. *Ans.* 28.8 ounces.

11. The specific gravity of lead being 11.35; of cork, .24; of fir, .45; how much cork must be added to 60 lbs. of lead, that the united mass may weigh as much as an equal bulk of fir?

Ans. 65.8527 lbs.

12. A cone, whose specific gravity is $\frac{1}{8}$, floats on water with its vertex downward; what part of the axis is immersed?

Ans. One-half.

13. A cone, having the same specific gravity as the above, floats with its vertex upward; how much of its axis is immersed?

Ans. 0.0436.

14. What is the weight of a chain of pure gold, which raises the water 1 inch in height, in a cubical vessel whose side is 3 inches? and suppose a chain of the same weight were adulterated with $14\frac{1}{2}$ ounces of silver; how much higher would it raise the water in the vessel? 1 ft. water = 911.458 oz. troy.

Ans. Weight = 91.35 oz.; height .133 in. more.

CHAPTER II.

LIQUIDS IN MOTION.

232. Depth and Velocity of Discharge.—From an aperture which is small, compared with the breadth of the reservoir, *the velocity of discharge varies as the square root of the depth.* For the pressure on a given area varies as the depth (Art. 207). If the area is removed, this pressure is a force which is measured by the momentum of the water; therefore the *momentum* varies as the depth (d). But momentum varies as the mass (q) multiplied by the velocity (v); hence $q\,v \propto d$. But it is obvious that q and v vary alike, since the greater the velocity, the greater in the same ratio is the quantity discharged. Therefore, $q^2 \propto d$, or $q \propto d^{\frac{1}{2}}$; also $v^2 \propto d$, or $v \propto d^{\frac{1}{2}}$.

Not only does the velocity vary as the square root of the depth of the orifice, but *it is equal to that acquired by a body falling through the depth.*

Let h = the height of the liquid above the orifice, and h' = the height of an infinitely thin layer at the orifice.

If this thin layer were to fall through the height h', under the action of its own weight or pressure, the velocity acquired would be $v' = \sqrt{2gh'}$ (Art. 28).

Denoting the velocity generated by the pressure of the entire column by v, we have, since velocity $\propto \sqrt{\text{depth}}$,

$$v : v' :: \sqrt{h} : \sqrt{h'}, \text{ or}$$
$$v : \sqrt{2gh'} :: \sqrt{h} : \sqrt{h'} :: \sqrt{2gh} : \sqrt{2gh'};$$
$$\therefore v = \sqrt{2gh}.$$

But $\sqrt{2gh}$ is also the velocity acquired in falling through the distance h (Art. 28).

From an orifice $16\frac{1}{12}$ feet below the surface of water, the velocity of discharge is $32\frac{1}{6}$ feet per second, because this is the velocity acquired in falling $16\frac{1}{12}$ feet; and at a depth *four* times as great, that is, $64\frac{1}{3}$ feet, the velocity will only be doubled, that is, $64\frac{1}{3}$ feet per second.

As the velocity of discharge at any depth is equal to that of a body which has fallen a distance equal to the depth, it is theoretically immaterial whether water is taken upon a wheel from a gate at the same level, or allowed to fall on the wheel from the top of the reservoir. In practice, however, the former is best, on account of the resistance which water meets with in falling through the air.

233. Descent of Surface.—When water is discharged from the bottom of a cylindric or prismatic vessel, the surface descends with a *uniformly retarded* motion. For the velocity with which the surface descends varies as the velocity of the stream, and therefore as the square root of the depth (Art. 232). But this is a characteristic of uniformly retarded motion, that it varies as the square root of the distance from the point where the motion terminates, as in the case of a body ascending perpendicularly from the earth.

The descent of the surface of water in a prismatic vessel has been used for measuring time. The *clepsydra*, or water-clock of the Romans, was a time-keeper of this description. The graduation must increase upward, as the odd numbers 1, 3, 5, 7, &c.; since, by the law of this kind of motion, the spaces passed over in equal times are as those numbers.

If a prismatic vessel is kept full, it discharges *twice* as much water in the same time as if it is allowed to empty itself. For the velocity, in the first instance, is *uniform;* and in the second it is *uniformly retarded*, till it becomes zero. We reason in this case, therefore, as in regard to bodies moving uniformly, and with motion uniformly accelerated from rest, or uniformly retarded till it ceases (Art. 25), that the former motion is *twice* as great as the latter.

234. Discharge from Orifices in Different Situations.—Other circumstances besides *area* and *depth* of the aperture are found to have considerable influence on the velocity of discharge. Observations on the directions of the filaments are made by introducing into the water particles of some opaque substance, having the same density as water, whose movements are visible. From such observations it appears that the particles of water descend in vertical lines, until they arrive within three or four inches of the aperture, when they gradually turn in a direction more or less oblique toward the place of discharge. This convergence of the filaments extends outside of the vessel, and causes the stream to diminish for a short distance, and then increase. The smallest section of the stream, called the *vena contracta*, is at a distance from the aperture varying from *one-half* of its diameter to the *whole*.

If water is discharged through a circular aperture in a thin plate in the bottom of the reservoir, and at a distance from the sides, as in Fig. 161, 1, the filaments form the vena contracta at a distance beyond the aperture equal to *one-half* of its diameter; the area of the section at the vena contracta is less than *two-thirds* (0.64) of the area of the aperture; and the quantity discharged is also about two-thirds of that obtained by calculation for the full size of the aperture.

FIG. 161.

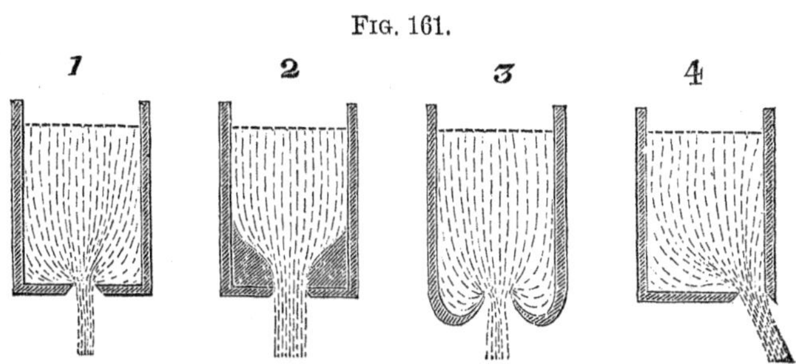

If the reservoir terminates in a short pipe or ajutage, whose interior is adapted to the curvature of the filaments, as far as to the vena contracta, or a little beyond, as in Fig. 161, 2, it is found the most favorable for free discharge, which in some cases reaches 0.98 of the theoretical discharge. The stream is smooth and pellucid like a rod of glass. The most unfavorable form is that in which the ajutage, instead of being external, as in the case just described, projects inward, as in Fig. 161, 3; the filaments in this case reach the aperture, some ascending, others descending, and therefore interfere with each other. Hence the stream is much roughened in its appearance, and the flow is only 0.53 of what is due to the size of the aperture and its depth.

When the aperture is through a thin plate, the contraction of the stream and the amount of discharge are both modified by the circumstance of being near one or more sides of the reservoir. There is little or no contraction on the side next the wall of the vessel, since the filaments have no obliquity on that side; and the quantity is on that account increased. The filaments from the opposite side also divert the stream a few degrees from the perpendicular (Fig. 161, 4).

235. Friction in Pipes.—As has just been stated, an ajutage extending to or slightly beyond the vena contracta, and adapted to the form of the stream, very much increases the quantity discharged; but beyond that, the longer the pipe, the more does it impede the discharge by friction. The friction varies directly as the length of the pipe, and inversely as its diameter. In order, therefore, to convey water at a given rate through a long pipe, it is necessary either to increase the head of water or to enlarge the pipe, so as to compensate for friction. If a given quantity of water is discharged per second by an aperture five inches in diameter, through the thin bottom of a reservoir, and we wish to discharge the same at the end of a horizontal pipe 150 feet long, and of the same diameter as the orifice, it will require *ten* times the head of water to accomplish it; or it may be done by the same head of water, if we use a pipe about 8 inches, instead of 5 inches, in diameter.

An aqueduct should be as *straight* as possible, not only to avoid unnecessary increase of length, but because the force of the stream is diminished by all changes of direction. If there must be change, it should be a gradual curve, and not an abrupt turn. When a pipe changes its direction by an *angle,* instead of a curve, there is a useless expenditure of force; a change of 90° requires that the head of water should be increased by nearly the height due to the velocity of discharge. For instance, if the discharge is *eight* feet per second (which is the velocity due to one foot of fall), then a right angle in the pipe requires that the head of water should be increased by nearly *one* foot, in order to maintain that velocity.

236. Jets.—Since a body, when projected upward with a certain velocity, will rise to the same height as that from which it must have fallen to acquire that velocity, therefore, if water issue from the side of a vessel through a pipe bent upward, it would, were it not for the resistance of the air and friction at the orifice, rise to the level of the water in the reservoir. If water is discharged from an orifice in any other than a vertical direction, it describes a parabola, since each particle may be regarded as a projectile (Art. 44).

If a semicircle be described on the perpendicular side of a vessel as a diameter, and water issue horizontally from any point, its *range*, measured on the level of the base, equals *twice the ordinate* of that point. For, the velocity with which the fluid issues from the vessel, being that which is due to the height $B\ G$ (Fig. 162), is $\sqrt{2\ g\ .\ B\ G}$ (Art. 28). But after leaving the orifice, it arrives at the horizontal plane in the time in which a body would fall freely through $G\ D$, which is $\sqrt{\frac{2\ G\ D}{g}}$. Since the horizontal motion is uniform, the space equals the product of the time by the velocity; that is, $D\ E = \sqrt{\frac{2\ G\ D}{g}} \times \sqrt{2\ g\ .\ B\ G} = 2\sqrt{B\ G\ .\ G\ D} =$

FIG. 162.

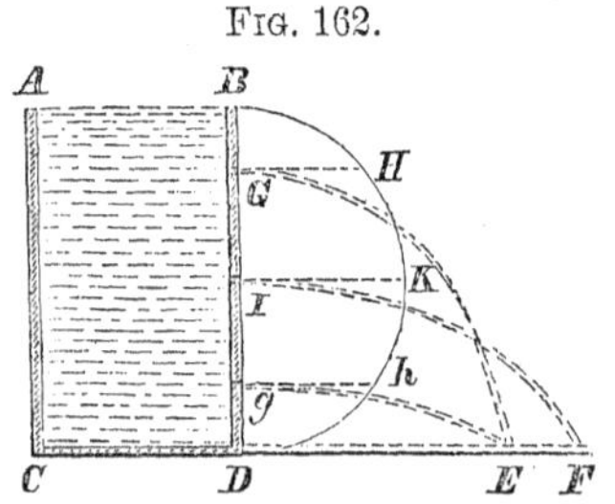

$2\,G\,H$, or twice the ordinate of the semicircle at the place of discharge.

The greatest range occurs when the fluid issues from the centre, for then the ordinate is greatest; and the range at equal distances above and below the centre is the same.

The remarks already made respecting pipes apply to those which convey water to the jets of fire-engines and fountains. If the pipe or hose is very long, or narrow, or crooked, or if the jet-pipe is not smoothly tapered from the full diameter of the hose to the aperture, much force is lost by friction and other resistances, especially in great velocities. If the length of hose is even *twenty* times as great as its diameter, 32 per cent. of height is lost in the jet, and more still when the ratio of length to diameter is greater than this.

237. Rivers.—Friction and change of direction have great influence on the flow of rivers. A *dynamical equilibrium*, as it is called, exists between gravity, which causes the descent, and the resistances, which prevent acceleration, beyond a certain moderate limit; so that, in general, the water of a river moves uniformly. The velocity in all parts of the same section, however, is not the same; it is greatest at that part of the surface where the depth is greatest, and least in contact with the bed of the stream.

To find the mean *velocity* through a given section, it is necessary to float bodies at various places on the surface, and also below it, to the bottom, and to divide the sum of all the velocities thus obtained, by the number of observations. To obtain the *quantity* of water which flows through a given section of a river, having determined the velocity as above, find next the area of the section, by taking the depth at various points of it, and multiplying the mean depth by the breadth. The quantity of water is then found by multiplying the area by the velocity.

The increased velocity of a stream during a freshet, while the stream is confined within its banks, exhibits something of the acceleration which belongs to bodies descending on an inclined plane. It presents the case of a river flowing upon the top of another river, and consequently meeting with much less resistance than when it runs upon the rough surface of the earth itself. The augmented force of a stream in a freshet arises from the simultaneous increase of the quantity of water and the velocity. In consequence of the friction of the banks and beds of rivers, and the numerous obstacles they meet with in their winding course, their velocity is usually very small, not more than three or four miles per hour; whereas, were it not for these impediments, it would become immensely great, and its effects would be exceedingly disastrous. A very slight declivity is sufficient for giving the running motion to water. The largest rivers in the world fall about five or six inches in a mile.

238. Hydraulic Pumps.—The most common pumps for raising water operate on a principle of pneumatics, and will be described under that subject.

In the *lifting-pump* the water is pushed up in the pump tube by a piston placed below the water-level. In the tube *A B* (Fig. 163) is a fixed valve *V*, a little below the water-level *L L*, while still lower is the piston *P*, in which there is a valve. Both of these valves open upward. The piston is attached to a rod, which extends downward to the frame *F F*. This frame can be moved up and down on the outside of the tube by a lever. When the piston descends, the water passes through its valve by hydrostatic pressure; and when raised, it pushes the water before it through the fixed valve, which then prevents its return. In this manner, by repeated strokes, the water can be driven to any height which the instrument can bear.

Fig. 163.

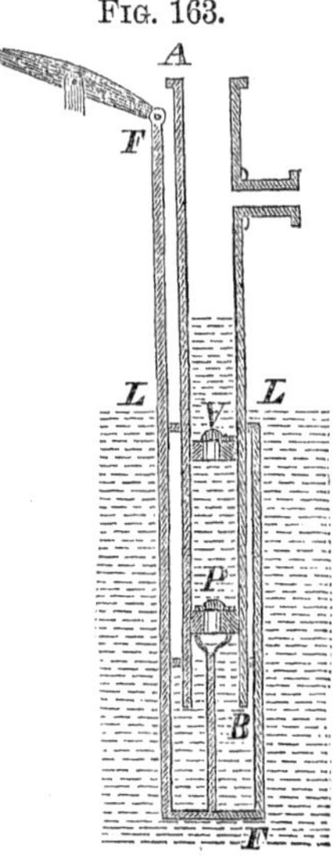

The *chain pump* consists of an endless chain with circular disks attached to it at intervals of a few inches, which raise the water before them in a tube, by means of a wheel over which the chain passes; the wheel may be turned by a crank. The disks cannot fit closely in the tube without causing too great resistance; hence, a certain velocity is requisite in order to raise water to the place of discharge,; and after the working of the pump ceases, the water soon descends to the level in the well.

239. The Hydraulic Ram.—When a large quantity of water is descending through an inclined pipe, if the lower extremity is suddenly closed, since water is nearly incompressible, the shock of the whole column is received in a single instant, and if no escape is provided, is very likely to burst the pipe. The intensity of the shock of water when stopped is made the means of raising a portion of it above the level of the head. The instrument for effecting this is called the *hydraulic ram*. At the lower end of a long pipe, *P* (Fig. 164), is a valve, *V*, opening downward; near it, another valve, *V'*, opens into the air-vessel, *A*; and from this ascends the pipe, *T*, in which the water is to be raised. As the valve V lies open by its weight, the water runs out, till its momentum at length shuts it, and the entire column is suddenly

FIG. 164.

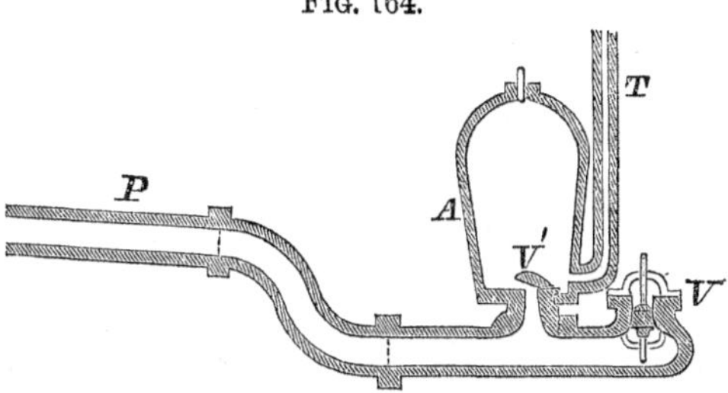

stopped; this impulse forces the water into the air-vessel, and thence, by the compressed air, up the tube T. As soon as the momentum is expended, the valve V drops, and the process is repeated.

240. Water-Wheels with a Horizontal Axis.—The *overshot wheel* (Fig. 165) is constructed with buckets on the circumference, which receive the water just after passing the highest point, and empty themselves before reaching the bottom. The weight of the water, as it is all on one side of a vertical diameter, causes the wheel to revolve. It is usually made as large as the fall will allow, and will carry machinery with a very small supply of water, if the fall is only considerable. The *moment* of each bucket-full constantly increases from a, where it is filled, to F, where its acting distance is radius, and therefore a maximum. From F downward the moment decreases, both by loss of water and diminution of acting distance, and becomes zero at L.

FIG. 165.

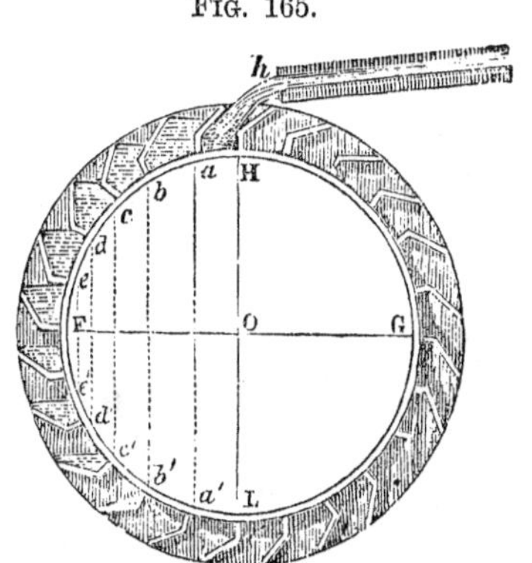

The *undershot wheel* (Fig. 166) is revolved by the momentum of running water, which strikes the float-boards on the lower side. When these are placed, as in the figure, perpendicular to the circumference, the wheel may turn either way; this is the construction adopted in tide-mills. When the wheel is required to turn only in one direction, an advantage is gained by placing

FIG. 166.

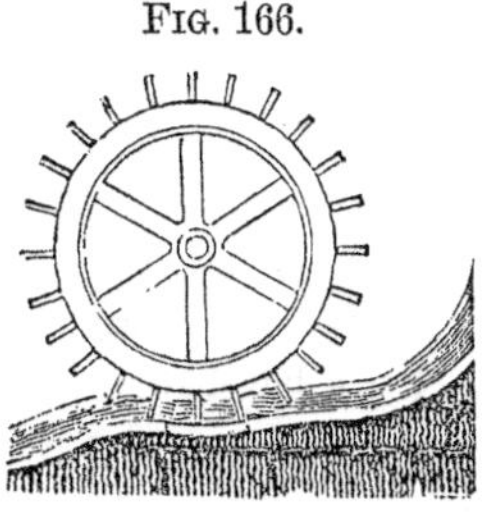

the floa -boards so as to present an acute angle toward the current, by which means the water acts partly by its weight, as in the overshot wheel. The undershot wheel is adapted to situations where the supply of water is always abundant.

In the *breast wheel* (Fig. 167) the water is received upon the float-boards at about the height of the axis, and acts partly by its weight, and partly by its momentum. The planes of the float-boards are set at right angles to the circumference of the wheel, and are brought so near the mill-course that the water is held and acts by its weight, as in buckets.

Fig. 167.

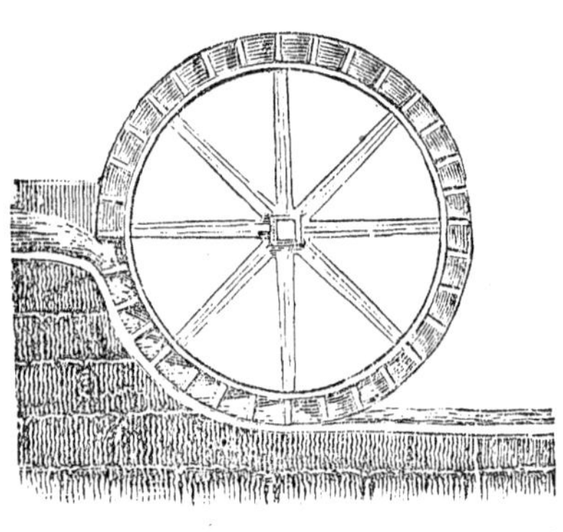

241. The Turbine.—This very efficient water-wheel, frequently called the French turbine, is of modern invention, and has received its chief improvements in this country. It revolves on a vertical axis, and surrounds the bottom of the reservoir from which it receives the water. The lower part of the reservoir is divided into a large number of sluices by curved partitions, which direct the water nearly into the line of a tangent, as it issues upon the wheel. The vanes of the wheel are curved in the opposite direction, so as to receive the force of the issuing streams at right angles. The horizontal section (Fig. 168) shows the lower part of the reservoir with its curved guides, *a*, *a*, *a*, and the wheel with its curved vanes, *v*, *v*, *v*, surrounding the reservoir; *D* is the central tube, through which the axis of the wheel passes. Fig. 169 is a vertical section of the turbine; but it does not present the guides of the reservoir, nor the vanes of the wheel. *C G*, *C G*, is the outer wall of the reservoir; *D*, *D*, its inner wall or tube; *F F*, the base, curved so as to turn the descending water gradually into a horizontal direction. The outer wall, which terminates at *G G*, is connected with the base and tube by the guides which are shown at *a*, *a*, in Fig. 168.

Fig. 168.

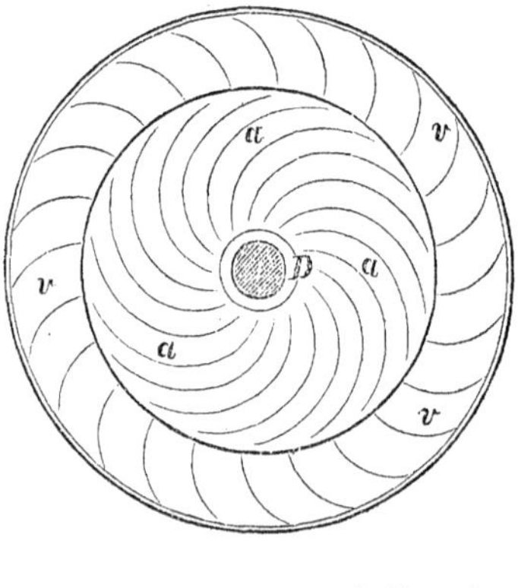

The lower rim of the wheel, *H, H,* is connected with the upper rim, *P, P,* by the vanes between them, *v, v* (Fig. 168), and to the

Fig. 169.

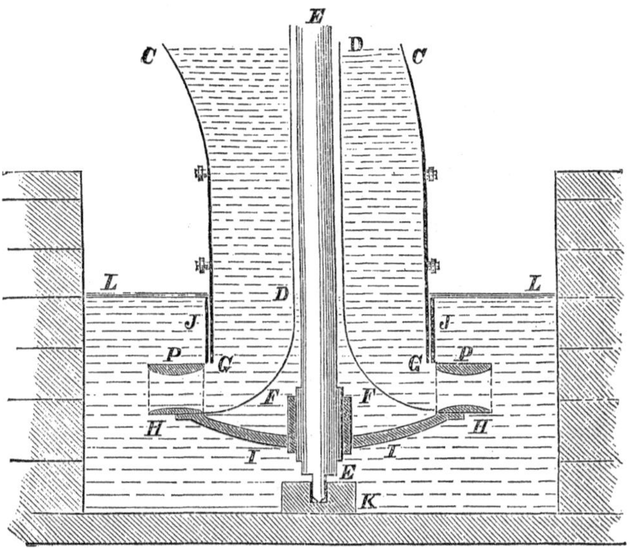

axis, *E, E,* by the spokes *I, I.* The gate, *J, J,* is a thin cylinder which is raised or lowered between the wheel and the sluices of the reservoir. The bottom of the axis revolves in the socket *K,* and the top connects with the machinery. As the reservoir cannot be supported from below, it is suspended by flanges on the masonry of the wheel-pit, or on pillars outside of the wheel. To prevent confusion in the figure, the supports of the reservoir and the machinery for raising the gate are omitted. By the curved base and guides of the reservoir, the water is conducted in a spiral course to the wheel, with no sudden change of direction, and thus loses very little of its force. The wheel usually runs below the level of the water in the wheel-pit, as represented in the figure, *L L* being the surface of the water. The reservoir is sometimes merely the extremity of a large tapering tube or supply pipe, bent from a horizontal to a vertical direction. In such a case, the tube *D D,* in which the axis runs, passes through the upper side of the supply pipe. The figure represents only the lower part.

242. Barker's Mill.—This machine operates on the principle of *unbalanced hydrostatic pressure.* It consists of a vertical hollow cylinder, *A B* (Fig. 170), free to revolve on its axis *M N,* and having a horizontal tube connected with it at the bottom. Near each end of the horizontal tube, at P and P', is an orifice, one on one side, and one on the opposite. The cylinder, being kept full of water, whirls in a direction opposite to that of the discharging streams from P and P'. This is owing to the fact that hydrostatic pressure is removed from the apertures, while on the interior of the tube, at points exactly opposite to them, are pressures which are now unbalanced, but which would be counteracted by the pressures at the apertures, if they were closed. The centrifugal force, after the machine is in rotation, has the effect to increase the pressure, and therefore the speed of rotation.

FIG 170.

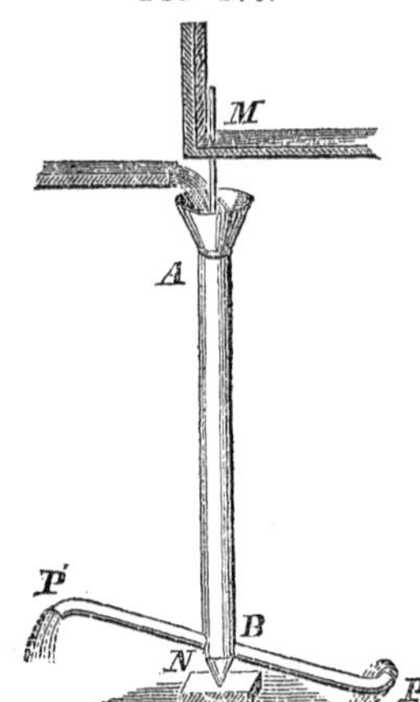

243. Resistance to Motion in a Liquid.—The resistance which a body encounters in moving through any fluid arises from the inertia of the particles of the fluid, their want of perfect mobility among each other, and friction. Only the first of these admits of theoretical determination. So far as the inertia of the fluid is concerned, the *resistance* which a surface meets with in moving perpendicularly through it *varies as the square of the velocity*. For the resistance is measured by the *momentum* imparted by the moving body to the fluid. And this momentum (m) varies as the product of the quantity of fluid set in motion (q), and its velocity (v); or $m \propto q\ v$. But it is obvious that the quantity displaced varies as the velocity of the body, or $q \propto v$; hence $m \propto v^2$. Therefore the resistance varies as the square of the velocity.

This proposition is found to hold good in practice, where the velocity is small, as the motions of boats or ships in water; but when the velocity becomes very great, as that of a cannon ball, the resistance increases in a much higher ratio than as the square of the velocity. Since action and reaction are equal, it makes no difference, in the foregoing proposition, whether we consider the body in motion and the fluid at rest, or the fluid in motion and striking against the body at rest.

Since the resistance increases so rapidly, there is a wasteful expenditure of force in trying to attain great velocities in navigation. For, in order to double the velocity of a steamboat, the force of the steam must be increased four fold; and in order to triple its velocity, the force must become nine times as great.

When the resistance becomes equal to the moving force, the body moves uniformly, and is said to be in a state of *dynamical equilibrium.* Thus, a body falling freely through the air by gravity does not continue to be accelerated beyond a certain limit, but is finally brought, by the resistance of the air, to a uniform motion.

244. Water Waves.—These are moving elevations of water, caused by some force which acts unequally on its surface. There are two very different kinds of waves, called, respectively, *waves of oscillation* and *waves of translation.* In the first kind the particles of water have a vibratory or reciprocating motion, by which the vertical columns are alternately lengthened and shortened. A familiar example of this kind is the *sea-wave.* In the waves of translation the particles are raised, transferred forward, and then deposited in a new place, without any vibratory movement.

245. Waves of Oscillation.—If a pebble be tossed upon still water, it crowds aside the particles beneath it, and raises them above the level, forming a wave around it in the shape of a ring. As soon as this ring begins to descend, it elevates above the level another portion around itself, and thus the ring-wave continues to spread outward every way from the centre. But in the meantime the water at the centre, as it rises toward the level, acquires a momentum which lifts it above that level. From that position it descends, and once more passes below the level, thus starting a new wave around it, as at first, only of less height. Hence, we see as the result of the first disturbance, a series of concentric waves continually spreading outward and diminishing in height at greater distances, until they cease to be visible. In Fig. 171 are represented three circular waves at one of the moments of time when the centre is lowest. The shaded parts are the basins or *troughs,* and the light parts, *c, c, c,* are the ridges or *crests.* Fig. 172 is a vertical section along the line *c, c,* through the centre of the system, corresponding to the momentary arrangement of Fig. 171. The central basin is at *b,* and the crests at *c, c, c.* A little later, when either crest has moved half way to the place of the next one, both figures will have become reversed; the centre will be a hillock, the troughs will be at *c, c,* and the crests at the middle points between them.

Fig. 171.

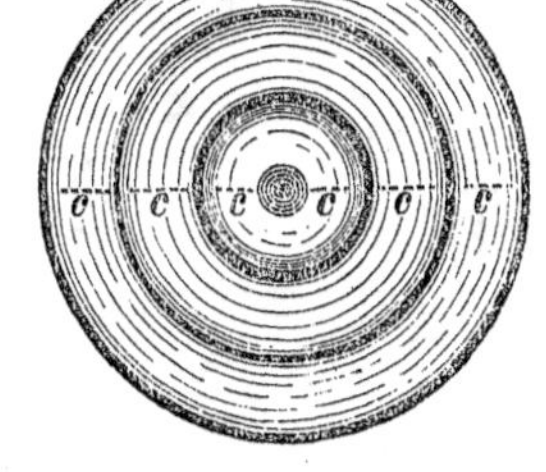

Fig. 172.

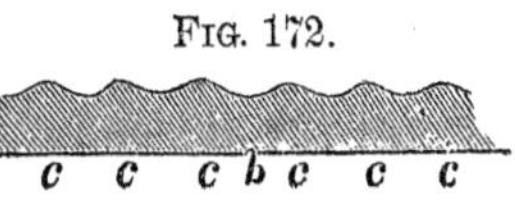

Except in the circular arrangement of the crests and troughs around a centre, the waves of the foregoing experiment illustrate common sea-waves. They constitute a system of elevations and depressions moving along the surface at right angles to the line of the wave-crest.

246. Phases.—In the cross-section (Fig. 172), where the waves are shown in profile, any particular part of the curve is called a *phase*. Different phases are generally unlike, both in elevation and in movement. The corresponding parts of different waves are called *like phases;* and those points in which the molecular motions are reversed are called *opposite phases.* The highest points of the crests of two waves are like phases; the highest point of the crest and the lowest point of the trough are opposite phases. Two points half way from crest to trough, one on the front of the wave, and the other on the rear of it, are also opposite phases, although they are at the same elevation; for they are moving in opposite directions. The *length* of a wave is the horizontal distance between two successive like phases.

247. Molecular Movements.—The water which constitutes a system of waves does not advance along the surface, as the waves themselves do; for a floating body is not borne along by them, but alternately rises and falls as the waves pass under it. Each particle of water, instead of advancing with the wave, oscillates about its mean place, alternately rising as high as the crest, and falling as low as the trough. Its path is the circumference of a vertical circle. Let *B B'* (Fig. 173) represent two successive troughs, and

Fig. 173.

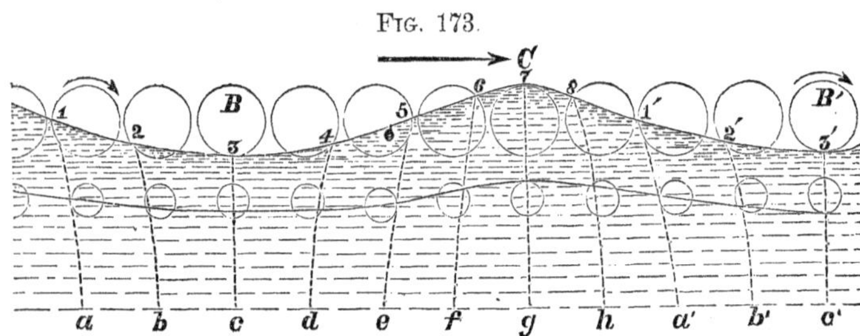

C the intervening crest; and for convenience suppose *a a'*, the wave length, to be divided into eight equal parts. The waves move in the direction of the straight arrow, while the particles of water revolve in the direction of the bent arrows. The points 1, 2, 3, &c., represent particles which, if the water were at rest, would be directly above the points *a*, *b*, *c*, &c. At the moment represented, 1 is at the extreme left of its revolution, 2 is at 45° below, 3 at the lowest point, &c. When the wave has advanced one-eighth of its length, 1 will have ascended 45°, 2 will have ascended to the extreme left, and each of the eight particles will have revolved one-eighth of the circumference shown in the figure. Then 4 will be at the bottom, and 8 at the top. Each particle of water on the front of the wave, from 1 to 3, and from 7 to 3′, is ascending; each one on the rear, from 3 to 7, is descending. It is plain that while the wave advances its whole length, that is, while the phase B is moving to B', each particle makes a complete revolution; 3′, which is now lowest, will be lowest again, having in the meantime occupied all other points of the circumference.

Particles *below* the surface, as far as the wave disturbance reaches, perform synchronous revolutions, but in smaller circles, as represented in the figure.

248. Form of Waves of Oscillation.—The sectional form of these waves is that of the inverted *trochoid,* a curve described by a point in a circle as it rolls on a straight line. The curvature of the crest is always greater than that of the trough, and the summit may possibly be a sharp ridge, in which case the section of the trough is a *cycloid,* the describing point of the rolling circle being on the circumference; the height of such waves is to their length as the diameter of a circle to the circumference. If waves are ever higher than about one-third of their length, the summits are broken into spray.

249. Distortion of the Vertical Columns.—Where the surface is depressed below its level, some of the water must be crowded laterally out of its place, and the vertical columns, being shorter, must necessarily be wider, at least in the upper part. So, too, where the surface is raised above its level, the lengthened columns must be narrower. In Fig. 173 these effects are made apparent as the necessary result of the revolutions of the particles. The dotted lines, 1 *a*, 2 *b*, 3 *c*, &c., were all vertical lines when the water was at rest. But now they are swayed, some to the right and some to the left, none being vertical, except under the highest and lowest points of the waves. Under the trough the lines are spread apart, and under the crest they are drawn together. The sectional figures 1 *a b* 2, 2 *b c* 3, &c., which would all be rectangular if the water were at rest, are now distorted in form, the upper parts being alternately expanded and contracted in breadth as the successive phases pass them.

250. Sea-Waves.—The waves raised by the wind rarely exhibit the precise forms above described, and the particles rarely revolve in exact circles, partly because there is scarcely ever a system of waves undisturbed by other systems, which are passing over the water at the same time, and partly because the wind, which was the original cause of the waves, acts continually upon their surfaces to distort and confuse them.

The *interference* of waves denotes, in general, the resultant system, which is produced by the combination of two or more separate systems. The joint effect of two systems is various, according as they are more or less unlike as to length of waves. But even if two systems are just alike, still the effect of interference will vary, according to the coincidence or the degree of discrepancy of their like phases. For instance, if two similar systems exactly coincide, phase for phase, the waves simply have double height; or, in general terms, there is double intensity in the wave motion. But if the phases of one system exactly coincide with the *opposite* phases of the other, then the water is nearly level, the crests of each system filling the troughs of the other. These two effects may be plainly seen in the intersections of ring-waves formed by dropping two pebbles on still water.

251. Waves of Translation.—The principal characteristics of the wave of translation are, that it is solitary—i. e., it does not belong to a system, like the other kind; and that its length and velocity both depend on the depth of the water. Where the water is deeper, the wave travels faster, and its length (measured in the direction of its progress) is longer. A wave of this character is started in a canal by a moving boat; and when the boat stops, it moves on alone. A grand example of this species is found in the tide-wave of the ocean. It is called the wave of translation because the particles of water are borne forward a certain distance while the wave is passing, and then remain at rest.

PART III.

PNEUMATICS.

CHAPTER I.

PROPERTIES OF GASES.—INSTRUMENTS FOR INVESTIGATION.

252. Gases Distinguished from Liquids.—The property of *mobility* of particles, which belongs to all fluids, is more remarkable in gases than in liquids.

While gaseous substances are compressed with ease, they are always ready to expand and occupy more space. This property, called *dilatability*, scarcely belongs to liquids at all. The force which gases show in expanding is called *tension*.

253. Change of Condition.—Liquids, and even solids, may be changed into the gaseous or aeriform condition by heating them sufficiently. By being cooled, they return again to their former state. In the gaseous form they are called *vapors*. And some substances which are ordinarily gases can be so far cooled, especially under great pressure, as to be reduced to the liquid or solid form. Those which have never been thus reduced are called *permanent gases*.

The mechanical properties of the gases may all be illustrated by experiments performed upon atmospheric air.

254. Mariotte's Law.—

At a given temperature, the volume of air is inversely as the compressing force.

An instrument constructed for showing this is called *Mariotte's tube*. The end *B* (Fig. 174) is sealed, and *A* open. Pour in small quantities of mercury, inclining the tube so as to let air in or out, till both branches are filled to the zero point. The air in the short branch now has the same tension as the external air, since they just balance each other. If mercury be poured in till the column in the short tube rises to *C*, the inclosed air is reduced to one-half of its original volume, and the column *A* in the long branch is found to be 29 or 30 inches above the level of *C*, according to the barometer at the time. Thus, *two* atmospheres, one of mercury, the other of air above it, have compressed the inclosed air into *one-half* its volume. If the tube is of sufficient length, let mercury be poured in again, till the air is compressed to one-third of its original space; the long column, measured from the level of the mercury in the short one, is now twice as high as before; that is, *three* atmospheres, two of mercury and one of air, have reduced the same quantity of air to *one-third* of its first volume. This law has been found to hold good in regard to atmospheric air up to a pressure of nearly thirty atmospheres.

Fig. 174.

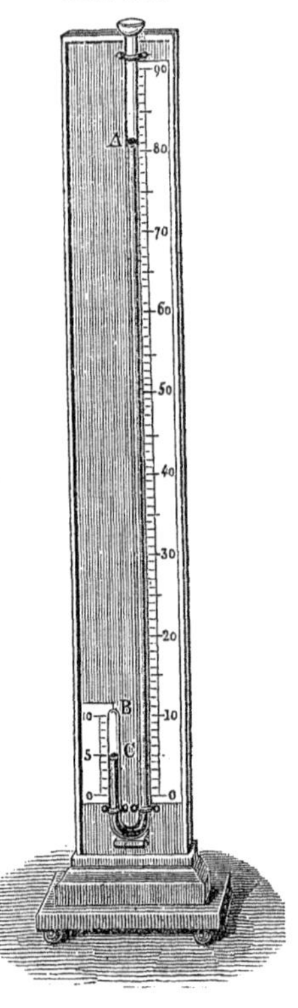

On the other hand, if the pressure on a given mass of air is diminished, its volume is found to increase according to the same law. When the pressure is *half* an atmosphere, the volume is *doubled;* when *one-third* of an atmosphere, the volume is three times as great, &c.

This law is found, however, not to be strictly applicable to all the gases. Some are compressed a little more, and others a little less, than Mariotte's law would require.

Since the tension of the inclosed air always balances the compressing force, and since the density is inversely as the volume, it follows from Mariotte's law that when the temperature is the same,

The tension of air varies as the compressing force; and
The tension of air varies as its density.

255. The Air-Pump.—This is an instrument by which nearly all the air can be removed from a vessel or receiver. It has a variety of forms, one of which is shown in Fig. 175. In the barrel *B* an air-tight piston is alternately raised and depressed by the lever, the piston-rod being kept vertical by means of a guide. The pipe *P* connects the bottom of the barrel with the brass plate *L*, on which rests the receiver *R*. The surface of the plate and the edge of the receiver are both ground to a plane. *G* is the gauge which indicates the degree of exhaustion. There are three valves, the first at the bottom of the barrel, the second in the piston, and the third at the top of the barrel. These all open upward, allowing the air to pass out, but preventing its return.

FIG. 175.

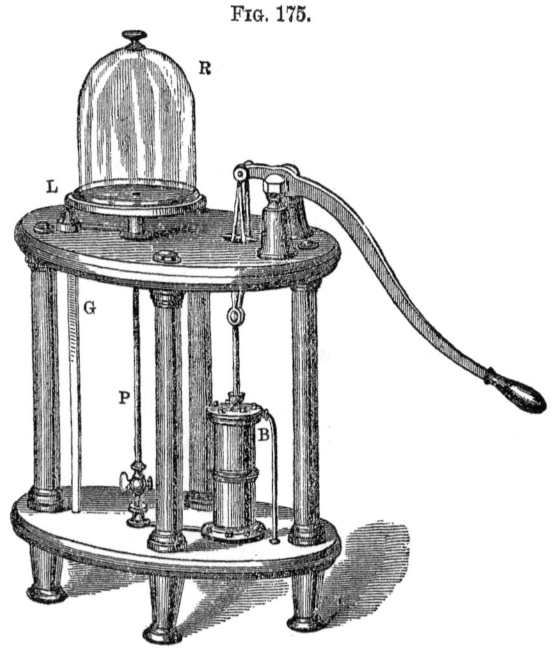

256. Operation.—When the piston is depressed, the air below it, by its increased tension, presses down the first valve, and opens the second, and escapes into the upper part of the barrel. When the piston is raised, the air above it cannot return, but is pressed through the third valve into the open air; while the air in the receiver and pipe, by its tension, opens the first valve, and diffuses itself equally through the receiver and barrel. Another descent and ascent only repeat the same process; and thus, by a succession of strokes, the air is nearly all removed.

FIG. 176.

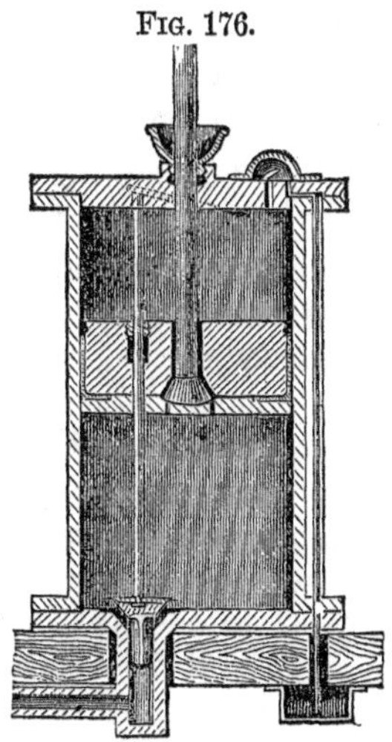

The exhaustion can be made more complete if the first and second valves are opened by the action of the piston and rod, rather than by the tension of the air. This method is illustrated by Fig. 176, a section of the barrel and piston. The first and second valves, as shown in the figure, are conical or *puppet* valves, fitting into conical sockets. The first has a long stem attached, which passes through the piston air-tight, and is pulled up by it a little way, till it is arrested by striking the top of the barrel. The second valve is a conical frustum on the end of the piston-rod. When the rod is raised, it shuts the valve before moving the piston; when it begins to descend, it opens the valve again before giving motion to the piston. The first valve is shut by a lever, which the piston strikes at the moment of its reaching the top. The oil which is likely to be pressed through the third valve is drained off by the pipe (on the right in both figures) into a cup below the pump.

257. Rate of Exhaustion.—The quantity removed, by successive strokes, and also the quantity remaining in the receiver, diminishes in the same geometrical ratio. For, of the air occupying the barrel and receiver, a barrel-full is removed at each stroke, and a receiver-full is left. If, for example, the receiver is *three* times as large as the barrel, the air occupies *four* parts before the descent of the piston; and by the first stroke *one-fourth* is removed, and *three-fourths* are left. By the next stroke, three-fourths as much will be removed as before ($\frac{1}{4}$ of $\frac{3}{4}$, instead of $\frac{1}{4}$ of the whole), and so on continually. The quantity left obviously diminishes also in the same ratio of three-fourths. In general, if b expresses the capacity of the barrel, and r that of the receiver and connecting-pipe, the ratio of each descending series is $\frac{r}{b+r}$. With a given barrel, the rate of exhaustion is obviously more rapid as the receiver is smaller. If the two were equal, ten strokes would rarefy the air more than a thousand times. For $(\frac{1}{2})^{10} = \frac{1}{1024}$.

As a term of this series can never reach zero, a complete exhaustion can never be effected by the air-pump; but in the best condition of a well-made pump, it is not easy to discover by the gauge that the vacuum is not perfect.

258. Experiments with the Air-Pump.—By the air-pump a great variety of experiments may be performed, illustrative of the mechanical properties of the air. The *obstruction* of the air being removed, light and heavy bodies are seen to fall with equal rapidity; a wheel with vanes perpendicular to the plane of rotation runs as freely as if they coincided with that plane, and water boils below blood-heat. The *weight* of a given volume of air is obtained by first weighing a vessel filled with air, and then empty. The *pressure* of air in every direction is rendered apparent by many striking effects, such as lifting weights, holding together the Magdeburg hemispheres, and throwing jets of water; also, by the *difference* of pressures on the upper and lower side, bodies are shown to weigh less in air than in a vacuum. And, finally, the *tension* or expansive force is exhibited by experiments equally numerous and interesting.

259. The Air Condenser.—While the air-pump shows the tendency of air to *dilate* indefinitely, as the compressing force is removed, another useful instrument, the *condenser*, exhibits the indefinite compressibility of air. Like the pump, it consists of a barrel and piston, but its valves, one in the piston and one at the bottom of the barrel, open downward. Fig. 177 shows the exterior of the instrument. If it be screwed upon the top of a strong receiver (Fig. 178), with a stop-cock connecting them, air may be forced in, and then secured by shutting the stop-cock. When the piston is depressed, its own valve is shut by the increased tension of the air beneath it, and the lower one opened by the same force. When the piston is raised, the lower valve is kept shut by the condensed air in the receiver, and that of the piston is opened by the weight of the outer air, which thus gets admission below the piston.

Fig. 177.

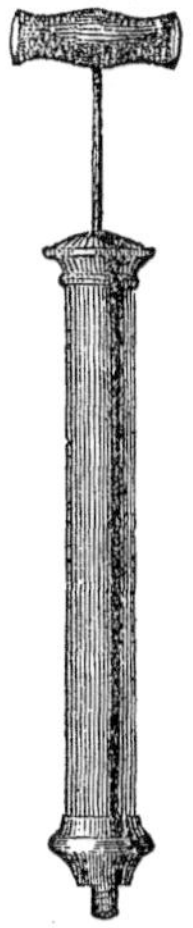

The quantity of air in the receiver increases at each stroke in an arithmetical ratio, because the same quantity, a barrel-full of common air, is added every time the piston is depressed. A small Mariotte's tube is attached to the receiver, to show how many atmospheres have been admitted.

Fig. 178.

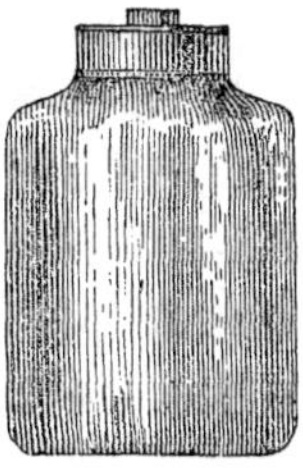

260. Experiments with the Air Condenser.—If the receiver be partly filled with water, and a pipe from the stop-cock extend into it, then when the condenser has been used and removed, and the stop-cock opened, a jet of water will be thrown to a height corresponding to the tension of the inclosed air. A gas-bag being placed in the condenser, then filled and shut, will become flaccid when the air around it is compressed. A thin glass bottle, sealed, will be crushed by the same force. By these and other experiments may be shown the effects of increased tension.

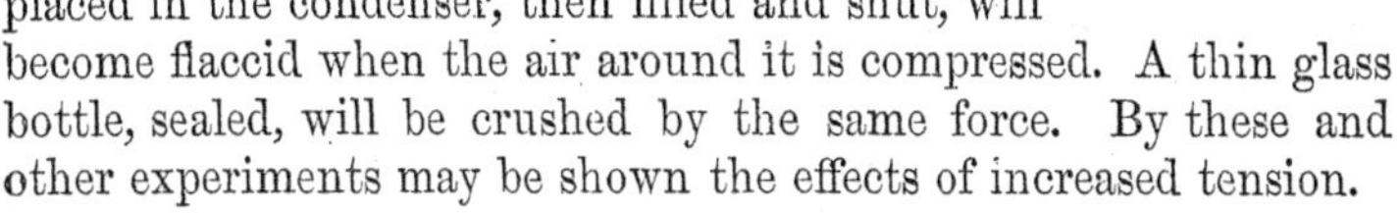

261. Torricelli's Experiment.—A glass tube *A B* (Fig. 179) about three feet long, and hermetically sealed at one end, is filled with mercury, and then, while the finger is held tightly on the open end, it is inverted in a cup of mercury. On removing the finger after the end of the tube is beneath the surface of the mercury, the column sinks a little way from the top, and there remains. Its height is found to be nearly thirty inches above the level of mercury in the cup. If sufficient care is taken to expel globules of air from the liquid, the space above the column in the tube is as perfect a vacuum as can be obtained. It is called the *Torricellian* vacuum, from Torricelli of Italy, a disciple of Galileo, who, by this experiment, disproved the doctrine that *nature abhors a vacuum*, and fixed the limits of atmospheric pressure.

Fig. 179.

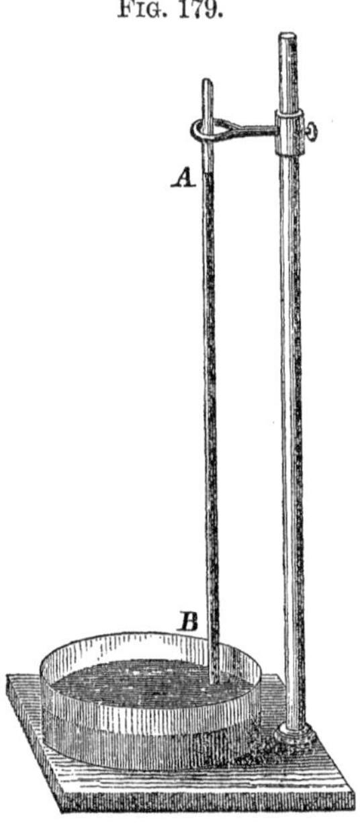

262. Pressure of Air Measured.—The column is sustained in the Torricellian tube by the pressure of air on the surface of mercury in the vessel; for the level of a fluid surface cannot be preserved unless there is an equal pressure on every part. Hence, the column of mercury on one part, and the column of air on every other equal part, must press equally. To determine, therefore, the pressure of air, we have only to weigh the column of mercury, and measure the area of the mouth of the tube. If this is carefully done, it is found that the weight of mercury is about 14.7 lbs. on a square inch. Therefore the atmosphere presses on the earth with a force of nearly 15 pounds to every square inch, or more than 2000 lbs. per square foot.

The specific gravity of mercury is about 13.6; and therefore the height of a column of water in a Torricellian tube should be 13.6 times greater than that of mercury, that is, about 34 feet. Experiment shows this to be true. And it was this significant fact, that *equal weights* of water and mercury are sustained in these circumstances, which led Torricelli to attribute the effect to a common force, namely, the pressure of the air.

263. Pascal's Experiment.—As soon as Torricelli's discovery was known, Pascal of France proposed to test the correctness of his conclusion, by carrying the apparatus to the top of a mountain, in order to see if less air above the instrument sustained the mercury at a less height. This was found to be true; the column gradually fell, as greater heights were attained. The experiment of Pascal also determined the relative density of mercury and air. For the mercury falls one-tenth of an inch in ascending 87.2 feet; therefore the weight of the one-tenth of an inch of mercury was balanced by the weight of the 87.2 feet of air. Therefore the specific gravities of mercury and air (being inversely as the heights of columns in equilibrium) are as ($87.2 \times 12 \times 10 =$) 10464 : 1. In the same way it is ascertained that water is 770 times as dense as air. These results can of course be confirmed by directly weighing the several fluids, which could not be done before the invention of the air-pump.

264. The Barometer.—When the Torricellian tube and basin are mounted in a case, and furnished with a graduated scale, the instrument is called a barometer. The scale is divided into inches and tenths, and usually extends from 26 to 32 inches, a space more than sufficient to include all the natural variations in the weight of the atmosphere. By attaching a vernier to the scale, the reading may be carried to hundredths and thousandths of an inch, as is commonly done in meteorological observations. By observing the barometer from day to day, and from hour to hour, it is found that the atmospheric pressure is constantly fluctuating.

As the meteorological changes of the barometer are all comprehended within a range of two or three inches, much labor has been expended in devising methods for magnifying the motions of the mercurial column, so that more delicate changes of atmospheric pressure might be noted. The inclined tube and the wheel barometer are intended for this purpose. A description of these contrivances, however, is unnecessary, as they are all found to be inferior in accuracy to the simple tube and basin.

265. Corrections for the Barometer.—

1. For *change of level in the basin.*—The numbers on the barometer scale are measured from a certain zero point, which is assumed to be the level of the mercury in the basin. If now the column falls, it raises the surface in the basin; and if it rises, it lowers it. If the basin is broad, the change of level is small, but it always requires a correction. To avoid this source of error, the bottom of the basin is made of flexible leather, with a screw underneath it, by which the mercury may be raised or lowered, till its surface touches an index that marks the zero point. This adjustment should always be made before reading the barometer.

2. For *capillarity.*—In a glass tube mercury is depressed by capillary action (Art. 227). The amount of depression is less as the tube is larger. This error is to be corrected by the manufacturer, the scale being put below the true height by a quantity equal to the depression.

There is a slight variation in this capillary error, arising from the fact that the rounded summit of the column, called the *meniscus*, is more convex when ascending than when descending. To render the meniscus constant in its form, the barometer should be jarred before each reading.

3. For *temperature.*—As mercury is expanded by heat and contracted by cold, a given atmospheric pressure will raise the column too high, or not high enough, according to the temperature of the mercury. A thermometer is therefore attached to the barometer, to show the temperature of the instrument. By a table of corrections, each reading is reduced to the height the mercury would have if its temperature was 32° F.

4. For *altitude of station.*—Before comparing the observations of different places, a correction must be made for altitude of station, because the column is shorter according as the place is higher above the sea level.

266. The Aneroid Barometer.—This is a small and portable instrument, in appearance a little like a large chronometer. The essential part of this barometer is a flat cylindrical metallic box, whose upper surface is corrugated, so as to be yielding. The box being partly exhausted of air, the external pressure causes the top to sink in to a certain extent; if the pressure increases, the surface descends a little more; if it diminishes, a little less. These small movements are communicated by a system of levers to an index on the graduated face of the barometer. The box and levers are concealed and protected within the outer case. As might be expected, its range is limited, and its indications not perfectly reliable; but for obtaining results in which accuracy is not essential, its lightness and convenient form and size recommend it, especially for portable uses.

267. Pressure and Latitude.—The mean pressure of the atmosphere at the level of the sea is very nearly 30 inches. But it is not the same at all latitudes. From the equator either northward or southward, the mean pressure increases to about latitude 30°, by a small fraction of an inch, and thence decreases to about 65°, where the pressure is less than at the equator, and beyond that it slightly increases. This distribution of pressures in zones is due to the great atmospheric currents, caused by heat in connection with the earth's rotation on its axis.

The amount of variation in barometric pressure is very unequal in different latitudes; and in general, the higher the latitude, the greater the variation. Within the tropics the extreme range scarcely ever exceeds one-fourth of an inch, while at latitude 40° it is more than two inches, and in higher latitudes even reaches three inches.

268. Diurnal Variation.—If a long series of barometric observations be made, and the mean obtained for each hour of the day, the changes caused by weather become eliminated, and the diurnal oscillation reveals itself. It is found that the pressure reaches a maximum and a minimum twice in 24 hours. The times of greatest pressure are from 9 to 10, and of least pressure from 3 to 4, both A. M. and P. M. In tropical climates this variation is very regular, though small; but in the temperate zones the irregular fluctuations of weather conceal it in a great degree.

This double oscillation is the mingled effect of heat and moisture, each of which alone would produce a single oscillation extending through the entire day.

In some countries of the torrid zone there is a regular *annual* oscillation of the barometer; but in the temperate zones this is scarcely perceptible.

269. The Barometer and the Weather.—The changes in the height of the barometer column depend directly on nothing else than the atmospheric pressure. But these changes of pressure are due to several causes, such as wind and changes of temperature and moisture.

The practice formerly prevailed of engraving at different points of the barometer scale several words expressive of states of weather, "fair, rain, frost, wind," &c. But such indications are worthless, being as often false as true; this is evident from the fact that the height of the column would be changed from one kind of weather to another by simply carrying the instrument to a higher or lower station.

No general system of rules can be given for anticipating changes of weather by the barometer, which would be applicable in different countries. Rules found in English books are of very little value in America.

Severe and extensive storms are almost always accompanied by a fall of the barometer while passing, and succeeded by a rise of the barometer.

270. Heights Measured by the Barometer.—Since mercury is 10464 times as heavy as air (Art. 263), if the barometer is carried up until the mercury falls one inch, it might be inferred that the ascent is 10464 inches, or 872 feet. This would be the case if the density were the same at all altitudes. But, on account of diminished pressure, the air is more and more expanded at greater heights. Besides this, the height due to a given fall of the mercury varies for many reasons, such as the temperature of the air, the temperature of the mercury, the elevation of the stations, and their latitude. Hence, the measurement of heights by the barometer is somewhat troublesome, and not always to be relied on. Formulæ and tables for this purpose are to be found in practical works on physics.

271. The Gauge of the Air-Pump.—The Torricellian tube is employed in different ways as a gauge for the air-pump, to indicate the degree of exhaustion. In Fig. 175 the gauge *G* is a tube about 33 inches long, both ends of which are open, the lower immersed in a cup of mercury, and the upper communicating with the interior of the receiver. As the exhaustion proceeds, the pressure is diminished within the tube, and the external air raises the mercury in it. A perfect vacuum would be indicated by a height of mercury equal to that of the barometer at the time.

Another kind of gauge is a barometer already filled, the basin of which is open to the receiver. As the tension of air in the receiver is diminished, the column descends, and would stand at the same level in both tube and basin, if the vacuum were perfect.

A modified form of the last, called the *siphon gauge*, is the best for measuring the rarity of the air in the receiver when the vacuum is nearly perfect. Its construction is shown by Fig. 180. The top of the column, *A*, is only 5 or 6 inches above the level of *B* in the other branch of the recurved tube. As the air is withdrawn from the open end *C*, the tension at length becomes too feeble to sustain the column; it then begins to descend, and the mercury in the two branches approaches a common level.

Fig. 180.

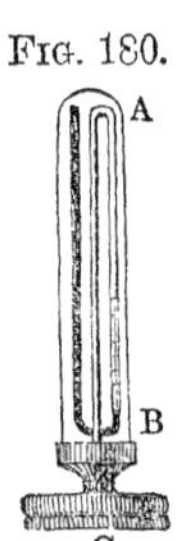

272. Buoyant Power of the Air.—If a large and a small body are in equilibrium on the two arms of a balance, and the whole be set under a receiver, and the air be removed, the larger body will preponderate, showing that it is really the heaviest. Their apparent equality of weight when in the air is owing to its buoyant power; for air, like water and all fluid substances, diminishes the apparent weight of an immersed body by just the weight of the displaced fluid. Hence, the larger the body, the more weight it loses.

It follows that if a body weighs less than the displaced air, it will rise just as light bodies do in water. It is in this way that balloons are made to ascend. By the use of a large volume of hydrogen, inclosed in a silk envelope, rendered air-tight by varnish, a car with several persons in it can be carried to a great height. The greatest height ever attained is about 23000 feet, or nearly 4.5 miles. The mercury of a barometer at that height falls to 12.5 inches.

CHAPTER II.

INSTRUMENTS WHOSE OPERATION DEPENDS ON THE PROPERTIES OF AIR.

273. Pneumatic Instruments.—Besides the apparatus described in the foregoing chapter, by the aid of which the properties of the air are discovered, there are several articles in common use whose utility depends more or less on the same properties, and which serve as good illustrations of the principles already presented.

274. The Bellows.—The simple or *hand-bellows* consists of two boards or lids hinged together, and having a flexible leather round the edges, and a tapering tube through which the air is driven out. In the lower board there is a hole with a valve lying on it, which can open inward. On separating the lids, the air by its pressure instantly lifts the valve and fills the space between them; but when they are pressed together, the valve shuts, and the air is compelled to escape through the pipe. The stream is intermittent, passing out only when pressure is applied.

The *compound bellows*, used for forges where a constant stream is needed, are made with two compartments. The partition *C T* (Fig. 181) is fixed, and has in it a valve *V′* opening upward. The lower lid has also a valve *V* opening upward, and the upper one is loaded with weights. The pipe *T* is connected with the upper compartment. As the lower lid is raised by the rod *A B*, which is worked by the lever *E B*, the air in the lower part is crowded through *V′* into the upper part, whence it is by the weights pressed through the pipe *T* in a constant stream. When the lower lid falls, the air enters the lower compartment by the valve *V*.

FIG. 181.

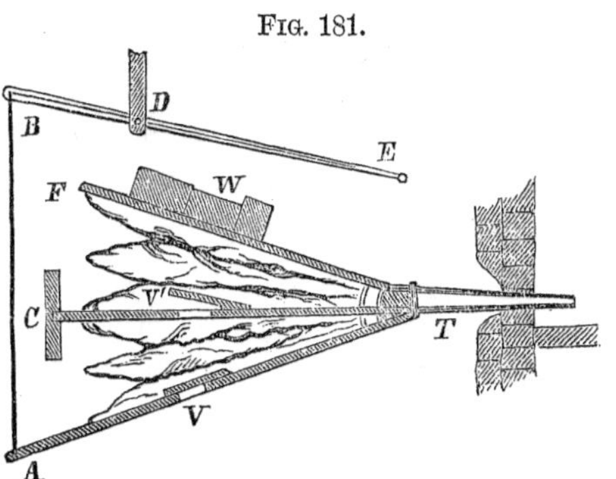

275. The Siphon.—If a bent tube *A B C* (Fig. 182) be filled, and one end immersed in a vessel of water, the liquid will be discharged through the tube so long as the outer end is lower than the level in the vessel. Such a tube is called a *siphon*, and is much used for removing a liquid from the top of a reservoir without disturbing the lower part. The height of the bend B above the fluid level must be less than 34 feet for water, and less than 30 inches for mercury. The reasons for the motion of the water are, that the atmosphere is able to sustain a column higher than $E\,B$, and that $C\,B$ is longer than $E\,B$. The two pressures on the highest cross-section B of the tube are unequal. For the atmospheric pressure at E is able to sustain 34 feet of water, and therefore at B exerts a pressure equal to $34 - E\,B$ toward the right. At C the air also presses upward with a force equal to 34 feet, and therefore at B it exerts a pressure to the left equal to $34 - C\,B$. Subtracting $34 - C\,B$ from $34 - E\,B$, we have the remainder, $C\,D$, for the excess of pressure to the right, or outward from the vessel. Therefore the water will flow with a velocity due to the weight of $D\,C$; hence, the velocity diminishes as the vessel empties.

FIG. 182.

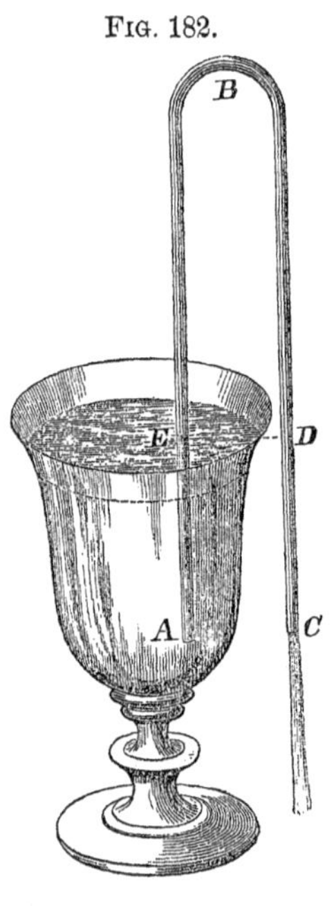

If the tube is small, it may be filled by suction, after the end A is immersed. If it is large, it may be inverted and filled, and then secured by stop-cocks, till the end is beneath the water.

276. Intermitting Springs.—Springs which flow freely for a time, and then cease for a certain interval, after which they flow again, are found in some cases to operate on the principle of the siphon. Suppose a reservoir or hollow in the interior of a hill, having a siphon-shaped outlet. It is obvious, upon hydrostatic principles, that no water will be discharged until the fluid has reached a level in the reservoir as high as the top of the bend in the outlet. Then it will begin to run out, and will continue to run until the water has descended to the level of the outlet; after which no more water will be discharged until enough has collected to reach the higher level, as before.

277. The Suction Pump.—The section (Fig. 183) exhibits the construction of the common suction pump. By means of a lever, the piston P is moved up and down in the tube $A\,V$. In the piston is a valve opening upward, and at the top of the pipe $H\ C$ is another valve, also opening upward. The latter must be at a less height than 34 feet above the water C. When the piston is raised, its valve is kept shut by the weight of air above, and the atmospheric pressure at C lifts a column of water $C\ H$ to such a height that its weight, added to the tension of the rarefied air, $H\ P$, equals 34 feet of water. When P descends, the air below is prevented from returning by the lower valve, and escapes through the piston. The piston being raised again, the water rises still higher, till at length it passes through the valve, and the piston dips into it; after this it is lifted directly to the discharge pipe S, without the intervention of the air.

FIG. 183.

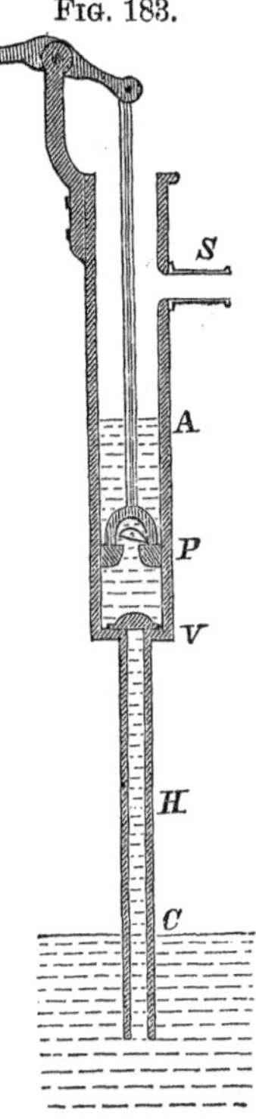

278. Calculation of the Force.—Let the whole atmospheric pressure be represented by 34 (its equivalent in feet of water), and the height of water, $C\ H$, by h. Since the tension of air in the tube, added to h, equals 34, therefore the tension $= 34 - h$; and this force is exerted upward on the lower side of the piston; while the downward pressure on the top of the piston $= 34$. The difference of the two $= h$, which is, therefore, the height of water, whose downward pressure is to be overcome. We arrive at the same result if the water is above the piston at A, when $h = A\ C$. For, in this case, the pressure upward on $P = 34 - P\ C$; while the downward pressure $= 34 + A\ P$; and the difference between them is $P\ C + A\ P = h$. Therefore, in every case, the force required to lift the piston and column of water is that which would be required to lift the same weight in any other way. The atmosphere has no other agency than to furnish a convenient mode of applying the force.

If $d =$ the diameter of the piston, in decimals of a foot, then $\frac{1}{4}\ \pi\ d^2 =$ its area; $\frac{1}{4}\ \pi\ d^2\ h =$ the cubic *feet* of water; and $\frac{1}{4}\ \pi\ d^2\ h \times 62.5 =$ the *pounds* of water.

279. The Forcing Pump.—The piston of the forcing pump (Fig. 184) is solid, and the upper valve V' opens into the side pipe $V'\ S$. In the ascent of the piston, the water is raised as in the suction pump; but in its descent, a force must be applied to press the water which is above V into the side pipe through V'.

Fig. 184.

As in the suction pump, the force expended is that required to lift $\frac{1}{4} \pi d^2 h \times 62.5$ pounds of water. But the two differ in this respect: in the suction pump the force is all expended in *raising* the piston; in the forcing pump the force is divided, and the column below P is lifted while the piston ascends, and that above P while it descends.

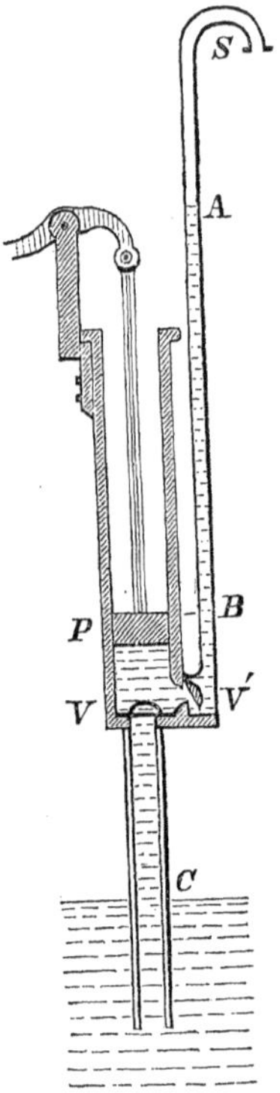

The piston is only one of many contrivances for producing rarefaction of air in a pump-tube; but since it is the most simple and most easily kept in repair, the piston-pump is generally preferred to any other.

280. The Fire-Engine.—This machine generally consists of one or more forcing pumps, with a regulating air-vessel, though the arrangement of parts is exceedingly varied. Fig. 185 will illustrate the principles of its construction. As the piston, P, ascends, the water is raised through the valve, V, by atmospheric pressure. As P descends, the water is driven through F into the air-vessel, M, whence by the condensed air it is forced out without interruption through the hose-pipe, L. The piston P' operates in the same way by alternate movements. The piston-rods are attached to a lever (not represented), to which the strength of several men can be applied at once by means of hand-bars called *brakes*.

The air-vessel may be attached to any kind of pump, whenever it is desired to render the stream constant.

Fig. 185.

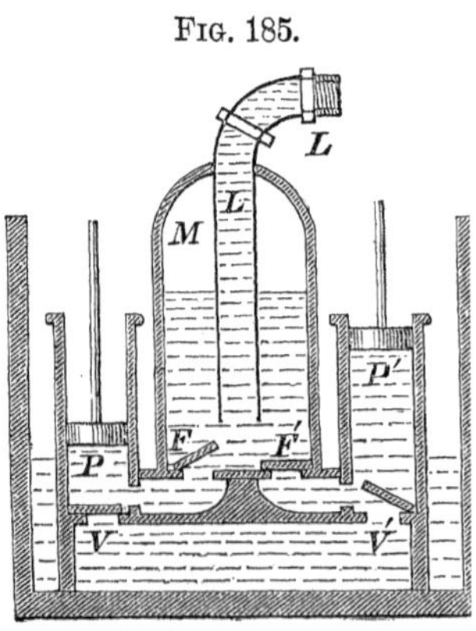

281. Hero's Fountain.—The condensation in the air-vessel, from which water is discharged, may be produced by the weight of a column of water. An illustration is seen in Hero's fountain, Fig. 186. A vertical column of water from the vessel, A, presses into the air-vessel, B, and condenses the air more or less, according to the height of A B.

From the top of this vessel an air-tube conveys the force of the compressed air to a second air-vessel, *C*, which is nearly full of water, and has a jet-pipe rising from it. Since the tension of air in *C* is equal to that in *B*, a jet will be raised which, if unobstructed, would be equal in height to the compressing column, *A B*.

Fig. 186.

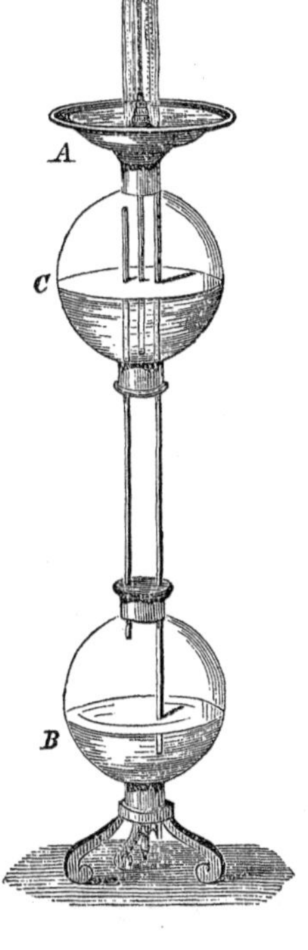

This plan has been employed to raise water from a mine in Hungary, and hence called "the Hungarian machine."

282. Manometers.—These are instruments for measuring the tension of gases or vapors. In one kind of manometer the law of Mariotte is employed. The tube *A B* (Fig. 187), closed at the top, has its open end beneath the surface of mercury in the closed cistern *C*. The vessel *D*, containing the gas or vapor whose tension is to be measured, communicates with the top of the cistern. If the mercury is at the same level in the cistern and tube, the pressure equals one atmosphere. As the tension in *D* increases, the column in *A B* rises, and compresses the air in the tube. The tension of the air in *A B* above the mercury, together with the weight of the mercury above the level in the cistern, is equal to the tension in *D*; so that the number (2) will not be in the middle point between (1) and the top, but somewhat below. A scale of atmospheres is calculated according to the proportions of the instrument, and placed by the side of the tube.

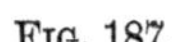

Fig. 187.

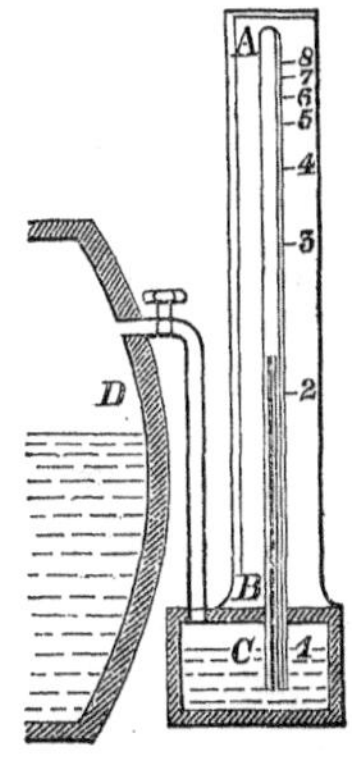

283. Apparatus for Preserving a Constant Level.—Let *A B* (Fig. 188) be a reservoir which supplies a liquid to the vessel *C D*; and suppose it is desired to preserve the level at the point *C* in the vessel, while the liquid is discharged from it irregularly or at intervals. This is accomplished by letting the discharge pipe *E* enter *C D* below the required level *C*, while the air is supplied to the reservoir only by a tube *F B*, which just reaches that level. So long as the liquid in $C D$ is below C, it is at a greater depth from the surface A than B is, and therefore the pressure is greater, and the liquid will run from E, and air enter at B. But when the vessel is filled to C, the hydrostatic pressures at C and B are equal; it is therefore impossible that the water should overcome the air at C and pass out, and that the air should at the same time overcome the water at B, and pass in. Hence, if G discharges more slowly than E, it is immaterial whether water is running from G or not; the vessel will remain always filled to the level $C B$.

FIG. 188.

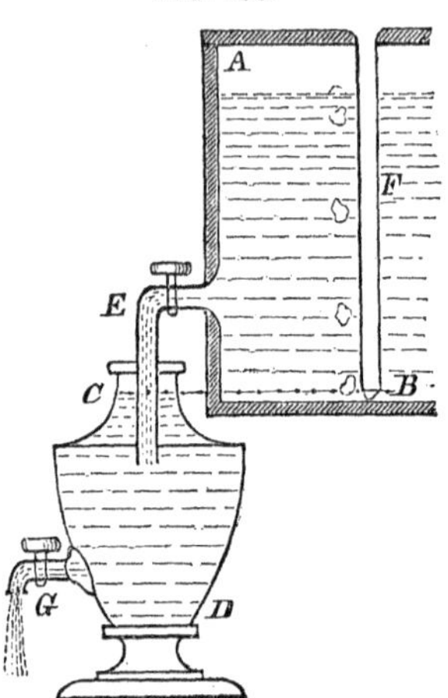

CHAPTER III.

THE ATMOSPHERE.—ITS QUANTITY, HEIGHT, AND MOTIONS.

284. Quantity of the Atmosphere.—Since the air sustains a column of mercury thirty inches high, the weight of the whole atmosphere is equal to that of a stratum of mercury thirty inches thick covering the globe. The thickness is relatively so small that the volume of the stratum may be reckoned as that of a parallelopiped, thirty inches in height, and having a base equal to the surface of the earth.

Letting R = the radius of the earth, and h = the depth of mercury, the earth's surface $= 4 \pi R^2$, and the volume of mercury $= 4 \pi R^2 h = 4 \times 3.14159 \times (3956 \times 5280)^2 \times 2.5$ cubic feet.

This multiplied by 62.5×13.6, the weight of a cubic foot of mercury, gives about 11,650,000,000,000,000,000 lbs. This is, therefore, the weight of the earth's atmosphere.

285. Virtual Height of the Atmosphere.—When two fluid columns are in equilibrium with each other, their heights are inversely as their specific gravities (Art. 221). The specific gravity of mercury is 10464 times that of the air at the ocean level. Therefore, if the air had the same density in all parts, its height would be found by the proportion,

$$1 : 10464 : : 2.5 : 26160 \text{ feet},$$

which is almost five miles. Hence, the quantity of the entire atmosphere of the earth is pretty correctly conceived of when we imagine it having the density of that which surrounds us, and reaching to the height of five miles.

286. Decrease of Density.—But the atmosphere is very far from being throughout of uniform density. The great cause of inequality is the decreasing weight of superincumbent air at increasing altitudes. The law of diminution of density, arising from this cause, is the following:

The densities of the air decrease in a geometrical as the altitudes increase in an arithmetical ratio. For, let us suppose the air to be divided into horizontal strata of equal thickness, and so thin that the density of each may be considered as uniform throughout. Let a be the weight of the whole column from the top to the earth, b the weight of the whole column above the lowest stratum, c that of the whole column above the second, &c. Then the weight of the lowest stratum is $a - b$, and the weight of the second is $b - c$, &c. Now the *densities* of these strata, and therefore their *weights* (since they are of equal thickness), are as the compressing forces; or,

$$a - b : b - c :: b : c;$$
$$\therefore a c - b c = b^2 - b c; \ \therefore a c = b^2;$$
$$\therefore a : b :: b : c;$$

in the same way, $b : c :: c : d$;

that is, the *weights* of the entire columns, from the successive strata to the top of the atmosphere, form a geometrical series; therefore, the *densities* of the successive strata, varying as the compressing forces, also form a geometrical series. If, therefore, at a certain distance from the earth, the air is twice as rare as at the surface of the earth, at twice that distance it will be four times as rare, at three times that distance eight times as rare, &c.

By barometric observations at different altitudes, it is found that at the height of three and a half miles above the earth the air is one-half as dense as it is at the surface. Hence, making an arithmetical series, with $3\frac{1}{2}$ for the common difference, to denote heights, and a geometrical series, with the ratio of $\frac{1}{2}$, to denote densities, we have the following:

Heights, $3\frac{1}{2}$, 7, $10\frac{1}{2}$, 14, $17\frac{1}{2}$, 21, $24\frac{1}{2}$, 28, $31\frac{1}{2}$, 35.
Densities, $\frac{1}{2}$, $\frac{1}{4}$, $\frac{1}{8}$, $\frac{1}{16}$, $\frac{1}{32}$, $\frac{1}{64}$, $\frac{1}{128}$, $\frac{1}{256}$, $\frac{1}{512}$, $\frac{1}{1024}$.

According to this law, the air, at the height of 35 miles, is at least a thousand times less dense than at the surface of the earth. It has, therefore, a thousand times less weight resting upon it; in other words, only one-thousandth part of the air exists above that height.

287. Actual Height of the Atmosphere.—The foregoing law, founded on that of Mariotte, cannot, however, be applicable except to moderate distances. If it were strictly true, the atmosphere would be unlimited. But that is impossible on a revolving body, since the centrifugal force must at some distance or other equal the force of gravity, and thus set a limit to the atmosphere; and that limit in the case of the earth is more than 20,000 miles high. The actual height of the atmosphere is doubtless far below this; for there can be none above the point where the repellency of the particles is less than their weight; and the repellency diminishes just as fast as the density, while the weight diminishes very slowly. The highest portions concerned in reflecting the sunlight are about 45 miles above the earth. But there is reason to believe that the air extends much above that height, probably 100 or 200 miles from the earth.

288. The Motions of the Air.—The air is never at rest. When in motion, it is called *wind.* The equilibrium of the atmosphere is disturbed by the unequal heat on different parts of the earth. The air over the hotter portions becomes lighter, and is therefore pressed upward by the cooler and heavier air of the less heated regions. And the motions thus caused are modified as to direction and velocity by the rotation of the earth on its axis.

289. The Trade Winds.—The most extensive and regular system of winds on the earth is known by the name of the *trade winds,* so called on account of their great advantage to commerce. They are confined to a belt about equal in width to the torrid zone, but whose limits are four or five degrees further north than the tropics.

In the northern half of this trade-wind zone the wind blows continually from the northeast, and in the southern half from the southeast. As these currents approach each other, they gradually become more nearly parallel to the equator, while between them there is a narrow belt of calms, irregular winds, and abundant rains.

The oblique directions of the trade winds are the combined effects of the heat of the torrid zone and the rotation of the earth. The cold air of the northern hemisphere tends to flow directly south, and crowd up the hot air over the equator. In like manner, the cold air of the southern hemisphere tends to flow directly northward. So that if the earth were at rest, there would be *north* winds on the north side of the equator, and *south* winds on the south side. But the earth revolves on its axis from west to east, and the air, as it moves from a higher latitude to a lower, has only so much eastward motion as the parallel from which it came.

Therefore, since it really has a less motion from the west than those regions over which it arrives, it has relatively a motion *from the east.* This motion from the east, compounded with the motion from the north on the north side of the equator, and with that from the south on the south side, constitutes the northeast and southeast tradewinds.

The limits of this system move a few degrees to the north during the northern summer, and to the south during the northern winter, but very much less than might be expected from the changes in the sun's declination.

In certain localities within the tropics the wind, owing to peculiar configurations of coast and elevations of the interior, changes its direction periodically, blowing six months from one point, and six months from a point nearly opposite. The *monsoons* of southern India are the most remarkable example.

290. The Return Currents.—The air which is pressed upward over the torrid zone must necessarily flow away northward and southward towards the higher latitudes, to restore the equilibrium. Hence, there are south winds in the upper air on the north side of the equator, and north winds on the south side. But these upper currents are also oblique to the meridians, because, having the easterly motion of the equator, they move faster than the parallels over which they successively arrive, so that a motion from the west is combined with the others, causing southwest winds in the northern hemisphere, and northwest in the southern. These motions of the upper air are discovered by observations made on high mountains, and in balloons, and by noticing the highest strata of clouds. It is to be borne in mind that although the atmosphere is more than 100 miles high, yet the lower half does not extend beyond three and a half miles above the earth (Art. 286).

291. Circulation Beyond the Trade Winds.—The upper part of the air which flows away from the equator cannot wholly retain its altitude, because of the diminishing space on the successive parallels. About latitude 30°, it is so much accumulated that it causes a sensible increase of pressure (Art. 267), and begins to descend to the earth. It is probable that some of the descending air still retains its oblique motion towards higher latitudes (for the prevailing winds of the northern temperate zone are from the southwest, and of the southern temperate zone from the northwest), while a part joins with the lower air which is moving towards the equator. Only so much of the rising equatorial mass can flow back to the polar regions as is needed to supply the comparatively small area within them. On account of the successive descent of the air returning from the equator, there is much less distinctness and regularity in the general circulation outside of the torrid zone than within it. Besides this, various local causes, such as mountain ranges, sea-coasts, and ocean currents, clear and cloudy skies, &c., mingle their effects with the more general circulation, and modify it in every possible way.

292. Land and Sea Breezes.—These are limited circulations over adjoining portions of land and water, the wind blowing from the water to the land in the day time, and in the contrary direction by night. When the sun begins to shine each day, it heats the land more rapidly than the water. Hence the air on the land becomes warmer and lighter than that on the water, and the surface current sets toward the land. By night the flow is reversed, because the land cools most rapidly, and the air above it becomes heavier than that over the water. These effects are more striking and more regular in tropical countries, but are common in nearly all latitudes.

293. A Current Through a Medium.—There are some phenomena relating to currents moving through a fluid, either of the same or a different kind, which belong alike to hydraulics and pneumatics; a brief account of these is presented here.

If a stream is driven through a medium, it carries along the adjoining particles by *friction* or *adhesion*. The experiment of Venturi illustrates this kind of action, as it takes place between the particles of water. A reservoir filled with water has in it an inclined plane of gentle ascent, whose summit just reaches the edge of the reservoir. A stream of water is driven up this plane with force sufficient to carry it over the top; but in doing so, it takes out continually some part of the water of the reservoir, and will in time empty it to the level of the lowest part of the stream. A stream of air through air produces the same effect, as may be shown by the flame of a lamp near the stream always bending toward it. In like manner, water through air carries air with it; when a stream of water is poured into a vessel of water, air is carried down in bubbles; and cataracts carry down much air, which as it rises forms a mass of foam on the surface. The strong wind from behind a high waterfall is owing to the condensation of air brought down by the back side of the sheet.

294. Ventilators.—If the stream passes across the end of an open tube, the air within the tube will be taken along with the stream, and thus a partial vacuum formed, and a current established. It is thus that the wind across the top of a chimney increases the draught within. To render this effect more uniformly successful, by preventing the wind from striking the interior edge of the flue, appendages, called *ventilators*, are attached to the chimney top. A simple one, which is generally effectual, consists of a conical frustum surrounding the flue, as in Fig. 189, so that the wind, on striking the oblique surface, is thrown over the top in a curve, which is convex upward. The same mechanical contrivance is much used for the ventilation of public halls and the holds of ships. A horizontal cover may be supported by rods, at the height of a few inches, to prevent the rain from entering.

FIG. 189.

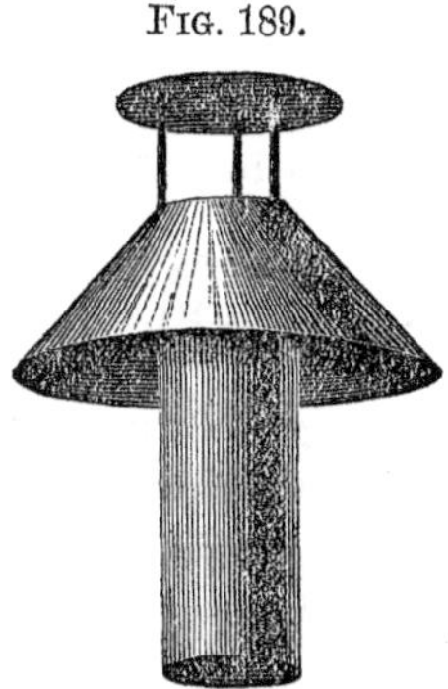

295. A Stream Meeting a Surface.—Though the moving fluid may be elastic, yet, when it meets a surface, it tends to *follow* it, rather than to rebound from it. This effect is partly due to adhesion, and partly to the resistance of the medium in which the stream moves. It will not only follow a plane or concave surface, but even one which is convex, provided the velocity of the current is not too great, or the curvature too rapid. A stream of air, blown from a pipe upon a plane surface, will extinguish the flame of a lamp held in the direction of the surface beyond its edge, while, if the lamp be held elsewhere near the stream, the flame will point toward the stream, according to Art. 293. Hence, snow is blown away from the windward side of a tight fence, and from around trees.

296. Diminution of Pressure on a Surface.—When a stream is thus moving along a surface, the fluid pressure on that surface is slightly diminished. This is proved by many experiments. If a curved vane be suspended on a pivot, and a stream of air be directed tangentially along the surface, it will move toward the stream, and may be made to revolve rapidly by repeating the blast at each half revolution. What is frequently called the *pneumatic paradox* is a phenomenon of the same kind. A stream of air is blown through the centre of a disk, against another light disk, which, instead of being blown off, is forcibly held near to it by the means. The pressure is diminished by all the radial streams along the surface contiguous to the other disk, and the full pressure on the outside preponderates. Another form of the experiment is to blow a stream of air through the bottom of a hemispherical cup, in which a light sphere is lying loosely. The sphere cannot be blown out, but, on the contrary, is held in, as may be seen by inverting the cup, while the blast continues. It appears to be for a reason of the same sort that a ball or a ring is sustained by a jet of water. It lies not on the *top*, but on the *side* of the jet, which diminishes the pressure on that side of the ball, so that the air on the outside keeps it in contact. The tangential force of the jet causes the body to revolve with rapidity. A ball can be sustained a few inches high by a stream of air.

297. Vortices where the Surface Ends.—As a current reaches the termination of the surface along which it was flowing, a *vortex* or whirl is likely to occur in the surrounding medium behind the edge of the surface. Vortices are formed on water, whose flow is obstructed by rocks; and often when the obstructing body is at a distance below the surface, the whirl which is established there is communicated to the top, so that the vortex is seen, while its cause is out of sight. There is a depression at the centre, caused by the centrifugal force; and if the rotation is rapid, a spiral tube is formed, in which the air descends to great depths. These are called *whirlpools*. In a similar manner whirls are produced in the air, when it pours off from a surface. The eddying leaves on the leeward side of a building in a windy day often indicate such a movement, though it may have no permanency, the vortex being repeatedly broken up and reproduced.

298. Vortices by Currents Meeting.—But vortices are also formed by counteracting currents in an open medium. When an aperture is made in the middle of the bottom of a vessel, as the water runs toward it, the filaments encounter each other, and usually, though not invariably, they establish a rotary motion, and form a whirlpool. Vortices are a frequent phenomenon of the atmosphere, sometimes only a few feet in diameter, in other instances some rods or even miles in width. The smaller ones, occurring over land, are called *whirlwinds;* over water, *water-spouts*. They probably originate in currents which do not exactly oppose each other, but act as a *couple* of forces, tending to produce rotation (Art. 54).

The burning of a forest sometimes occasions whirlwinds, which are borne away by the wind, and maintain their rotation for miles. As the pressure in the centre is diminished by the centrifugal force, substances heavier than air, as leaves and spray, are likely to be driven up in the axis, and floating substances, as cloud, will for the same reason descend. The rising spray and the descending cloud frequently mark the progress of a vortex in the air, as it moves over a lake or the ocean. Such a phenomenon is called a water-spout.

For a notice of cyclones, see Part VIII, on Heat.

PART IV.

SOUND.

CHAPTER I.

NATURE AND PROPAGATION OF SOUND.

299. Sound.—Vibrations.—The impression which the mind receives through the organ of hearing is called *sound.* But the same word is constantly used to signify that *progressive vibratory movement* in a medium by which the impression is produced, as when we speak of the velocity of sound.

This is one of the several modes of motion mentioned in Art. 4. The vibrations constituting sound are comparatively slow, and are often perceived by sight and by feeling as well as by hearing. For these reasons, the true nature of sound is investigated with far greater ease than that of light, electricity, &c. It is not difficult to discover that *vibrations* in the medium about us are essential to hearing; and these vibrations are always traceable to the body in which the sound originates. A body becomes a source of sound by producing an impulse or a series of impulses on the surrounding medium, and thus throwing the medium itself into motion. A single sudden impulse causes a *noise,* with very little continuance; an irregular and rapid succession of impulses, a *crash,* or *roar,* or *continued noise* of some kind; but if the impulses are rapid and perfectly equidistant, the effect is a *musical sound.* In most cases of the last kind the impulses are vibrations of the body itself; and whatever affects these vibrations is found to affect the sound emanating from it; and if they are destroyed, the sound ceases.

If we rub a moistened finger along the edge of a tumbler nearly full of water, or draw a bow across the strings of a viol, we can procure sounds which remain undiminished in intensity as long as the operation by which they are excited is continued. In both cases the vibrations are visible; those of the tumbler are plainly seen as crispations on the water to which they are communicated; the string appears as a broad shadowy surface. If a wire or light piece of metal rests against a bell or glass receiver, when ringing, it will be made to rattle. If sand be strewed on a horizontal plate while a bow is drawn across its edge, the sand will be agitated, and dance over the surface, till it finds certain places where vibrations do not exist. Near an organ-pipe the tremor of the air is perceptible, and pipes of the largest size jar the seats and walls of an edifice. Every species of sound may be traced to impulses or vibrations in the sounding body.

300. Sonorous Bodies.—Two qualities in a body are necessary, in order that it may be sonorous. It must have a form favorable for vibratory movements, and sufficient strength of elasticity.

The favorable *forms* are in general rods and plates, rather than very compact masses, like spheres and cubes; because the particles of the former are more free to receive lateral movements than those of the latter, which are constrained on every side. But even a thin lamina may have a form which allows too little freedom of motion, such as a spherical shell, in which the parts mutually support each other. If the shell be divided, the hemispheres are bell-shaped and very sonorous.

The *elasticity* of some materials is too imperfect for continued vibration; thus lead, in whatever form, has no sonorous quality. In other cases, where the elasticity is nearly *perfect*, yet it is a *feeble* force, and hence the vibrations are slow and inaudible. Thus india-rubber is quite elastic, but its force is feeble, and occasions but little sound.

301. Air as a Medium of Sound.—There must not only be a vibrating body, as a *source* of sound, but a medium for its *communication* to the organ of hearing. The ordinary medium is air. Let a bell mounted with a hammer and mainspring, so as to continue ringing for several minutes, be placed on a thick cushion under the receiver of an air-pump. The cushion, made of several thicknesses of woolen cloth, is necessary to prevent communication through the metallic parts of the instrument. As the process of exhaustion goes on, the sound of the bell grows fainter, and at length ceases entirely. From this experiment we learn that sound cannot be propagated through a vacant space, even though it be only an inch or two in extent; and also that air conveys sound more feebly as it is more rare. The latter is proved by the faintness of sounds on the tops of high mountains. Travelers among the Alps often observe that at great elevations a gun can be heard only a small distance. The fact that meteoric bodies are sometimes heard when passing over at the height of 40 or 50 miles does not conflict with the above statements; for the velocity of meteors is vastly greater than any other velocities which occur within the earth's atmosphere. On the other hand, when air has more than the natural density, it conveys sound with more intensity, and therefore to a greater distance. In a diving-bell sunk to a considerable depth a whisper is painfully loud.

302. Velocity of Sound in Air.—Sound occupies an appreciable time in passing through air. This is a fact of common observation. The flash of a distant gun is seen before the report is heard. Thunder usually follows lightning after an interval of many seconds; but if the electric discharge is quite near, the lightning and thunder are almost simultaneous. If a person is hammering at a distance, the perceptions of the blows received by the eye and the ear do not generally agree with each other: or if in any case they do agree, it will be observed that the first stroke seen is inaudible, and the last one heard is invisible; for it requires just the time between two strokes for the sound of each to reach us. Many careful experiments were made in the 18th century to determine the velocity of sound; but as the temperature was not recorded, they have but little value. During the present century, the velocity has been determined by several series of observations in different countries, and all reduced for temperature to the freezing-point. The agreement between them is very close, and the mean of all is 1090 feet per second at 32° F.

303. Velocity as Affected by the Condition of the Air and the Quality of the Sound.—

Temperature affects the velocity of sound; the latter is increased about one foot (0.96 ft.) for each degree of rise in the temperature. Therefore, in most New England climates, the velocity of sound varies more than 100 feet during the year on this account. Probably the celebrated experiments of Derham, in London, 1708, who made the velocity 1142 feet, were performed in the heat of summer.

Wind of course affects the velocity of sound by the addition or subtraction of its own velocity, estimated in the same direction, because it transfers the medium itself in which the sound is conveyed. This modification, however, is only slight, for sound moves ten times faster than wind in the most violent hurricane.

But other changes in the condition of the air produce little or no effect. Neither pressure, nor moisture, nor any change of weather, alters the *velocity* of sound, though they may affect its *intensity*, and therefore the distance at which it can be heard. Falling snow and rain *obstruct* sound, but do not *retard* it.

All *kinds* of sound—the firing of a gun—the blow of a hammer—the notes of a musical instrument, or of the voice, however high or low, loud or soft, are conveyed at the same rate. That sounds of different pitch are conveyed with the same velocity was conclusively proved by Biot, in Paris, who caused several airs to be played on a flute at one end of a pipe more than 3000 feet long, and heard the same at the other end distinctly, and without the slightest displacement in the order of notes, or intervals of silence between them.

304. The Calculated Velocity.—For several years there was a large unexplained difference between the calculated velocity of sound and the actual velocity as determined by experiment. While the latter is, as already stated, 1090 feet per second at the freezing-point, calculation gave 916 feet. The difference was at length explained by La Place, who ascertained that it arises from the heat developed in the air by the compression which it undergoes. The calculations previously made regarded the elasticity as varying with the density alone, according to Mariotte's law, assuming that the temperature remained unchanged. But it is a well-known fact that when air is compressed, a part of its latent heat becomes sensible, and raises its temperature. If the condensation is gradual, the heat is radiated or conducted off, especially if in contact with other bodies; but the heat developed in the propagation of sound has little opportunity to escape, and, though without continuance, it augments the elasticity of the air, so as to add 174 feet to the velocity of sound in it.

305. Diffusion of Sound.—Sound produced in the open air tends to spread equally in all directions, and will do so whenever the original impulses are alike on every side. But this is rarely the case. In firing a gun, the first impulse is given in one direction, and the sound will have more intensity, and be heard further in that direction than in others. It is ascertained by experiment, that a person speaking in the open air can be equally well heard at the distance of 100 feet directly before him, 75 feet on the right and left, and 30 feet behind him; and therefore an audience, in order to hear to the best advantage, should be arranged within limits having these proportions. But, as will be seen hereafter, this rule is not applicable to the interior of a building.

Sound is also heard in certain directions with more intensity, and therefore to a greater distance, if an obstacle prevents its diffusion in other directions. On one side of an extended wall sound is heard further than if it spread on both sides; still further, in an angle between two walls; and to the greatest distance of all, when confined on four sides, and limited to one direction, as in a long tube. The reason in these several cases is obvious; for a given force can produce a given amount of motion; and if the motion is prevented from spreading to particles in some directions, it will reach more distant ones in those directions in which it does spread. Speaking-tubes confine the movement to a slender column of air, and therefore convey sound to great distances, and are on this account very useful in transmitting messages and orders between remote parts of manufacturing edifices and public houses.

306. Nature of Acoustic Waves.—The vibrations of a medium in the transmission of sound are of the kind called *longitudinal;* that is, the particles vibrate longitudinally with regard to the movement of the sound; whereas, in water-waves, the particle-motion is partly transverse to the wave-motion (Art. 247). If, for example, sound is passing from A to B (Fig. 190), the par-

Fig. 190.

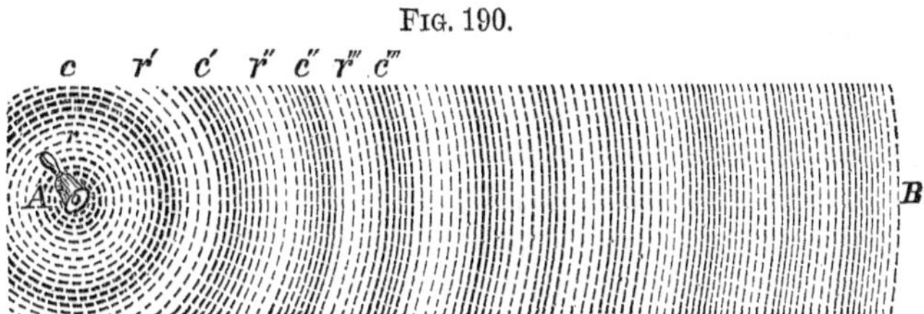

ticles just about A are (at the moment represented) in a state of condensation; around this condensed centre is a rarefied portion, then a condensed portion, &c., as marked by the letters r, c, r', c', r'', &c. From r to c the particles are advancing; so likewise from r' to c', and from r'' to c''. But from c to r', from c' to r'', &c., they are rebounding. The condensed wave near B has advanced from A, and others have followed it at equal intervals; and between these waves of condensation are waves of rarefaction, which in like manner spread outward from the centre A. And yet no one particle has any other motion than a small vibration back and forth in the line, near its original place of rest. The *amplitude* is the distance through which a particle vibrates. The intensity or loudness of sound depends on the amplitude.

In water-waves we distinguish carefully between the motion of the *wave* and the motion of the *water* which forms the wave; so here, the wave-motion is totally different from the motion of the air itself. The wave, i. e. the state of condensation and subsequent rarefaction, travels swiftly forward; but the masses of air, which suffer these condensations and rarefactions, simply tremble in the line of that motion.

Since the motion is propagated in all directions alike, the entire system of waves around the point where sound originates consists of spherical strata of air alternately condensed and rarefied. As the quantity set in motion in these successive layers increases with the square of the distance, the amount of motion communicated to each particle must diminish in the same ratio. Hence, the intensity of sound varies inversely as the square of the distance.

Fig. 190 is a section of a system of spherical waves around the source *A*.

A *ray* of sound is any one of the radii of the sphere whose centre is the source of sound. The vibratory motion is propagated along each of the rays.

307. Other Gaseous Bodies, as Media of Sound.—Let a spherical receiver, having a bell suspended in it, be exhausted of air, till the bell ceases to be heard; then fill it with any gas or vapor instead of air, and the bell will be heard again. By means of an organ-pipe blown by different gases, it can be learned with what velocity sound would move in each kind of gas experimented upon, because the pitch of a given pipe depends upon the velocity of the waves, as will be seen hereafter. In hydrogen sound is exceedingly feeble, but moves nearly three times as fast as in air. Momentary development of heat by compression produces, in all gaseous bodies, the effect of increasing the velocity of sound.

308. Liquids as Media.—Many experimenters have determined the circumstances of the propagation of sound in water. Franklin found that a person with his head under water could hear the sound of two stones struck together at a distance of more than half a mile. In 1826, Colladon made many careful experiments in the water of Lake Geneva. The results of these and other trials are principally the following:

1. Sounds produced in the air are very faintly heard by a person in water, though quite near; and sounds originating under water are feebly communicated to the air above, and in positions somewhat oblique are not heard at all.

2. Sounds are conveyed by water with a velocity of 4700 feet per second, at the temperature of 47° F., which is more than four times as great as in air. The calculated and the observed velocity of sound in water agree so nearly with each other, that there appears to be no appreciable effect arising from heat developed by compression.

3. Sounds conveyed in water to a distance, lose their sonorous quality. For example, the ringing of a bell gives a succession of short sharp strokes, like the striking together of two knife-blades. The musical quality of the sound is noticeable only within 600 or 700 feet. In air, it is well known that the contrary takes place; the blow of the bell-tongue is heard near by, but the continued musical note is all that affects the ear at a distance.

4. Acoustic *shadows* are formed; that is, sound passes the edges of solid bodies nearly in straight lines, and does not turn around them except in a very slight degree. In this respect, sound in water resembles light much more than it does sound in air.

To enable the experimenter to hear distant sounds without placing himself under water, Colladon pressed down a cylindrical tin tube, closed at the bottom, thus allowing the acoustic pulses in the water to strike perpendicularly on the sides of the tube. In this way, the faintest sounds were brought out into the air. It appears to be true of sound as of light, that it cannot pass from a denser to a rarer medium at large angles of incidence, but suffers nearly a total reflection.

309. Solids as Media.—Solid bodies of high elastic energy are the most perfect media of sound which are known. An iron rod—as, for instance, a lightning-rod—will convey a feeble sound from one extremity to the other, with much more distinctness than the air. If the ears are stopped, and one end of a long wire is held between the teeth, a slight scratch or blow on the remote end will sound very loud. The sound in this case travels through the wire and the bones of the head to the organ of hearing. The *stethoscope*, an instrument used by physicians for determining whether the lungs or heart have a diseased or healthy action, illustrates the conduction of sound by solids. The instrument is a tubular rod of wood, one end of which is pressed upon the chest of the patient, while the ear is applied to the other. The movements of the vital organs are thus distinctly heard, and the character of those movements readily distinguished. The sound of earthquakes and volcanic eruptions is transmitted to great distances through the solid earth. By laying the ear to the ground, the tramp of cavalry may be heard at a much greater distance than through the air.

310. Velocity in Solids.—Structure.—The velocity of sound in cast iron is about 11000 feet per second—ten times greater than in air. This was determined by Biot, in his experiments on some aqueduct pipes in Paris, already alluded to. A blow upon one end was brought to an observer at the other through two channels, and seemed to be two blows. One sound traveled in the air within the tube, the other in the iron itself of which the pipe was made. From the observed interval of time between the two sounds, and the known velocity of sound in air, the velocity in iron is readily calculated. The pitch of sound produced by rods and tubes of different materials, when vibrating longitudinally, enables us to determine with tolerable accuracy the velocity of propagation in those substances respectively.

In one important particular solids differ from fluids, namely, in the fixed relations of the particles among themselves. These relations are usually different in different directions; hence, sound is likely to be transmitted more perfectly in some directions through a given solid than in others. The scratch of a pin at one end of a stick of timber seems loud to a person whose ear is at the other end. The sound is heard more perfectly in the direction of the grain than across it. In crystallized substances it is unquestionably true that the vibrations of sound move with different speed and with different intensity in the line of the axis, and in a line perpendicular to it.

311. Mixed Media.—In all the foregoing statements it has been supposed that the medium was homogeneous; in other words, that the material, its density, and its structure, continue the same, or nearly the same, the whole distance from the source of sound to the ear. If abrupt changes occur, even a few times, the sound is exceedingly obstructed in its progress. When the receiver is set over the bell on the pump plate, the sound in the room is very much weakened, though the glass may not be one-eighth of an inch in thickness, and is an excellent conductor of sound. The vibrations of the internal air are very imperfectly communicated to the glass, and those received by the glass pass into the air again with a diminished intensity. If a glass rod extended the whole distance from the bell to the ear, the sound would arrive in less time, and with more loudness, than if air occupied the whole extent. For a like reason, walls, buildings, or other intervening bodies, though good conductors of sound themselves, obstruct the progress of sound in the air. This explains the fact mentioned in Art. 308, that sound in air is heard faintly in water, and *vice versa*. When the texture of a substance is loose, having many alternations of material, it thereby becomes unfit for transmitting sound. It is for this reason that the bell-stand, in the experiment just referred to, is set on a cushion made of several thicknesses of loose flannel, that it may prevent the vibrations from reaching the metallic parts of the pump. The waves of sound, in attempting to make their way through such a substance, continually meet with new surfaces, and are reflected in all possible directions, by which means they are broken up into a multitude of crossing and interfering waves, and are mutually destroyed. A tumbler, nearly filled with water, will ring clearly; but if filled with an effervescing liquid, it will lose all its sonorous quality, for the same cause as before. The alternate surfaces of the liquid and gas, in the foam, confuse the waves, and deaden the sound.

CHAPTER II.

REFLECTION, REFRACTION, AND INFLECTION OF SOUND.

312. Reflection of Sound.—Sound is reflected from surfaces in accordance with the common law of reflection in the case of elastic bodies; that is,

The angle of incidence equals the angle of reflection, and the two angles are on opposite sides of the perpendicular to the reflecting surface.

Suppose sound to emanate from *A* (Fig. 191), and meet the plane surface *B D*. The particles of air in the ray *A B* vibrate back and forth in that line, and those contiguous to *B* will, after striking the surface, rebound on the line *B G*, as an elastic ball would do (Art. 103), and propagate their motion along that line. The angle of incidence *A B F* equals the angle of reflection *F B G*, and the two angles are on opposite sides of *F B*, which is perpendicular to the reflecting surface *B D*. If *G B* be produced backward, it will meet the perpendicular *A E* at *C*, as far behind *B D* as *A* is before it. In like manner, every ray of sound after reflection proceeds as if from *C*, and the successive waves are situated as represented by the dotted lines in the figure. From the point *E* the reflection is directly back in the line *E A*.

Fig. 191.

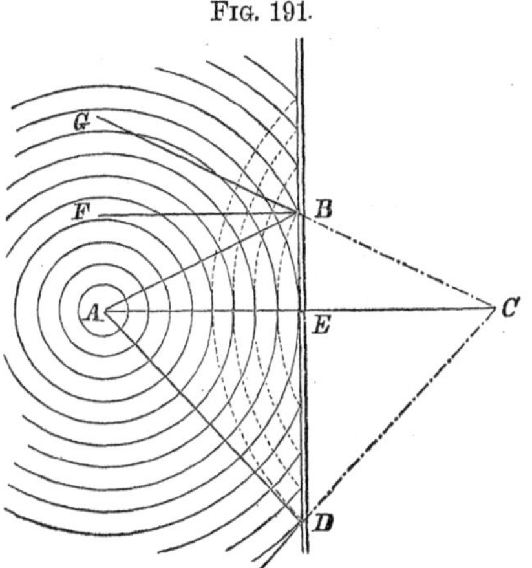

313. Echoes.—When sound is so distinctly reflected from a surface that it seems to come from another source, it is called an *echo*. Broad and even surfaces, such as the walls of buildings and ledges of rock, often produce this effect. According to the law (Art. 312), a person can hear the echo of his own voice only by standing in a line which is perpendicular to the echoing surface. In order that one person may hear the echo of another's voice, they must place themselves in lines making equal angles with the perpendicular.

The interval of time between a sound and its echo enables one to judge of the distance of the surface, since the sound must pass over it twice. Thus, if at the temperature of 74° the echo of the speaker's voice reaches him in two seconds after its utterance, the distance of the reflecting body is about 1130 feet, and in that proportion for other intervals. And he can hear a distinct echo of as many syllables as he can pronounce while sound travels twice the distance between himself and the echoing surface.

314. Simple and Complex Echoes.—When a sound is returned by one surface, the echo is called *simple*; it is called *complex* when the reflection is from two or more surfaces at different distances, each surface giving one echo. Thus, a cannon fired in a mountainous region is heard for a long time echoed on all sides, and from various distances.

A complex echo may also be produced by two parallel walls, if the hearer and the source of sound are both situated between them. The firing of a pistol between parallel walls a few hundred feet apart has been known to return from 30 to 40 echoes before they became too faint to be heard. The rolling of thunder is in part the effect of reverberation between the earth and the clouds. This is made certain by the observed fact that the report of a cannon, which in a level country and under a clear sky is sharp and single, becomes in a cloudy day a prolonged roar, mingled with distant and repeated echoes. But the peculiar inequalities in the reverberations of thunder are doubtless due in part to the irregularly crinkled path of the electric spark. A discharge of lightning occupies so short a time, that the sound may be considered as starting from all points of its track at once. But that track is full of large and small curves, some convex and some concave to the ear, and at a great variety of distance; and all points which are at equal distances would be heard at once. Hence, the original sound comes to the hearer with great irregularity, loud at one instant and faint at another. These inequalities are prolonged and intensified by the echoes which take place between the clouds and the earth.

315. Concentrated Echoes.—The divergence of sound from a plane surface continues the same as before, that is, in spherical waves, whose centre is at the same distance behind the plane as the real source is in front. But concave surfaces in general produce a concentrating effect. A sound originating in the centre of a hollow sphere will be reflected back to the centre from every point of the surface. If it emanates from one focus of an ellipsoid, it will, after reflection, all be collected at the other focus. So, if two concave paraboloids stand facing each other, with their axes coincident, and a whisper is made at the focus of one, it will be plainly heard at the focus of the other, though inaudible at all points between. In the last case the sound is twice reflected, and passes from one reflector to the other in parallel lines. All these effects are readily proved from the principle that the angles of incidence and reflection are equal.

The speaking-trumpet and the ear-trumpet have been supposed by many writers to owe their concentrating power to multiplied reflections from the inner surface. But a part of the effect, and sometimes the whole, is doubtless due to the accumulation of force in one direction, by preventing lateral diffusion, till the intensity is greatly increased.

Concave surfaces cause all the curious effects of what are called *whispering galleries*, such as the dome of St. Paul's, in London. In many of these instances, however, there seems to be a continued series of reflections from point to point along the smooth concave wall, which all meet simultaneously (if the curves are of equal length) at the opposite point of the dome; for the whisperer places his mouth, and the hearer his ear, close to the wall, and not in a focus of the curve. The *Ear of Dionysius* was probably a curved wall of this kind in the dungeons of Syracuse. It is said that the words, and even the whispers, of the prisoners were gathered and conveyed along a hidden tube to the apartment of the tyrant. The sail of a ship when spread, and made concave by the breeze, has been known to concentrate and render audible to the sailors the sound of a bell 100 miles distant. A concave shell held to the ear concentrates such sounds as may be floating in the air, and is suggestive of the murmur of the ocean.

316. Resonance of Rooms.—If a rectangular room has smooth, hard walls, and is unfurnished, its reverberations will be loud and long-continued. Stamp on the floor, or make any other sudden noise, and its echoes passing back and forth will form a prolonged musical note, whose pitch will be lower as the apartment is larger. This is called the *resonance* of the room. Now, let furniture be placed around the walls, and the reverberations will be weakened and less prolonged. Especially will this be the case if the articles be of the softer kinds, and have irregular surfaces. Carpets, curtains, stuffed seats, tapestry, and articles of dress have great influence in destroying the resonance of a room. The appearance of an apartment is not more changed than is its resonance by furnishing it with carpet and curtains. The blind, on entering a strange room, can, by the sound of the first step, judge with tolerable accuracy of its size and the general character of its furniture.

The reason why substances of loose texture do not reflect sound well, is essentially the same as what has been stated (Art. 311) for their not transmitting well; they are not homogeneous—the waves are reflected in all directions by successive surfaces, interfere with each other, and are destroyed.

317. Halls for Public Speaking.—In large rooms, such as churches and lecturing halls, all echoes which can accompany the voice of the speaker syllable by syllable, are useful for increasing the volume of sound; but all which reach the hearers sensibly later, only produce confusion. It is found by experiment that if a sound and its echo reach the ear within *one-sixteenth* of a second of each other, they seem to be one. Hence, this fraction of time is called the *limit of perceptibility*. Within that time an echo can travel about 70 feet more than the original sound, and yet appear to coincide with it. If an echoing wall, therefore, is within 35 feet of the speaker, each syllable and its echo will reach every hearer within the limit of perceptibility. The distance may, however, be increased to 40 or even 50 feet without injury, especially if the utterance is not rapid. Walls intended to aid by their echoes should be smooth, but not too solid; plaster on lath is better than plaster on brick or stone; the first echo is louder, and the reverberations less. Drapery behind the speaker deprives him of the aid of just so much echoing surface. A lecturing hall is improved by causing the wall behind the speaker to change its direction, on the right and left of the platform, at a very obtuse angle, so as to exclude the rectangular corners from the room. The voice is in this way more reinforced by reflection, and there is less resonance arising from the parallelism of opposite walls. Paneling, and any other recesses for ornamental purposes, may exist in the reflecting walls without injury, provided they are not curved. The ceiling should not be so high that the reflection from it would be delayed beyond the limit of perceptibility. Concave surfaces, such as domes, vaults, and broad niches, should be carefully avoided, as their effect generally is to concentrate all the sounds they reflect. An equal diffusion of sound throughout the apartment, not concentration of it to particular points, is the object to be sought in the arrangement of its parts.

As to distant parts of a hall for public speaking, the more completely all echoes from them can be destroyed, the more favorable is it for distinct hearing. It is indeed true that if a hearer is within 35 feet of a wall, however remote from the speaker, he will hear a syllable, and its echo from that wall, as one sound; but to all the audience at greater distances from the same wall, the echoes will be perceptibly retarded, and fall upon subsequent syllables, thus destroying distinctness. The distant walls should, by some means, be broken up into small portions, presenting surfaces in different directions. A gallery may aid in effecting this; and the seats of the gallery and of the lower floor may rise rapidly one behind another, so that the audience will receive directly much of the sound which would otherwise go to the remote wall, and be reflected. Especially should no large and distant surfaces be *parallel* to nearer ones, since it is between parallel walls that prolonged reverberation occurs.

318. Refraction of Sound.—It has been ascertained by experiment that sound, like light, may be *refracted*, or bent out of its rectilinear course by entering a substance of different density. If a large convex lens be formed of carbonic acid gas, by inclosing it in a sphere of thin india-rubber, a feeble sound, like the ticking of a watch, produced on one side, will be concentrated to a focal point on the other. In this case, the several diverging rays of sound are refracted toward each other on entering the sphere, and still more on leaving it, so that they are converged to a focus.

319. Inflection of Sound.—If air-waves are allowed to pass through an opening in an obstructing wall, they are not entirely confined within the radii of the wave-system produced through the opening, but spread with diminished intensity in lateral directions. The particles near the edges of the opening, as *B* and *C* (Fig. 192) may be considered as sources of sound; and if they be made centres of concentric spheres, whose radii are equal to the length of the wave, *B b*, or *C c*, and its multiples, then these spherical surfaces will represent the lateral systems of waves which are diffused on every side of the direct beam, *B D*, *C E*. But the sound is in general more feeble as the distance from *B D*, or *C E*, is greater, and in certain points is destroyed by interference. This spreading of sound in lateral directions is called the *inflection* of sound.

FIG. 192.

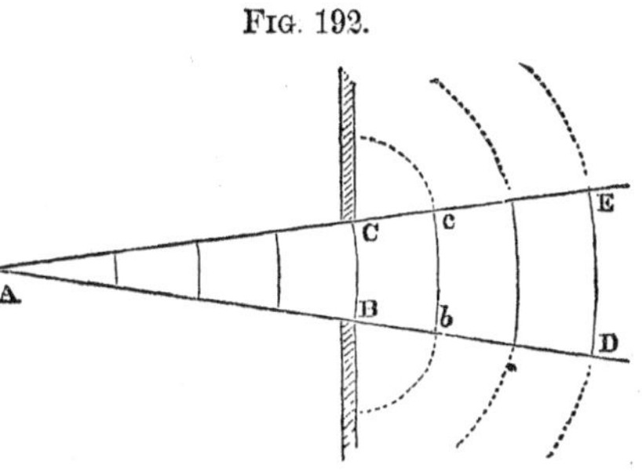

What is true of all sides of an opening is of course true whenever sound passes by the side of an obstacle. Instead of being limited by lines almost straight drawn from the source, as light is in the formation of a shadow, it bends round the edge, and is heard, though more feebly, behind the intervening body. It has been already noticed (Art. 308) that in water there is little or no inflection of sound.

CHAPTER III.

MUSICAL SOUNDS AND MODES OF PRODUCING THEM.

320. The Vibrations in Musical Sounds.—When the impulses of a sounding body upon the air are equidistant, and of sufficient frequency, they produce what is termed a musical sound. In most cases these impulses are the isochronous vibrations of the body itself, but not necessarily so; it is found by experiment that blows or pulses, of any species whatever, if they are more than about 15 or 20 per second, and possess the property of *isochronism*, cause a musical tone. For example, the snapping of a stick on the teeth of a metallic wheel would seem as unlikely as anything to produce a musical sound; but when the wheel is in rapid motion, the succession causes a pure musical note. Equidistant *echoes* often produce a musical sound, as when a person stamps on the floor of a rectangular room, finished, but unfurnished (Art. 316). So, on a walk by the side of a long baluster fence, a sudden sharp sound, like the blow of a hammer on a stone, brings back a tone more or less prolonged, resembling the chirp of a bird. It is occasioned by successive equidistant echoes from the balusters of the fence. A flight of steps will sometimes produce the same effect, the tone being on a lower key than that from the fence, as it should be.

321. The Pitch of Musical Sounds.—What is called the *pitch* of a musical sound, or its degree of acuteness, is owing entirely to its rate of vibration. Other qualities of sounds are due to other and often unknown circumstances; but rapidity of vibration is the only condition on which the pitch depends. In comparing one musical sound with another, if the number of vibrations per second is greater, the sound is more acute, and is said to be of a *higher pitch;* if the vibrations are fewer per second, the sound is graver, or of a *lower pitch.*

322. The Monochord.—If a string of uniform size and texture is stretched on a box of thin wood, by means of a pulley and weight, the instrument is called a *monochord,* and is useful for studying the laws of vibrations in musical sounds. The sound emitted by the vibrations of the whole length of the string is called its *fundamental* sound.

If the string be drawn aside from its straight position, and then released, one component of the force of tension urges every particle back towards its place of rest; but the string passes beyond that place, on account of the momentum acquired, and deviates as far on the other side; from which position it returns, for the same reason as before, and continues thus to vibrate till obstructions destroy its motion. By the use of a bow, the vibrations may be continued as long as the experimenter chooses.

The pitch of the fundamental sound of musical strings is found by experience to depend on three circumstances; the *length* of the string—its *weight* or quantity of matter—and its *tension.* The tone becomes more acute as we increase the tension, or diminish either the length or the weight. The operation of these several circumstances may be seen in a common violin. The pitch of any one of the strings is raised or lowered by turning the screw so as to increase or lessen its tension; or, the tension remaining the same, higher or lower notes are produced by the same string, by applying the fingers in such a manner as to shorten or lengthen the string which is vibrating; or, both the tension and the length of the string remaining the same, the pitch is altered by making the string larger or smaller, and thus increasing or diminishing its weight.

A string is said to make a *single vibration* in passing from the extreme limit on one side to the extreme limit on the other; a *double vibration* is the motion across and back again to the original position. Independently of calculation, it is easy to see that, with a given weight per inch, and a given tension, the string will vibrate *slower*, if *longer*, since there is more matter to be moved, and only the same force to move it; and for a similar reason, the length and tension being given, it will also vibrate *slower*, if *heavier*. On the other hand, if length and weight are given, it will vibrate *faster*, if the *tension* is *greater;* because a greater force will move a given quantity at a swifter rate.

323. Time of a Single Vibration.—The mathematical formula for the time of a vibration is the following, in which $T=$ the time of a single vibration; $l=$ the length of the string in inches; $w=$ the weight of one inch of the string; $t=$ the tension in lbs.; and $g=$ the force of gravity $= 386$ inches $= 32\frac{1}{6}$ feet (Art. 28);

$$T = l\left(\frac{w}{gt}\right)^{\frac{1}{2}}.$$

The constant factor, g, being omitted, the variation may be expressed thus:

$$T \propto \frac{l\sqrt{w}}{\sqrt{t}}; \text{ that is,}$$

The time of a vibration varies as the length of the string multiplied by the square root of its weight per inch, and divided by the square root of its tension.

As the *distance* of the string from its quiescent position does not form an element of the algebraic expression for the time of a vibration, it follows that the time is independent of the amplitude. Hence, as in the pendulum, the vibrations of a string, fixed at both ends, are performed in equal times, whether the amplitude of the vibrations be greater or smaller. It is on this account that the pitch of a string does not alter, when left to vibrate till it stops. The excursions from side to side grow less, and therefore the sound more feeble, till it ceases; but the *rate* of vibration, and therefore the pitch, remains the same to the last. This property of *isochronism*, independent of extent of excursion, is common to sounding bodies generally, and is owing to what may be called *the law of elasticity, that the restoring force, acting on any particle, varies directly as its distance from the place of rest.* For example, each particle of the string, if removed twice as far from its place of rest, is urged back by a force twice as great, and therefore returns in the same time.

324. The Number of Vibrations in a Given Time.—The greater is the length of one vibration, the less will be the number of vibrations in a given time; that is, if N represents the number, $N \propto \frac{1}{T}$; but as $T \propto \frac{l\sqrt{w}}{\sqrt{t}}$, $\therefore N \propto \frac{\sqrt{t}}{l\sqrt{w}}$. If t and w are constant, $N \propto \frac{1}{l}$; if l and t are constant, $N \propto \frac{1}{\sqrt{w}}$; and if l and w are constant, $N \propto \sqrt{t}$; that is,

1. *The number of vibrations varies inversely as the length.*
2. *The number of vibrations varies inversely as the square root of the weight of the string.*
3. *The number of vibrations varies as the square root of the tension.*

Thus, the number of vibrations in a second may be doubled, either by *halving* the length of the string or by making its weight *one-fourth as great*, or, finally, by making its tension *four times as great.*

325. Vibrations of a String in Parts.—The monochord may be made to vibrate in parts, the points of division remaining at rest; and this mode of vibration may even coexist with the one already described. Of course the sound produced by the parts will be on a higher pitch, since they are shorter, while the tension and the weight per inch remain unaltered. It is a noticeable fact that the parts are always such as will exactly measure the whole without a remainder. Hence the vibrating parts are either halves, thirds, fourths, or other aliquot portions. The sounds produced by any of these modes of vibration are called *harmonics*, for a reason which will appear hereafter. Suppose a string (Fig. 193) to

Fig. 193.

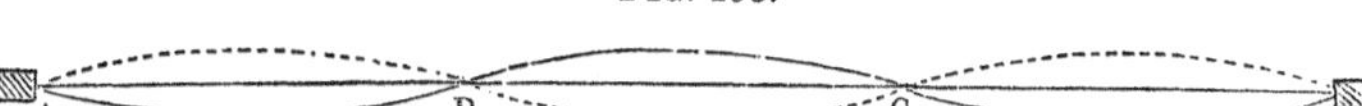

be stretched between A and B, and that it is thrown into vibration in three parts. Then while $A\ D$ makes its excursion on one side, $D\ C$ will move in the opposite direction, and $C\ B$ the same as $A\ D$; and when one is reversed, the others are also, as shown by the dotted line. In this way D and C are kept at rest, being urged toward one side by one portion of string, and toward the opposite by the next portion. But the string may at the same time vibrate as a whole; in which case D and C will have motion to each side of their former places of rest, while relatively to them the three portions will continue their movements as before. The points C and D are called *nodes;* the parts $A\ D$, $D\ C$, and $C\ B$, are called *ventral segments.* By a little change in the quickness of the stroke, the bow may be made to bring from the monochord a great number of harmonic notes, each being due to the vibrations of certain aliquot parts of the string. By confining a particular point, however, at the distance of $\frac{1}{2}$, $\frac{1}{3}$, or other simple fraction of the whole from the end, the particular harmonic belonging to that mode of division may be sounded clear, and unmingled with the others.

326. Vibrations of a Column of Air.—When a musical sound is produced by a pipe of any kind, it is the column of inclosed air which must be regarded as the sounding body. A condensed wave is caused, by some mode of excitation, to travel back and forth in the pipe, followed by a rarefied portion; and these waves affect the surrounding air much in the same way as do the alternate excursions of a string. That it is the air, and not the pipe itself, which is the source of sound, is proved by using pipes of various materials—the most elastic and the most inelastic—as glass, wood, paper, and lead; if they are of the same form and size, the tone in each case has the same pitch.

In order to examine the manner in which the air-columns in

pipes perform their vibrations, it is convenient to consider them in three classes:

1st. Pipes which are closed at both ends.

2d. Those which are closed at one end and open at the other.

3d. Those which are open at both ends.

327. Both Ends of the Pipe Closed.—Suppose the ends of the pipe, *A C B* (Fig. 194), to be closed, and an impulse in some way to be communicated at the centre, *C*; then the motion of the column will consist of a constant and regular fluctuation of the whole mass to and fro within the pipe, the air being always condensed in one half, while it is rarefied in the other. While the condensed pulse moves from *B* to *C*, the point of rarefaction runs from *A* to *C*, where they pass each other; hence, at the middle of the pipe there is no change of density, since every degree of condensation is at that point met by an equal degree of rarefaction of the other half of the general wave. At the extremities, *A* and *B*, there is alternately a *maximum* of condensation and of rarefaction, each being reflected and returning, to meet again at *C*. Fig. 195 shows the air in a state of condensation at *A*, and of rarefaction at *B*. At all points between the centre and the ends there is alternate condensation and rarefaction, but in a less degree according to the distance from the ends.

Fig. 194.

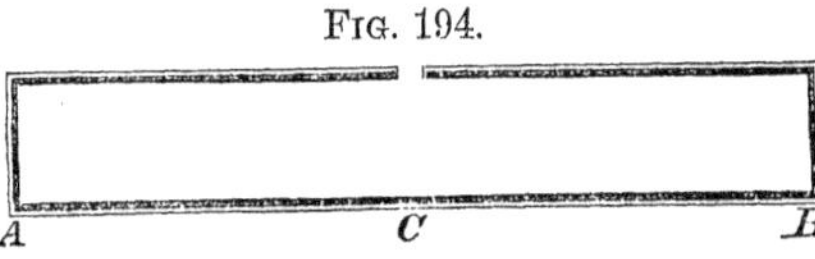

Fig. 195.

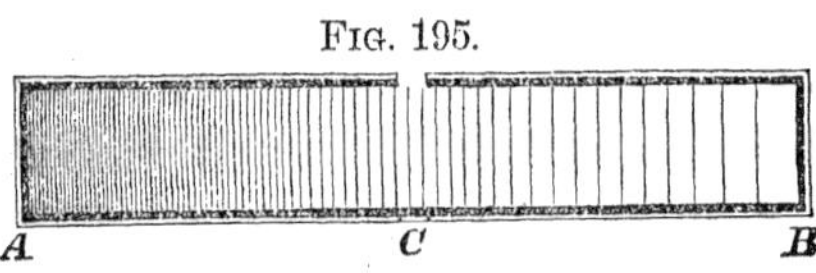

On the other hand, the *excursions* of the particles are greatest at *C*, and nothing at *A* and *B*, where all motion is prevented by the fixed stoppers by which the pipe is closed. Between the ends and the centre, the amplitude of vibration is greater, as the distance from the centre is less.

The pitch of such a pipe will be lower, as the pipe is longer, because the waves have a greater distance to travel between the successive reflections, and hence there will be a smaller number per second. So also, lowering the temperature lowers the pitch, since the wave then travels more slowly, and suffers fewer reflections in a second.

328. One End of the Pipe Closed, the other Open.—If, while the column *A B* is vibrating as a whole, an aperture is made at the centre, or even if the pipe is divided there, so that the aperture extends entirely round it, this will not interrupt the oscillation already described, because there is neither rarefaction nor condensation at the point *C*, and hence no tendency there to lateral motion. The means employed for exciting vibrations may therefore be applied at the open section. Let the pipe *A B* (Fig. 196), remaining stopped at *A* and *B*, be divided at the middle, *C*; and let the half pipe *B C* be removed, while the exciting cause remains at *C* (Fig. 194), then the vibrations in *A C* will still continue, and the pitch be unaltered. For now the condensed pulse, on reaching *C*, will be returned to *A* by the vibrating disk or spring which excites it, and will make a second reflection at *A* at the same instant as it would have done at *B* in the whole pipe *A B*; thus the same movements are performed now in *one* half which were before performed alternately in the *two*. Hence it is that a pipe with only one end closed, and a pipe of twice its length, with both ends closed, give the same pitch.

FIG. 196.

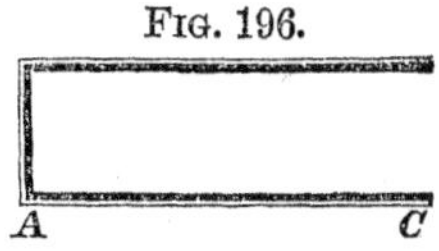

329. Both Ends of the Pipe Open.—When both ends of a pipe are open, it may still produce a musical tone, by having a node in the centre of it, thus forming two pipes like the one last described. When the vibration is established in such a pipe, the pulses from the ends move simultaneously toward *C* (Fig. 197), and again from it after reflection. Thus *C* is a fixed point, where the greatest condensation and rarefaction occur alternately, like *A* in Fig. 194. It therefore has the same pitch as *A C* alone, stopped at *C* and open at *A*. If a solid partition be inserted at *C*, it causes no change of pitch.

FIG. 197.

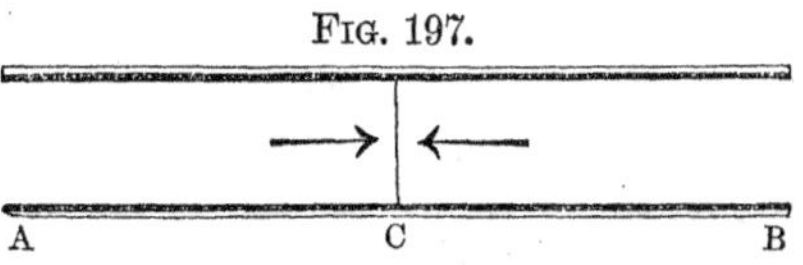

Such a pipe can produce no sound, except by the formation of at least one node.

330. The Second Kind of Pipe is the Elementary Form.—In comparing with each other the three kinds of pipe which have been described, it is observable that the *first* kind (stopped at both ends), and the *third* kind (open at both ends), is each a double pipe of the *second* kind (open at one end, and stopped at the other). For, if two pipes of the second kind be placed with their open ends together, as we have seen, they form one of the first kind, and there is no change of pitch. Again, if the two be placed with the closed ends in contact, they form a pipe of the third class; since the partition may remain or be removed, without affecting the mode of vibration. Hence, a pipe open at both ends, and one of the same length closed at both ends, each yields the same fundamental note as a pipe of half their length, open only at one end.

331. Vibrations of a Column of Air in Parts.—The same is true of a column of air as of a string, that it may vibrate in parts; and also, that two or more modes of vibration may coexist in the same column.

The *first* and *third* kinds of pipe can divide so that the whole and the vibrating segments have the ratios of $1 : \frac{1}{2} : \frac{1}{3}$: &c.; these ratios in the closed pipe are shown in Figs. 194, 198, and 199; and in the open pipe in Figs. 197 and 200. In Fig. 198 the pipe is divided into two equal parts, in each of which the vibrations take place in the same manner as in the whole, Fig. 194. Condensations run simultaneously from *A* and *B* to the middle point *C*, and thence back to *A* and *B*. When *C* is condensed, *A* and *B* are rarefied; and when *A* and *B* are condensed, *C* is rarefied. Those three points have no amplitude, but the greatest changes in density. But the points midway between have the greatest amplitude, and no change of density. As the waves run over the parts in half the time that they would over the whole, the pitch is raised accordingly. In this mode of vibrating, the opening where the vibrations are excited cannot be at *C*, where the node is formed.

Fig. 198.

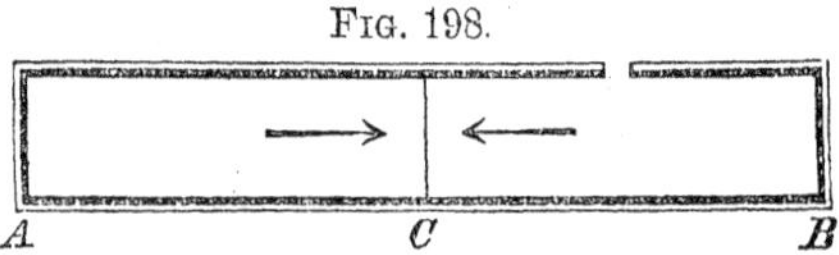

In Fig. 199 are shown *three* vibrating segments. *B* and *D* are condensed at one moment, *A* and *E* at another.

Fig. 199.

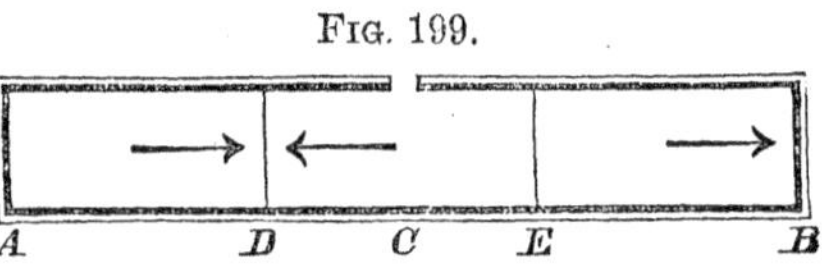

In the third kind, as already stated (Art. 329), there must be at least one node. When there are two, it is apparent by Fig. 200 that they must be one-fourth of the length from each end, in order that the three parts may vibrate in unison; for the middle part is a complete segment, like the pipe *A B* (Fig. 194), while the ends are half segments, like the pipe *A C* (Fig. 196). If there were three nodes, there would be two complete segments between them, and two half segments at the ends. It is evident that the lengths of the half segments, being $\frac{1}{2}$, $\frac{1}{4}$, $\frac{1}{6}$, &c., are as 1, $\frac{1}{2}$, $\frac{1}{3}$, &c., of the whole pipe; therefore the rates of vibra-

Fig. 200.

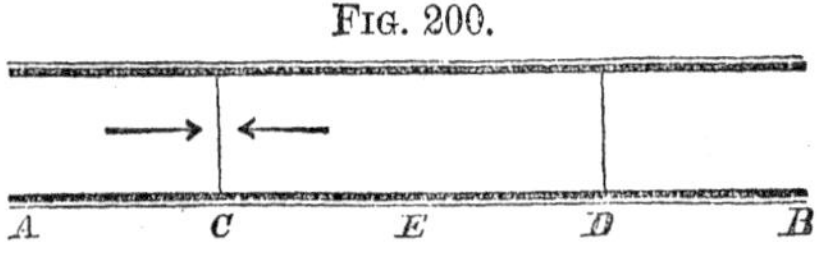

tion (being inversely as the lengths) are as the numbers 1, 2, 3, &c.

In the second kind of pipe the ratios of length for successive modes of vibration are $1 : \frac{1}{3} : \frac{1}{5}$, &c. The simplest division is by one node, a third of the length from the open end, as in Fig. 201. Then *C D*, a half segment, and *A D*, a complete segment, have the same rate of vibration. If there were two nodes, one must be a fifth from the open end, while the other divides the remainder into two complete segments. Therefore, in the several modes of vibration of the second kind of pipe, the half segments, being 1, $\frac{1}{3}$, $\frac{1}{5}$, &c., of the whole length, the rates of vibration in them are as the odd numbers 1, 3, 5, &c.

FIG. 201.

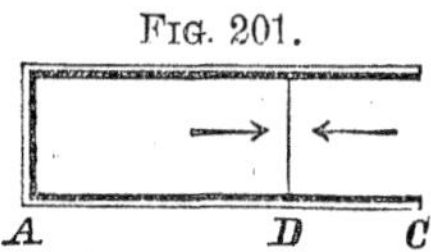

332. Modes of Exciting Vibrations in Pipes.—There are two methods of making the air-column in a pipe to vibrate: one by a stream of air blown across an orifice in the pipe, the other by an elastic plate called a reed. A familiar example of the first is the *flute*. A stream of air from the lips is directed across the *embouchure*, so as just to strike the opposite edge; this causes a wave to move through the tube. The stream of air, like a spring, vibrates so as to keep time with the movement of the wave to and fro, while at each pulse it renews that movement, and makes the sound continuous. For higher notes, the stream must be blown more swiftly, that by its greater elastic force, it may be able to conform to the more rapid vibration of the column. A large proportion of the pipes of an organ are made to produce musical tones essentially in the same way as the flute, and are called *mouth-pipes*.

Fig. 202 shows the construction of the mouth-pipe of an organ; *o b* is the mouth; and as the stream of air issues from the channel *i*, it starts a wave in the pipe, and then the stream itself vibrates laterally past the lip *b*, keeping time with the successive returns of the wave in the pipe. The pipe is attached to the wind-chest by the foot *P*.

FIG. 202.

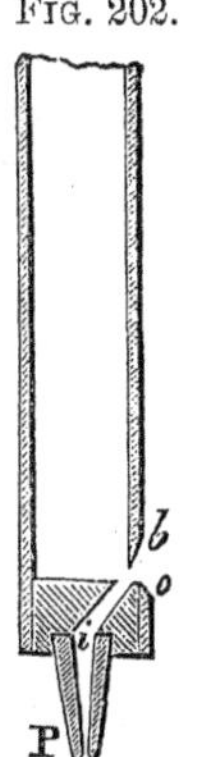

The *clarinet* is an example of vibrations in an air-column by a reed. In that instrument the reed is often made of wood; when the air is blown past its edge into the tube, the reed is thrown into vibration, and by it the column of air. The strength of elasticity in the reed should be such that its vibrations will keep time with the excursions of the wave in the column. What are called the reed pipes of the organ are constructed on the same principle, but the reeds are metallic. An example is seen in Fig. 203, which represents a model of the reed pipe, made to show the vibrations through the glass walls at *E*. A chimney, *H*, is usually attached, sometimes of a form (as in the figure) to increase the loudness of the sound, and sometimes of a different form, for softening it.

Fig. 203.

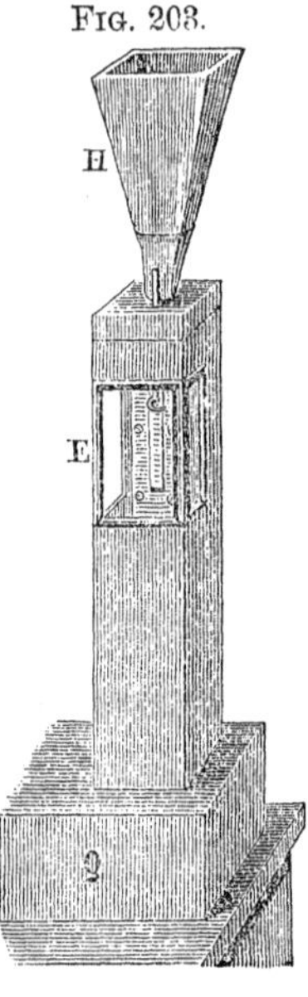

333. Vibrations of Rods and Laminæ.—A plate of metal called a reed is much used for musical purposes in connection with a column of air, as already stated. Except in such connection, the sounds of wires and laminæ are generally too feeble to be employed in music. But their vibrations have been much studied, on account of the interesting phenomena attending them.

334. Wires.—If one end of a steel wire is fastened in a vise and vibrated, while a thin blade of sunlight falls across it, the path of the illuminated point may be traced. It is not ordinarily a circular arc about the fixed point as a centre, but some irregular figure; and frequently the point describes two systems of ellipses, the vibrations passing alternately from one system to the other several times before running down. If the structure of the wire were the same in every part across its section, and if the fastening pressed equally on every point around it, the orbit of each particle would be a series of ellipses, whose major axes are on the same line. If, moreover, there was no obstruction to the motion, and the law of elasticity could obtain perfectly, it would vibrate in the same elliptic orbit forever, the force toward the centre being directly as the distance. It is easy to cause the wire, in the experiment just described, to vibrate also in parts; in which case each atom, while describing the elliptic orbit, will perform several smaller circuits, which appear as waves on the circumference of the larger figure.

335. Chladni's Plates.—If a square plate of glass or elastic metal, of uniform thickness and density, be fastened by its centre in a horizontal position, and a bow be drawn on its edge, it will emit a pure musical tone; and by varying the action of the bow, and touching different points of the edge with the finger, a variety of sounds may be obtained from it. The plate necessarily vibrates in parts; and the lowest pitch is produced when there are two nodal lines parallel to the sides, and crossing at the centre, thus dividing the plate into four square ventral segments. The position of the nodal lines, and the forms of the segments, are beautifully exhibited by sprinkling writing-sand on the plate. The particles will dance about rapidly till they find the lines of rest, where they will presently be collected. For every new tone the sand will show a new arrangement of nodal lines; and as two or more modes of vibration may coexist in plates, as well as in strings and columns of air, the resultant nodes will also be rendered visible. Again, by fastening the plate at a different point, still other arrangements will take place, each distinguishable by the position of its nodal lines and the pitch of its musical note. The form of the plate itself may also be varied, and each form will be characterized by its own peculiar systems. Chladni, who first performed these interesting experiments, delineated and published the forms of *ninety* different systems of vibration in the square plate alone.

If a fine light powder, as lycopodium (the pollen of a species of fern), be scattered on the plate, it is affected in a very different manner from heavy sand. It will gather into rounded heaps on those portions of the segments which have the greatest amplitude of vibration; the particles which compose the heaps performing a continual circulation, down the sides of the heaps, along the plate to the centre, and up the axis. If the vibration is violent, the heaps will be thrown up from the plate in little clouds over the portions of greatest motion. The cause of this singular effect was ascertained by Faraday, who found that in an exhausted receiver the phenomenon ceased. It is due to a circulation of the air, which lies in contact with a vibrating plate. The air next to those parts which have the greatest amplitude is at each vibration thrown upward more powerfully than elsewhere, and surrounding particles press into its place, and thus a circulation is established; and a fine light powder is more controlled by these atmospheric movements than by the direct action of the plate.

336. Bells.—If a thin plate of metal takes the form of a cylinder or bell, its fundamental note is produced when each ring of the material changes from a circle to an ellipse, and then into a second ellipse, whose axis is at right angles to that of the former, as seen in Fig. 204. It thus has four ventral segments and four nodal lines, the latter lying in the plane of the axis of the bell or cylinder. If the rings which compose the bell were all detached from one an-

Fig. 204.

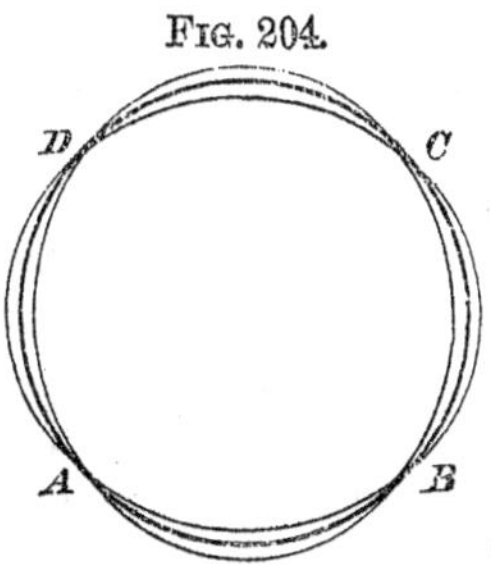

other, they would have different rates of vibration according to their diameter, and hence would produce tones of various pitch; but, being bound together by cohesion, they are compelled to keep the same time, and hence give but one fundamental tone. But a bell, especially if quite thin, may be made to emit a series of harmonic sounds by dividing up into a greater number of segments. It is obvious that the number of nodes must always be *even*, because two successive segments must move in opposite directions in one and the same instant; otherwise the point between them could not be kept at rest, and therefore would not be a node. Besides the principal tone of a church-bell, one or two subordinate sounds on a different pitch may usually be detected. A glass bell, suitably mounted for the lecture-room, will yield *ten* or *twelve* harmonics, by means of a bow drawn on its edge.

337. The Voice.—The vocal organ is complex, consisting of a cavity called the *larynx*, and a pair of membranous folds like valves, having a narrow opening between them; this opening, called the *glottis*, admits the air to the larynx from the wind-pipe below. The edges of these valves are thickened into a sort of cord, and for this reason the apparatus is called the *vocal cords.* In the act of breathing, the folds of the glottis lie relaxed and separate from each other, and the air passes freely between them, without producing vibration. But in the effort to form a vocal sound, they approach each other, and become tense, so that the current of air throws them into vibration. These vibrations are enforced by the consequent vibrations in the air of the larynx above; and thus a fullness of sound is produced, as in many musical instruments, in which a reed, and the air of a cavity, perform synchronous vibrations, and emit a much louder sound than either could do alone. If two pieces of thin india-rubber be stretched across the end of a tube, with their edges parallel, and separated by a narrow space, as represented in Fig. 205, the arrangement will give an idea of the larynx and glottis of the vocal organ. If air be forced through, a sound is produced, whose pitch depends on the size of the tube and the tension of the valves.

Fig. 205.

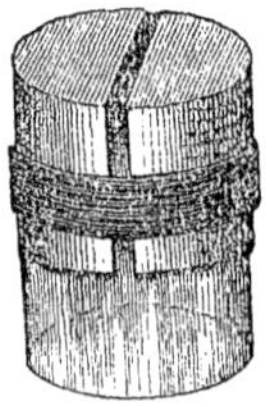

The natural key of a person's voice depends on the length and weight of the vocal cords, and the size of the larynx. The yielding nature of all the parts, and the ability, by muscular action, to change the form and size of the cavity and the tension of the valves, give great variety to the pitch, and the power of adjusting it with precision to every shade of sound within certain limits. No instrument of human contrivance can be brought into comparison with the organ of voice. After the voice is formed by its appropriate organ, it undergoes various modifications, by means of the palate, the tongue, the teeth, the lips, and the nose, before it is uttered in the form of articulate speech.

338. The Organ of Hearing.—The principal parts of the ear are the following:

1. The *outer ear*, *E a* (Fig. 206), terminating at the membrane of the tympanum, *m*.

FIG. 206.

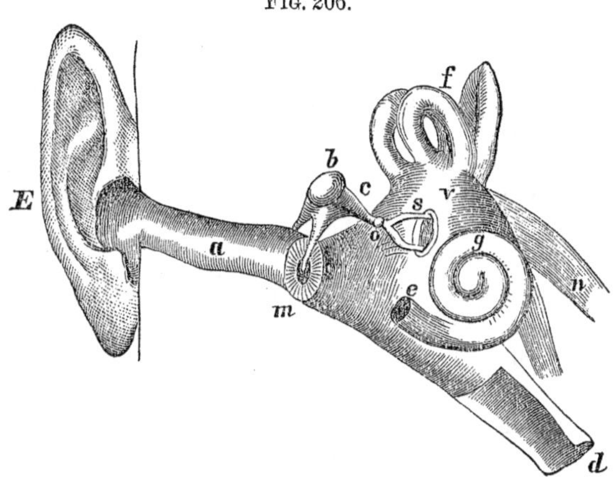

2. The *tympanum*, a cavity separated from the outer ear by a membrane, *m*, and containing a series of four very small bones (*ossicles*), *b*, *c*, *o*, and *s*, severally called, on account of their form, the *hammer*, the *anvil*, the *ball*, and the *stirrup*. The figure represents the walls of the tympanum as mostly removed, in order to show the internal parts. This cavity is connected with the back part of the mouth by the *Eustachian tube*, *d*.

3. The *labyrinth*, consisting of the *vestibule*, *v*, the *semicircular canals*, *f*, and the *cochlea*, *g*. The latter is a spiral tube, winding two and a half times round. The parts of the labyrinth are excavated in the hardest bone of the body. The figure shows only its exterior. There are two orifices through the bone which separates the labyrinth from the tympanum, the round orifice, *e*, passing into the cochlea, and the oval orifice, *s*, leading to the vestibule. These orifices are both closed by a thin membrane. The ossicles of the tympanum form a chain which connects the centre of the membrane, *m*, with that which closes the oval orifice. The labyrinth is filled with a liquid, in various parts of which float the fibres of the auditory nerve.

By the form of the outer ear, the waves are concentrated upon the membrane of the tympanum, thence conveyed through the chain of bones to the membrane of the labyrinth, and by that to the liquid within it, and thus to the auditory nerve, whose fibres lie in the liquid.

CHAPTER IV.

MUSICAL SCALES.—THE RELATIONS OF MUSICAL SOUNDS.

339. Numerical Relations of the Notes.—To obtain the series of notes which compose the common scale of music, it is convenient to use the monochord. Calling the sound, which is given by the whole length of the string, the *fundamental*, or *key note*, of the scale, we measure off the following fractions of the whole for the successive notes, namely: $\frac{8}{9}, \frac{4}{5}, \frac{3}{4}, \frac{2}{3}, \frac{3}{5}, \frac{8}{15}, \frac{1}{2}$. If the whole, and these fractions, are made to vibrate in order, the ear will recognize the sounds as forming the series called the *gamut*, or *diatonic scale*. And the interval between the fundamental and each of the others is named according to its distance inclusively. Thus, the interval from the whole ($= 1$) to $\frac{8}{9}$, is called the *second*; from 1 to $\frac{4}{5}$, the *third*, &c.; therefore, from 1 to $\frac{1}{2}$, the *eighth*, or *octave*. Now, as the number of vibrations varies inversely as the length of the string, the numbers corresponding to the notes respectively, are expressed by the same fractions inverted, 1, $\frac{9}{8}, \frac{5}{4}, \frac{4}{3}, \frac{3}{2}, \frac{5}{3}, \frac{15}{8}$, 2. Reducing these to a common denominator, and using the numerators (since they have the same ratios), we have the following series, 24, 27, 30, 32, 36, 40, 45, 48, to express in the simplest manner the relative numbers of vibrations in the notes of the scale, however produced. The sounds represented by these numbers are not arbitrarily chosen to form the scale, but they are demanded by the ear, and constitute the basis of the music of all ages and nations.

340. Interval of the Second.—In examining the relation of each two successive numbers in the foregoing series, we find three different ratios. Thus,

27 : 24, 36 : 32, and 45 : 40, is each as 9 : 8.
30 : 27 . . . . . and 40 : 36, ... 10 : 9.
32 : 30 . . . . . and 48 : 45, ... 16 : 15.

Therefore, of the seven intervals, called *the second*, in the diatonic scale, there are three equal to $\frac{9}{8}$, two equal $\frac{10}{9}$, and two others equal to $\frac{16}{15}$. Each of the first five is called a *major second*, or a *tone*; each of the last two is called a *minor second*, or a *semitone*.

One of the larger tones exceeds one of the smaller by $\frac{81}{80}$; for $\frac{9}{8} \div \frac{10}{9} = \frac{81}{80}$. This small interval, $\frac{81}{80}$, is called a *comma*, and is employed as a measuring unit in estimating the relations of intervals. The minor second, though called a semitone, is in fact more than half of either kind of major second.

341. Repetition of the Scale.—The eighth note of the scale so much resembles the first in sound, that it is regarded as a repetition of it, and called by the same name. Beginning, therefore, with the half string, where the former series closed, let us consider the sound of that as the fundamental, and take $\frac{8}{9}$ of it for the second, $\frac{4}{5}$ of it for the third, &c.; we then close a second series of notes on the quarter-string, whose sound is also considered a repetition of the former fundamental. Each fraction of the string used in the second scale is obviously half of the corresponding fraction of the whole string, and therefore its note an octave above the note of that. This process may be repeated indefinitely, giving the *second octave, third octave,* &c. Ten or eleven octaves comprehend all sounds appreciable by the human ear; the vibrations of the extreme notes of this entire range have the ratio of $1 : 2^{10}$, or $1 : 2^{11}$; that is, $1 : 1024$, or $1 : 2048$. Hence, if 16 vibrations per second produce the lowest appreciable note, the highest varies from 16,000 to 33,000. It was ascertained by Dr. Wollaston that the highest limit is different for different ears; so that when one person complains of the piercing shrillness of a sound, another maintains that there is no sound at all. The lowest limit is indefinite for a different reason; the sounds are heard by all, but some will recognize them as low musical tones, while others only perceive a rattling or fluttering noise. Few musical instruments comprehend more than six octaves, and the human voice has only from one to three, the male voice being in pitch an octave lower than the female.

342. Modes of Naming the Notes.—There is one system of names for the notes of the scale, which is fixed, and another which is movable. The first is by the seven letters, *A, B, C, D, E, F, G.* The notes of the second octave are expressed by the same letters, in some way distinguished from the former. The best method is to write by the side of the letter the numeral expressing that index of 2, which corresponds to the octave: as A_2, A_3, &c., in the octaves above; $A_{\bar{1}}$, $A_{\bar{2}}$, in those below.

The second mode of designation is by the syllables, *do, re, mi, fa, sol, la, si.* These express merely the *relations* of notes to each other, *do* always being the fundamental, *re* its second, *mi* its third, &c. In the natural scale, *do* is on the letter *C, re* on *D,* &c.; but by the aid of interpolated notes, the scale of syllables may be transferred, so as to begin successively with every letter of the fixed scale.

343. The Chromatic Scale.—Let the notes of the diatonic scale be represented (Fig. 207) by the horizontal lines, *C, D,* &c.; the distance from *C* to *D* being a tone, from *D* to *E* a tone, *E* to *F* a semitone, &c. It will be observed that the fundamental, *C,* is so situated that there are *two* whole tones above it, before a semitone occurs, and then *three* whole tones before the next semitone. *C* is therefore the letter to be called by the syllable *do,* in order to bring the first semitone between the 3d and 4th, and the other semitone between the 7th and 8th, as the figure represents them. Now, that we may be able to transfer the scale of relations to every part of the fixed scale (which is necessary, in order to vary the character of music, without throwing it beyond the reach of the voice), the whole tones are bisected, and two semitone intervals occupy the place of each. The dotted lines in the figure show the places of the interpolated notes, which, with the original notes of the diatonic scale, divide the whole into a series of semitones. This is called the *chromatic* scale. The interpolated note between *C* and *D* is written *C*♯ (*C* sharp), or *D*♭ (*D* flat), and so of the others. As the whole tones lie in groups of *twos* and *threes,* so the new notes inserted are grouped in the same way. This explains the arrangement of the *black keys* by twos and threes alternately in the key-board of the organ and piano-forte. The white keys compose the diatonic scale, the white and black keys together, the chromatic scale. It is obvious that on the chromatic scale any one of the twelve notes which compose it may become *do,* or the fundamental note, since the required series, 2 tones, 1 semitone, 3 tones, 1 semitone, can be arranged to succeed each other, at whatever note we begin the reckoning. This change, by which the fundamental note is made to fall on different letters, is called the *transposition* of the scale.

FIG. 207.

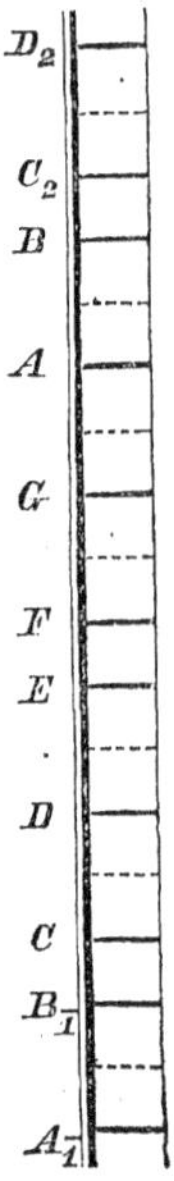

D_2
C_2
B
A
G
F
E
D
C
$B_{\bar{1}}$
$A_{\bar{1}}$

344. Chords and Discords.—When two or more sounds, meeting the ear at once, form a combination which is agreeable, it is called a *chord;* if disagreeable, a discord. The disagreeable quality of a discord, if attended to, will be perceived to consist in a certain roughness, or harshness, however smooth and pure the simple sounds which are combined. On examining the combinations, it will be found that if the vibrations of two sounds are in some very simple relations, as 1 : 2, 1 : 3, 2 : 3, 3 : 4, &c., they produce a chord; and the lower the terms of the ratio, the more perfect the chord. On the other hand, if the numbers necessary to express the relations of the sounds are large, as 8 : 9, or 15 : 16, a discord is produced. It appears that concordant sounds have *frequent* coincidences of vibrations. If, in two sounds, there is coincidence at every vibration of each, then the pitch is the same, and the combination is called *unison*. If every vibration of one coincides with every alternate vibration of the other, the ratio is 1 : 2, and the chord is the *octave*, the most perfect possible. The *fifth* is the next most perfect chord, where every second vibration of the lower meets every third of the higher, 2 : 3. The *fourth*, 3 : 4, the *major third*, 4 : 5, the *minor third*, 5 : 6, and the *sixth*, 3 : 5, are reckoned among chords; while the *second*, 8 : 9, and the *seventh*, 8 : 15, are harsh discords. What is called the *common chord* consists of the 1st, 3d, and 5th, combined, and is far more used in music than any other. *Harmony* consists of a succession of chords; or rather, of such a succession of combined sounds as is pleasing to the ear; for discords are employed in musical composition, their use being limited by special rules. Many combinations, which would be too disagreeable for the ear to dwell upon, or to finish a musical period, are yet quite necessary to produce the best effect; and without the relief which they give, perfect harmony, if long continued, would satiate.

345. Temperament.—This is a term applied to the small errors introduced into the notes, in tuning an instrument of fixed keys, in order to adapt the notes equally to the several scales. If the tones were all equal, and if semitones were truly half tones, no such adjustment of notes would be needed; they would all be exactly correct for every scale. Representing the notes in the scale whose fundamental is C by the numbers in Art. 339, we have,

$$\begin{array}{cccccccccc} C, & D, & E, & F, & G, & A, & B, & C_2, & D_2, & E_2, \&c. \\ 24, & 27, & 30, & 32, & 36, & 40, & 45, & 48, & 54, & 60, \&c. \end{array}$$

Now suppose we wish to make D, instead of C, our key-note; then it is obvious that E will not be exactly correct for the second on the new scale. For the fundamental to its second is as 8 to 9; and 8 : 9 : : 27 : 30.375, instead of 30. Therefore, if D is the key-note, we must have a new E, slightly above the E of the original scale. So we find that A, represented by 40, will not serve to be the 5th in the new scale; since 2 : 3 : : 27 : 40.5, which is a little higher than A (= 40). After adding these and other new notes, to render the intervals all exactly right for the new key of D, if we proceed in the same manner, and make E (= 30) our key-note, and obtain its second, third, &c., exactly, we shall find some of them differing a little, both from those of the key of C, and also of the key of *D*. Using in this way all the twelve notes of the chromatic scale in succession for the fundamental, it appears that several different *E's*, *F's*, *G's*, &c., are required, in order to make each scale perfect. In instruments, whose sounds cannot be modified by the performer, like the organ and piano-forte, as it is considered impossible to insert all the pipes or strings necessary to render every scale perfect, such an *adjustment* is made as to distribute these errors equally among all the scales. For example, *E* is not made a perfect *third* for the key of *C*, lest it should be too *imperfect* for a *second* in the key of *D*, and for its appropriate place in other scales. It is this equal distribution of errors among the several scales which is called *temperament*. The errors, when thus distributed, are too small to be observed by most persons; whereas, if an instrument was tuned perfectly for any one scale, all others would be intolerable.

The word temperament, as above explained, has no application except to instruments of fixed keys, as the organ and piano-forte; for, where the performer can control and modify the notes as he is playing, he can make every key perfect, and then there are no errors to be distributed. The flute-player can roll the flute slightly, and thus humor the sound, so as to cause the same fingering to give a precisely correct *second* for one scale, a correct *third* for another, and so on. The player on the violin does the same, by touching the string in points slightly different. The organs of the voice, especially, can be adjusted to make the intervals perfect on *every* scale. In these cases there is no *tempering*, or dividing of errors among different scales, but a *perfect adjustment* to each scale, by which all error is avoided.

346. Harmonics.—The fact has been mentioned that a string, or a column of air, may vibrate in parts, even while vibrating as a whole. It only remains to show the musical relations of the sounds thus produced. When a string vibrates in parts, it divides into halves, thirds, fourths, or other *aliquot* parts. Now, a half-string produces an *octave* above the whole, making the most perfect chord with it. The third of a string being two-thirds of the half-string, produces the *fifth* above the octave, a very perfect chord. The quarter-string gives the *second octave;* the fifth part of it, being $\frac{4}{5}$ of the quarter, gives the *major third* above the second octave; and the sixth part, being $\frac{2}{3}$ of the quarter, gives the *fifth* above the second octave. Thus, all the simpler divisions, which are the ones most likely to occur, are such as produce the best chords; and it is for this reason that the sounds are called *harmonics*. The same is true of air-columns and bells. The Æolian harp furnishes a beautiful example of the harmonics of a string. Two or more fine smooth cords are fastened upon a box, and tuned, at suitable intervals, like the strings of a violin; and the box is placed in a narrow opening, where a current of air passes. Each string at different times, according to the intensity of the breeze, will emit a pure musical note; and, with every change, will divide itself in a new mode, and give another pitch, while it will frequently happen that the vibrations of different divisions will coexist, and their harmonic sounds mingle with each other.

347. Overtones.—But the parts into which a sounding body divides do not always harmonize with the whole. For instance, $\frac{1}{9}$ or $\frac{1}{11}$ of a string is discordant with the fundamental. The word *harmonics* is not, therefore, applicable except to a very few of the many possible sounds which a body may produce. The word *overtone* is used to express in general any sound whatever, given by a part of a sounding body. A string may furnish 20 or 30 overtones, but only a small number of them would be harmonics.

348. Quality of Tone.—Even when the pitch of two sounding bodies is the same, the ear almost always distinguishes one sound from the other by certain qualities of tone peculiar to each. Thus, if the same letter be sounded by a flute and the string of a piano, each note is easily distinguished from the other. Two church-bells may be upon the same key, and yet one be agreeable, and the other harsh to the ear. While these great diversities may to some extent be due to circumstances not yet discovered, still it is certain that they in no small degree arise from the vibrations of various parts mingling with those of the fundamental sound. A long monochord can, by varying the mode of exciting the vibrations, be made to yield a great variety of sounds, while there is perceived in them all the same fundamental undertone which determines the pitch. If the string be struck at the *middle*, then no node can be formed at that point; hence, the mixed sound will contain no overtones of the $\frac{1}{2}$, $\frac{1}{4}$, $\frac{1}{6}$, $\frac{1}{8}$, $\frac{1}{10}$, or other *even* aliquot parts of the string; for all such would require a node at the middle. But if struck at *one-third* of its length from the end, then the overtones, $\frac{1}{2}$, $\frac{1}{4}$, &c., may exist, but not those of $\frac{1}{3}$, $\frac{1}{6}$, $\frac{1}{9}$, or any other parts whose node would fall at $\frac{1}{3}$ of the length from the end.

For reasons which are mostly unknown, some sounding bodies have their fundamental accompanied by harmonic overtones, and others by overtones which are discordant. And this is one cause of the agreeable or unpleasant quality of the sounds of different bodies.

349. Communication of Vibrations.—The acoustic vibrations of one body are readily communicated to others, which are near or in contact. We have already noticed that the vibrations of a reed will excite those of a column of air in a pipe. If two strings, which are adapted to vibrate alike, are fastened on the same box, and one of them is made to sound, the other will sound also more or less loudly, according to the intimacy of their connection. The vibrations are communicated partly through the air, and partly through the materials of the box. So, if a loud sound is uttered near a piano-forte, several strings will be thrown into vibration, whose notes are heard after the voice ceases. The noticeable fact in all such experiments is, that the vibrations thus communicated from one body to another cause sounds which *harmonize* with each other, and with the original sound. For the rate of vibration will either be identical, or have those simple relations which are expressed by the smallest numbers. Let a person hold a pneumatic receiver or a large tumbler before him, and utter at the mouth of it several sounds of different pitch; and he will probably find some one pitch which will be distinctly reinforced by the vessel. That particular note, which the receiver by its size and form is adapted to produce, will not be called forth by a sound that would be discordant with it. The melodeon, seraphine, and instruments of like character, owe their full and brilliant notes to reeds, each of which has its cavity of air adapted to vibrate in unison with it. It sometimes happens that the second body, vibrating as a whole, would not harmonize with the first, and yet will give the same note by some mode of division. Thus it is that all the various sounds of the monochord, and of the strings of the viol, are reinforced by the case of thin wood upon which they are stretched. The plates of wood divide by nodal lines into some new arrangement of ventral segments for every new sound emitted by the string. In like manner, the pitch of the tuning-fork, and all the rapid notes of a music-box, are rendered loud and full by the table, in contact with which they are brought. The extended material of the table is capable of division into a great variety of forms, and will always give a sound in unison with the instrument which touches it.

350. One System of Vibrations Controlling Another.—If two sounding bodies are nearly, but not precisely on the same key, they will sometimes, when brought into close contact, be made to harmonize perfectly. The vibrations of the more powerful will be communicated to the other, and control its movements so that the discordance, which they produce when a few inches apart, will cease, and concord will ensue. Two diapason pipes of an organ, tuned a quarter-tone or even a semitone from unison, so as to jar disagreeably upon the ear, when one inch or more asunder, will be in perfect unison, if they are in contact through their whole length. Even the slow oscillations of two watches will influence each other; if one gains on the other only a few beats in an hour, then, if they are placed side by side on the same board, they will beat precisely together.

351. Crispations of Fluids.—Among the numerous acoustic experiments illustrating the communication of vibrations, none are more beautiful than those in which the vibrations of glass rods are conveyed to the surface of a fluid. Let a very shallow pan of glass or metal be attached to the middle of a thin bar of wood, three or four feet long, and resting near its ends on two fixed bridges; let water be placed in the pan, and a long glass rod standing in it, or on the wood, be vibrated longitudinally, by drawing the moistened fingers down upon it; the liquid immediately shows that the vibrations are communicated to it. The surface is covered with a regular arrangement of heaps, called *crispations,* which vary in size with the pitch of sound, which is produced by the same vibration. If the pitch is higher, they are smaller, and may be readily varied from three or four inches in diameter to the fineness of the teeth of a file. Crispations of the same character are also formed in clusters on the water in a large tumbler or glass receiver, when the finger is drawn along its edge; every ventral segment of the glass produces a group of hillocks by the side of it on the surface of the water.

352. Interference of Waves of Sound.—Whenever two sounds are moving through the air, every particle will, at a given instant, have a motion which is the resultant of the two motions which it would have had if the sounds were separate. These motions may conspire, or they may oppose each other. The word *interference* is used in scientific language to express the *resultant effect,* whatever it may be. The *beats,* which are frequently heard in listening to two sounds, indicate the points of maximum condensation produced by the union of the condensed parts of both systems of waves. And the sounds are considered discordant when these beats are just so frequent as to produce a disagreeable fluttering or rattling. If too near or too far apart for this, they are regarded practically as concordant. And when the beats are too close to be perceived separately, yet the peculiar adjustment of condensations of one system with those of the other, according as *one* wave measures *two,* or *two* waves measure *three,* or *four* measure *five,* &c., is at once distinguished by the ear, and recognized as the chord of the *octave,* the *fifth,* the *third,* &c. When a sound and its octave are advancing together, there are instants in which any given particle of air is impressed with two opposite motions, and other alternate moments when both motions are in the same direction. For the waves of the highest sound are half as long as those of the lowest; hence, while every *second* condensation of the former coincides with *every* condensation of the latter, the alternate ones of the former must be at the points of greatest rarefaction of the latter; and this cannot occur without opposite movements of the particles. If two simultaneous sounds have the same pitch, i. e., the same length of wave, they ordinarily *run together*, so that like phases in the two systems are coincident, and the compound sound (called *unison*) simply has twice the loudness of one of them alone. But, by a delicate mode of experiment, one of these two systems of waves, having equal lengths, and equal intensities, may be made to fall half a wave behind the other, in which case opposite phases coincide, and the two sounds destroy each other. Thus, two sounds produce *silence*, on the same principle that two systems of water-waves may produce *level water*.

353. Number and Length of Waves for Each Note.— Though the vibrations of any musical note are too rapid to be counted, yet the number may be ascertained in several ways. One of the readiest methods is by means of a little instrument called the *siren*, invented by De La Tour. The pulses are produced by streams of air driven through holes in a revolving wheel. The revolutions of the wheel are recorded by machinery, and the number of vibrations in each revolution is known from the number of holes through which the air rushes. When such a velocity of revolution is given as to produce the required pitch, then the revolutions of the index per minute may be counted, and the number of *vibrations* in the same time will be known, and therefore the number per second. In this and other ways it is ascertained that the numbers corresponding to the letters of the scale are the following:

$$\begin{cases} C, & D, & E, & F, & G, & A, & B, & C_2, & D_2, \\ 128, & 144, & 160, & 170\frac{2}{3}, & 192, & 213\frac{1}{3}, & 240, & 256, & 288. \end{cases}$$

The highest note of the above series, D_2, 288, is the lowest on the common or *D*-flute. There is not, however, a perfect agreement of pitch in different countries, and among different classes of musicians. Accordingly, *C*, which is given above as corresponding to 128 vibrations per second, has several values, varying from 127 to 131.

To find the *length* of acoustic waves for any given pitch, we have only to divide the velocity of sound in one second by the number of vibrations which reach the ear in the same length of time. For example, at the temperature of 60°, sound travels 1118 feet per second; therefore the length of waves of low *D* on the flute = 1118 ÷ 288 = nearly *four* feet. The waves of the lowest musical note are about 70 feet long; and of the highest, less than half an inch.

354. Acoustic Vibrations Visibly Projected.—The vibrations of heavy tuning-forks can be magnified and rendered distinctly visible to an audience by projecting them on a screen. The fork being constructed with a small metallic mirror attached near the end of one prong, a sunbeam reflected from the mirror will exhibit all the movements of the fork greatly enlarged on a distant wall; and if the fork is turned on its axis, the luminous projection will take the form of a waving line. And by the use of two forks, all the phenomena of interference may be rendered as distinct to the eye as they are to the ear.

PART V.

MAGNETISM.

CHAPTER I.

THE MAGNET AND ITS PROPERTIES.

355. The Magnet.—Fragments of iron ore are sometimes found which strongly attract iron; and bars of steel are artificially prepared which exhibit the same property. These bodies are called *magnets;* the ore is the *natural magnet,* commonly called *lodestone;* the prepared steel bar is an *artificial magnet.*

Another property distinguishes the magnet, namely, that when properly suspended on a pivot, it assumes a certain definite direction with regard to the earth. This property of the magnet is called its *polarity.*

356. The Attraction Between a Magnet and Iron.—The magnetic property which is likely to be first noticed is the attraction of iron. If a lodestone or a bar magnet be rolled in iron filings (Fig. 208), there are two opposite points to which the

Fig. 208.

filings attach themselves in thick clusters, arranged in diverging filaments. These opposite points of greatest action are called *poles.* The attraction diminishes from the poles towards the central parts; and about in the middle between them there is little or none; this is called the *neutral point.* The straight line from one pole to the other is called the *axis.*

The mutual attraction between a magnet and iron is shown by bringing a piece of iron toward either pole of the magnetic needle; the needle instantly turns so as to bring its pole as near as possible to the iron (Fig. 209). On the other hand, an iron needle being suspended in like manner, the same movement takes place, when either pole of a magnet is brought near to it.

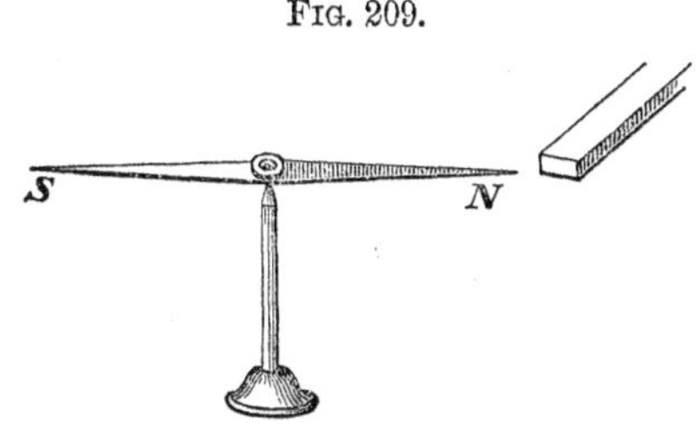

Fig. 209.

Nickel, cobalt, and sometimes manganese, exhibit the same magnetic properties in some degree. But these exist in comparatively small quantities, and therefore by magnetic bodies are usually intended only iron and steel.

357. Polarity.—If a light magnet is delicately suspended on a pivot at the neutral point, as in Fig. 210, it is called a *magnetic needle.* When thus placed and left to itself, it oscillates for a time, and finally settles with its axis in a certain fixed direction, which in most places is nearly north and south. The end which points in a northerly direction is called the *north pole;* the other, the *south pole.* These poles are usually marked on the larger magnets by the letters N and S, so that they may be instantly distinguished. If a magnetic needle has simply a mark or stain on one end, that end is understood to be the north pole.

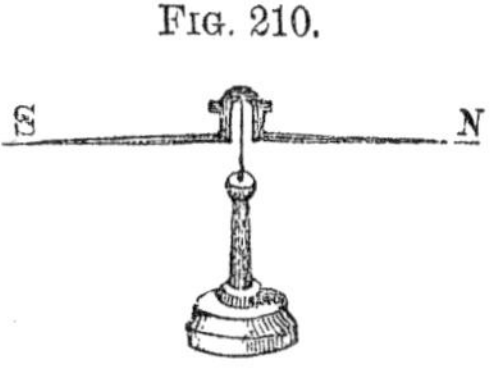

Fig. 210.

358. Action of Magnets on Each Other.—While either pole of a magnet attracts and is attracted by a piece of iron, it is otherwise when the pole of one magnet is brought near the pole of another. There is attraction in some cases, and repulsion in others. If the magnets are properly marked, and one of them suspended so as to move freely, it is readily discovered that the law of action is the following:

Poles of the same name repel, and those of contrary name attract each other.

Thus, the pole S of the magnet (Fig. 211) repels s of the needle, and attracts n; and if the magnet were inverted, and the pole N brought near to n, the latter would be repelled, and s be attracted.

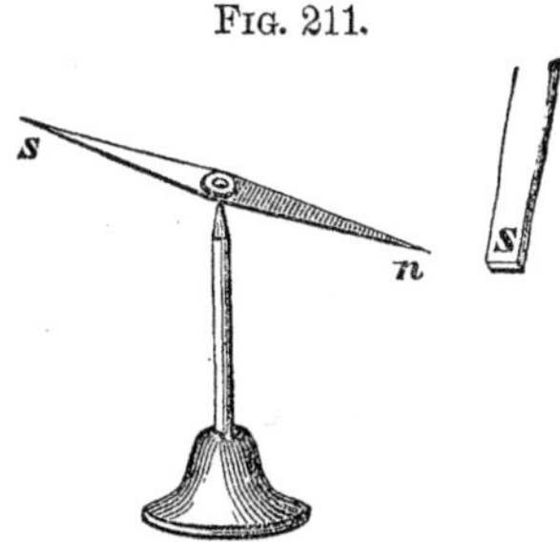

Fig. 211.

359. Magnetic Induction.—When a bar of iron is brought near to the pole of a magnet, though attraction is the phenomenon first observed, as stated (Art. 356), yet it is readily proved that this attraction results from a change which is previously produced in the iron. It becomes a *magnet* through the influence of the magnet which is near it. That end of the iron bar which is placed near one pole of a magnet becomes a pole of the opposite name, and the remote end a pole of the same name. Hence, according to the law (Art. 358), the poles which are contiguous attract each other, because they are unlike. The influence by which the iron becomes a magnet is called *induction*. A magnet, when brought near to a piece of iron, *induces* upon the iron the magnetic condition, without any loss of its own magnetic properties. This influence is more powerful according as the two are nearer to each other; it is, therefore, greatest when the two bars are in contact.

That the iron is truly a magnet for the time being is proved by bringing a needle near to its remote end; one pole is attracted, and the other repelled. If the iron had not been changed into a magnet, each pole of the needle would be attracted by it (Art. 356).

360. Successive Inductions.—Let a bar of iron, *B* (Fig. 212), be suspended from the south pole of the magnet *A*; then the upper end of *B* is a north pole, and the lower end a south pole. Now, as *B* is a magnet, it will induce the magnetic state on another bar, *C*, when brought in contact; and, as before, the poles of opposite name will be contiguous. Therefore, the upper end of *C* is north, and the lower end south. *D* is also a magnet by the inductive power of *C*. Thus, there is an indefinite series of inductions, growing weaker, however, from one to another, as the number is greater, and as the bars are longer.

Fig. 212.

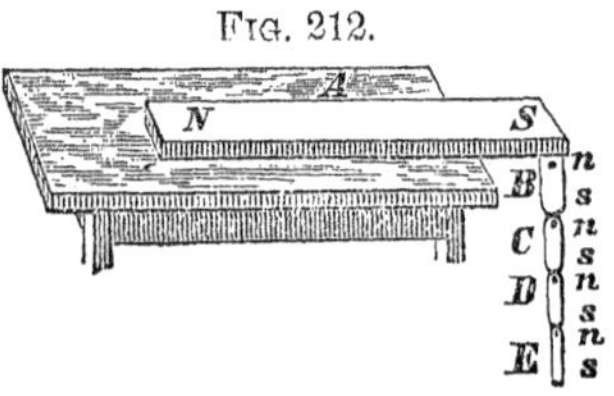

The filaments of iron filings which attach themselves to the pole of a magnet (Art. 356) are so many series of small magnets formed in the same manner as just described. Every particle of iron is a complete magnet, having its poles so arranged that the opposite poles of two successive particles are always contiguous.

361. Reflex Influence.—When a magnet exerts the inductive power upon a piece of iron which is near it, *its own magnetic intensity is increased*. The end of the piece of iron contiguous to the pole of the magnet is no sooner endued with the opposite polarity than it reacts upon the magnet, and increases its intensity; so that, if fragments of iron are attached to a magnet, as many as it will sustain, then after a time another may be added, and again another, till there is a very sensible increase of its original power.

Hence, too, the force of attraction of the dissimilar poles of two magnets is greater than the force of repulsion of the similar poles; because, when the poles are unlike, each acts inductively on the other to develop its poles more fully; but when they are alike, the influence which they reciprocally exert tends to make them unlike, and of course to diminish their repulsive force.

An extreme case of this diminution of repulsive force occurs when the like poles of two very unequal magnets are brought into contact. The small magnet immediately clings to the large one, as though the poles were unlike; and if examined, it is found that they are unlike. The powerful magnet has in an instant reversed the poles of the weak one by its strong inductive power, the latter not having force enough to diminish sensibly the strength of the other.

362. Double Induction.—The effects of *two* inductions at once on a bar of iron are various.

1. The bar may become a single magnet of double strength.
2. It may consist of two distinct magnets.
3. It may have no magnetic power at all.

The first case is illustrated by bringing the north pole of a magnet to one end of the iron, and the south pole of another magnet to the other end. Each magnet will form two poles by induction, and it is evident that the two pairs of poles will coincide. Even one magnet produces the same effect when laid by the side of a bar of iron of the same length.

To show the second effect, apply one pole of a magnet to the middle of the iron bar; then an opposite pole is formed at the middle, and a like pole at each end, each half of the bar being a separate magnet. The same effect is produced by bringing the like poles of two magnets in contact with the ends of the bar; for both ends will be of the opposite kind, and the middle of the same kind, as the poles applied. If a pole is applied to the middle of a star of iron, the extremity of each ray is a pole of the same kind; if to the middle of a circle of iron, the same polarity is found at every point of the circumference.

As an example of the third case, suspend a bar of iron from the pole of a magnet, and then bring the opposite pole of an equal magnet to the point of contact; the two poles induced by one are contrary to the two induced by the other, and they are found to be completely neutralized.

This last case shows that two opposite and equal magnetic poles formed at the same point destroy each other.

363. Coercive Force.—If in the several experiments on iron bars, which have been already described, pure annealed iron is used, the effects take place instantly; and when the magnet is removed, they as suddenly disappear. But if the iron is hard, magnetic poles are developed in it slowly; and when they have been developed, the iron returns also slowly to its neutral condition.

That property of hard unannealed iron which obstructs the development of magnetism in it, and which hinders its return to a neutral state, is called the *coercive force.* In iron which is pure and well annealed there seems to be no coercive force. It appears in a slight degree in iron not carefully prepared, and increases with its hardness; it is great in tempered steel, which is a compound of iron and carbon, and greatest of all in steel, which is tempered to the utmost hardness.

It is, therefore, difficult to make a strong magnet of a steel bar by ordinary induction, unless it is quite thin; but after the development has once been made, the bar becomes a permanent magnet, and may by care be used as such for years.

364. Change in the Coercive Force.—The coercive force is weakened by any cause which excites a tremulous or vibratory motion among the particles of the steel. This happens when the bar is struck by a hammer, so as to produce a ringing sound, which indicates that the particles are thrown into a vibratory motion. The passage of an electric discharge through a steel bar under the influence of a magnet, overcomes the coercive force for the time being, and permanent magnetism is developed. Heat produces the same effect; and hence a steel bar is conveniently magnetized by heating it to redness, placing it under a powerful inductive influence, and then hardening it by sudden cooling. The coercive force is thus neutralized by heat, till the development takes place, when it is restored, and the bar is a permanent magnet.

A magnet, however, loses its power by the same means as, during the process of induction, were used to develop it. Accordingly, any mechanical concussion or rough usage impairs or destroys the power of a magnet. By falling on a hard floor, or by being struck with a hammer, it is injured. Heat produces a similar effect. A boiling heat weakens, and a red heat totally destroys the magnetism of a needle.

365. Magnetism not Transferred, but only Developed.—This is strikingly proved by the fact that if a magnet be divided, even at the neutral point, where there is no sign of magnetism, the parts instantly become complete magnets, two unlike poles manifesting themselves at the place of fracture. Both polarities seem to exist at every point, and are developed wherever the bar is divided. If each part is divided again, the same phenomenon is repeated, and so on indefinitely. There is, therefore, no *transfer* of magnetism from one point to another, any more than from one bar to another, but only an excitation of what existed in every part of the body before. Both the north and the south pole must be conceived as latent at every point of a piece of iron or steel; and when the piece is magnetized, either north or south polarity is developed more or less fully in all parts except the neutral point.

It is not necessary that the particles should be united by cohesion in a solid bar. A magnet can be formed by filling a brass tube with iron filings and sand, or by forming a rod of cement mixed with filings, and then subjecting them to inductive influence. Fig. 213 will give an idea of the probable structure of every magnet. Each particle of it is a complete magnet, the like poles of all are turned the same way, and unlike poles are therefore contiguous to each other, and each acts inductively on the next.

FIG. 213.

S N

366. Magnetic Intensity and Distance.—The law of the magnetic force is the following:

The intensity of the magnetic force, whether attraction or repulsion, varies inversely as the square of the distance.

The law in the case of the repulsion of like poles is readily proved by Coulomb's *torsion balance*, which is figured and described in Art. 402, under Electricity. The angle of torsion is used as a measure of the repulsion, and it is found that the wire must be twisted through four times as large an angle to bring the poles to one-half the distance, and nine times as large an angle to bring them to one-third the distance, &c., the force increasing as the square of the distance diminishes.

To prove the law for the attraction of opposite poles, the vibrations of a needle are counted, when it is placed at different distances from a magnet. The square of the number made in a given time is a measure of the attractive force, just as the square of the number of vibrations of a pendulum is a measure of the force of gravity (Art. 170).

In each of these experiments, the magnetic influence of the earth upon the needle must be eliminated, in order to obtain a correct result.

367. Equilibrium of a Needle Near a Magnet.—If a small needle, free to revolve, be placed near the pole of a magnet, so that its centre is in the axis of the magnet produced, it will place itself in the line of that axis. For suppose that *N S* (Fig. 214) is a large magnetic bar, and *n s* a small needle suspended near the north pole of the magnet, with its centre in the axis of the bar produced at *a*; it will be seen that the action of the pole of the magnet is such as to bring the needle into a line with the magnet. The action of the pole *N* upon the needle tending to give it this direction (since it repels *n* and attracts *s*), is equal to the sum of its actions upon both poles. The pole *S*, by repelling *s*, and attracting *n*, tends to reverse this position, but, on account of greater distance, its force is less than that of *N*.

FIG. 214.

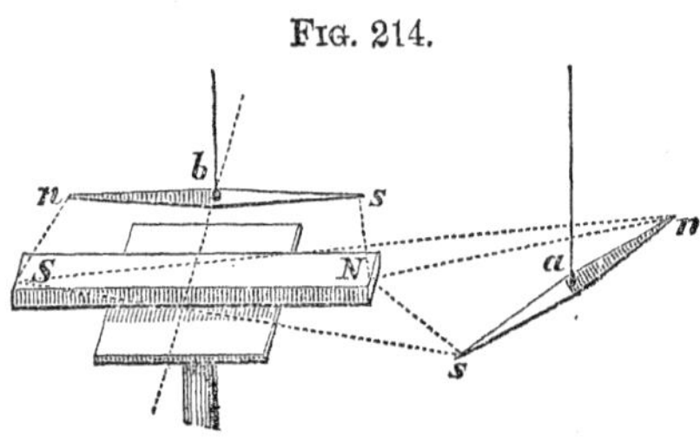

If the centre of the needle is in a line perpendicular to the bar at its middle point, the needle will be in equilibrium when parallel to the bar with its poles in contrary order. Thus, supposing the needle to be suspended at *b*, it will be seen that the actions of both poles of the magnet conspire to move *n* to the left, and *s* to the right; and as these forces are equal, equilibrium takes place only when the needle is parallel to the bar.

At intermediate points the needle will assume all possible inclinations to the axis of the bar, each position being determined by the resultant of the four forces which act on the needle. In Fig. 215 are indicated some of the positions which the needle takes

FIG. 215.

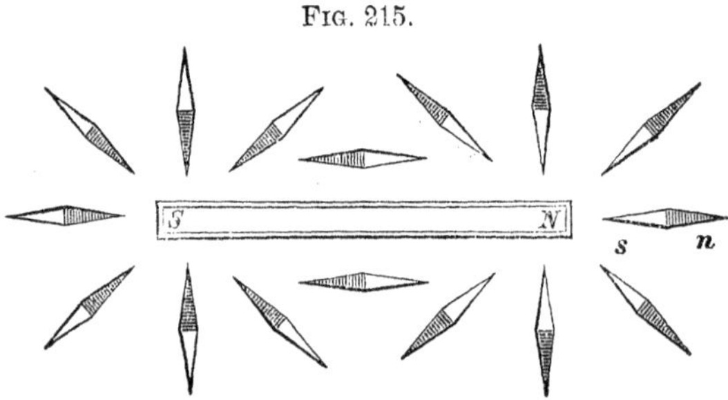

in being carried round the magnet. While it goes once round the magnet, it makes two revolutions on its own axis.

It is to be observed that in all positions the needle tends, as a whole, to move toward the bar, since the attractions always exceed the repulsions.

368. Magnetic Curves.—All the foregoing cases are shown at once by iron filings strewn on paper or parchment, which is stretched on a frame and placed near a magnet. Let the paper be slightly jarred, while the magnet lies parallel to it, either above or below, and all the inclinations of the needle will be represented by the particles of iron arranged in curves from pole to pole (Fig. 216). Near the poles of the magnet the filings stand up on the

FIG 216.

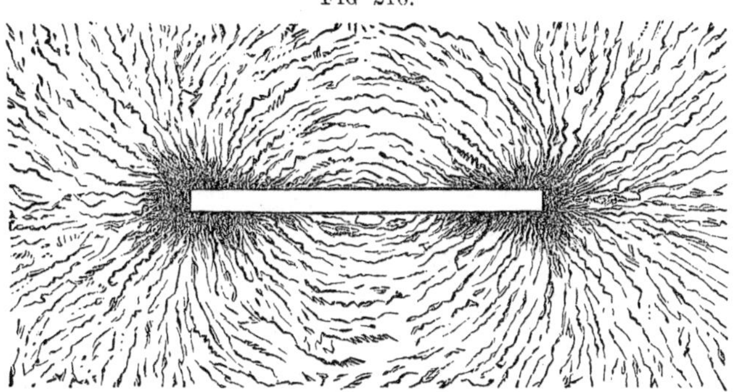

paper at various inclinations. These are the extremities of still other curves, which would be formed in all possible planes passing through the axis of the magnet, provided the filings could float suspended in the air, while the magnet is placed in the midst of them. These are called *magnetic curves.*

When the magnet is *below* the paper, the particles move away from the area over the poles, as in Fig. 216; but when it is *above,* they gather in a cluster under each pole. This singular difference arises from the force of gravity acting on the filaments, which are raised up on the paper, and which lean, in the former case, *from* each other, and in the latter, *toward* each other.

CHAPTER II.

RELATIONS OF THE MAGNET TO THE EARTH.

369. Declination of the Needle.—When the needle is balanced horizontally, and free to revolve, it does not generally point exactly north and south; and the angle by which it deviates from the meridian is called the *declination.* A vertical circle coincident with the direction of the needle at any place is called the *magnetic meridian*. As the angle between the magnetic and the geographical meridians is generally different for different places, and also varies at different times in the same place, the word *variation* expresses these *changes* in declination, though it is much used as synonymous with declination itself.

370. Isogonic Curves.—This name is given to a system of lines imagined to be drawn through all the points of equal declination on the earth's surface. We naturally take as the standard line of the system that which connects the points of *no declination*, or the isogonic of 0° (Fig. 217). Commencing at the north

Fig. 217.

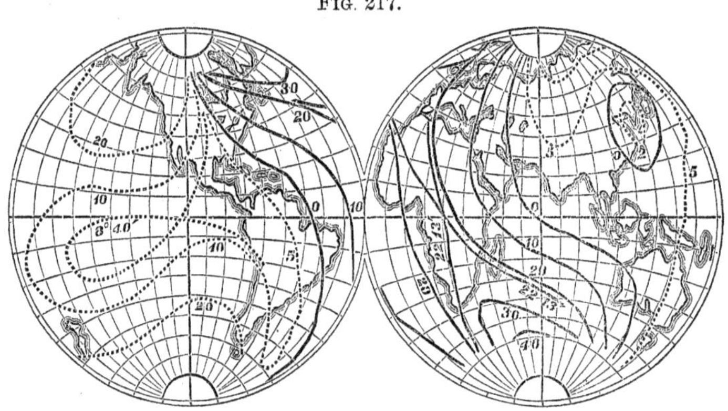

pole of dip, about Lat. 70°, Lon. 96°, it runs in a general direction E. of S., through Hudson's Bay, across Lake Erie, and the State of Pennsylvania, and enters the Atlantic Ocean on the coast of North Carolina. Thence it passes east of the West India Islands, and across the N. E. part of South America, pursuing its course to the south polar regions. It reappears in the eastern hemisphere, crosses Western Australia, and bears rapidly westward across the Indian Ocean, and then pursues a northerly course across the Caspian Sea to the Arctic Ocean. There is also a detached line of no declination, lying in eastern Asia and the Pacific Ocean, returning into itself, and inclosing an oval area of 40° N. and S. by 30° E. and W. Between the two main lines of no declination in the Atlantic hemisphere, the declination is *westward*, marked by continued lines, in Fig. 217; in the Pacific hemisphere, outside of the oval line just described, it is *eastward*, marked by dotted lines. Hence, on the American continent, in all places east of the isogonic of 0°, the north pole of the needle declines westward, and in all places west of it, the north pole declines eastward; on the other continent this is reversed, as shown by the figure.

Among other irregularities in the isogonic system, there are two instances in which a curve makes a wide sweep, and then intersects its own path, while those within the loop thus formed return into themselves. One of these is the isogonic of 8° 40′ E., which intersects in the Pacific Ocean west of Central America; the other is that of 22° 13′ W., intersecting in Africa.

In the northeastern part of the United States the declination has long been a few degrees to the west, with very slow and somewhat irregular variations.

371. Secular and Annual Variation.—The declination of the needle at a given place is not constant, but is subject to a slow change, which carries it to a certain limit on one side of the meridian, when it becomes stationary for a time, and then returns, and proceeds to a certain limit on the other side of it, occupying two or three centuries in each vibration. At London, in 1580, the declination was 11¼° E.; in 1657, it was 0°; after which time the needle continued its western movement till 1814, when the declination was 24½° W.; since then the needle has been moving slowly eastward. The entire secular vibration will probably last more than three centuries. The average variation from 1580 to 1814 was 9′ 10″ annually. But like other vibrations, the motion is slowest toward the extremes.

There has also been detected a small *annual* variation, in which the needle turns its north pole a few minutes to the east of its mean position between April and July, and to the west the rest of the year. This annual oscillation does not exceed 15 or 18 minutes.

372. Diurnal Variation.—The needle is also subject to a small *daily* oscillation. In the morning the north end of the needle has a variation to the east of its mean position greater than at any other part of the day. During winter this extreme point is attained at about 8 o'clock, but as early as 7 o'clock in the summer. After reaching this limit it gradually moves to the west, and attains its extreme position at about 3 o'clock in winter, and 1 o'clock in summer. From this time the needle again returns eastward, reaching its first position about 10 P. M., and is almost stationary during the night. The whole amount of the diurnal variation rarely exceeds 12 minutes, and is commonly much less than that. These diurnal changes of declination are connected with changes of *temperature*, being much greater in summer than in winter. Thus, in England the mean diurnal variation from May to October is 10 or 12 minutes, and from November to April, only 5 or 6 minutes.

373. Dip of the Needle.—A needle first balanced on a horizontal axis, and then magnetized and placed in the magnetic meridian, assumes a fixed relation to the horizon, one pole or the other being usually depressed below it. The angle of depression is called the *dip* of the needle. Fig. 218 represents the *dipping needle*, with its adjusting screws and spirit-level; and the depression may be read on the graduated scale. After the horizontal circle *m* is leveled by the foot-screws, the frame *A* is turned horizontally till the vertical circle *M* is in the magnetic meridian. For north latitudes, the north end of the needle is depressed, as *a* in the figure.

FIG 218

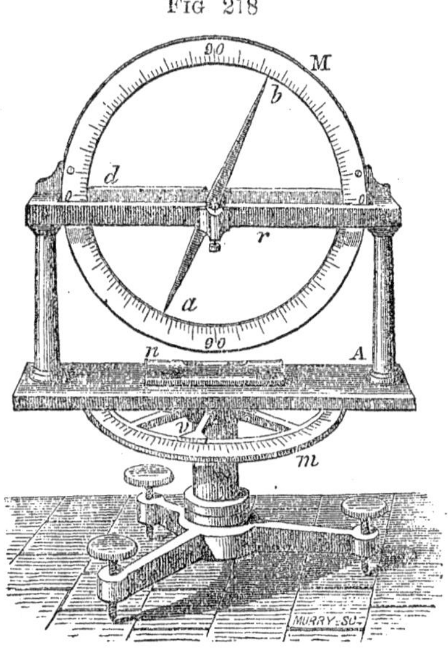

374. Isoclinic Curves.—A line passing through all points where the dip of the needle is nothing, i. e. where the dipping needle is horizontal, is called the *magnetic equator of the earth.* It can be traced in Fig. 219 as an irregular curve around

FIG. 219.

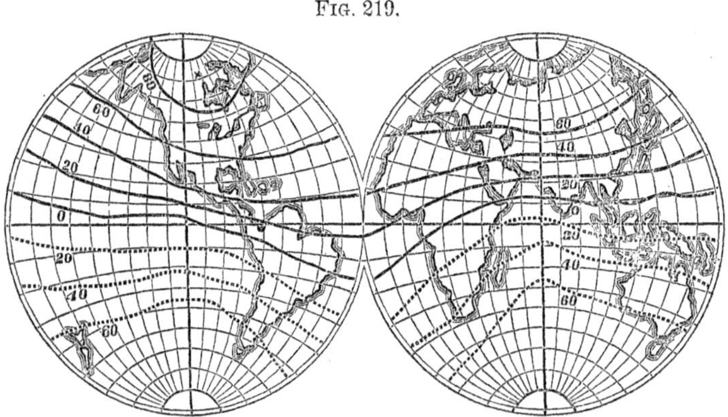

the earth in the region of the equator, nowhere departing from it more than about 15°. At every place north of the magnetic equator the north pole of the needle descends, and south of it the south pole descends; and, in general, the greater the distance, the greater is the dip. Imagine now a system of lines, each passing through all the points of equal dip; these will be nearly parallel to the magnetic equator, which may be regarded as the standard among them. These magnetic parallels are called the *isoclinic curves;* they somewhat resemble parallels of latitude, but are inclined to them, conforming to the oblique position of the magnetic equator. In the figure, the broken lines show the dip of the south pole of the needle; the others, that of the north pole. The points of greatest dip, or dip of 90°, are called the *poles of dip.* There is one in the northern hemisphere, and one in the southern. The north pole of dip was found, by Capt. James C. Ross, in 1831, to be at or very near the point, 70° 14′ N.; 96° 40′ W., marked x in the figure. The south pole is not yet so well determined.

At the poles of dip the horizontal needle loses all its directive power, because the earth's magnetism tends to place it in a vertical line, and, therefore, no component of the force can operate in a horizontal plane. The isogonic lines in general converge to the two dip-poles; but, for the reason just given, they cannot be traced quite to them.

The dip of the needle, like the declination, undergoes a variation, though by no means to so great an extent. In the course of 250 years, it has diminished about five degrees in London. In 1820 it was about 70°, and diminishes from two to three minutes annually.

Since the dip at a given place is changing, it cannot be supposed that the poles are fixed points; they, and with them the entire system of isoclinic curves, must be slowly shifting their locality.

375. Magnetic Intensity of the Earth.—The force exerted by the magnetism of the earth varies in different places, being generally least in the region of the equator, and greatest in the polar regions. The ratio of intensity in different places is measured by the number of vibrations which the needle makes in a given time. In the discussion of the pendulum, it was proved (Art. 170) that gravity varies as the square of the number of vibrations. For the same reason, the magnetic force at any place varies as the square of the number of vibrations of the needle at that place.

376. Isodynamic Curves.—After ascertaining, by actual observation, the intensity of the magnetic force in different parts of the earth, lines are supposed to be drawn through all those points in which the force is the same; these lines are called isodynamic curves, represented in Fig. 220. These also slightly resemble parallels of latitude, but are more irregular than the isoclinic lines. There is no one standard equator of minimum intensity, but there are two very irregular lines surrounding the earth in the equatorial region, in some places almost meeting each

FIG. 220.

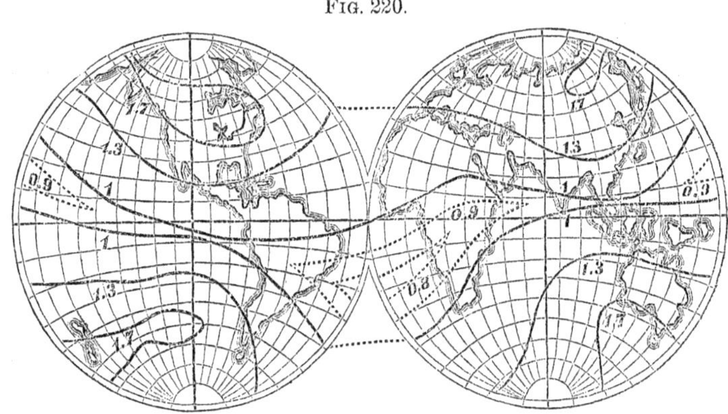

other, and in others spreading apart more than two thousand miles, on which the magnetic intensity is the same. These two are taken as the standard of comparison, because they are the lowest which extend entirely round the globe. The intensity on them is therefore called *unity*, marked 1 in the figure. In the wide parts of the belt which they include—lying one in the southern Atlantic, and the other in the northern Pacific oceans—there are lines of lower intensity which return into themselves, without encompassing the earth. In approaching the polar regions, both north and south, the curves, retaining somewhat the form of the unit lines, are indented like an hour-glass, as those marked 1.7 in the figure, and at length the indentations meet, forming an irregular figure 8; and at still higher latitudes, are separated into two systems, closing up around two poles of maximum intensity. Thus there are on the earth four poles of maximum intensity, two in the northern hemisphere and two in the southern. The American north pole of intensity is situated on the north shore of Lake Superior. The one on the eastern continent is in northern Siberia. The ratio of the least to the greatest intensity on the earth is about as 0.7 to 1.9; that is, as 1 to $2\frac{5}{7}$. In the figure, intensities less than 1 are marked by dotted lines.

377. Magnetic Charts.—These are maps of a country, or of the world, on which are laid down the systems of curves which have been described. But for the use of the navigator, only the isogonic lines, or lines of equal declination, are essential. There are large portions of the globe which have as yet been too imperfectly examined for the several systems of curves to be accurately mapped. It must be remembered, too, that the earth is slowly but constantly undergoing magnetic changes, by which, at any given place, the declination, dip, and intensity are all essentially altered after the lapse of years. A chart, therefore, which would be accurate for the middle of the nineteenth century, will be, to some extent, incorrect at its close.

378. Magnetic Observatories.—In accordance with a suggestion of Humboldt, in 1836, systematic observations have been since made upon terrestrial magnetism, in various parts of the world, in order to deduce from them the laws of its changes. Buildings have been erected without any iron in their construction, to serve as magnetic *observatories;* and the most delicate *magnetometers* have been devised and used for detecting minute oscillations both in the horizontal and vertical planes. By these means has been discovered a class of phenomena called *magnetic storms*, in which the needle suffers numerous and rapid disturbances, sometimes to the extent of several degrees; and it is a remarkable and interesting fact that these disturbances occur at the same absolute time in every part of the earth.

379. Aurora Borealis.—This phenomenon is usually accompanied by a disturbance of the needle, thus affording visible indications of a magnetic storm; but the contrary is by no means generally true, that a magnetic storm is accompanied by auroral light. The connection of the aurora borealis with magnetism is manifested not only by the disturbance of the needle, but also by the fact that the streamers are parallel to the dipping needle, as is proved by their apparent convergence to that point of the sky to which the dipping-needle is directed. This convergence is the effect of perspective, the lines being in fact straight and parallel.

380. Source of the Earth's Magnetism.—If a needle is carried round the earth from north to south, it takes approximately all the positions in relation to the earth's axis which it assumes in relation to a magnetic bar, when carried round it from end to end (Art. 367). At the equator it is nearly parallel to the axis, and it inclines at larger and larger angles as the distance from the equator increases; and in the region of the poles, it is nearly in the direction of the axis. The earth itself, therefore, may be considered a magnet, since it affects a needle as a magnet does, and also induces the magnetic state on iron. But it is necessary, on account of the attraction of opposite poles, to consider the northern part of the earth as being like the south pole of a needle, and the southern part like the north pole. To avoid this, the words *boreal* and *austral* are applied to the two magnetic states, and the *boreal magnetism* is the name given to that development found in the northern hemisphere, and the *austral magnetism* to that in the southern. Hence, it becomes necessary, in using these names for a magnet, to reverse their order, and to speak of its north pole as exhibiting the austral, and its south pole the boreal magnetism.

Modern discoveries in electro-magnetism and thermo-electricity furnish a clew to the hypothesis which generally prevails at this day. Attention has been drawn to the remarkable agreement between the *isothermal* and the *isomagnetic* lines of the globe. The former descend in crossing the Atlantic Ocean toward America, and there are two poles of maximum cold in the northern hemisphere. The isoclinic and the isodynamic curves also descend to lower latitudes in crossing the Atlantic westward; so that, at a given latitude, the degree of *cold*, the magnetic *dip*, and the magnetic *intensity*, is each considerably greater on the American than on the European coast. This is only an instance of the general correspondence between these different systems of curves. It has likewise been noticed (Art. 372) that the needle has a movement diurnally, varying westward during the middle of the day, and eastward at evening, and that this oscillation is generally much greater in the hot season than the cold. It is obvious, therefore, that the development of magnetism in the earth is intimately connected with the temperature of its surface. Hence it is supposed that the heat received from the sun excites electric currents in the materials of the earth's surface, and these give rise to the magnetic phenomena.

381. Formation of Permanent Magnets.—Needles and small bars may be more or less magnetized by the following methods, the reasons for which will be readily understood:

1. A feeble magnetism may be developed in a steel bar, by causing it to ring while held vertically. The earth's influence upon it, however, is stronger if it is held, not precisely vertical, but leaning in a direction parallel to the dipping needle. The inductive influence of the earth explains the fact often noticed, that rods of iron or steel that have stood for many years in a position nearly vertical, as, for instance, lightning-rods, iron pillars, stoves, &c., are found somewhat magnetic, with the north pole downward.

2. A needle may be magnetized by simply suffering it to remain in contact with the pole of a strong magnet, or better, between the opposite poles of two magnets.

3. Place the needle across the opposite poles of two parallel

magnets, while a bar of soft iron connects the other two poles. Thus, removing one of the keepers, *A*, *B*, from the ends of the magnets (Fig. 221), put the needle in its place, being careful that the end of the needle marked for *north* is adjacent to the *south* pole of the magnet.

Fig. 221.

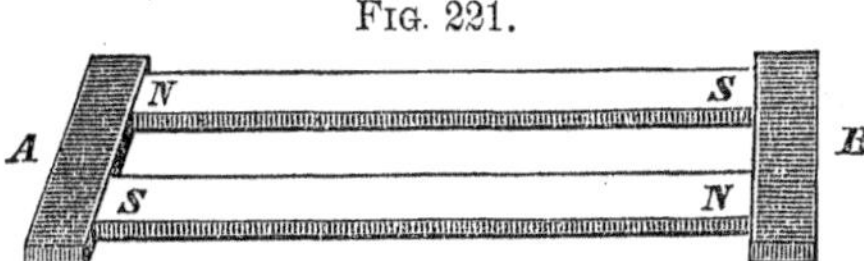

4. In order to take advantage of the earth's inductive influence, along with that of steel magnets, place the needle parallel to the dipping needle, and draw the south pole of one magnet over the lower half, and the north pole of another over the upper half, with repeated and simultaneous movements.

None of these methods, however, are of great practical value at the present day, since the galvanic circuit affords a far readier and more efficient means of magnetizing bars.

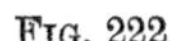
Fig. 222.

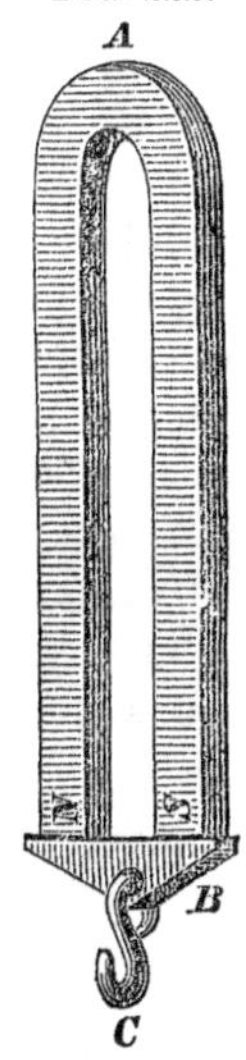

The *horse-shoe magnet*, sometimes called the *U-magnet* (Fig. 222), is for many purposes a very convenient form, and originated in the practice of *arming* the lodestone; that is, furnishing it with two pieces of soft iron, which are confined by brass straps to the poles of the stone, and project below it, so that a bar and weight may be attached. When a magnet has this form, both poles may be applied to a body at once. The *U*-magnet, *A N S*, being suspended, and the keeper, *B*, made of soft iron, being attached to the poles, weights may be hung upon the hook *C*, to show the strength.

382. The Declination Compass.—This instrument consists of a magnetic needle suspended in the centre of a cylindrical brass box covered with glass; on the bottom of the box within is fastened a circular card, divided into degrees and minutes, from 0° to 90° on the several quadrants. On the top of the box are two uprights, either for holding sight-lines or for supporting a small telescope, by which directions are fixed. The quadrants on the card in the box are graduated from that diameter which is vertically beneath the line of sight.

When the axis of vision is directed along a given line, the needle shows how many degrees that line is inclined to the magnetic meridian. In order that the angle between the line and the geographical meridian may be found, the declination of the needle for the place must be known.

383. The Mariner's Compass.—In the mariner's compass (Fig. 223) the card is made as light as possible, and attached to the needle, so that the north and south points marked on the card always coincide with the magnetic meridian. The index, by which the direction of the ship is read, consists of a pair of vertical lines, diametrically opposite to each other, on the interior of the box. These lines, one of which is seen at *a*, are in the plane of the ship's keel. Hence, the degree of the card which is against either of the lines shows at once both the angle with the magnetic meridian and the quadrant in which that angle lies.

FIG. 223.

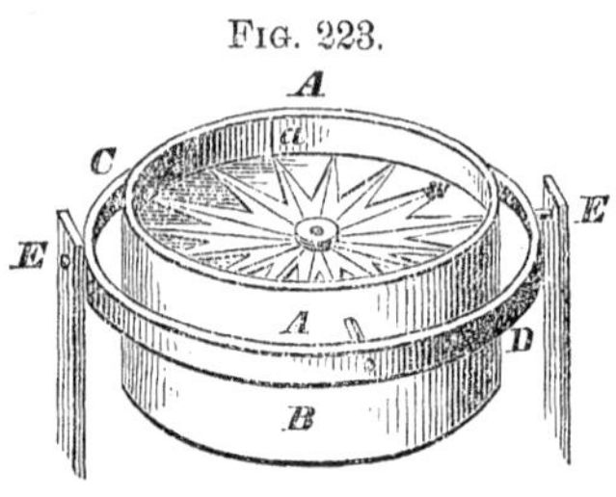

In order that the top of the box may always be in a horizontal position, and the needle as free as possible from agitation by the rolling of the ship, the box, *B*, is suspended in *gimbals*. The pivots, *A*, *A*, on opposite sides of the box, are centred in the brass ring, *C*, *D*, while this ring rests on an axis, which has its bearings in the supports, *E*, *E*. These two axes are at right angles to each other, and intersect at the point where the needle rests on its pivot. Therefore, whatever position the supports, *E*, *E*, may have, the box, having its principal weight in the lower part, maintains its upright position, and the centre of the needle is not moved by the revolutions on the two axes.

On account of the dip, which increases with the distance from the equator, and is reversed by going from one hemisphere to the other, the needle needs to be loaded by a small adjustable weight, if it is to be used in extensive voyages to the north or south. In north latitudes the south end must be heaviest; in south latitudes, the north end.

384. The Needle Rendered Astatic.—Though magnetic intensity increases at greater distances from the equator, yet the directive power of the compass grows more feeble in approaching the poles of dip, because the horizontal component constantly diminishes, and at the poles becomes zero (Art. 374). A needle in such a situation, in which the earth's magnetism has no influence to give it direction, is called *astatic*. The compass needle is astatic at the north and south poles of dip. And the dipping needle may be rendered astatic at any place by setting its plane of rotation perpendicular to its line of dip at that place; for then there will remain no component of the magnetic force in the only plane in which the needle is at liberty to move.

The needle may also be made astatic at any place by holding a magnet at such a distance, and in such a position, as to neutralize the earth's influence. Or, if a wire, suspended vertically by a thread, pass through the centres of two needles, whose poles point in opposite directions, each needle will be astatic. The needles in Fig. 224, with like poles in opposite directions, are slipped tightly upon the wire *b c*, which is suspended by the thread *a b*, free from torsion. This method of liberating a magnetic needle from the earth's influence is of great use in electro-magnetism.

FIG. 224.

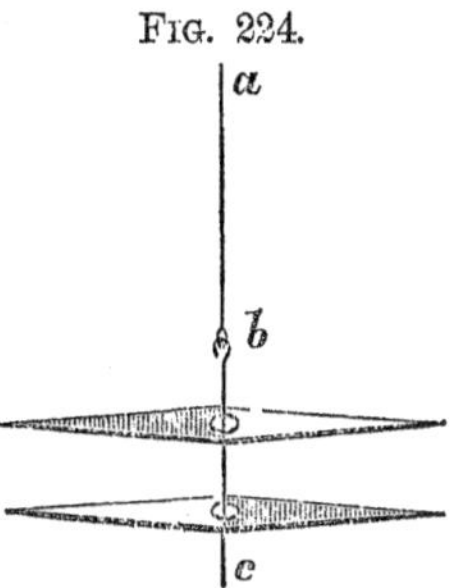

385. Theory of Magnetism.—The nature of the agency called magnetism is unknown. Much of the language employed by writers on the subject implies that there exist in iron, steel, &c., two imponderable *fluids*, called the *austral* and *boreal magnetisms;* that these fluids attract each other, and are ordinarily mingled and neutralized, so that no magnetic phenomena appear; and that in every magnet the two fluids have been separated by the inductive influence of the earth or of another magnet, one fluid manifesting itself at one pole, and the other at the other pole. As science advances, however, these views seem more and more crude and unsatisfactory. Magnetism is now regarded by many as one of those modes of *molecular motion* which are so difficult of investigation. If it is a mode of motion, then it may manifest itself as a force, as we know it does. It will be seen in the discussion of Electro-magnetism that there is a most intimate connection between magnetism and electricity, so much so that the former is generally considered as only a particular form in which the latter is developed.

Magnetism differs from the other molecular agencies—electricity, light, and heat—in producing no direct effect on any of our senses. We witness its direct effects only in the *motion* which it gives to certain kinds of matter, such as iron and steel.

16

PART VI.

FRICTIONAL OR STATICAL ELECTRICITY.

CHAPTER I.

ELEMENTARY PHENOMENA.

386. Definitions.—The name *Electricity,* from the Greek word for *amber,* is given to a peculiar agency, which is the cause of a variety of phenomena, such as attracting and repelling light bodies, producing light, heat, sound, and chemical decomposition, and, when concentrated in its action, violently rending or exploding bodies. Lightning and thunder are an example of its intense action.

Frictional electricity is so called because generally excited by friction, and to distinguish this form of development from the *galvanic electricity* which is excited by chemical means. The former is often called *statical,* and the latter *dynamical* electricity.

Bodies are said to be electrically *excited* when they show signs of electricity by some action performed upon them, as friction, for example. They are said to be *electrified* when they receive electricity by communication.

Conductors are bodies which transmit electricity freely; *non-conductors* are those which do not transmit it at all, or only very imperfectly. A body is said to be *insulated* when in contact only with non-conductors, so that electricity is retained in it.

387. Electroscopes.—The feeblest indication of electricity is usually *attraction* or *repulsion;* and instruments prepared for showing these effects are called *electroscopes.* The word *electrometer,* though sometimes used in the same sense, is more properly defined to be an instrument for measuring the quantity of electricity.

The *pendulum electroscope* (Fig. 225) consists of a glass standard, supported by a base, and bent into a hook at the top, from which is suspended a pith ball by a fine silk thread.

The *gold-leaf electroscope* consists of two narrow strips of gold-leaf, *n*, *n* (Fig. 226), suspended within a glass receiver, *B*, from a

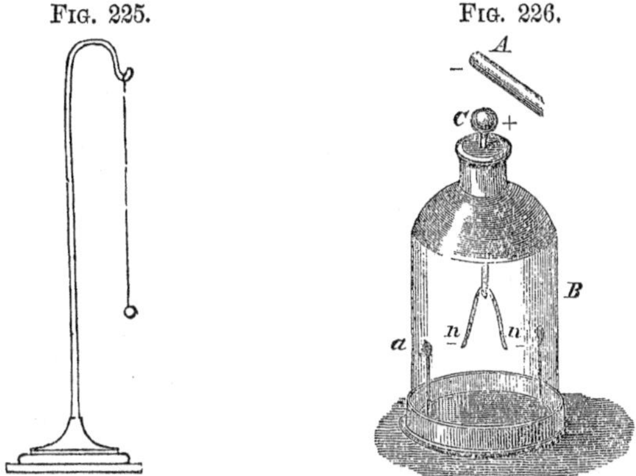

FIG. 225. FIG. 226.

metallic rod which passes through the top and terminates in a ball, *C*. A metallic base is cemented to the receiver, and strips of tin-foil, *a*, are attached to the inside, reaching to the base. When an electrified body is brought near the knob *C*, the gold leaves separate, or, if separated, collapse, or separate more, according to circumstances.

Certain modifications are convenient for some purposes. One is, a metallic wire with a ball on the top, having a thread and pith ball hanging by the side of it; and another, two threads with pith balls suspended together below a conductor, as in Fig. 231.

388. Common Indications of Electricity.—Though the frictional or statical electricity may be developed in several ways, pressure, evaporation, &c., the method generally employed is friction. By this means it can be excited in a greater or less degree in all substances, and from some it may be easily and abundantly obtained.

If amber, sealing-wax, or any other resinous substance, be rubbed with dry woolen cloth, fur, or silk, and then brought near the face, the excited electricity disturbs the downy hairs upon the skin, and thus causes a sensation like that produced by a cobweb. When the tube is strongly excited, it gives off a spark to the finger held toward it, accompanied by a sharp snapping noise. A sheet of writing-paper, first dried by the fire, and then laid on a table and rubbed with india-rubber, becomes so much excited as to adhere to the wall of the room or any other surface to which it is applied. As the paper is pulled up slowly from the table by one edge, a number of small sparks may be seen and heard on the under side of the paper. In dry weather, the brushing of a garment causes the floating dust to fly back and cling to it.

If an *iron* or *brass* rod be held in the hand and rubbed with silk, the rod shows no sign of electricity. It will be seen hereafter that the electricity excited in the rod is conveyed away by the conducting quality of the metal and the human body.

389. The Two Electrical States.—When friction has taken place between two bodies, they are found in electrical conditions, which in some remarkable particulars are *unlike each other.* These two electrical states are usually called the *positive* and the *negative,* terms which were employed by Franklin in his theory of *one* electric fluid, to indicate that the excited body has either more or less electricity than belongs to it in its common unexcited condition. Du Fay, in his theory of *two* kinds of electricity, uses the words *vitreous* and *resinous* to distinguish them, vitreous corresponding to the positive, and resinous to the negative. But it is very common to use Du Fay's *theory,* and to apply Franklin's *terms,* positive and negative, to the two kinds of electricity.

If in any case only one electricity is discovered when friction causes development, it is to be understood that the other is diffused through some large conductor, so as to be imperceptible. The earth is the great reservoir, in which any amount of electricity may be diffused and lost sight of.

390. Nature of Electricity.—The real nature of electricity is unknown. Though it is in most treatises spoken of as a *fluid,* of exceeding rarity, and more rapid in its movements than light, yet the prevailing belief at the present day is, that it is a peculiar mode of *vibratory motion,* either in the luminiferous ether which is imagined to fill all space, or else in the ordinary matter constituting the bodies and media about us, or in both of these. Electricity is brought to view by friction, by heat, and by other agencies which are calculated to cause *movements* in matter, rather than to bring new kinds of matter to light. It is undoubtedly one of the forms of *force,* into which other forces may be transformed. But until a more definite *wave-theory* or *force-theory* can be constructed than exists at present, it is comparatively easy to give to the learner an intelligible description of electrical phenomena by using the language of the *two-fluid theory* of Du Fay. In trying to give a statement of observed facts without the use of these hypothetical terms, it is necessary to employ in their stead tedious circumlocutions, which only confuse the mind of the learner.

391. Du Fay's Theory.—According to this theory, the two fluids are imagined to inhere in all kinds of matter, combined with each other and neutralized. In this condition, they afford no evidence of their existence. But they can in several ways be *separated* from each other; and when thus separated, they give rise to electrical phenomena.

392. The Two States Developed Simultaneously.—If bodies are rubbed together, the two electricities are separated, and one body is electrified positively, the other negatively. For example, glass rubbed with silk is itself positive, and the silk is negative. But the same substance does not always show the same kind of electricity, since that depends frequently on the substance against which it is rubbed. Dry woolen cloth rubbed on smooth glass is negative, but on sulphur it is positive. The following table contains a few substances, arranged with reference to this. Any one of them, rubbed with one that follows it, is positively electrified itself, and the other negatively:

1. Fur of a cat.	7. Silk.
2. Smooth glass.	8. Gum lac.
3. Flannel.	9. Resin.
4. Feathers.	10. Sulphur.
5. Wood.	11. India-rubber.
6. Paper.	12. Gutta-percha.

According to the above table, silk rubbed on smooth glass is negatively excited; but rubbed on sulphur, it is excited positively. It is sometimes found, however, that the previous electrical condition of one of the bodies will invert the order stated in the table. For example, if silk, having been rubbed on smooth glass, and therefore being negative, should then be rubbed on resin, it would probably retain its negative state, and the resin become positively electrified, contrary to the order of the table.

The mechanical condition of the surface sometimes changes the order of the two electricities. Thus, if glass is ground, so as to lose its polish, it is likely to be negative when rubbed with silk; but the excitation of rough glass is very feeble.

393. Mutual Action.—Bodies electrified *in different ways attract*, and *in the same way repel* each other. Thus, if an insulated pith ball, or a lock of cotton, be electrified by touching it with an excited glass tube, it will immediately recede from the tube, and from all other bodies which are charged with the positive electricity, while it will be attracted by excited sealing-wax, and by all other bodies which are negatively electrified. If a lock of fine long hair be held at one end, and brushed with a dry brush, the separate hairs will become electrified, and will repel each other. In like manner, two insulated pith balls, or any other light bodies, will repel each other when they are electrified the same way, and attract each other when they are electrified in different ways.

Hence it is easy to determine whether the electricity developed in a given body is positive or negative; for, having charged the electroscope with excited glass, then all those bodies which, when excited, attract the ball, are negative, while all those which repel it are positive.

394. Conduction.—Electricity passes through some bodies with the greatest facility; through others with difficulty, or scarcely at all; and others still have a conducting power intermediate between the two. As the conducting quality exists in different substances in all conceivable degrees, it is impossible to draw a dividing line between them, so as to arrange all conductors on one side, and all non-conductors on the other. The following brief table contains some of the more important of the two classes; the first column in the order of *conducting* power, the second in the order of *insulating* power:

Conductors.	Insulators.
The metals,	Lac, amber, the resins,
Charcoal,	Paraffine,
Plumbago,	Sulphur,
Water, damp snow,	Wax,
Living vegetables,	Glass, precious stones,
Living animals,	Silk, wool, hair, feathers,
Smoke, steam,	Paper,
Moist earth, stones,	Air, the gases,
Linen, cotton.	Baked wood.

When air is rarefied, its insulating power is diminished, and the further the rarefaction proceeds, the more freely does electricity pass. Hence, we might expect that it would pass with perfect freedom through a complete vacuum. It is found, however, that in an absolute vacuum electricity cannot be transmitted at all.

395. Modes of Insulating.—Solid insulating supports are usually made of glass; and, in order to improve their insulating power, they are sometimes covered with shell-lac varnish. Insulating threads for pith balls, or cords for suspending heavier bodies, are made of silk. The best insulator for suspending any very small weight is a single fiber of silk, a hair, or a fine thread of gum lac. In order to perform electrical experiments, the air must be dry, or no care whatever relating to apparatus can insure success; and therefore, in a room occupied by an audience, especially if the weather is damp, it is necessary to dry the air artificially by fires. If the air were a good conductor, it is probable that no facts in this science would ever have been discovered.

396. Communication and Influence.—The *sphere of communication* is the space within which a spark may pass from an electrified body, in any direction. It is sometimes called the striking distance. The *sphere of influence* is the space within which the power of attraction of an electrified body extends every way, beyond the sphere of communication. A glass tube strongly excited will give motion to the gold-leaf electroscope at the distance of several feet, although a spark could not pass from the tube to the cap of the electroscope at a greater distance than a few inches. The electricity which a body manifests by being brought towards an excited body, without receiving a spark from it, is said to be acquired by *induction*. The principle of induction resembles that noticed in magnetism, and will be discussed in connection with the Leyden jar.

CHAPTER II.

ELECTRICAL MACHINES.—LAW OF FORCE.—MODE OF DISTRIBUTION.

397. The Plate Machine.—In order that glass may be conveniently subjected to friction for the development of electricity, it is made in the form of a circular *plate,* and mounted on an axis, which is supported by a wooden frame, and revolved by a crank, while *rubbers* press against its surface. Fig. 227 represents one of the many forms which have been adopted. The crank, *M,* gives rotary motion to the plate, *P*, which is pressed by the rubbers, *F*, *F*; this pressure is equalized by their being placed at top and bottom, and on both sides of the glass. The prime conductor, *C C,* is made of hollow brass, and supported by glass pillars. The extremities terminate in two bows, which pass around the edges of the plate, and present to it a few sharp points, to facilitate the passage of electricity. But all other parts are carefully rounded in cylindrical and spherical forms, without edges or points, as these tend to dissipate the electricity. The glass, as it revolves from the rubbers to the points of the prime conductor, is protected by silk covers, to prevent the electricity from escaping into the air. The rubbers are made of soft leather, attached to a piece of wood or metal, and from time to time are rubbed over with an *amalgam* of zinc, tin, and mercury, or with the bi-sulphuret of tin, which is one of the best exciters on glass. The diameter of the plate varies from $1\frac{1}{2}$ to 3 feet; but in some of the largest it is 6 feet, and two plates are sometimes mounted on one axis.

FIG. 227.

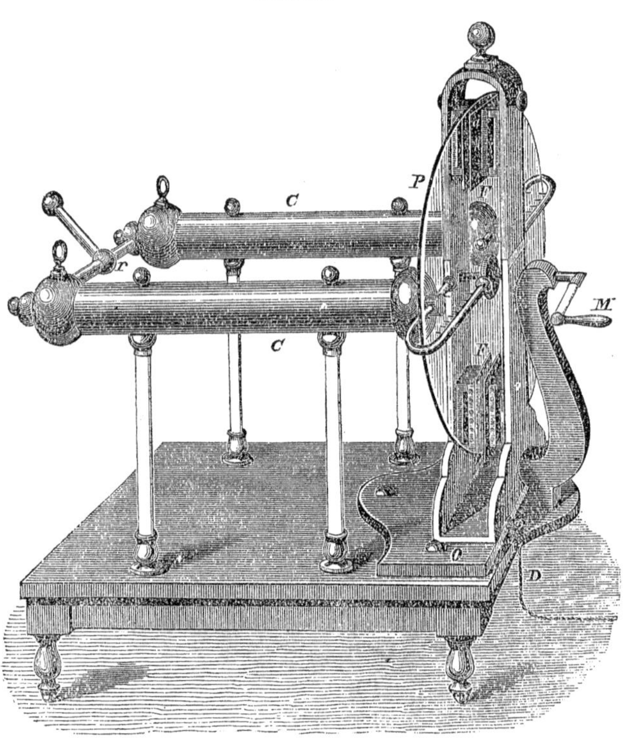

To give free passage of the negative electricity from the rubbers to the earth, a chain, *D*, may be attached to the wooden support, while its other end lies on the floor.

398. The Cylinder Machine.—In many electrical machines of the smaller sizes, a hollow cylinder is employed, having a length considerably exceeding its diameter. In the cylinder machine, the rubber is applied to one side, and the prime conductor receives the fluid from the opposite. The rubber is usually mounted on a glass pillar, so that it can be insulated, whenever it is desired

399. The Hydro-Electric Machine.—It was discovered in 1840 that a steam-boiler electrically insulated gave out sparks, and that the steam issuing from it was also electrified. Hence resulted the construction of *the hydro-electric machine.* It consists of a boiler mounted on glass pillars, and furnished with a row of jet-pipes and a metallic plate, against which the steam strikes. The prime conductor, to which the steam-plate is attached, is electrified positively, and the boiler itself negatively. Professor Faraday ascertained that the electricity in this case is developed, not by evaporation or condensation, but by the friction of watery particles in the jet-pipes. That the machine may act with energy, it was found necessary to make the interior of the jet-pipes angular, and quite irregular.

In connection with the subject of induction will be described a machine of still more recent invention, and known as the *induction machine.*

400. The Quadrant Electrometer.—In order to measure the intensity of electricity in the prime conductor, there is set upon it, whenever desired, a *quadrant electrometer* (Fig. 228). This consists of a pillar, *d*, about six inches high, having a graduated semicircle, *c*, attached to one side, and a delicate rod and ball, *a*, suspended from the centre of the semicircle. As the conductor becomes electrified, the rod is repelled from the pillar, and the arc passed over indicates rudely the degree of electrical intensity.

Fig. 228.

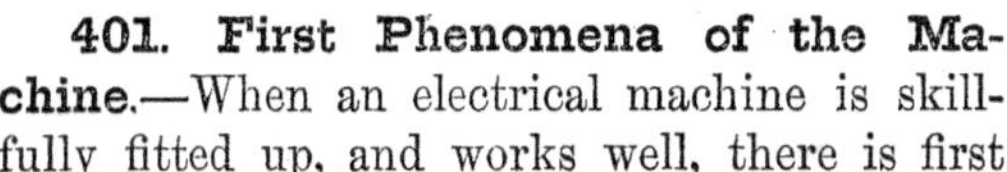

401. First Phenomena of the Machine.—When an electrical machine is skillfully fitted up, and works well, there is first perceived, on turning it, a crackling sound; and then, on bringing the knuckles toward the prime conductor, a brilliant spark leaps across, causing a sharp pricking sensation. If the room be darkened, brushes of pale light are seen to dart off continually from the most slender parts of the prime conductor, with a hissing or fluttering noise, while circles of light snap along the glass between the rubbers and the edges of the covers. When electricity is escaping plentifully from the machine, a person standing near also perceives a peculiar odor, which is that of *ozone*, and which seems always to accompany the development of electricity.

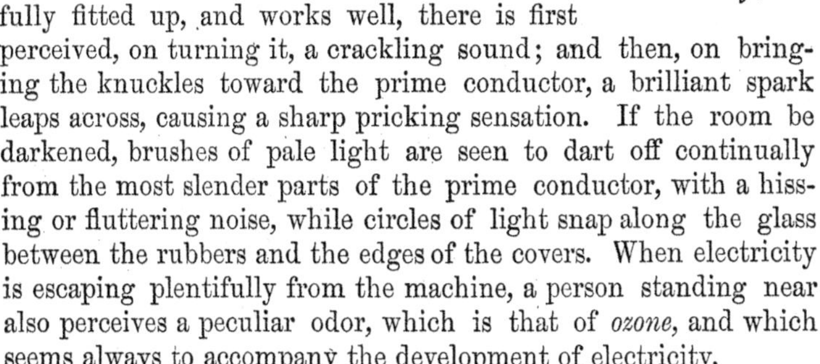

Therefore, at least *four* of the senses are directly affected by this remarkable agency, while magnetism affects none of them.

The phenomena of repulsion of like and attraction of unlike electricities, are well shown by the machine. A skein of thread or a tuft of hair, suspended from the prime conductor, will, as soon as the plate is revolved, spread into as wide a space as possible, by the repellency of the fibers which are electrified alike. Melted sealing-wax is thrown off in fine threads, and dropping water is diverged into delicate filaments. Even air, on those parts of the prime conductor which are most strongly charged, becomes so self-repellent as to fly off in a stream of wind, which is plainly felt.

On the other hand, light bodies, when brought toward the machine while in action, instantly fly to the prime conductor; for that is positive, but the nearer sides of the other bodies are made negative by induction.

The difference between substances as to their conducting quality is readily perceived by setting the quadrant electrometer on the prime conductor, raising the index by turning the plate, and then touching the prime conductor with the remote end of the body to be tried. If an iron rod, or even a fine iron wire, be thus applied, the index will fall instantly; a long dry wooden rod will cause it to descend slowly, while a glass rod will produce no effect at all. These experiments show that iron is a perfect conductor, wood an imperfect conductor, and glass a non-conductor.

402. Coulomb's Torsion Balance.—When a long fine wire is stretched by a small weight, its elasticity of torsion is a very delicate force, which is successfully employed for the measurement of other small forces. When such a wire is twisted through different angles, the *force* of torsion is found to vary as the *angle* of torsion; it is therefore easy to measure the force which is in equilibrium with torsion. The torsion balance is represented in Fig. 229. The needle of lac, *n o*, is suspended by a very fine wire from a stem at the top of the tube *d*. The cap of the tube, *e*, is a graduated circle, whose exact position is marked by the index, *a*. The stem from which the wire hangs is held in place in the centre of the cap by friction, but can be turned round so as to place the needle in any direction desired. At the end of the lac needle is a small disk of brass-leaf, *n*, and by its side a gilt ball, *m*, connected with the handle, *r*, by the glass rod, *i*. This apparatus is suspended in the glass cylinder, covered with a glass plate, on the centre of which the tube *d* is fastened. There is a graduated circle around the cylinder on the level of the needle.

Fig. 229.

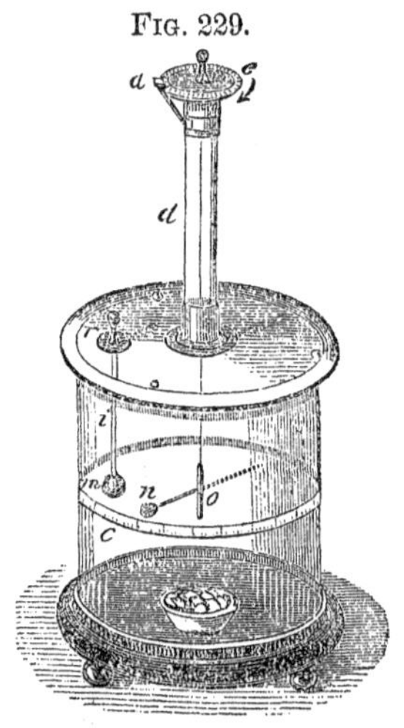

403. Law of Electrical Force as to Distance.—Adjustment is now made by turning the stem so that, while the wire is in its natural condition, the disk, *n*, touches the ball, *m*, and is at zero, and the index at top also at zero on the circle *e*. Let a minute charge of electricity be communicated to *m*, and it will repel *n*, and cause it, after a few oscillations, to settle at a certain distance—suppose, for instance, at 36°. The circle *e* is now turned in the opposite direction, until the needle is brought within 18° of the ball *m*. In order to bring it thus near, the index has to be turned 126°, which added to the 18°, makes the whole torsion 144°, or *four times* as great as before. Therefore, at *one-half* the distance there is *four times* the repulsion. In like manner, it is found that at *one-third* the distance there is *nine times* the repulsion. Hence, the law,

Electrical repulsion varies inversely as the square of the distance.

In a manner somewhat similar to the foregoing, it was conclusively proved by Coulomb that electrical *attraction* obeys the same law of distance, though there is more practical difficulty in performing the experiments. But if the electrified body *m* is placed outside of the circle described by *n*, so that the latter is allowed to vibrate both to the right and left, the square of the *number* of vibrations in a given time becomes a measure of the attractive force, as in the case of the pendulum (Art. 170).

404. Waste of Electricity from an Insulated Body.—In making accurate investigations like the foregoing, in which considerable time is necessarily occupied, a difficulty arises from the loss of the electrical charge. The *first* and most obvious source of waste is the moisture in the air, which conducts away the fluid; but this may be nearly avoided by setting into the cylinder a cup of dry lime, or other powerful absorbent of moisture, as represented in the figure. A *second* is the imperfect insulation afforded by even the most perfect non-conductors. A *third* is the mobility of the air, whose particles, when they have touched the electrified body, and become charged, are repelled, taking away with them the charge they have received. The loss in these ways is very slight, when the charge is small, and allowance can be made for it with a good degree of accuracy. But when bodies are highly charged, they lose their electricity at a rapid rate.

405. An Electrical Charge Lies at the Surface.—This is proved in many ways. A *hollow* ball, no matter how thin, will receive as large a quantity of electricity as a *solid* one. Hence it is that the prime conductor of the electrical machine, and metallic articles of electrical apparatus generally, are made of sheet brass, for the sake of lightness.

Let a metallic ball, supported on a glass pillar, be charged with either kind of electricity. Then apply to it two thin metallic hemispheres, by means of insulating handles. If they now be quickly removed from the ball, all the electricity which was previously on the ball is found on the hemispheres.

Let a dish, *a* (Fig. 230), be made of two brass rings and cambric sides and bottom, with an insulating handle, *b*, attached to the larger ring. If this vessel be charged with electricity, the charge is found on the outside; turn it over quickly, so as to throw it the other side out, and the charge is instantly found on the outside again, and none on the inside. It may be inverted several times with the same result, before the charge becomes too feeble to be perceived.

FIG. 230.

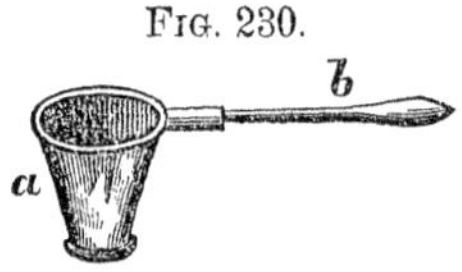

If cavities are sunk into a solid conductor, no sensible quantity of electricity is found at the bottom of such cavities. In experiments of this kind, Coulomb found his torsion balance (Fig. 229) of great service. A *proof plane*, as he termed it—that is, a small piece of gilt paper cemented upon the end of a slender rod of lac, was first touched to that part of an electrified body which was to be examined, and then applied to the ball of the instrument. The distance to which the needle was repelled indicated the intensity of electricity at the point in question. The charge taken from the bottom of an abrupt cavity was never sufficient to move the needle.

Another proof that the charge occupies only the outside surface is that the intensity diminishes as the *surface* is enlarged, while the *mass* of the conductor remains the same. A metallic ribbon rolled upon an insulated cylinder may be unrolled, and thus the surface enlarged to any extent. An electroscope standing on the instrument will fall as the ribbon is unrolled, and rise when it is again rolled up.

406. Distribution of a Charge on the Surface.—Developed electricity resides at the surface of a body, as we have seen, but is not uniformly diffused over it, except in the case of the sphere. In general, the more *prominent* the part, and the more *rapid its curvature*, the more intensely is the fluid accumulated there.

In a long slender rod, nearly the whole charge is collected at the extremities. On the surface of an ellipsoid it is found to be arranged according to a very simple law, namely: *the quantity of the charge at each point varies as the diameter through that point.* But the tendency to escape increases at a more rapid rate, and varies as the *square* of the diameter. Hence it is that electricity is so rapidly dissipated from *points*, which may be regarded as the extremities of ellipsoids indefinitely elongated. If the surface of a body is partly convex and partly concave, the distribution is still more unequal; nearly all the charge collects on the convex parts; and if the concavities are deep or abrupt, like those mentioned in Art. 405, no sign of electricity is discovered in them.

407. Rotation by Unbalanced Pressure.—As the electric charge on the surface of a body presses outward in all directions, wherever it escapes from a point, there the pressure is removed; consequently, on the opposite part there is *unbalanced* pressure. Therefore, if the body is delicately suspended, and one or more points are directed tangentially, the unbalanced pressure will cause rotation in the opposite direction, just as Barker's mill rotates by the unbalanced pressure of water. Electrical wheels and orreries are revolved in this way.

A windmill may also be revolved by the stream of air issuing from a stationary point attached to the prime conductor (Art. 401).

408. The Charge Held on the Surface by Atmospheric Pressure.—The mutual repellency, which drives the particles asunder till they reach the surface of the conductor, tends to make them escape in all directions from that surface; and it is the air alone which prevents. For if one extremity of a charged and insulated conductor extends into the receiver of an air-pump, the charge is dissipated by degrees, as the receiver is exhausted; and when the exhaustion is as complete as possible, the most abundant supply from the machine fails to charge the conductor. As the atmospheric pressure is limited to about 15 lbs. per square inch, so the amount of charge is limited which can be retained on a conductor of given form. Hence the reason for the well-known fact that the prime conductor receives all the charge which it is capable of retaining in one or two turns of the machine. All that is gained over and above this, by continuing to turn, flies off through the air.

CHAPTER III.

ELECTRICITY BY INDUCTION.—LEYDEN JAR.

409. Elementary Experiment.—When an electrified body is placed near one which is unelectrified, but not within the sphere of communication, the natural electricities of the latter are decomposed, one being attracted toward the former, the other repelled from it (Art. 396). Thus the ends become electrified by the *influence* of the first body, without *receiving* any electricity from it. Let *A* (Fig. 231) be charged with positive electricity,

FIG. 231.

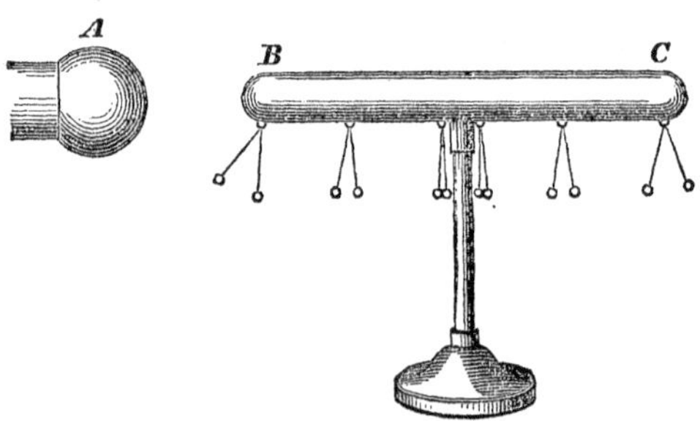

and let the insulated conductor, *B C*, be furnished with several electroscopes, as represented. Those nearest the ends will diverge most, and the others less according as they are nearer the centre, where there is no sign of electricity. By taking off small quantities with the proof-plane, and testing them, it is found that negative electricity occupies the end nearest to *A*, and positive the remote end. Remove the bodies to a distance from each other, and *B C* returns to its unelectrified condition; bring them near again, and it is electrified as before. As this electrical state is *induced* upon the conductor by the electrified body in its vicinity, without any *communication* of electricity, it is said to be electrified by *induction*. If *A* is first charged with the negative electricity, the two electricities of *B C* will be arranged in reversed order; the positive will be attracted to the nearest end, the negative repelled to the farthest.

Electrical induction is exactly analogous to *magnetic* induction; the opposite kind is developed at the nearer end, and the like kind at the remote end.

410. Successive Actions and Reactions.—If A is itself an insulated conductor, the foregoing is not the entire effect; for a reflex influence is exerted by the electricity in the nearer end of the conductor. Let *A* have a positive charge, as at first. After the negative electricity is attracted to the nearer end of *B C*, it in turn attracts the positive charge of *A*, and accumulates it on the nearest side, leaving the remote side less strongly charged than before. This is shown by electroscopes attached to the opposite sides of *A*. The charge of *A*, being now nearer, will exert more power on *B C*, separating more of its original electricities, and thus making the nearest end more strongly negative and the remote end more strongly positive than before; and this new arrangement of fluids in *B C* causes a second reaction upon *A*, of the same kind as the first. Thus an indefinite diminishing series of adjustments takes place in a single moment of time.

411. Division of the Conductor.—Suppose that before the experiment begins, *B C* is in two parts with ends in contact; the entire series of mutual actions takes place as already described. Now, while *A* remains in the vicinity, let the parts of *B C* be separated; then the negative electricity is secured in the nearest half, and the positive in the other. And if *A* is now removed, the positive charge is diffused over the more distant half. Thus each kind of electricity can be completely separated from the other by means of induction.

Here we find a marked difference between magnetism and frictional electricity. The electricities may be secured in their separate state, one in one conductor, the other in another. In magnetism this is not possible; for when an iron bar is magnetized, and then broken, each kind of magnetism is found in each half of the bar. At the point of division both polarities exist, and as soon as the bar is broken, they manifest themselves there as strongly as at the extremities.

412. Effect of Lengthening the Conductor.—If the conductor, *B C*, is *lengthened*, the accumulation on the adjacent parts of the two bodies is somewhat increased. The positive electricity which, at the remote end of the shorter conductor, operated in some degree by its repulsion to prevent accumulation on the nearest side of *A*, is now driven to a greater distance; and therefore a larger charge will come from the remote to the nearer side of *A*, which in turn attracts more negative to the nearer end of *B C*, and thus a new series of actions and reactions takes place in addition to the former. To obtain the greatest effect from this cause, the conductor, *B C*, is connected with the earth—that is, it is *uninsulated;* then the positive part of its decomposed electricities is driven to the earth, and entirely disappears, and the negative part is attracted to the nearer end; so that, when the series of adjustments is completed, the remote end of the conductor is in the neutral state. This experiment is performed by touching the finger to the conductor, after it has become electrified by induction. The electroscope nearest to *A* instantly rises a little higher, and the distant ones collapse.

If the original charge in *A* was *negative* instead of *positive*, the foregoing experiments are in all particulars the same, except that the order of the two fluids is reversed.

413. Disguised Electricity.—The electricity which occupies the surface of the prime conductor, or any other body electrified in the ordinary way, and which is kept from diffusing itself in every direction only by the pressure of the air (Art. 408), is called *free* electricity; for it will instantly spread over the surface of other conductors, when they are presented, and therefore will be lost in the earth, the moment a communication is made. But the electricity which is accumulated by the inductive influence is not free to diffuse itself; the same attractive force which has condensed it still holds it as near as possible to the original charge; and if we touch the electrified body with the hand, the electricity does not pass off; it is therefore called *disguised* electricity. In this respect the two fluids on the contiguous sides of *A* and *B C* are alike; either may be touched, or in any way connected with the earth, but, unless communication is made between them, or unless they are both allowed to pass to the earth, they hold each other in place by their mutual attraction, and show none of the phenomena of free electricity.

414. A Series of Conductors.—If another insulated conductor, *D*, is placed near to the remote end of *B C*, and *A* is charged positively, then that extremity of *B C* nearest to *D* is inductively charged with positive, as already stated. Hence, the electricities of *D* are separated, the negative approaching *B C*, and the positive withdrawing from it; there is therefore the same arrangement of fluids in both bodies, but a less intensity in *D* than in *B C*. For, on account of distance, the positive is not so intensely accumulated at the remote end of *B C* as in the original body *A*, and therefore a less force operates on *D* than on *B C*. The same effects are produced in a less and less degree in an indefinite series of bodies; and the shorter they are, the more nearly equal will be the successive accumulations. The same facts were noticed in a series of magnets.

415. An Electrified Body Attracts an Unelectrified Body.—This fact, which is the first to be noticed in observing electrical phenomena (Art. 401), is explained by induction. If *B C* is light, and delicately suspended, a consequence of the arrangement of fluids already described is, that *B C* will move toward *A*. For, according to the law of distance (Art. 403), the negative in the nearer part is attracted more strongly than the positive in the remote part is repelled; hence the body yields to the greater force, and moves toward *A*. That the attracted fluid does not leave the body, *B C*, behind, and go to *A*, is owing to the fact, noticed in Art. 408, that the body and the electricity are confined to each other by atmospheric pressure.

416. The Inductive Action Greatly Increased.—In the experiments as now described, the inductive influence is feeble, and the accumulation of electricities very small; for the bodies present toward each other only a limited extent of area, and they are necessarily as much as four or five inches distant, in order to prevent the fluid from passing across. By giving the bodies such a form that a large extent of surface may be equidistant, and then interposing a solid non-conductor, as glass, between them, so that the distance may be reduced to one-eighth of an inch or less, it is easy to increase the attracting and repelling forces many thousands of times. Let a glass plate, *C D* (Fig. 232), supported on a base, have attached to the middle of each side a rectangular piece of tin-foil. This is called a Franklin plate. Let *A* be connected with one coating, and *B* with the other. If, now, *A* forms a part of the prime conductor of an electrical machine, and *B* has communication with the earth, we are prepared to notice the remarkable phenomena of the Leyden jar. If the amount of surface and the thickness of glass are the same, the particular form of the instrument is immaterial; but, for most purposes, a vessel or jar is more convenient than a pane of glass of equal surface, and is generally employed for electrical experiments.

Fig. 232.

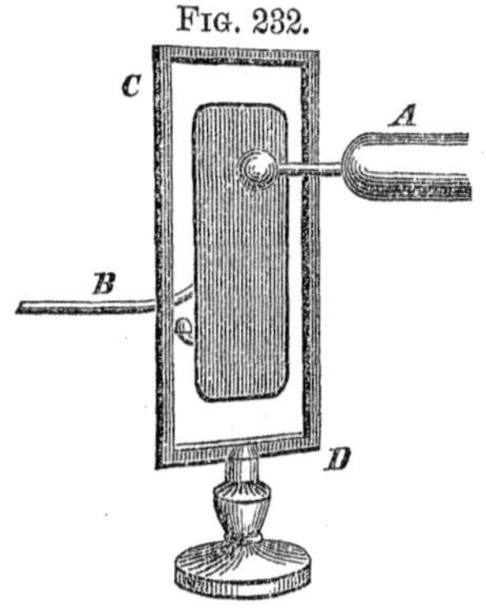

417. The Leyden Jar.—This article of electrical apparatus consists of a glass jar (Fig. 233), coated on both sides with tin-foil, except a breadth of two or three inches near the top, which is sometimes varnished for more perfect insulation. Through the cork passes a brass rod, which is in metallic contact with the inner coating, and terminates in a ball at the top.

Fig. 233.

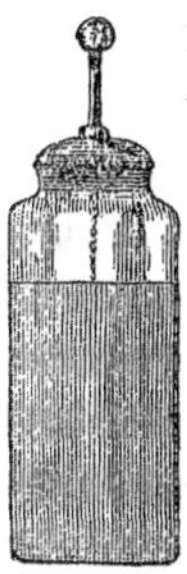

On presenting the knob of the jar near to the prime conductor of an electrical machine, while the latter is in operation, a series of sparks passes between the conductor and the jar, which will gradually grow more and more feeble, until they cease altogether. The jar is then said to be *charged*. If now we take the *discharging-rod*, which is a curved wire, terminated at each end with a knob, and insulated by glass handles (Fig. 234), and apply one of the knobs to the outer coating of the jar, and bring the other to the knob of the jar, a flash of intense brightness, accompanied by a sharp report, immediately ensues. This is the *discharge* of the jar.

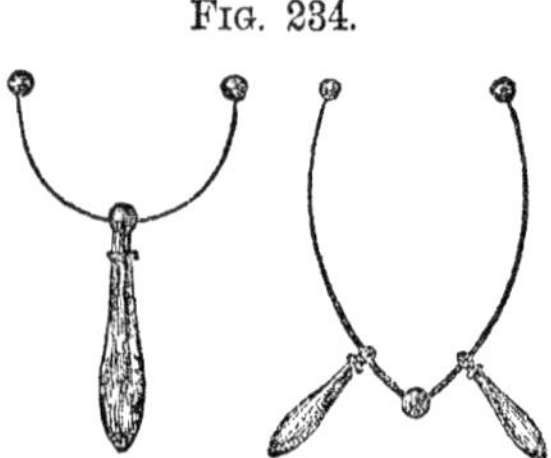

FIG. 234.

If, instead of the discharging-rod, a person applies one hand to the outside of the charged jar, and brings the other to the knob, a sudden *shock* is felt, convulsing the arms, and when the charge is heavy, causing pain through the body. The shock produced by electricity was first discovered accidentally by persons experimenting with a charged phial of water. This occurred in Leyden, and led to the construction and name of the Leyden jar.

418. Theory of the Leyden Jar.—This instrument accumulates and condenses great quantities of electricity on its surfaces, upon the principle of mutual attraction between unlike electricities, one of which is furnished by the machine, the other obtained from the earth by induction. First, suppose the outer coating insulated; a spark of the positive electricity passes from the prime conductor to the inner coating, which tends to repel the positive from the outer coating; but as the latter cannot escape, it remains to prevent, by its counter-repulsion, any addition to the charge of the inside, and thus the process stops. But now connect the outer coating with the earth, and immediately some of its positive electricity, repelled by the charge on the inside, passes off, while its negative is attracted close upon the glass, and gives room for the accession of more from the earth. The slight condensation of negative upon the outside, by its attraction, condenses the positive of the inner coating, and allows a second spark to pass in from the prime conductor. This produces the same effect as the first, and a second addition of negative is made to the outer coating, the latter being obtained from the earth as before. These actions and reactions go on in a diminishing series, till there is a great accumulation of the two electricities, held by mutual attraction as near each other as possible, on opposite sides of the glass. The jar in this condition is said to be charged.

If the positive electricity is on the inner coating, the jar is said to be *positively* charged; if on the outside, *negatively* charged.

419. The Spontaneous Discharge.—This occurs when the quantities accumulated are so great that their attraction will cause them to fly together with a flash and report over the edge of the jar. If the glass is soiled or damp, the fluids may pass over and mingle with only a hissing noise, in which case it is impossible for the jar to be highly charged.

If the glass is clean and dry, and especially if varnished with gum lac, a charge may not wholly disappear for days, or even weeks.

420. Series of Jars.—The same amount of electricity from the prime conductor which is required to charge one jar will charge an indefinite series, the strength of the charge being less and less from the first to the last. This case is analogous to the series of conductors (Art. 414). Insulate a series of jars, *A*, *B*, *C*, &c., and connect the inner coating of *A* with the prime conductor, and its outer coating with the inner coating of *B*, the outer of *B* with the inner of *C*, and so on. Then, as *A* is charged, the positive electricity of its outer coating, instead of passing to the earth, goes to the inside of *B*, and that on the outside of *B* to the inside of *C*, &c., while that on the outside of the last in the series passes to the earth. Thus each jar is charged positively by the inductive influence of the preceding, just as a series of magnets is formed with poles in the same order by a succession of magnetic inductions.

421. Division of a Charge in any Given Ratio.—If one of two jars be charged, and the other not, and if the inner coatings be brought into communication, and also the outer coatings, the charge of the first jar is instantly diffused over the two, with a report like that of a discharge. In this way a charge may be halved, or divided in any other ratio, according to the relative surfaces of the jars.

The self-repellency of each fluid tends to diffuse it over a greater surface, and they will be thus diffused if allowed to remain within each other's attracting influence; but *one* of the fluids will not be spread over the coatings of another jar, unless opportunity is given for *both* to do it.

An experiment somewhat resembling the foregoing is this: charge two equal jars, one positively, the other negatively, and insulate them both. If the two knobs be connected by a conductor, the electricities, notwithstanding their strong attraction, will not unite; for each is held disguised by that on the other side of the glass. But if the outer coatings are first connected, then, on joining the knobs, the jars are both discharged at once.

422. Use of the Coatings.—If a jar is made with a wide open top, and the coatings movable, then, after charging the jar and removing the coatings, very little of the electricities adheres to the latter, but nearly the whole remains on the glass. The same mutual attraction which condensed them at first still holds them there after the coatings are removed. When they come to be replaced, the jar can be discharged as usual. But the coatings are necessary in charging, to diffuse the electricity over those parts of the glass which they cover, and also in discharging, to conduct off the whole charge at once.

423. The Free Portion of an Electrical Charge.—Either kind of electricity is said to be *free* when it remains on a body only because held by the pressure of the air; but if held by the attraction of the opposite kind, it is said to be disguised (Art. 413). Nearly all the electricity of a charged jar is disguised, but not the whole.

The moment after a jar is charged there is a small quantity of free electricity on the coating to which the fluid was furnished in charging, but not on the other. But after the jar has stood charged some minutes, a little is free on both coatings. If the charged jar be upon an insulating stand, and the finger brought to one coating, a slight spark is taken off; if it be touched again immediately, there is no spark, for the free electricity all escaped by the first contact. Let the finger now be brought to the other coating, and a spark flies from that. Immediately afterward a second spark can be taken from the first coating, and so on alternately for hundreds of times usually before the charge wholly disappears. What is removed at each contact is the free part of the charge, which always appears *alternately* on the two coatings. If a small electroscope be connected with each coating, the fluid alternately set free is indicated to the sight. The electroscope on the coating which is touched instantly falls, and the other rises.

424. Explanation of this Phenomenon.—The positive electricity which is conveyed to the inner coating, in charging a jar, attracts to the outer coating from the earth a quantity of the negative fluid which is a little *less* than itself. This is because of the thickness of the glass. If it were infinitely thin, the negative would be just equal to the positive, and they would neutralize each other, and both be perfectly disguised. But as the glass has some thickness, the positive exceeds the negative, and disguises it. Now if the jar, after being charged, is insulated, it is obvious that the negative charge on the outer coating cannot disguise all the positive (which is more than itself), but only a quantity a little less than itself. Hence there must be a little of the positive on the inner coating in a free state. By touching the knob, we allow this free portion to pass off, and there is left less of the positive in the inner coating than there is of the negative in the outer. Therefore, *all* the negative cannot now be disguised, but a slight quantity is liberated and ready to pass off as soon as touched. And thus, by alternate contacts, the process of discharge goes on, the series being longer as the glass is thinner, because then the two quantities are more nearly equal.

425. Electrical Vibrations and Revolutions.—If two jars be charged in opposite ways, and a figure made of pith be suspended between the knobs by a long thread, it will be attracted by that knob whose action on it happens to be greatest. As soon as it touches, it is charged with that kind and repelled, and of course attracted by the other knob, which is in the opposite state; thus it vibrates between them, causing a very slow discharge of both jars. In this case, the outside of the jars not being insulated, the electricity, which is slowly set free on the outside, passes off, and therefore there is always some free electricity on the knob to be imparted to the vibrating figure.

The electricity of the prime conductor will also cause vibrations, without the use of a jar. Suspend from it a metallic disk horizontally a few inches above another which is connected with the earth; then if a glass cylinder surround the two disks so as to prevent escape, a number of pith balls between the disks will continue to vibrate up and down so long as the machine is in action. Each ball lying on the lower disk, being electrified by induction in the opposite way from the upper one, springs up to it, and then, being charged in the same way, is repelled.

In a similar manner a chime of bells may be rung, orreries revolved, &c.

426. Residuary Charge.—If a jar stand charged a few minutes, and after the discharge remain some minutes more, then a *second*, and possibly a *third*, discharge can be made; but these are usually very slight. The electricity remaining after the first discharge is called the *residuary charge*. The larger the jar, and the more intense the charge, the larger is this residuum. It is probably explained as follows: The charge, at first limited to the coatings, gradually diffuses itself on the uncoated glass for a little distance, according to the intensity of the charge and the length of time the jar remains charged. At the first discharge, only the electricity which is in contact with the coating is taken off, and that which lies on the uncoated glass slowly diffuses itself back again, and is conducted over the whole coated surface; so that, after the lapse of a minute or two, a sensible discharge occurs on applying the rod a second time.

427. The Electric Battery.—Leyden jars are made of various sizes, from a half-pint to one or two gallons. But when a great amount of surface is needed, it is more convenient, and, in case of fracture by violent discharge, more economical, to connect several jars, so that they may be used as one. Four, nine, twelve, or even a greater number of jars, are set in a box (Fig. 235), whose interior is lined with tinfoil, so as to connect all the outer coatings together. Their inner coatings are also connected, by wires joining all the knobs, or by a chain passing round all the stems. Care is necessary in discharging batteries, that the circuit is not too short and too perfect, since the violence of discharge is liable to perforate the jars. A chain, three or four feet long in the circuit, will generally prevent the accident.

Fig. 235.

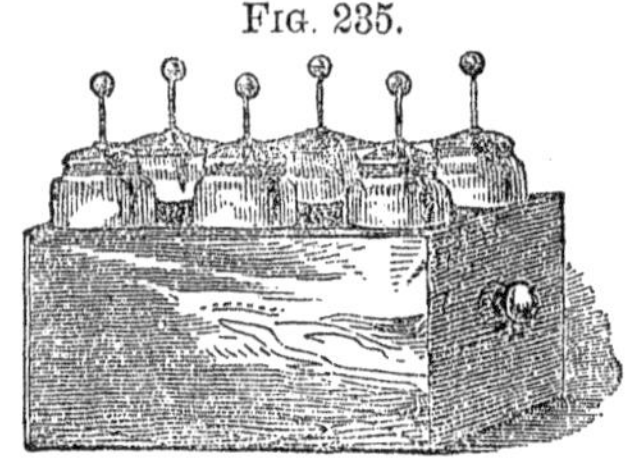

428. Different Routes of Discharge.—If two or more circuits are opened at once between the two coatings of a charged jar or battery, the discharge will take one or another, or divide between them, according to circumstances. If the circuits are alike except in length, the discharge will follow the *shorter*. If they differ only in conducting quality, the electricities will take the *best conductor*. If the circuits are interrupted, and in all respects alike, except that the conductors of one are pointed at the interruptions, and of the others not pointed, the discharge will follow the line which has *pointed conductors*. If the circuits are *very attenuated* (as very fine wire, or threads of gold-leaf), the charge is liable to divide among them.

429. Discharging Electrometers.—These are instruments contrived for measuring the charge in the act of discharging the jar. Fig. 236 represents Lane's discharging electrometer. *D* is a

Fig. 236. Fig. 237.

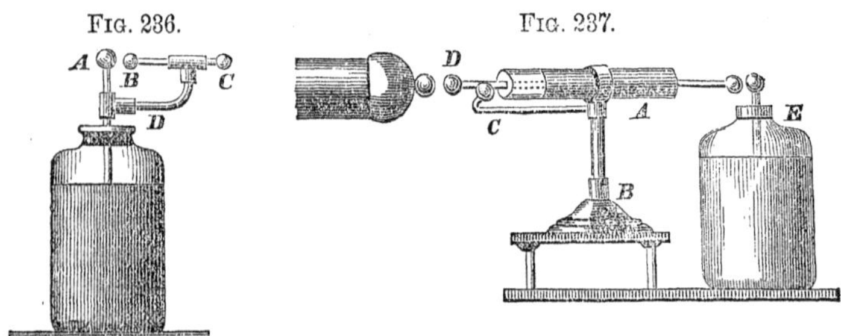

rod of solid glass, which holds the metallic rod and balls *B C*. This rod, being in a horizontal position at the height of the knob *A*, can be placed at any desired distance from it. Then, if the circuit through which the charge is to be sent extends from the rod *B C* to the external coating, the interval of air between *A* and *B* is all which prevents discharge; and as soon as the charge is increased, till its tension is sufficient to leap that interval, the discharge will take place. The greater that space is, of course the greater the charge must be, before it will pass across. If *A* and *B* are in contact, no charge at all will collect.

The *unit jar* is used to measure the charge of another jar, by conveying to it successive equal charges of its own. *A B* (Fig. 237) is the instrument, consisting of a small open jar, placed horizontally on an insulating stand, *B*. From the metallic part of the support, the bent rod and ball, *C*, come near to *D*, the rod of the inner coating, and can be turned so as to increase or diminish its distance. Let the knob, *D*, be near the prime conductor, and that of the outer coating near the top of the jar, *E*, which is to be charged, the outside of the latter being in communication with the earth. While *A* is charging, the positive electricity of its outer coating goes to the inner coating of the large jar, and partially charges it. Presently the unit jar discharges spontaneously across the distance between *D* and *C*. It is then in the neutral condition, as at first, and the process is repeated till *E* is charged with the requisite number of units.

430. The Effect of a Point Presented to an Electrified Body.—It has been noticed (Art. 406) that a pointed wire attached to the prime conductor wastes the charge very quickly, because of the accumulation at the point. The conductor loses its charge just as quickly by presenting a pointed rod *toward* it. For the induced electricity of the rod and person holding it is in like manner accumulated at the point, and readily escapes to mingle with and neutralize its opposite in the prime conductor. Thus the charge disappears at once. In a similar way is to be explained the use of the points on the prime conductor presented to the glass plate. When the two electricities are separated at the surface of contact between the plate and rubbers, the plate is positively electrified. This positive charge acts inductively on the prime conductor, attracting the negative kind to the points, where it passes off and neutralizes what is on the plate, and leaves a positive charge on the prime conductor.

431. The Gold-Leaf Electroscope.—The principle of induction explains the construction of some other instruments besides the Leyden jar; as the gold-leaf electroscope, the electrical condensers, and the electrophorus. The first has been already described (Art. 387); we have only to explain its operation by the principle of induction. Let a body positively electrified be brought within a few feet of the knob. It attracts the negative from the leaves into the knob, and repels the positive from the knob into the leaves; they are thus electrified alike, and repel each other. If the charged body is brought so near that the leaves touch the conductors, which are placed on the sides of the cylinder, and discharge their induced electricity to them, then they collapse. After this, they will diverge again, whether the electrified body is brought still nearer, or withdrawn; if brought nearer, they diverge by means of a new portion of positive, repelled from the knob; if withdrawn, they diverge by the return of negative electricity from the knob, which is no longer neutralized by the positive, since the latter has been discharged to the earth.

432. The Electrical Condenser.—Instruments called by this name are intended for the accumulation of electricity from some feeble source, until it may be rendered sensible. The most delicate is the gold-leaf condenser. Suppose the gold-leaf electroscope to have a disk, *A*, instead of a knob on the top (Fig. 238). Another disk, *B*, is furnished with an insulating handle, and between the disks is placed the thinnest possible non-conductor, as a film of varnish. Bring the finger in contact with the under-side of *A*, to connect it with the earth. Then bring to the upper side of *B* the source of feeble electricity (as, for example, a piece of copper, after being touched to a piece of zinc), the small quantity of electricity imparted to *B* induces an equal amount of the opposite in *A*, drawn in from the earth. After the disks have touched each other again, a second contact upon *B* repeats the action; and when this has been done a great number of times, there are condensed on the two sides of the varnish small charges which are held in that state by induction. As yet, the gold-leaves are at rest; but on removing the finger from *A*, and taking up *B* by the insulating handle, the electricity condensed in *A* is set free, flows down to the leaves and repels them, thus rendering the accumulation perceptible.

Fig. 238.

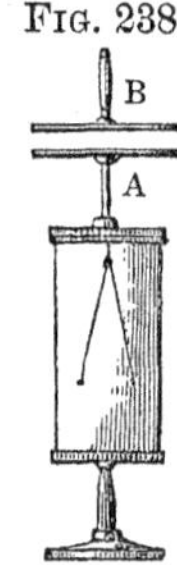

433. The Electrophorus.—This is a very simple *electrical machine* for giving the spark. It consists of a circular cake of resin in a wooden base, *A* (Fig. 239), and a metallic disk, *B*, having a glass handle. Excite the resin by fur or flannel; set the disk *B* upon it, and touch the latter with the finger. The disk now has a disguised charge of positive electricity, drawn in by the negative charge of the plate through the finger. On lifting the disk by the handle, its charge is set free, and may be taken

Fig. 239.

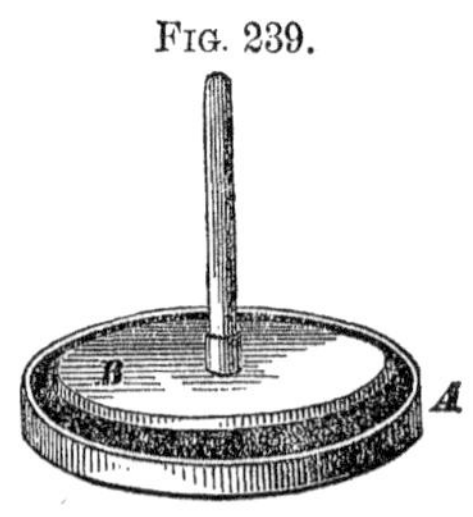

off in a brilliant spark. Set the disk down again, touch it, and lift it, and the same thing occurs, even hundreds or thousands of times, and, after standing for hours, is ready to operate still in the same way.

This case of inductive action seems at first perplexing, because there is no glass plate, no film of varnish, no non-conductor of any kind, between the two opposite electricities of the resin and disk. Why then do they not at once mingle, and neutralize each other? It is simply because the resin is an *excited* body, the negative electricity having been developed upon it by friction. When we touch the finger to the disk, the positive that enters does meet the negative, and neutralize it for the time being; but on separating the plate and disk, the electricities also separate, as is always the case when an electric and the rubber are removed from each other after friction. Thus one charge on the resin may be made to induce any number of successive charges upon the disk.

434. The Induction Machine.—This instrument (Fig. 240), known also as the Holtz machine, from the name of the inventor,

FIG. 240.

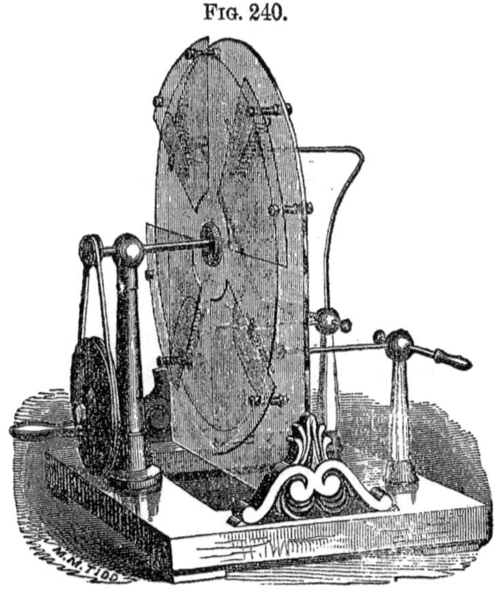

develops electricity with great rapidity without friction, except as it is employed for a moment at first to electrify one of the sectors, by the side of which the plate revolves. This electrified sector acts inductively on the successive portions of the revolving plate, without losing sensibly its own electric charge, just as the electrophorus plate induces charge after charge on the metallic disk, without loss to itself.*

435. The Leichtenberg Figures.—When a spark of electricity is laid upon a non-conductor, it will, by its own self-repellency, extend itself a little distance along the surface. The *Leichtenberg figures* furnish a visible illustration of this fact, and also show that the two fluids diffuse themselves in very different forms. Lay down sparks of positive electricity from the knob of the Leyden jar upon a plate of resin, and near them some sparks of negative electricity. Then blow upon the plate the mingled powders of sulphur and red-lead. The sulphur, by the agitation of passing through the air, will be electrified negatively, and attracted therefore by the positive sparks; the red-lead, positively electrified, will be attracted by the negative. Thus the spots on which the electricities are placed will appear in their exact forms by means of the colored powders attached to them. The positive resemble stars, or rather a group of crystals shooting out from a nucleus; the negative spots are circles with smooth edges; and the size of the electrified spots in each case depends on the quantity of electricity in the spark.

CHAPTER IV.

EFFECTS OF ELECTRICAL DISCHARGES.

436. Variety of Effects.—Some of the effects of electrical discharges have been incidentally noticed in the foregoing chapters. The bright light, the sharp sound, and the great suddenness of the transmission, are remarkable phenomena in every discharge of a Leyden jar or battery. The various effects may be classified as *luminous*, *mechanical*, *chemical*, and *physiological*.

437. Luminous Effects.—Light is seen only when electricity is discharged in considerable quantities through an obstructing medium. Hence, no light is perceived when it flows through a good conductor, unless of very small diameter. But if there is the least interruption, or if the conductor is reduced to a very slender form, then light appears at the interruption, and at those parts which are too small to convey the electricity. Thus,

---

* The Holtz machine has been greatly improved by Mr. E. S. Ritchie, of Boston. The figure presents this improved form. Physicists are not fully agreed as to the mode of explaining all the phenomena of this machine.

the discharge of a battery through a chain gives a brilliant scintillation at every point of contact between the links.

438. Modifications of the Light.—The length, color, and form of the electric spark vary with the nature and form of the conductors between which it passes, and with the quality of the medium interposed between them.

Electrical sparks are more brilliant in proportion as the substances *between* which they occur are better conductors. A spark received from the prime conductor upon a large metallic ball is short, straight, and white; on a small ball it is longer, and crooked; received on the knuckle, a less perfect conductor, the middle part is purplish; on wood, ice, a wet plant, or water, it is red.

From a point positively electrified, the electricity passes in the form of a faint brush or pencil of rays; a point connected with the negative side exhibits a luminous star.

When electricity passes through rarefied air, the light becomes faint, and is generally changed in color. The electrical spark, which in common air is interrupted, narrow, and white, becomes, as the rarefaction proceeds, continuous, diffused, and of a violet color, which tint it retains as long as it can be seen. If a battery is discharged through a tube several feet long, nearly exhausted of air, the whole space is filled with a rich purple light. The sparks from the machine, conveyed through the same tube, exhibit flashings and tints exceedingly resembling the Aurora Borealis.

The *Geissler tubes* are tubes of complex forms, and containing a slight trace of some gas or vapor, which show various colors and intensities of electric light, according to the kind of gas, the diameter of the parts, and the quality of the glass. The electricity is conveyed into the tubes by platinum wires sealed into their extremities.

Various colors are obtained by sending charges through different substances. An egg is bright crimson; the pith of cornstalk, orange; fluor-spar, green; and loaf-sugar, white and phosphorescent.

439. Luminous Figures.—Metallic conductors, if of sufficient size, transmit electricity without any luminous appearance, provided they are perfectly continuous; but if they are separated in the slightest degree, a spark will occur at every separation. On this principle, various devices are formed, by pasting a narrow band of tinfoil on glass, in the required form, and cutting it across with a penknife, where we wish sparks to appear. If an interrupted conductor of this kind be pasted round a glass tube in a spiral direction, and one end of the tube be held in the hand, and the other be presented to an electrified conductor, a coil of brilliant points surrounds the tube. Words, flowers, and other complicated forms, are also produced nearly in the same manner, by a suitable arrangement of interruptions in a narrow line of tinfoil, running back and forth on a plate of glass.

440. Mechanical Effects.—Powerful electric discharges through imperfect conductors produce certain mechanical effects, such as perforating, tearing, or breaking in pieces, which are all due to the sudden and violent repulsion between the electrified particles.

A discharge through the air is supposed to perforate it. If the air through which the spark is passed lies partially inclosed between two bodies which are easily moved, the force by which the air is rent will drive them asunder. Thus, a little block may be driven out from the foundation of a miniature building, and the whole be toppled down. But this enlargement of inclosed air is best seen in Kinnersley's *air thermometer* (Fig. 241). As the spark passes between the knobs in the large tube, the air confined in it is suddenly driven asunder, so as to press the water which occupies the lower part two or three inches up the tube, as represented. As soon as the discharge has occurred, the water quietly returns to its level. The sharp sound which is produced by the discharge of a Leyden jar is due to the sudden compression of the air, and also to the collapse which immediately succeeds.

Fig. 241.

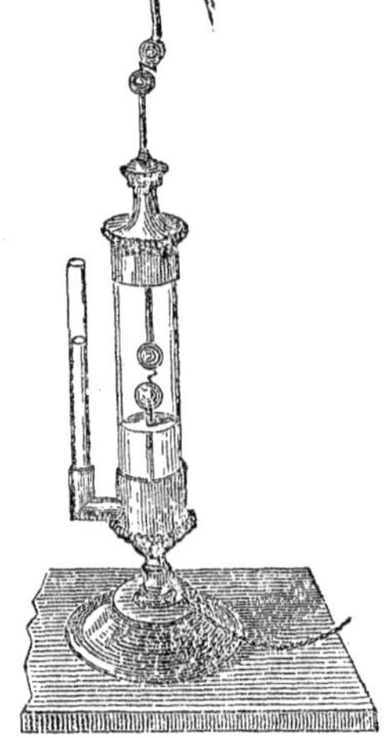

The path of the electric spark through the air, when short, is straight; but if more than about four inches long, is usually *crinkled.* This is supposed to arise from the condensation of the air before it, by which it is continually turned aside.

When the charge is passed through a thick card, or the cover of a book, a hole is torn through it, which presents the rough appearance of a bur on each side. By means of the battery, a quire of strong paper may be perforated in the same manner; and such is the velocity with which the fluid moves, that if the paper be freely suspended, not the least motion is communicated to it. Pieces of hard wood, of loaf-sugar, and brittle mineral substances, are split in two, or shivered to pieces, by an intense charge of a battery. But good conductors of much breadth are not thus affected. The charge, as it is transmitted, passes over the whole body, instead of being concentrated in any one line. But if liquids which are good conductors are closely confined on every side, they show that a violent expansion is produced by a discharge. Thus, when a charge is sent through water confined in a small glass tube or ball, the glass is shattered to pieces; and mercury in a thick capillary tube is expanded with a force sufficient to splinter the glass.

441. Chemical Effects.—These are various: combustion of inflammable bodies; oxydation, fusion, and combustion of metals; separation of compounds into their elements; reunion of elements into compounds.

Ether and alcohol may be inflamed by passing the electric spark through them; phosphorus, resin, and other solid combustible bodies, may be set on fire by the same means; gunpowder and the fulminating powders may be exploded, and a candle may be lighted. Gold-leaf and fine iron wire may be burned, by a charge from the battery. Wires of lead, tin, zinc, copper, platinum, silver, and gold, when subjected to the charge of a very large battery, are burned, and converted into oxides.

The same agent is also capable of restoring these oxides to their simple forms. Water is decomposed into its gaseous elements, and these elements may again be reunited to form water. By passing a great number of electric charges through a confined portion of air, the oxygen and nitrogen are converted into nitric acid. The ozone which is almost always perceived in connection with electrical experiments is to be considered as one of the chemical effects of electricity.

Galvanic electricity is a form of this agent much better adapted than frictional electricity to produce chemical as well as magnetic effects.

442. Physiological Effects.—The shock experienced by the animal system, when the charge of a jar passes through it, has been already mentioned.

A slight charge of the Leyden jar, passed through the body from one hand to the other, affects only the fingers or the wrists; a stronger charge convulses the large muscles of the arms; a still greater charge is felt in the breast, and becomes somewhat painful. The charge of a large battery is sufficient to destroy life, if it be sent through the vital organs. By connecting the chains which are attached to the jar with insulating handles, it is easy to pass shocks through any particular joint, muscle, or other part of the body, as is frequently done for medical purposes.

The charge may be passed through a great number of persons at the same time. Hundreds of individuals, by joining hands, have received the shock at once, though there is more difficulty in passing a charge of given intensity as the number is increased.

If the spark is taken by a person from the prime conductor, the quantity is not sufficient, unless the conductor is of extraordinary size, to produce what is called the shock; a pricking sensation in the flesh where the spark strikes, and a slight spasm of the muscle, is all that is noticeable. A person may make his own body a part of the prime conductor by standing on an *insulating stool*—that is, a stool having glass legs, and touching the conductor of the machine. This occasions no sensation at all, except what arises from the movement of the hair, in yielding to the repellency of the fluid. If another person takes the spark from him, the prick is more pungent, as the quantity is larger than in the prime conductor alone.

443. Velocity of Electricity.—This is so great that no appreciable time is occupied in any case of discharge. When we seem to see lightning move *from* the cloud *to* the earth, we find that such a progress is imagined, not perceived; for, by a little effort, we can just as well learn to see it pass from the earth to the cloud.

Wheatstone a few years since devised an ingenious method of measuring the time in which electricity passes over a wire only half a mile long. The wire was so arranged that three interruptions, one near each end, and one in the middle of the wire, were brought side by side. When the discharge of a jar was transmitted, the sparks at these interruptions were seen by reflection in a swiftly revolving mirror. An exceedingly small difference of time between the passage of those interruptions could be easily perceived by the *displacement* of the sparks as seen in the whirling mirror. The amount of observed displacement and the known rate of revolution of the mirror, would furnish the interval of time occupied by the electricity in passing from one interruption to the next. By a series of experiments, Wheatstone arrived at the conclusion *that, on copper wire, one-fifteenth of an inch in diameter, electricity moves at the rate of* 288,000 *miles per second*, a velocity much greater than that of light.

Galvanic electricity moves very much slower. Its rate on iron wire, of the size usually employed for telegraph lines, is about 16,000 miles per second.

CHAPTER V.

ATMOSPHERIC ELECTRICITY.—THUNDER STORMS.

444. Electricity in the Air.—The atmosphere is always more or less electrified, sometimes positively, sometimes negatively. This fact is ascertained by several different forms of apparatus. For the lower strata, it is sufficient to elevate a *metallic rod* a few feet in length, pointed at the top, and insulated at the bottom. With the lower extremity is connected an electroscope, which indicates the presence and intensity of the electricity. For experiments on the electricity of higher portions, a kite is employed, with the string of which is intertwined a fine metallic wire. The lower end of the string is insulated by fastening it to a support of glass, or by a cord of silk. If a cloud is near the kite, the quantity of electricity conveyed by the string may be greatly increased, and even become dangerous. Cavallo received a large number of severe shocks in handling the kite-string; and Richman, of Petersburgh, was killed by a discharge of electricity which came down the rod which he had arranged for his experiments, but which was not provided with a conductor near by it, for taking off extra charges.

The electricity of the atmosphere is most developed when hot dry weather succeeds a series of rainy days, or the reverse; and during a single day, the air is most electrical when dew is beginning to form before sunset, or when it begins to exhale after sunrise. In clear, steady weather, the electricity is generally positive; but in falling or stormy weather, it is frequently changing from positive to negative, and from negative to positive.

445. Thunder-Storms.—Thunder-clouds are, of all atmospheric bodies, the most highly charged with electricity; but all single, detached, or insulated clouds, are electrified in greater or less degrees, sometimes positively and sometimes negatively. When, however, the sky is completely overcast with a uniform stratum of clouds, the electricity is much feebler than in the single detached masses before mentioned. And, since fogs are only clouds near the surface of the earth, they are subject to the same conditions: a driving fog, of limited extent, is often highly electrified.

Thunder-storms occur chiefly in the hottest season of the year, and after mid-day, and are more frequent and violent in warm than in cold countries. They never occur beyond 75° of latitude—

seldom beyond 65°. In the New England States they usually come from the west, or some westerly quarter.

The storm itself, including everything except the electrical appearances, is supposed to be produced in the same manner as other storms of wind and rain; and the electricity is developed by the rapid condensation of watery vapor, and by friction. Electricity is not to be regarded as the *cause*, but as a *consequence* or *concomitant* of the storm. But the precipitation of vapor must be sudden and copious, since when the process is slow, too much of the electricity evolved would escape to allow of the requisite accumulation. Also, if a storm-cloud is of great extent, it is not likely to be highly electrified, because the opposite electricities, which may be developed in different parts of it, have opportunity to mingle and neutralize; and points of communication with the earth will here and there occur. Clouds of rapid formation, violent motion, and limited extent, are therefore most likely to be thunder-clouds.

446. Lightning.—When a cloud is highly charged, it operates inductively on other bodies near it, such as other clouds, or the earth. Hence, discharges will occur between them. Lightning passes frequently between two clouds, or even between two parts of the same cloud, in which opposite electricities are so rapidly developed that they cannot mingle by conduction. But, in general, the discharges of lightning take place between the electrified cloud and the earth, whose nearer part is thrown into the opposite electrical state by induction. It is supposed that, in some instances, a discharge occurs between two distant clouds by means of the earth, which constitutes an interrupted circuit between them. The crinkled form of the path of lightning is explained in the same way as that of the spark from the machine, and the thunder is caused by the simultaneous rupture and collapse of air in all parts of the line of discharge. The words *chain-lightning*, *sheet-lightning*, and *heat-lightning*, are supposed not to indicate any real differences in the lightning itself, but only in the circumstances of the person who observes it. If the crinkled line of discharge is seen, it is *chain* or *fork* lightning; if only the light which proceeds from it is noticed, it is *sheet*-lightning; if, in the evening, the thunder-storm is so far distant that the cloud cannot be seen, nor the thunder heard, but only the light of its discharges can be discerned in the horizon, it is frequently called *heat*-lightning.

447. Identity of Lightning and Electrical Discharges.—Franklin was the first to point out the resemblances between the phenomena of lightning and those of frictional electricity. He was also the first to propose the performance of electrical experiments by means of electricity drawn from the clouds. The points of resemblance named by Franklin were these: 1. The crinkled form of the path. 2. Both take the most prominent points. 3. Both follow the same materials as conductors. 4. Both inflame combustible substances. 5. They melt metals in attenuated forms. 6. They fracture brittle bodies. 7. Both have produced blindness. 8. Both destroy animal life. 9. Both affect the magnetic needle in the same manner. In 1752, he obtained electricity from a thunder-cloud by a kite, and charged jars with it, and performed the usual electrical experiments.

448. Lightning-Rods.—Franklin had no sooner satisfied himself of the identity of electricity and lightning than, with his usual sagacity, he conceived the idea of applying the knowledge acquired of the properties of the electric fluid so as to provide against the dangers of thunder-storms. The conducting power of metals, and the influence of pointed bodies, to transmit the fluid, naturally suggested the structure of the lightning-rod. The experiment was tried, and has proved completely successful; and probably no single application of scientific knowledge ever secured more celebrity to its author.

Lightning-rods are often constructed of wrought iron, about three-fourths of an inch in diameter. The parts may be made separate, but, when the rod is in its place, they should be joined together so as to fit closely, and to make a continuous surface, since the fluid experiences much resistance in passing through links and other interrupted joints. At the bottom the rod should terminate in two or three branches, going off in a direction from the building, and descending to such a depth that they will reach permanent moisture. At top the rod should be several feet higher than the highest parts of the building. It is best, when practicable, to attach it to the chimney, which needs peculiar protection, both on account of its prominence and because the products of the combustion, smoke, watery vapor, &c., are conductors of electricity. For a similar reason, a kitchen chimney, being that in which the fire is kept during the season of thunder-storms, requires to be especially protected. The rod is terminated above in one or more sharp points; and as these points are liable to lose their sharpness, and have their conducting power impaired by rust, they are protected from corrosion by being covered with gold-leaf or silver-plate. Rods may be made of smaller size than above described; but if so, there should be a proportionally greater number. It is well to connect with the rods, and with the earth, all extended conductors upon or within the building, such as metallic coverings of roofs, water conductors, bundles of bell-wires, &c.; in order that large discharges may have opportunity to divide, and take several circuits, without doing injury at the non-conducting intervals.

449. In what way Lightning-Rods Afford Protection.—Lightning-rods are of service, not so much in receiving a discharge when it comes, as in diminishing the number of discharges in their vicinity. They continually carry on a silent communication between the two electricities, which are attracting each other, one in the cloud, the other in the earth; so that a village well furnished with rods has few discharges of lightning in it. All tall pointed objects, like spires of churches and masts of ships, exert a similar influence, though in a less degree, because not so good conductors.

During a thunder-storm, or immediately after it, if a person can be near the top of a high rod, he will sometimes hear the hissing sound of electricity escaping from it, as from a point attached to the prime conductor of a machine. In the same circumstances, if it were quite dark, he would probably see the brush or star of light on the point. The statement of Cæsar in his Commentaries, "that the points of the soldiers' darts shone with light in the night of a severe storm," probably refers to the visible escape of electricity from the weapons as from lightning-rods.

450. Protection of the Person.—Silk dresses are sometimes worn with the view of protection, by means of the insulation they afford. They cannot, however, be deemed effectual unless they completely envelop the person; for if the head and the extremities of the limbs are exposed, they will furnish so many avenues as to render the insulation of the other parts of the system of little avail. The same remark applies to the supposed security that is obtained by sleeping on a feather bed. Were the person situated *within* the bed, so as to be entirely enveloped by the feathers, they would afford some protection; but if the person be extended on the surface of the bed, in the usual posture, with the head and feet nearly in contact with the bedstead, he would rather lose than gain by the non-conducting properties of the bed, since, being a better conductor than the bed, the charge would pass through him in preference to that. If the bedstead were of iron, its conducting quality would probably be a better protection than the insulating property of the feathers, since, by taking the charge itself, it would keep it away from the person. So, a man's garments soaked with rain have been known to save his life, being a better conductor than his body. Animals under trees are peculiarly exposed, because the trees by their prominence are liable to be the channels of communication for the electric discharge, and the animal body, so far as it reaches, is a better conductor than the tree. Tall trees, however, situated near a dwelling-house, furnish a partial protection to the building, being both better conductors than the materials of the house, and having the advantage of superior elevation.

451. How Lightning Causes Damage.—The word *strike*, which is used with reference to lightning, conveys no correct idea of the nature of the movement of electricity, or of the injury which it causes. One kind of electricity, developed in a cloud, causes the other to be accumulated by induction in the part of the earth nearest to it. These electricities strongly attract each other; consequently, that in the earth presses upward into all prominent conducting bodies toward the other; and, if those bodies are numerous, high, the best of conductors, and terminated by points, the electricity will flow off from them abundantly, and mingle with its opposite in the air above; and thus discharges are in a great degree prevented. But if these channels for silent communication are not furnished, the quantity of electricity will increase, till the strength of attraction becomes so great that the fluid will break its way through the air, usually from some prominent object, as a building or tree, and thus the union of the two electricities takes place. The building or tree in this case is said to be *struck by lightning;* it is rent, or otherwise injured, by the great quantity of electricity which passes violently through it, in an inconceivably short space of time. The effects produced are exactly like those caused by discharges of the electrical battery, on a greatly enlarged scale. The charge of a large battery, taken through the body in the usual way, would prostrate a person by the violence of the shock; but the same charge, if allowed to occupy a few seconds in passing by means of a point, would not be felt at all.

Fulgurites are tubes of silicious matter formed in the ground, where lightning has struck in sandy soil, and melted the sand around its path.

PART VII.

DYNAMICAL ELECTRICITY.

CHAPTER I.

THE GALVANIC CURRENT, AND APPARATUS FOR PRODUCING IT.

452. Electricity Developed by Chemical Action.—In a glass vessel (Fig. 242) containing a mixture of one part of sulphuric acid and seven or eight parts of water, put two plates, one of copper, *C*, and the other of zinc, *Z*, to each of which is soldered a copper wire. On bringing the extreme ends of the wires together, a feeble flow of electricity will take place through the wires, the plates, and the liquid. This is called the *galvanic or voltaic current* of electricity. It is developed by the chemical action of the acid on the metals; and this condition of electricity is called galvanic or voltaic, from Galvani and Volta, two Italian philosophers, who made the first discoveries of importance in this branch of science. It is also called *dynamical* electricity, for reasons to be mentioned hereafter.

Fig. 242.

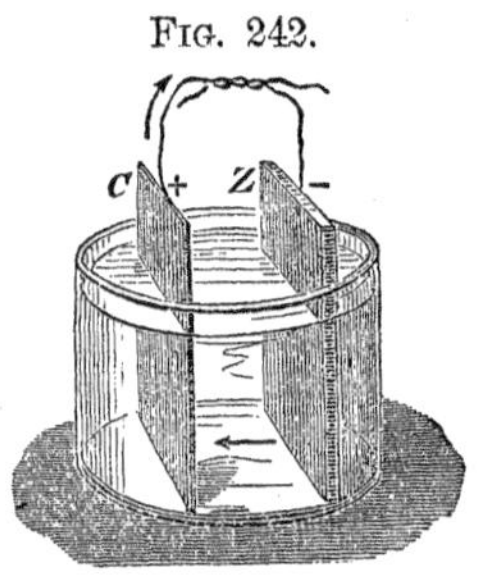

453. Definitions.—An *element* or *cell* is a jar containing any arrangement of substances for the purpose of obtaining the galvanic electricity. A *battery* is a number of elements properly connected with each other.

The *poles* or *electrodes* of a cell or battery are the extremities of the wires where the electricities appear.

The *circuit* is the path or conductor provided for the flow of the current—that is, the liquid, the plates, and the wires. The circuit is said to be *closed* when the wires are joined, so that there is a flow of the current; when they are separated, the current ceases, and the circuit is said to be *broken*, or to be *open*.

454. The Essential Parts of an Element.—An element must consist of *two unlike substances* (they are generally two different metals), *separated by continuous moisture.*

Volta's original battery, called the *dry pile*, consisting of alternate disks of copper, zinc, and paper, was no real exception, since the paper absorbed sufficient moisture from the atmosphere.

455. The Cell of Two Fluids.—An element of copper, zinc, and dilute acid, already described, soon loses its efficiency. Improved batteries, by which a constant flow of electricity may be maintained for a considerable length of time, are those in which two liquids are employed, and generally some other substance than copper for one of the metals. The liquids must be separated by some porous substance, which shall prevent them from mingling, and at the same time, being saturated by the liquid, shall not interrupt the necessary moist communication between the metals.

456. Constant Batteries.—Batteries composed of cells containing two liquids are called *constant,* because their action continues for so long a time without sensible abatement. Among the best of these is *Grove's battery,* one element of which is shown in Fig. 243, which represents a glass jar containing a hollow cylinder of zinc, which has a narrow opening on one side from top to bottom, that the liquid in which it is placed may circulate freely within it. Within the zinc is a cylindrical cup of porous earthenware, and within that is suspended a lamina of platinum. One of the circuit wires is in metallic communication with the zinc, and the other with the platinum, by means of the binding screws at the top. The earthen cup is now filled with strong nitric acid, while the space outside of it, in which the zinc is placed, contains dilute sulphuric acid. It is necessary to amalgamate the surface of the zinc with mercury, in order to prevent the action of the acid when the circuit is broken.

FIG. 243.

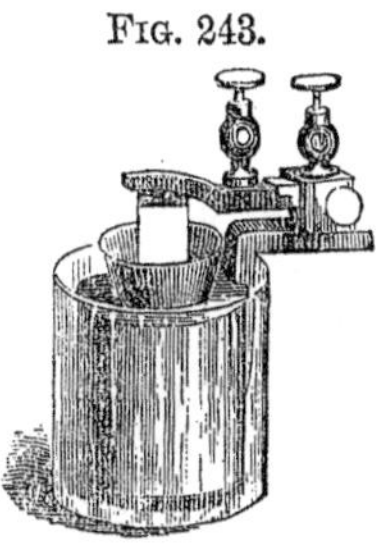

Bunsen's battery is the same as Grove's, except that in it a cylinder of carbon is used instead of a leaf of platinum, on account of the expense of the latter. It is very generally employed in telegraphy. Fig. 244 is a Bunsen battery of ten cells.

457. Direction of the Current.—Both positive and negative electricities are furnished by a galvanic battery. In one of copper and zinc, the former is found at the extremity of the wire connected with the copper plate, which extremity is therefore called the *positive* electrode. For a corresponding reason, the other electrode is the negative one. On the supposition that electricity is a fluid (a hypothesis which is now discarded, though the convenient terms which it gave rise to, as *current*, *flow*, &c., are

FIG. 244.

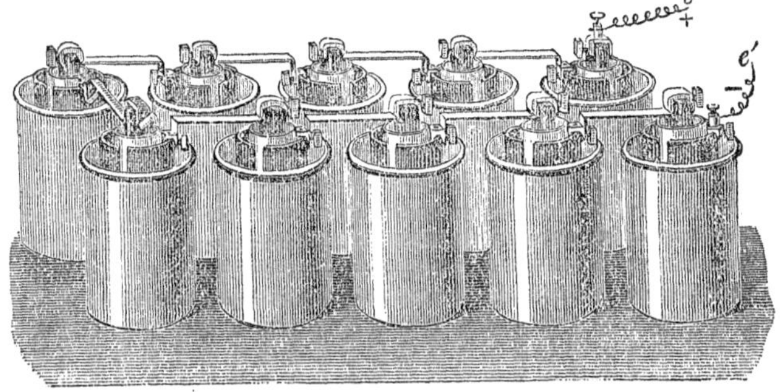

retained), there are manifestly two currents, flowing in opposite directions. For the sake of convenience, only the *positive* one is spoken of as *the current*. The direction in which this passes through the wires is *from the copper to the zinc*.

458. Galvanic and Frictional Electricity Compared.—The electricities furnished by chemical action and by friction are undoubtedly the same in kind. But they differ in that the former is produced in greater *quantity*, while the latter is in a state of greater *intensity*, or *tension*. This will be understood by referring to heat. The *quantity* of heat in a warm room is vastly greater than that in the flame of a lamp; yet the former is agreeable, while the latter, if touched, causes severe pain by its greater intensity. In a similar manner, a quantity of galvanic electricity may pass through the body without harm, which, if it possessed the intensity of frictional electricity, would instantly destroy life.

The word *tension*, or *intensity*, expresses the degree of force exerted by electricity in overcoming a given obstacle, as a break in a circuit.

(1) From this difference in quantity and intensity results a very great difference in *continuance of action*. This is indicated by the terms *dynamical* and *statical*. Galvanic electricity, being produced in prodigious quantities and with very feeble tension, may flow in a steady, gentle stream for many hours, and is hence called *dynamical*. While frictional electricity, being small in quantity and intense in action, darts through an opposing medium instantaneously, and with great violence. What motion it has is therefore

merely incidental to its passage from one state of *rest* to another. Hence the propriety of the term *statical.*

(2) Again, owing to its low tension, galvanic electricity will traverse many thousands of feet of wire rather than pass through the thin covering of silk with which the wire is insulated, and which would be but a slight obstacle in the path of frictional electricity.

(3) Analogous to the latter is its inability to pass from one conductor to another in its immediate vicinity. In order to establish the flow of a current, the electrodes must first be brought into actual contact, or exceedingly near to each other. They may then be separated more or less, according to the intensity of the battery, without interrupting the current.

459. Actual Amount.—Comparisons have been made of the actual quantities of electricity obtained by chemical action and by friction. Faraday has shown that to decompose one grain of water into its constituent elements, oxygen and hydrogen, requires an amount of frictional electricity equal to the charge of a Leyden battery with a metallic surface of *thirty-two acres,* equal to a very powerful flash of lightning. But by a galvanic current, the same result is accomplished in three minutes and forty-five seconds. From this some idea may be formed of the vast quantity of electricity produced during the steady flow for several hours of a Grove or Bunsen battery.

460. Quantity and Tension Regulated.—It may be stated in general that quantity increases with the surface of metal, and intensity with number of elements. Thus, from an element which presents two square feet of surface of metal to the action of the acids, we obtain a greater *quantity* of electricity than from one whose metallic surface is one square foot, but no increase of tension. On the other hand, from two elements, each of one square foot of surface, we find greater *tension,* but no increase in quantity.

461. Manner of Connecting the Elements of a Battery.—When *quantity* of electricity is desired, all the plates of the same name in the several elements should be united by connecting wires, as, for example, all the zinc plates together, and all the copper together. The battery thus becomes substantially a single large cell, and is called a *quantity battery.*

When *tension* is sought for, the zinc of one cell should be joined to the copper of the next, and so on through the series. A battery thus formed is called an *intensity battery.*

462. Effects.—The presence of a galvanic current is indicated by certain chemical, physical, physiological, or magnetic effects.

An example of the first is the decomposition of water, already mentioned.

A physical effect is the production of light. When the electrodes are brought together, and then separated, a spark is produced of varying intensity and duration.

The shock which is felt when the electrodes are held in the hands, and which affects more or less of the person, is a physiological effect.

The magnetic properties of a current will be spoken of hereafter.

463. Size of Battery for Required Results.—The physical or physiological results obtained from a single element of ordinary size of any kind are quite limited. No shock can be obtained from the direct current of a single cell. But a smart one is given by fifty Bunsen cells. It is felt only at the instant of closing or breaking the circuit. A shock from a battery of several hundred cells would affect the system painfully, if not dangerously. Nine hundred cells of copper, zinc, and dilute acid, furnish an arch of flame between the electrodes six inches in length. Brilliant results are also obtained from twenty Grove cells.

Such magnetic and chemical results as require a current of low intensity and small quantity may readily be obtained from a single cell. Such are electrotyping or the deflection of the magnetic needle.

CHAPTER II.

ELECTRO-MAGNETISM.

464. Helices.—A wire bent in a spiral, as in Fig. 245, is called a *coil* or *helix*. If the wire is coiled in the direction of the thread of a common or right-hand screw (Art. 136), it is called a *right-hand helix;* if in the direction of the thread of a left-hand screw, it is called a *left-hand helix.* Without referring to the screw, the distinction between the right and left hand helix may be described thus: When a person looks at a helix in the direction of its length, if the wire, as it is traced *from*

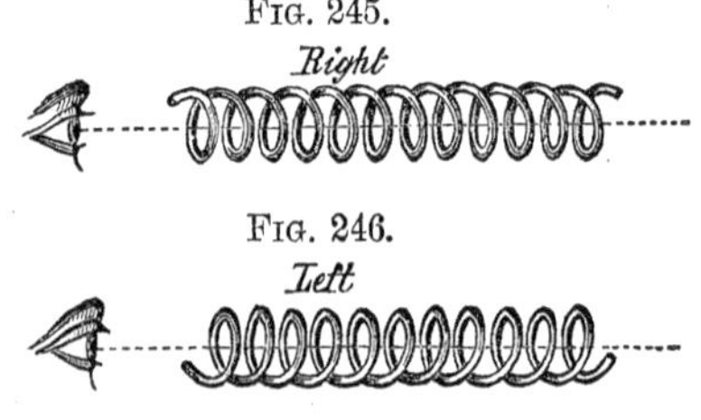

Fig. 245.

Fig. 246.

him, winds from the left *over* to the right, it is a right-hand helix (Fig. 245); if from the right over to the left, a left-hand helix (Fig. 246).

465. The Solenoid.—Let a helix be constructed as in Fig. 247, in which the ends are turned back through the coil, metallic contact being avoided throughout; this is called a *solenoid*—that is, a *tubular* or *channel-shaped* magnet. Next, let the electrodes *p* and *n* of a battery be furnished with sockets, one vertically above the other, in which the two ends of the helix wire are placed. The solenoid is then free to turn nearly a whole revolution around a vertical axis, at the same time that a current is passing through it. The helix is supposed to be a left-hand one, and is so connected with the battery that the current passes through it from *N* to *S*, and therefore around it from right over to left.

FIG. 247

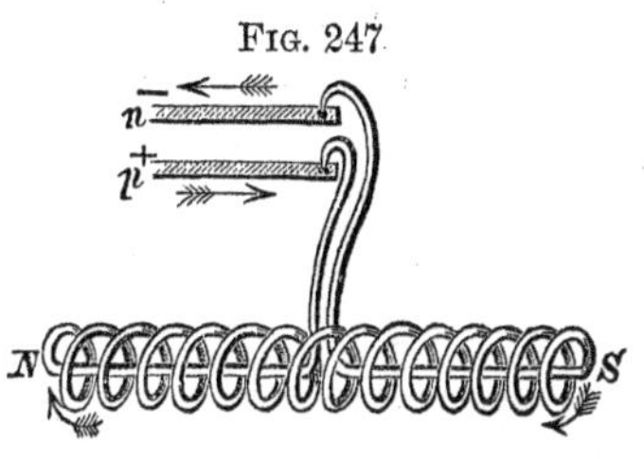

While the current flows, the following phenomena may be observed:

1. If a magnet be brought near it, *N* will be attracted by the south pole, and *S* by the north pole. If, instead of a magnet, another solenoid be presented to it, whose corresponding extremities are *N'* and *S'*, *N* and *S'* will attract each other, as also *S* and *N'*.

2. If not disturbed, the coil will place itself lengthwise in the direction of the magnetic meridian, with the extremity *N* toward the north, and *S* toward the south.

3. If a bar of iron be placed within it, the bar will become a magnet, having its north pole at *N*, and its south pole at *S*.

If a right-hand helix had been employed, all these phenomena would have been reversed.

466. Ampere's Theory of Magnetism.—In these experiments a coil is found to act the same as a magnet whose north and south poles are at *N* and *S* respectively. We therefore deduce the following:

1. A helix traversed by a galvanic current is a magnet the position of whose poles depends on the direction of the current.

2. Conversely, a magnet, like a coil, *may* be conceived to owe its magnetic properties to currents of electricity which traverse it.

This is the theory of Ampère, and is the one generally received, notwithstanding some objections to it.

In the helix a single current is present. But in a magnet we must conceive of an infinite number of currents, the circuit of each being confined to an individual molecule. Fig. 248 represents a magnet according to this theory, and *N* and *S* (Fig. 249) show the extremities of the north and south poles on a larger scale. The arrows on the convex surface show the general direction of all the currents—that is, of those portions of them nearest the surface, where magnetism is in fact developed—and may therefore represent them all.

Fig. 248.

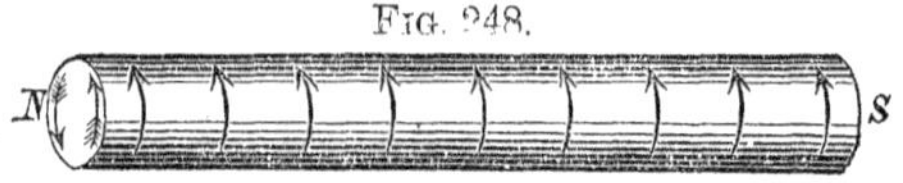

Fig. 249.

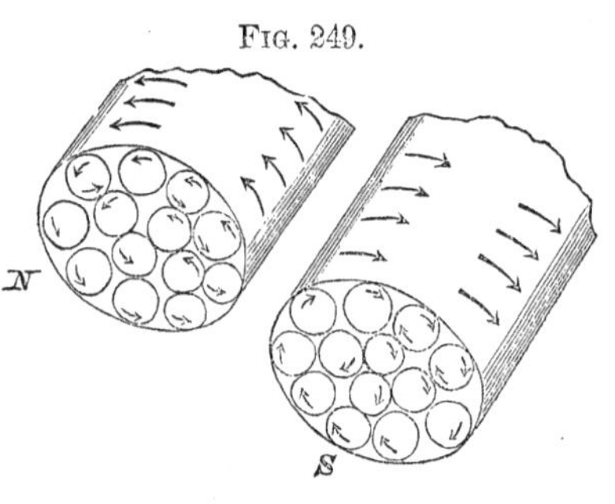

Since *S* is the south pole of the magnet, as supposed to be seen by an observer looking at it in the direction of its axis, it follows that when a magnet is in its normal position, that is, with its *north pole pointing northward*, its currents circulate from *west over to east*, and therefore from left over to right if the observer is also looking northward. In like manner, it is evident that to a person looking along the length of a magnet, from its north toward its south pole, the currents circulate from the right over to the left.

These supposed currents of the magnet are so small that we cannot take cognizance of them directly. But on the basis of Ampère's theory, we may substitute for them the large and manageable current of a helix. Then, by determining experimentally the causes of magnetic phenomena in the case of the latter, we may assign the same causes to like phenomena of the magnet.

467. Mutual Action of Currents.—

1. If galvanic currents flow through parallel wires in the *same direction*, they *attract* each other; if in *opposite directions*, they *repel* each other. These effects are shown in Fig. 250, where *A B*, *A′ B′*, turn toward each other, while *C D*, *C′ D′*, turn away from each other.

Fig. 250.

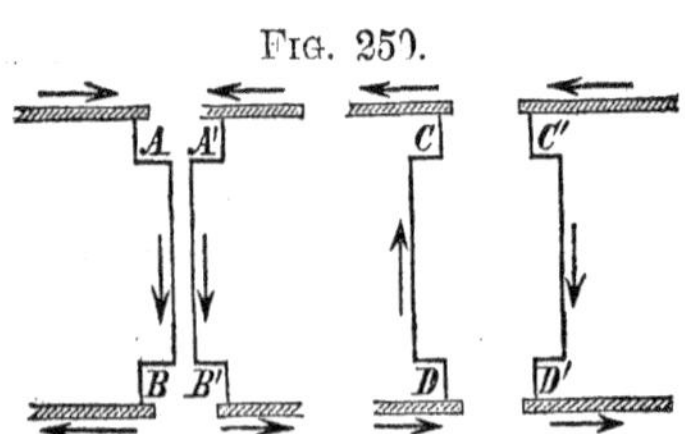

Hence, when a current flows through a loose and flexible helix, each turn of the coil attracts the next, since the current moves in the same direction through them all. In this way, a coil suspended above a cup of mercury, so as to just dip into the fluid, will vibrate up and down as long as a current is supplied. The weight of the helix causes its extremity to dip into the mercury below it; this closes the circuit, the current flows through it, the spirals attract each other, and lift the end out of the mercury; this breaks the circuit, and it falls again, and thus the movement is continued.

2. If currents flow through two wires near each other, which are free to change their directions, the wires tend to become parallel to each other, with the currents flowing in the same direction. Thus, two circular wires, free to revolve about vertical axes, when currents flow through them, place themselves by mutual attractions in parallel planes, as in Fig. 251, or in the same plane, as in Fig. 252. In the latter case, we must consider the parts of the two circuits which are nearest to each other as small portions of the dotted straight lines, *c d* and *e f*.

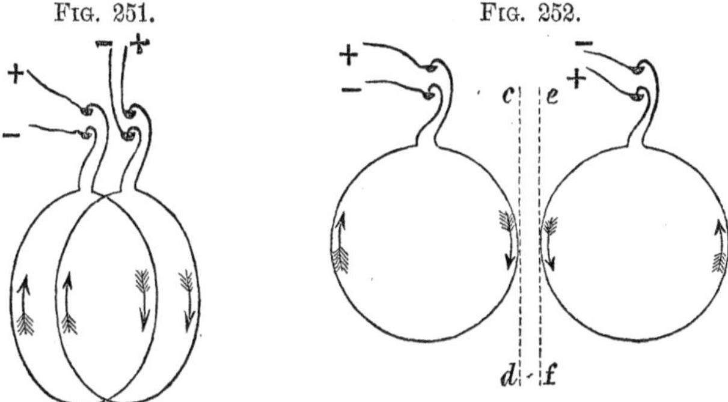

Fig. 251. Fig. 252.

It appears, therefore, that *galvanic currents, by mutual attractions and repulsions, tend to place themselves parallel to each other in such a manner that the flow is in the same direction.*

Supposing the same to hold true of the molecular currents of magnets, this single law will satisfactorily account for the phenomena of magnetic polarity.

In the following articles these phenomena are considered in the order in which they are mentioned in Art. 465.

468. Relations of Currents and Magnets to Each Other (1. Art. 465).—It should be constantly borne in mind that *when the north pole of a magnet turns toward a person, its currents circulate from his right over to his left.*

1. When two solenoids, suspended as in Fig. 247, or when a solenoid and a magnet, or two magnets, are brought near each other, poles of different names attract, and those of the same name

repel. For, when the magnets suspended from A and B (Fig. 253) are in the same line, it is seen that the currents are parallel and

FIG. 253.

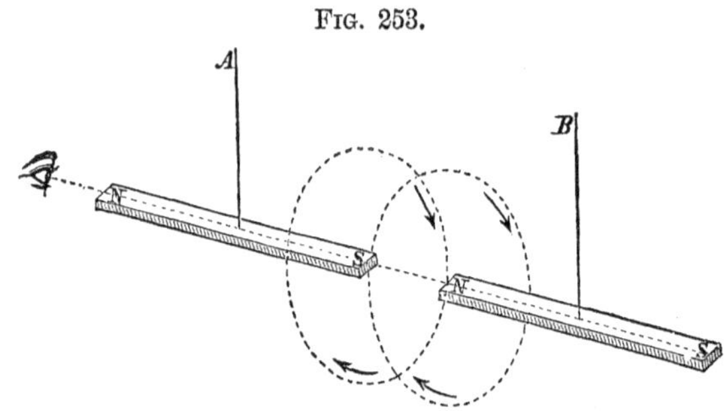

flow in the same direction in all the corresponding parts; and in Fig. 254, where they hang side by side, the nearer parts of the

FIG. 254. FIG. 255.

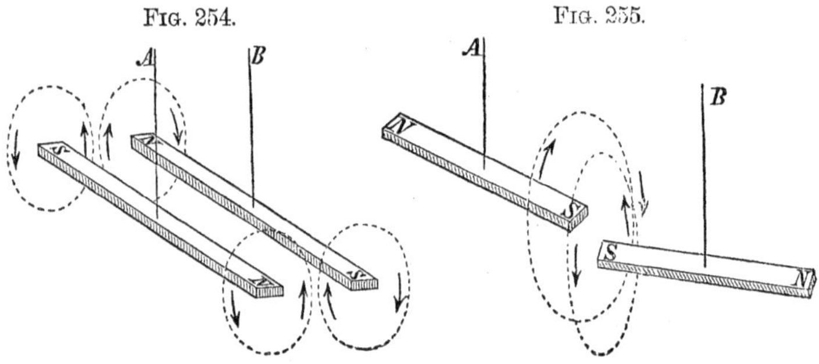

currents are parallel and flow in the same direction. While in Fig. 255, where like poles are contiguous, the corresponding parts of the currents flow in opposite directions.

2. When a magnet is suspended within a loop through which a current flows, if free to move it will place itself at right angles to the plane of the circuit, with the north pole pointing toward a person, when the current passes from his right over to his left (Fig. 256). Therefore, if the circuit is in a horizontal plane, the magnet turns its north pole downward, if the current flows as in Fig. 257, or upward if the current is reversed.

FIG. 256.

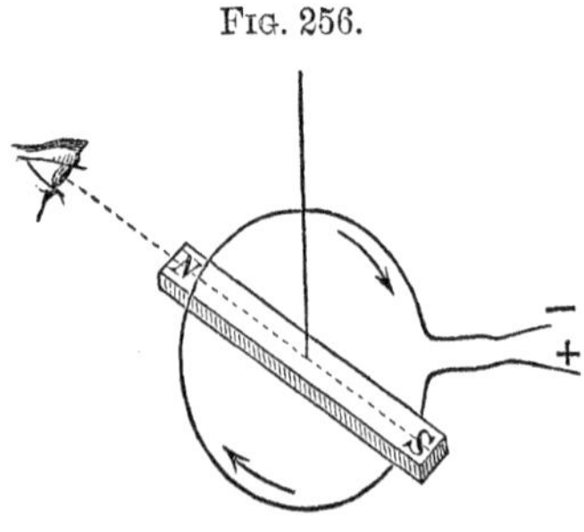

3. When a magnet is brought near a closed circuit wire, as + – (Fig. 258), it will place itself tangentially to a circle, $x\ y\ z$,

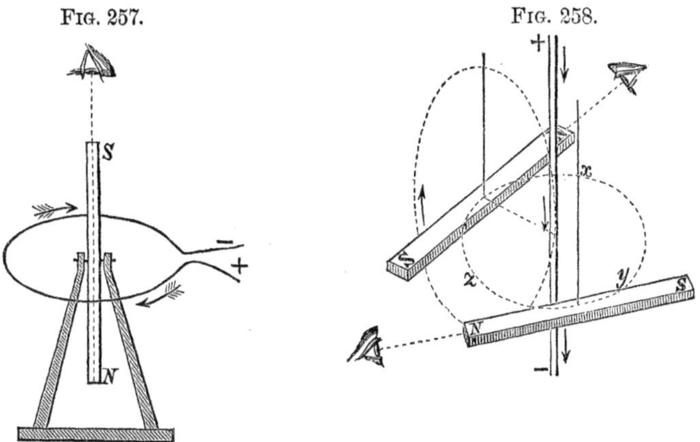

FIG. 257. FIG. 258.

whose centre is in the wire, and its plane perpendicular to it. The part of the wire nearest to the magnet may be considered as a small portion of a loop around it, as in Fig. 256. This tangential relation is maintained on all sides of the circuit, it being everywhere true that when the north pole is directed to a person, the current *descends on the left,* as if it had passed from the right over to the left.

Comparing Figs. 257 and 258, it is evident that the current and the magnet may change places without disturbing their relative directions, it being understood that the *current flows* in the same direction in which the *north pole points.*

469. The Galvanometer.—Advantage is taken of the directive influence of a current on a magnet in the construction of the galvanometer (Fig. 259). When the coil consists of many convolutions of wire, a very feeble current passing through will deflect the needle from its north and south direction, and the amount of deflection serves as a measure of the galvanic force. Hence the name of the instrument. To render it still more sensitive, a second smaller needle, with poles reversed, attached to the same vertical wire, makes the first nearly astatic with relation to the earth. In making such a coil, the wire must be carefully insulated. This is generally done by winding it with silk thread. In the figure, the galvanometer is repre-

FIG. 259.

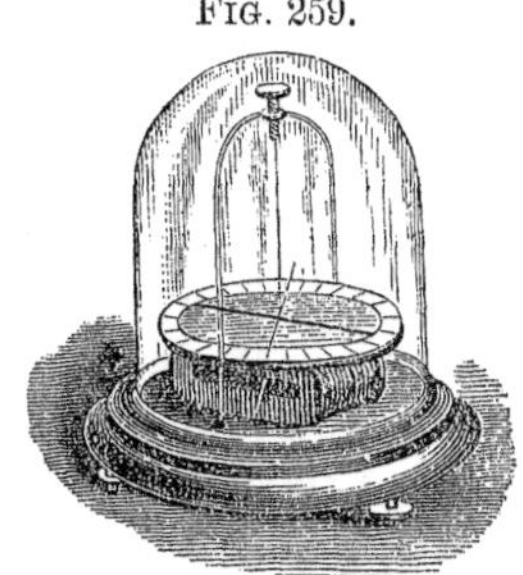

sented as covered by a bell-glass. The coil is seen beneath the graduated circle; the deflected needle projects as a white line from within the coil, and directly above it is the needle, which nearly neutralizes the earth's influence upon it.

470. Polarity with Respect to the Earth (2. Art. 465).— It is believed that currents of electricity are constantly traversing the earth's crust, passing around it from east to west, and making the earth itself a magnet, with boreal magnetism developed at the north pole, and austral at the south pole. Thus the earth may be taken as the standard magnet, and both it and the currents around it control the polarity of the needle. For, as in Fig. 260, in order

Fig. 260.

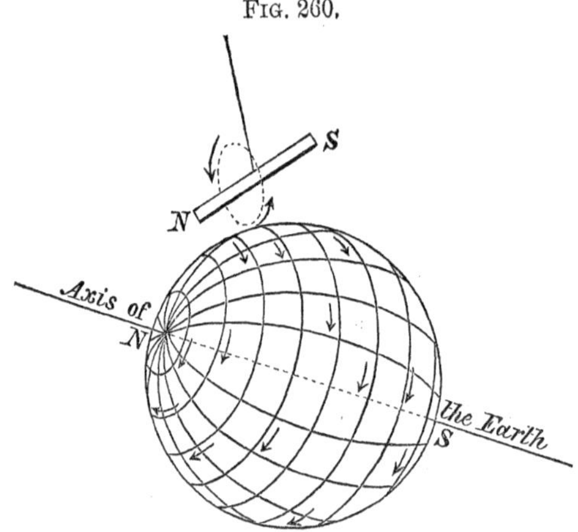

that the current of the magnet may be parallel with the adjacent terrestrial current, and in the same direction with it, since the latter passes from east to west, the lower side of the former must also pass from east to west. But in order that this may be the case, the north pole of the magnet must point northward, and this it does when free to obey the directive influence of the earth.

At first view, the earth currents from east to west seem to be in the wrong direction; for that is from left over to right, to a person to whom the north pole points. This, however, is explained by recollecting that the magnetism of the north pole of the earth is the same as that of the south pole of a magnet (Art. 380). For convenience, that end of a needle which *points* north is called the north pole; but by the law of attraction between opposite poles, it must be unlike the north pole of the earth. Therefore, the rule for the direction of currents around a magnet must be reversed when applied to the earth.

The existence of currents traversing the earth's crust has been variously accounted for. The strong analogy between them and those of thermo-electricity points to the heat of the sun as at least a very probable cause.

471. Thermo-Electricity.—Let a number of bars of bismuth (*b*) and antimony (*a*) be soldered together as in Fig. 261. Now if the flame of a candle be carried around so as to warm the outer joints, a current of electricity will pass through the circuit from left to right, and will influence a needle near it just as any other current would do, flowing in the same direction. Furthermore, it is only while the metals are unequally heated that the current flows. We may therefore suppose that the terrestrial current may be caused, in part at least, by the unequal heating of the heterogeneous substances composing the earth's crust, as the sun's heat is alternately poured upon and withdrawn from them once in every diurnal revolution.

Fig. 261.

472. Magnetic Induction by Currents (3. Art. 465).—Ampère accounted for the phenomena of magnetic induction by supposing that galvanic currents circulate through the molecules of all bodies, but in different directions, so that they mutually neutralize each other. That in a few substances, such as steel and iron, it is possible to control these currents and cause them all to flow in the same direction; and that when this is done, the phenomena of polarity ensue.

Supposing this to be the correct explanation, the effect of a galvanic current (and in fact of any method of magnetizing) is simply, by repulsion and attraction, to produce uniformity of direction among these magnetic currents.

473. The Permanent and Temporary Magnet.—When a current of sufficient strength is passed around a bar of well-tempered steel, a *permanent* magnet of considerable power may be obtained.

With *soft* iron, the result is a *temporary* magnet, which retains its magnetic properties only while the current is in motion. In either case the poles are always in the position which those of a needle would voluntarily assume if placed in the same relation to the current.

474. The U-Magnet.—Let a piece of soft iron, in the form of a horseshoe or the letter U (Fig. 262), be wound with a coil of insulated copper wire whose extremities, *W* and *w*, are dipped in cups of mercury, in which are also dipped the electrodes + and − of a battery. When all the wires are in metallic communication, the circuit is closed, and the current passing around the iron makes it a magnet; and since to a person looking along the length of the helix the current passes from right over to left, the north pole is at *N*, and the south pole at *S*. As soon as the circuit is broken by lifting out of the mercury any one of the wires, the weight which was previously sustained will fall, showing that the iron is no longer a magnet.

Fig. 262.

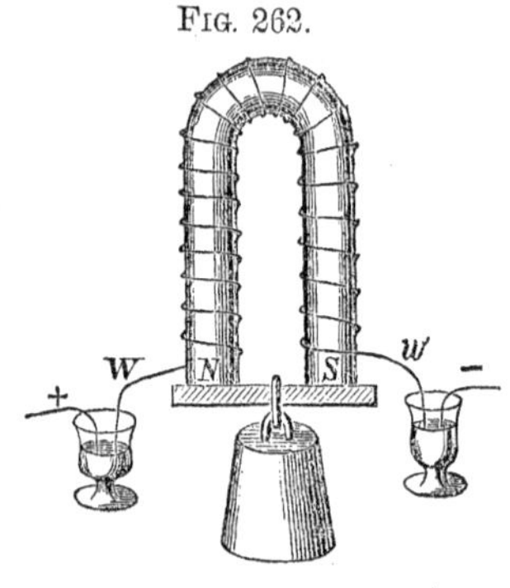

475. Helices.—The form of coil or helix generally employed is shown in Fig. 263. Many hundreds or even thousands of feet of insulated wire are wound around two bobbins, and through the centre of each passes a branch of the U-shaped iron; or, more frequently the central cores of iron are separate pieces, joined by a third one across two of the ends, and thus a U-magnet of modified form is obtained. By employing a fine wire coiled many times around the bobbins, a magnet of very great power may be formed, considering the weakness of the battery which furnishes the current. A magnet formed by the use of a small Bunsen cell has been known to lift five hundred pounds, and with twenty Grove cells can be made to sustain a weight of three tons.

Fig. 263.

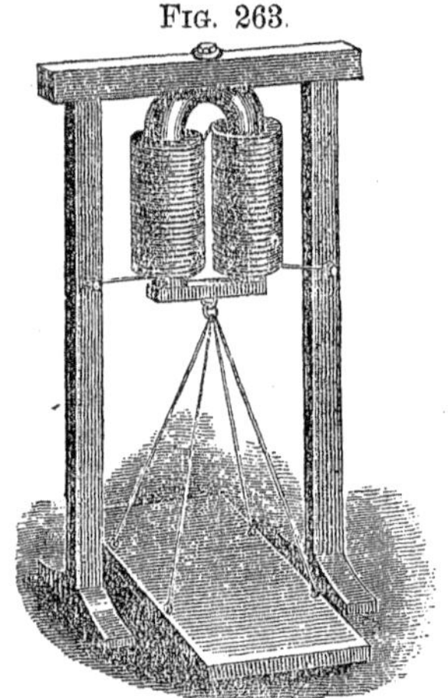

CHAPTER III.

INDUCED CURRENTS.

I. Currents Induced by Currents.

476. Experiment.—Place a copper wire, *a b* (Fig. 264), near *p n*, the circuit wire of a battery. The current flowing through *p n* acts inductively on *a b*, decomposing its natural electricity. According to the general law of induction, the positive electricity of

Fig. 264.

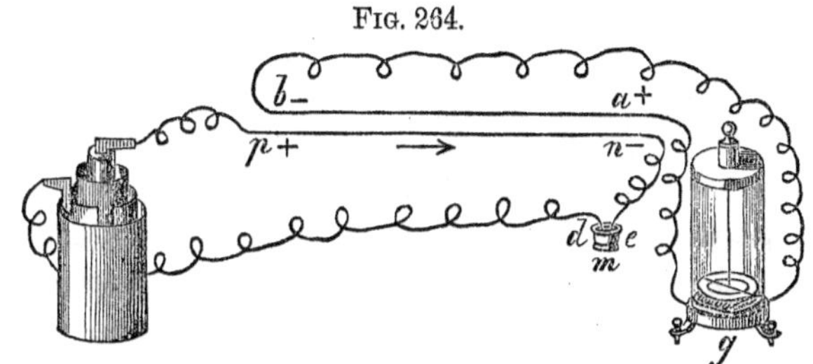

a b is attracted in the direction of *n*, and the negative in the direction of *p*. In this experiment let the following facts be noticed:

(1) The decomposition of the natural electricity of *a b* occurs *at the instant of closing* the circuit *p n*.

(2) While the circuit *remains closed*, the current passing through it *does not induce a current through a b*. There is probably, however, an accumulation of positive electricity in the direction of *b*, and of negative in the direction of *a*; for

(3) When the circuit is broken, the natural electrical equilibrium of *a b* is *instantly restored*, and no further signs of a current can be detected until the circuit is again closed.

The circuit is conveniently closed by dipping the end of the wire, *e* or *d*, in the cup of mercury *m*, and broken by removing it from the cup.

477. The Induced and Inducing Currents.—The sudden decomposition of the natural electricity mentioned in (1) of the preceding Article involves a momentary *flow* of the two electricities in opposite directions; that is, a *current* is made to traverse the wire from *a* to *b*, that being the direction of the flow of the positive fluid (Art. 457). And the restoration of equilibrium mentioned in (3) involves a *reversal* of this flow; that is, it produces a current which passes from *b* to *a*.

These two currents in $a\ b$ are called *induced currents;* and the one in $p\ n$, to which they owe their origin, is called the *inducing current.* The presence, direction, and duration of the induced currents are indicated by the galvanometer g.

The terms *flow, current, accumulation of electricity,* &c., when applied to the whole wire, have no significance except as they indicate the *resultants* of those electrical disturbances which are probably confined to the individual molecules of the wire.

478. Characteristics of Induced Currents.—It is obvious that induced currents differ materially from the current of a battery which is uniform in direction and constant in intensity for an appreciable length of time. The following are the distinctive features of induced currents:

(1) Induced currents are *instantaneous.*

(2) They result from *interruptions* of the inducing current.

(3) On *closing* the circuit, the direction of the resulting induced current is *opposite* to that of the inducing current.

(4) On *breaking* the circuit, the induced and inducing currents are in the *same* direction.

479. Inducing and Induced Currents in one Wire.—We have thus far considered only the inductive influence of the current on a wire exterior to its circuit. But the circuit-wire $p\ n$ itself possesses its share of natural electricity, as well as $a\ b$.

This is believed to exist independently of the galvanic current passing through it, and to be decomposed by that current. In order, therefore, to produce the preceding results with a single wire, let the circuit-wire be coiled as in Fig. 265. Each spire is now acted upon inductively by the galvanic current passing through the adjacent spires in the manner already described for separate wires. In addition to this mutual inductive influence of the several spires on each other, it is probable that the natural electricity of every portion of the wire is still further decomposed by the galvanic current passing through it. For it is a noticeable fact that when a very long circuit-wire is employed, induced currents are obtained even though it be so nearly straight that no one portion can act inductively on another.

FIG. 265.

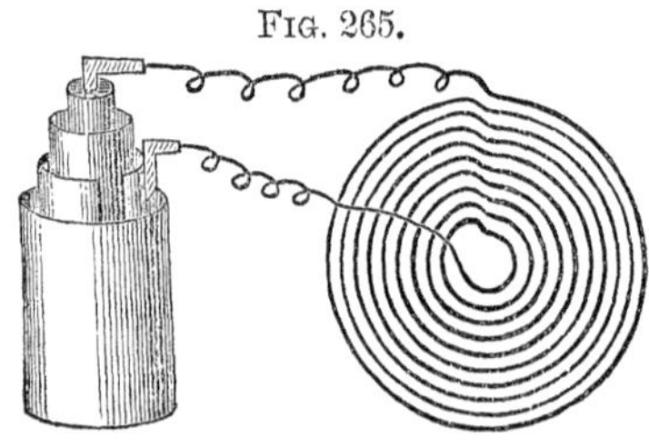

The result of these several inductive actions is that when the circuit is closed and broken, regular induced currents are generated in it. *And since these coexist for an instant of time with*

the inducing current, and pass through the same electrodes with it, it follows—

(1) That when the circuit is *closed,* the inducing current is partially neutralized, and has its intensity diminished by the induced current which flows in a direction contrary to its own; and

(2) That when it is *opened,* the induced current having now the same direction as the inducing current, reinforces it and augments its intensity.

480. Mode of Naming Circuits and Currents.—The phenomena of induced currents were discovered by Faraday in 1832, and to him we owe the foregoing explanation of them. The following terms now in use were also introduced by him:

The *inducing* current is called the *primary current,* and the wire it traverses the *primary wire.* Currents *induced* in the primary wire are called *extra currents;* the one obtained on *closing* the circuit is the *inverse extra current;* the one on *opening* it is the *direct extra current* (Art. 478, 3, 4).

A wire exterior to the primary, as *a b* in Fig. 264, is a *secondary* wire, and the currents induced in it are *secondary currents.*

481. Currents Induced in Coils.—Instead of straight wires or loose spirals, compact coils of carefully insulated wire are employed. Thus all parts of the wire are brought much nearer to each other, and the inductive influence is far more energetic. Indeed, without a coil, the presence of induced currents can generally be detected only with a delicate galvanometer. The following experiments show the effects of coils:

(1) Around a hollow wooden bobbin, *b* (Fig. 266), coil about 100 feet of No. 16 insulated copper wire. Let this be made a part of the circuit of a battery, as shown in the figure. This circuit is of course closed when *m* and *n* touch each other. Now if *m* and *n* be held one in each hand and then separated, the body of the operator becomes a part of the circuit, and the primary current, not having sufficient intensity to pass through it, ceases. But the direct extra current passes through, producing a shock. When the wires are brought together again, the primary and inverse extra currents pass through the metallic circuit, and no shock is felt.

FIG. 266.

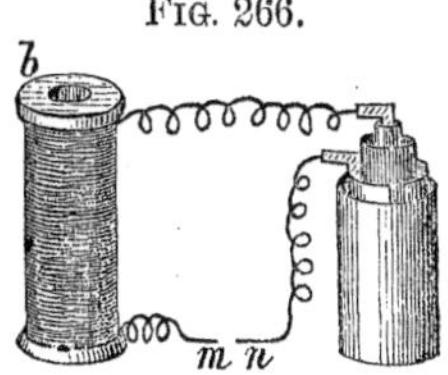

The more rapid the rate at which *m* and *n* are brought together and separated, the more decided are the results obtained.

To produce the most marked effect, attach a coarse file to one end, as *m*, and hold it in one hand while *n* is drawn rapidly over the ridges of its surface with the other.

(2) Fig. 267 represents the same coil as Fig. 266, with the addition of a bundle of soft iron wires, *w*, inserted in the hollow bobbin.

Fig. 267.

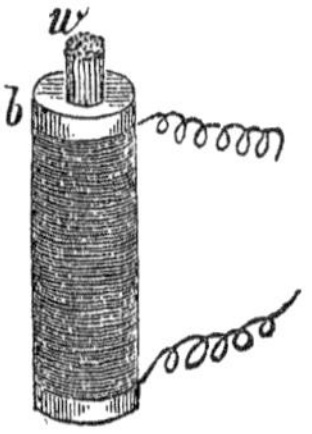

When the circuit is closed, these wires are magnetized—that is, the Ampèrean currents supposed to reside in them are made to circulate in the same direction as the battery current (Art. 472).

And since the appearance and disappearance of these magnetic currents are simultaneous with the appearance and disappearance of the primary current, they augment the effects of the latter, and the resulting extra currents are of greater intensity.

The effect of soft iron in the primary coil is an observed fact, and the above is the way in which Faraday accounted for it on the basis of Ampère's theory.

(3) Let the primary coil and bundle of wires of the preceding figure be placed within a secondary coil, *d* (Fig. 268), from which it is carefully insulated. This secondary coil should be made of wire much greater in length and smaller in diameter than that of which the primary coil is made. For instance, let it consist of 1500 feet of No. 35 insulated copper wire. When the ends of this wire, *h*, *h'*, are held one in each hand, every time the primary circuit is interrupted, a secondary current traverses the secondary circuit of which the person forms a part. The resulting shocks will be quite appreciable, though the primary current be produced by only a single small cell.

Fig. 268.

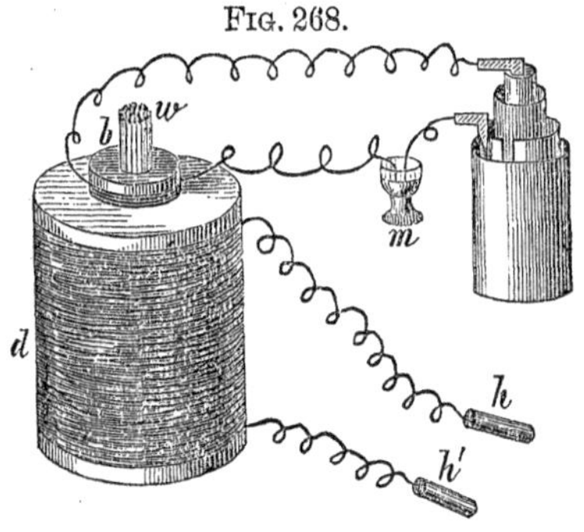

As in the first experiment, the effect on the person will become more marked as the interruptions increase in frequency.

In the third experiment, the magnetic currents of the iron core add their inductive influence, as already explained, to that of the primary current, thus increasing the intensity of the secondary currents.

The effect of the *extra currents* also should not be overlooked. As these traverse the primary coil, alternating with each other in direction, they materially modify the effects of the primary and magnetic currents, which are uniform in direction.

482. Ruhmkorff's Coil.—The celebrated Ruhmkorff coil (Fig. 269) is not essentially different from the one just described, except in having (1) an arrangement for producing a continued and rapid succession of interruptions in the primary current, (2) a *commutator*, or key, by which when desired the primary current may be stopped or its direction reversed, and (3) a *condenser* to neutralize the effects of the extra currents. The condenser consists of sheets of tin-foil insulated by oiled silk. They are placed out of sight, in the base of the apparatus, and are so connected with the primary wire that the extra currents pass into them. Owing to this diversion of these interfering currents the efficiency of the coil is very much increased.

FIG. 269.

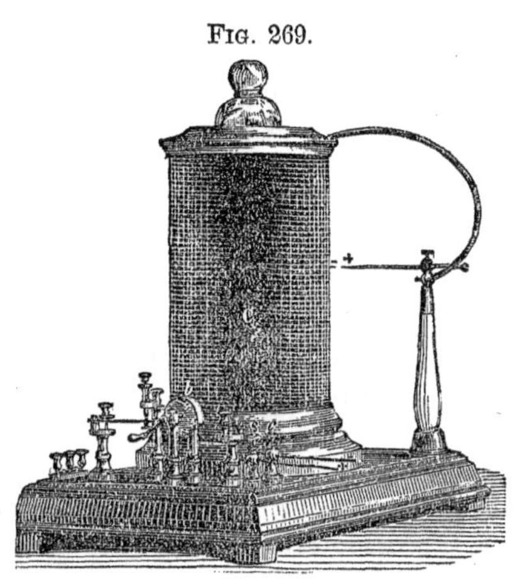

483. Power of the Ruhmkorff Coil.—The efficiency of a Ruhmkorff coil depends largely on *complete insulation;* and, in different coils, varies greatly with the *length and fineness* of the *secondary wire.*

To secure insulation, the wires are (as usual) wound with silk thread, then each individual coil around the axis is separated from the succeeding one by a layer of melted shellac, and lastly a cylinder of glass is placed between the primary and secondary coils.

With regard to the secondary coil, one of the largest size sometimes contains *sixty miles* of the finest copper wire. With such an apparatus, though the primary current be produced by only two or three Bunsen cells, the secondary currents are of such intensity that sparks *eighteen inches* in length, and of great brilliancy, may be obtained; and indeed, all the tension effects of a large electrical machine, as well as the quantity effects of a powerful galvanic battery, may be reproduced.

Great care should be taken in handling an induction coil of this size, for the shock resulting from its discharge through the body would be dangerous, and might possibly prove fatal.

484. One Coil moved into, and out of, another.—In all that has preceded, the interruptions of the primary current have been supposed to take place *instantaneously*. If these interruptions are *gradual*, the resulting induced currents remain the same in direction as before, but vary in intensity and duration.

Thus, if the primary coil *c* (Fig. 270) be made to fit loosely in the secondary coil *d*, and then be moved up and down (the primary circuit remaining closed), it will be found—

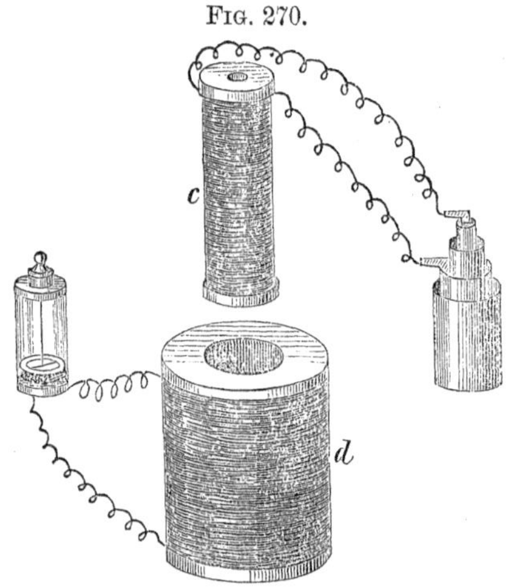

Fig. 270.

(1.) That each insertion and removal of it corresponds, the one to a gradual *closing*, the other to a gradual *opening* of the primary circuit—the result of the former being an *inverse* secondary current, of the latter a *direct* secondary current.

And since a continuous motion of the primary coil produces a continuous series of instantaneous secondary currents with no appreciable interval between them, it will be found—

(2.) That the secondary currents are *continuous* in effect as long as the motion of the primary coil is continuous;

(3.) That their intensity varies with the rate of motion of the primary coil, diminishing or increasing as that is moved slowly or rapidly; from which it follows—

(4.) That they cease whenever the primary coil is brought to a state of rest in any position.

485. Changes of Intensity in the Primary Current.—All the results just mentioned may be obtained if, instead of changing the position of the primary coil, as above, it remain at rest while a corresponding series of variations be produced in the primary *current*,—an *increase* of intensity in that corresponding to an *insertion* of the coil, and a *decrease* to a *removal* of it.

II. Currents Induced by Magnets.

486. Magneto-electricity.—Faraday reasoned that if currents could induce magnetism, a magnet ought to induce currents.

This he found to be the case, and thus discovered a new branch of physical science, to which he gave the name of magneto-electricity.

If a magnet be used instead of the primary coil in Fig. 270, all the phenomena mentioned in Art. 484 may be reproduced.

Thus, with the coil and magnet in Fig. 271, we obtain the following results:

Fig. 271.

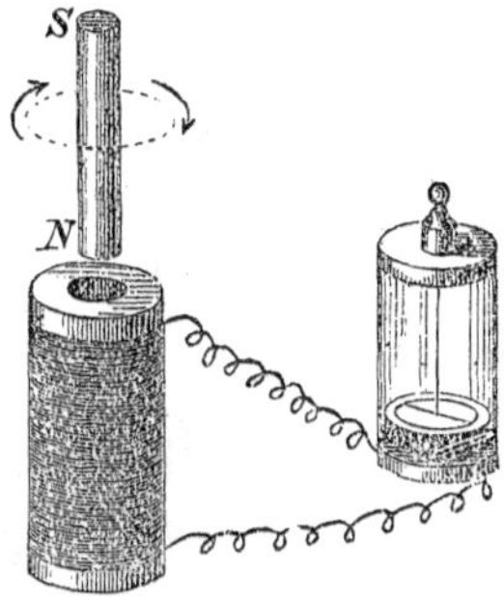

(1.) When the magnet is alternately inserted in and withdrawn from the coil, the latter is traversed by induced currents alternating with each other in direction.

(2.) These currents are continuous while the magnet is in motion.

(3.) Their intensity diminishes or increases as the magnet moves slowly or rapidly.

(4.) They cease when the motion of the magnet ceases.

487. Explanation of the foregoing Phenomena.—On the basis of Ampère's theory, the correspondence of these phenomena with those of Art. 484 can readily be accounted for. For the magnet may be considered a true primary coil, its magnetic currents corresponding to the primary current in Fig. 270. With regard to them, the induced currents are regular inverse and direct secondaries; for in any given case they will be found to have the same direction as those induced by a primary current whose direction corresponds with the *supposed* direction of the magnetic currents.

It will be seen at once what a strong argument is here furnished in favor of Ampère's theory.

Fig. 272.

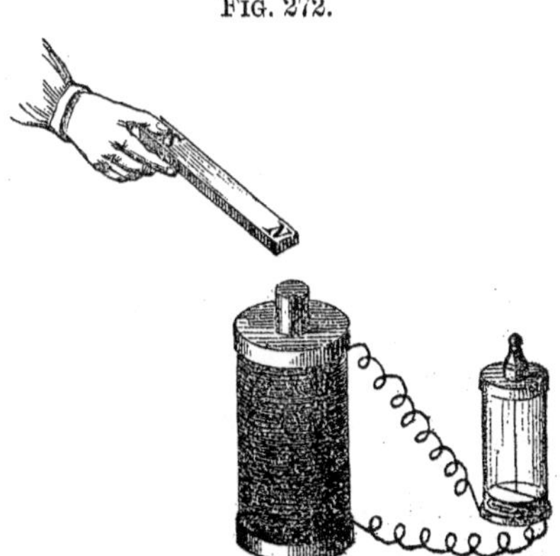

488. An Iron Core, changing its Magnetic Intensity. — Replace the magnet (Fig. 271) by a bar of soft iron inserted in the coil, and let a magnet be alternately brought near this, and removed from it, as in Fig. 272. The same results will be obtained as in the preceding series of experiments. The proximity of the magnet induces magnetism in the soft iron, and its motions to and fro produce variations in this induced magnetism corresponding precisely with the varying intensity of the primary current mentioned in Art. 485, and, as might be expected, the results are the same.

In this experiment, it is obviously immaterial whether the coil be at rest and the magnet be moved, or the magnet be at rest and the coil be moved. The latter method is adopted in some magneto-electric machines.

489. Clarke's Magneto-electric Machine.—In front of the poles of the U-magnet, *A* (Fig. 273), is revolved the *armature*,

FIG. 273.

consisting of the two *bobbins*, *B*, *B'*, which are coils of fine wire with cores of soft iron. These cores are joined to each other and to the axis of rotation by the bar of soft iron, *V*, and motion is communicated by the multiplying wheel and band at *W*.

As one of the bobbins passes before a *north* pole while the other is passing before a *south* pole, the resulting induced currents are *relatively* of contrary directions; but as one of the coils is always right-handed and the other left-handed, the currents passing through them at any given instant have the same *absolute* direction, so that the two coils act as one.

Fig. 274 shows the poles of the fixed magnet, and the direction in which the armature revolves. The maximum magnetization of the soft iron cores occurs when the bobbins are directly in front of *N* and *S*. While they move through the first and third quadrants they are *losing* their magnetism, and while moving through the second and fourth they are *acquiring* that of the *contrary kind*. The resulting induced currents will thus be direct and inverse *to contrary kinds of magnetism*, and will therefore have *the same absolute direction*. But as the bobbins pass from the second quadrant to the third, and from the fourth to the first, they lose the magnetism just acquired, and the induced currents change from inverse to direct *with reference to the same kind of magnetism*, and therefore become *reversed* in absolute direction.

FIG. 274.

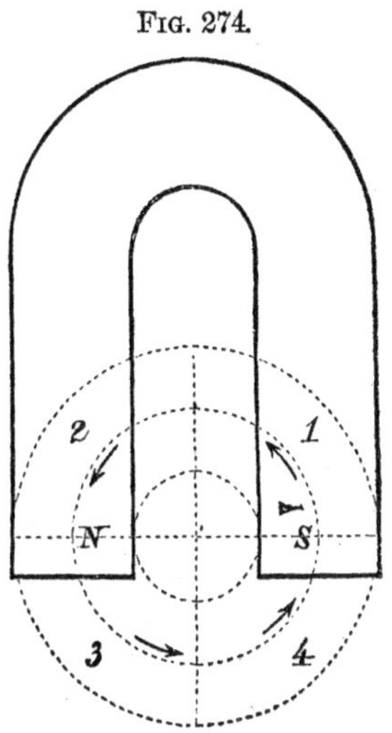

Hence the semi-revolutions of the armature on opposite sides of a line joining the poles of the permanent magnet produce currents of contrary directions.

490. The Commutator.—Fig. 275 is an enlarged view of the outer end of the axis (Fig. 273), and shows the *commutator*, or

FIG. 275.

arrangement by means of which the contrary currents just mentioned are made to furnish one—or rather a series of currents—flowing in the same direction.

Two pieces of brass, *m* and *n*, are insulated from each other by being fastened to an ivory ring, *i*, around the axis. They are so connected with the wires of the coils that at any given instant both coils present to *m* one kind of polarity, and to *n* the other; that is, they are made the *poles* of the two coils acting as one. Against m and n press two springs, b and c, which are brought into communication with each other through the plates P and P', and the handles h and h', whenever the latter are joined. When b and c press against m and n respectively, it will be seen that the circuit is complete. Let us suppose that the induced current is passing through it in the direction indicated by the arrows. When the armature has revolved through 180° from its present position, m and n will have changed places—*but they will also have changed polarities* (Art. 489). Therefore n presents to b the same polarity which m did, and hence there is no change in the direction of the current through b and c.

491. Effects of Rapid Revolution.—The intensity of the induced currents of this machine, as also the rapidity with which they succeed each other, is regulated by the rate of revolution of the armature. When this is rapidly revolved, they produce all the effects of a single voltaic current, so that the apparatus may be used as a galvanic battery with h and h' for its electrodes. At the same time its physiological effects are most remarkable, the shocks becoming unendurable when it is revolved with great rapidity. The shocks are more powerful when a third spring, a (Fig. 274), is attached to the plate P', near c.

492. Large Machines.—In large magneto-electric machines of this kind, increased efficiency is obtained in two ways:

First, by multiplication of magnets. In Nollet's machine, constructed in 1850, 192 magnetized steel plates are so combined as to make 40 powerful U-magnets. These are arranged in eight rows around the circumference of a large iron frame inside of which revolve sixty-four bobbins.

Second, by multiplication of currents. In Wild's machine, constructed on this plan, the induced currents first obtained, instead of being directly utilized, are passed through the coils of a large electro-magnet. Before the poles of this a second armature revolves, and the resulting induced currents are far more powerful than the first. These may in turn be made to magnetize a second electro-magnet before the poles of which a third armature revolves, &c.

Currents of very great intensity may be obtained from either of these machines. The motive power employed is generally a steam-engine of from one to fifteen-horse power.

CHAPTER IV.

PRACTICAL APPLICATIONS.

493. Classification.—The applications of Galvanic electricity in the arts and sciences, as well as in the affairs of every day life, are eminently *practical.* They may be classified according to the way in which are utilized those molecular forces whose resultant is known as *the current.*

It may be stated in general that these applications are made either *within the circuit,* or *exterior to it.*

Within the circuit. Here the electrical force is employed (I) *directly,* or (II) by being first made to produce the effects *light* and *heat,* which are then applied as desired.

Without the circuit. Here it is employed *indirectly* (III) by being made to reappear as mechanical motion through the intervention of the kindred force, magnetism.

Examples will be given of each.

I. Direct Applications of the Current.

494. Electrolysis.—When a current is passed through a binary compound (*i. e.,* one containing two elements), the compound is decomposed, one of its elements appearing at the positive electrode, the other at the negative.

For instance, water, consisting of the two gases oxygen and hydrogen, is thus decomposed.

In the bottom of the dish *D* (Fig. 276), partly filled with water, are fastened *p* and *n,* the platinum electrodes of a battery. Over these are placed two tubes, *O* and *H,* full of water. On closing the circuit, oxygen rises from *p* into *O,* and hydrogen from *n* into *H.*

Fig. 276

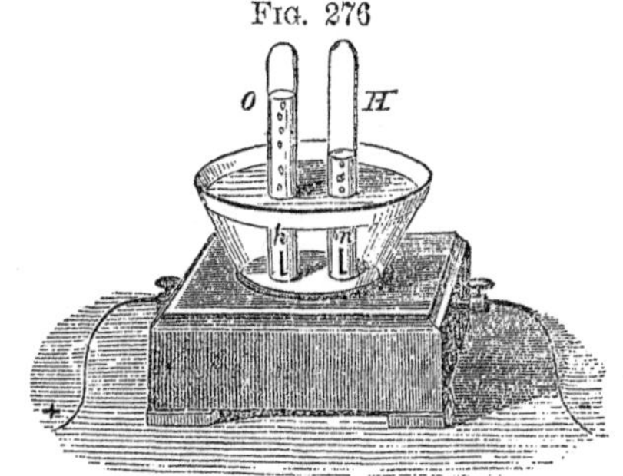

Electrolysis is of the utmost importance in chemistry. Thus, the preceding experiment gives a correct analysis of water, and if oxygen had been previously unknown, would have been the means of its discovery. In this way were discovered several of the metallic elements.

No less important are its applications in the arts. For when a solution of a metallic salt is subjected to the action of the current, it is decomposed and a permanent film of the metal is deposited on any suitable material placed so as to receive it. The process is then called *electro-metallurgy*, or *electro-plating*.

495. Electro-plating.—The bath (Fig. 277) contains a saturated solution of blue vitriol (sulphate of copper). In this is suspended by wires from the metallic rod *D* a plate of copper *C*, and from *B* (also metallic) the cast of a medal *m*, which is to be coated with copper. Connect *D* with the positive electrode of a battery and *B* with the negative. The current passing through the solution removes from it particles of copper and deposits them on *m*. Those taken from the liquid are replaced by others taken from *C*, which is thus gradually wasted away, and the solution is kept saturated.

FIG. 277.

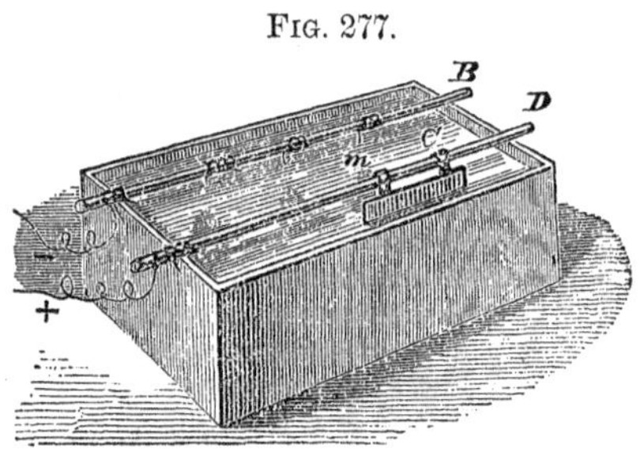

If the bath contains a solution of gold, and *C* is replaced by a piece of gold and *m* by a silver cup, the cup will be *electro-gilded*. *Electro-silvering* is an analogous process.

To produce in any case a firm and even coating, the process must be allowed to proceed *slowly* by the employment of a weak current. On a small scale a single cell is sufficient. In large establishments a magneto-electrical machine turned by steam has been successfully and economically used.

496. Electrotyping.—By taking proper precautions, the copper film deposited on *m* may be *removed*, and its surface will be found to present an exact fac-simile of the medal of which *m* is an impression. Therefore if *m* is an impression of the *type* from which a page is printed, when the copper has been removed and stiffened by melted lead (or some alloy) poured over its under surface, it may be used in the printing-press instead of the type. It is then called an *electro-type plate*, and when not in use may be preserved indefinitely for succeeding editions, while the type of which it is a copy can be distributed and used for other purposes.

497. Medicinal Applications.—The shocks produced by the passage of interrupted currents through the system have already been alluded to. In certain ailments these shocks, when properly applied, have been known to produce beneficial results. On the other hand, great injury has resulted from their misapplication. Hence they should be employed as remedies only under the direction of a reliable physician.

The familiar medical magneto-electrical machine, which comes compactly stored in a box ten inches long and about four inches square at the end, does not differ essentially from Clarke's (Art. 489), except in lacking the *commutator*, so that its currents pass through the body alternating with each other in direction.

II. Applications of Electric Light and Heat.

498. Light by the Electric Current.—The electric light may be advantageously employed for brilliant illumination on special occasions; also where a strong penetrating light is needed, as in light-houses, or for signals between ships. For exhibiting to an audience magnified images of small objects (as with the projecting microscope) it has no superior; and to the physical experimenter the various colors it assumes on passing through highly rarefied gases of different kinds are of great interest. But as yet it cannot compete with gas-light for ordinary illumination.

To obtain the most brilliant effects carbon electrodes must be employed, and as these are constantly changing in length they must be kept at a uniform distance apart by machinery. The flame is not straight, but curved, as in Fig. 278, and is called the *voltaic arc.* To obtain it, the electrodes must first be made to touch each other. With 92 Bunsen elements the light has been found to possess more than one-third the intensity of direct sunlight.

Fig. 278.

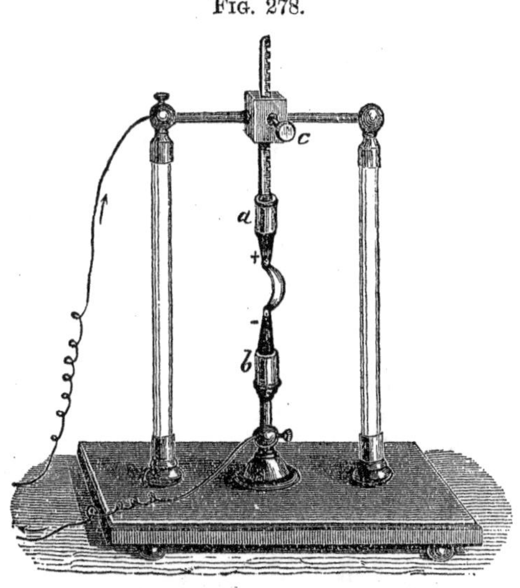

499. Heat by the Electric Current.—The heat as well as the light of the voltaic arc is intense. In the laboratory it is employed to deflagrate and vol-utilize refractory substances. When the lower electrode is hollowed out in the form of a cup, a piece of platinum (one of the least fusible of metals) placed in it is melted like wax in a candle, and a diamond, the hardest of known substances, is burnt to a black cinder. To produce either of these results, a battery of great power must be used.

Metals may also be deflagrated by being made part of the circuit in the form of very fine wires. They are thus employed to spring mines in time of war, or in blasting rocks. In Fig. 279, *B* is a box full of fulminating powder, and *w* is a very fine platinum

FIG. 279.

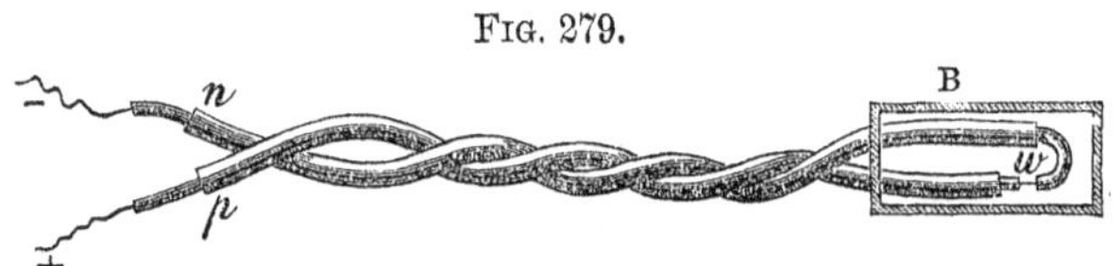

wire, about $\frac{3}{8}$ in. in length, fastened to *p* and *n*, which are insulated copper wires extending to a battery situated at any convenient distance. *B* and its contents are the fuse which is inserted in the powder to be fired. When a moderately strong current passes through *w*, it is heated sufficiently to ignite the fuse, and the powder explodes.

III. MECHANICAL APPLICATIONS.

500. Made through the Medium of Induced Magnetism. —As has been seen in preceding experiments, magnetism may be induced in a piece of steel or iron by the current passing through a circuit which is near it. Hence induced magnetism and its applications are results obtained *outside of the circuit.*

This magnetism may be utilized *directly.* For compass and galvanometer needles are ordinarily made by placing a steel needle in a helix through which a current is sent (Art. 473).

But its most numerous and important applications are in the way of *mechanical movements.* All these are modifications of the simple rising and falling of the armature of a U-magnet, mentioned in Art. 474. Thus, the armature *A* (Fig. 280) is limited in its fall by the metallic base *B*, so that it is within the influence of *M* the next time that it becomes a magnet. Hence, when the circuit is closed and broken at *n p*, the

FIG. 280.

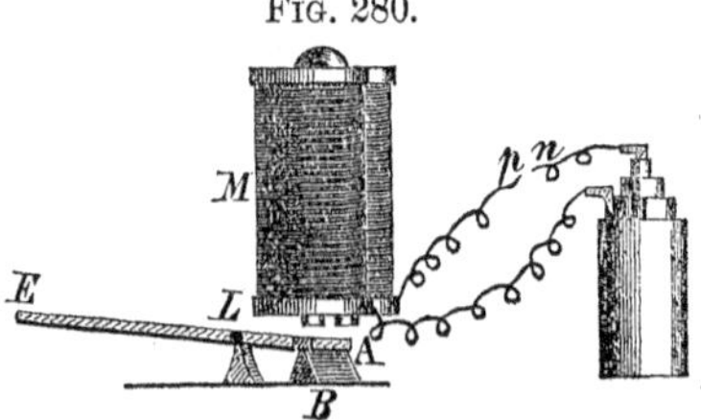

end *A* of the lever *L* rises and falls. It is evident that the corresponding motions of the end *E* may be applied in a variety of ways. A few of these are described in the following articles.

501. Electro-magnetic Engine.—*E* may be attached to a vertical arm, and that to the crank of a fly-wheel (Fig. 281), and the interruptions of the current may be made automatic by connecting *p* with *L*, and *n* with *B*. When *A* rests on *B* the circuit is closed; *M* becomes a magnet, and *A* is attracted by it; but as soon as *A* rises from *B* the circuit is opened, *M* no longer attracts it, and it falls back, only to close the circuit again and repeat the same movements as before. The tendency of this is to produce a rotary motion in the fly-wheel, and the apparatus involves the principle of a single-acting engine. With a second electro-magnet, and a somewhat different arrangement of parts, an actual double-acting engine may be constructed.

FIG. 281

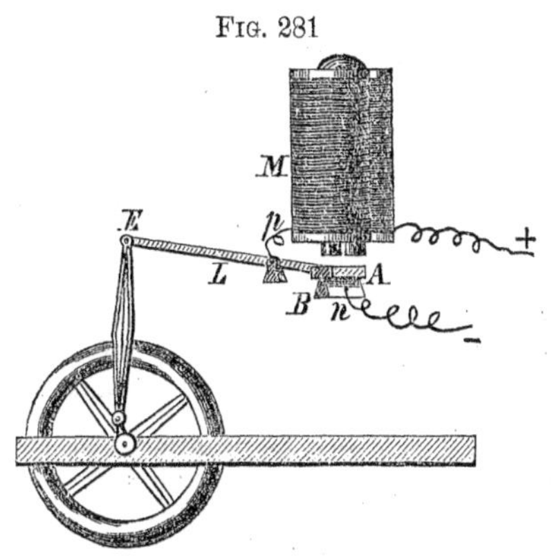

Various forms of this engine have been constructed and exhibited as curiosities, or used where expense was not regarded. But it cannot compete with the steam-engine as long as zinc and acids cost so much more than coal.

502. Electro-magnetic Telegraph.—To our countryman, Prof. S. F. B. Morse, is due the credit of the erection of the first telegraph line in the United States. It extended from Baltimore to Washington, and went into operation in 1844.

Communication in various ways by means of electricity between places a few miles apart was not unknown in Europe before that time, and several ingenious systems have appeared since. One of these is Wheatstone's, which is commonly used in England. But the Morse system has been very generally preferred on account of its greater simplicity and efficiency, and it is now widely used in the United States and on the continent of Europe, where it is known as the *American system*. The principle of its operation is as follows:

Let *E* of Fig. 280 be furnished with a *style e* (Fig. 282) directly over which is the groove on the surface of a solid brass roller *c*. Between *c* and *e* is the long paper ribbon *R R*. Also let *A* be placed *above* M and be furnished with a spring s to raise it as far as the screw i allows when it is not attracted by M. When the circuit is closed, A is attracted and e rises and forces the paper

FIG. 282.

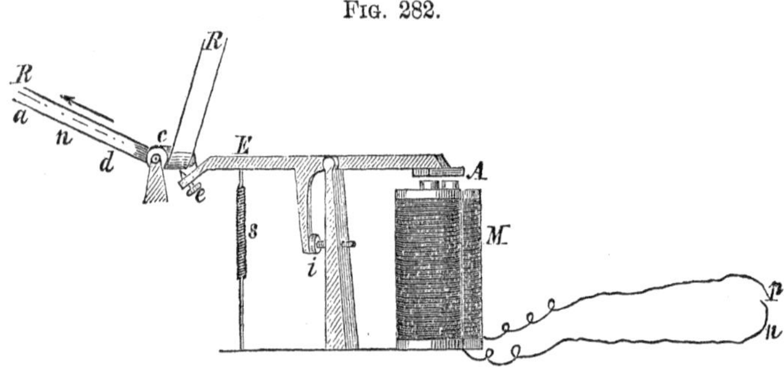

into the groove, producing a slight elevation on its upper surface. The ribbon is pulled along at a uniform rate in the direction of the arrow by clockwork (not shown in the figure), so that when the circuit remains closed for a little time, a *dash* is marked on the paper by e; when it is closed and instantly opened, the result is a *dot*—or rather a *very short dash*. *Spaces* are left between these whenever the circuit is opened. Combinations of these *dots*, *dashes*, and *spaces*, all carefully regulated in length, compose the letters of the alphabet. Spaces are also left between the letters, and longer ones between words.

By lengthening the circuit wire, it is evident that the person who sends the message at n p, and the one who receives it at E, may be miles apart, and the transmission will be almost instantaneous owing to the rapid passage of the current.

The essential parts of this system, or indeed of any system, are a *communicator* at n p, an *indicator* at E, and a *wire* extending from one to the other.

503. The Connecting Wire.—It was at first supposed that a complete metallic circuit was necessary, hence a return wire was

FIG. 283.

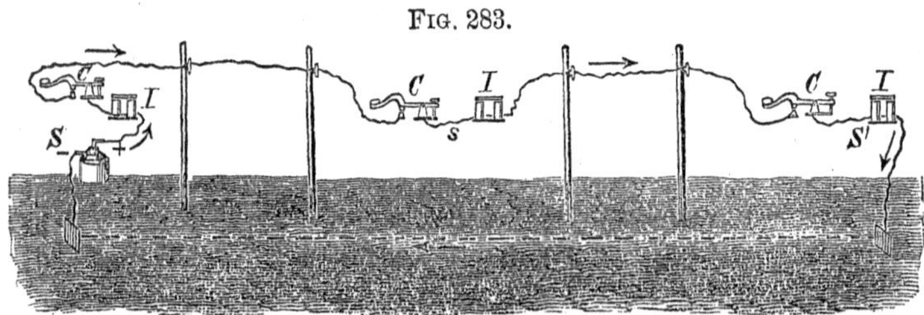

employed. But this was rejected when it was found that the earth could be used as a part of the circuit, as shown in Fig. 283.

S and *S'* are the terminal stations, and *s* is one of the way stations which may occur anywhere along the line. At every station both a communicator, *C*, and indicator, *I*, are introduced into the circuit, so that messages can be both sent and received.

504. The Communicator.—This consists of a *lever*, *l* (Fig. 284), and *anvil*, *a*, both of brass, and insulated from each other.

FIG. 284.

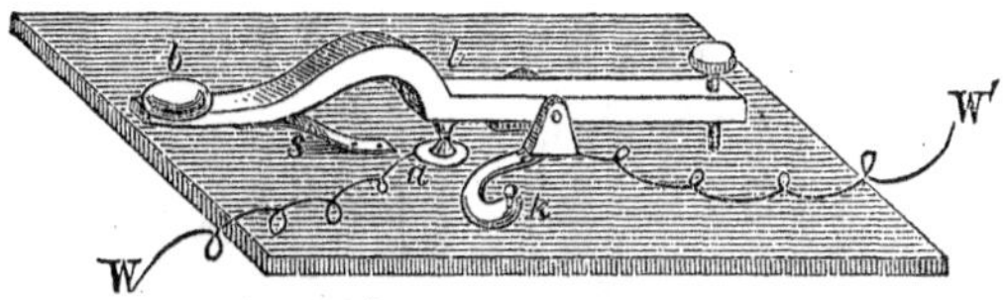

The anvil connects with the line wire *W*, and *l* with the rest of the circuit through *W'*, and *W' W''* of the next figure. (See also Fig. 283.) The end of *l* is depressed by the finger of the operator on the insulating button *b*, and is raised by the spring *s* when the pressure is removed. The former movement closes the circuit, the latter opens it, and by a succession of these the message is sent.

When the communicator is not in use, the brass bar *k* hinged to the base of *l* is pressed into contact with *a*. This closes the circuit for other stations on the line, and hence *k* is called the *circuit closer*. The whole apparatus is called *the key*.

505. The Indicator.—This consists of two parts, (1st) the *relay*, and (2d) either the *register*, or the *sounder*.

The first part, called the *relay*, or the *relay magnet* (Fig. 285), consists of an *electro-magnet*, an *armature*, a *lever*, and a *spring*, the same as in Fig. 282, except that the electro-magnet is horizontal, and the other parts correspond in position. The tension of the spring *s* is regulated by the screw and milled head *h*, and *M* is adjusted by a similar screw (between *W'* and *W''* in the figure), which slides it along the grooved way *X*. One end of the coil wire passes out through *W'* to *W'* of Fig. 284. The other end connects at *W''* with one pole of the battery if it is at *S* (Fig. 283), with the earth if it is at *S'*, or with the line wire to the next station if it is at *s*.

The reason for introducing the relay is this: The current from the preceding station has become too feeble to cause indentation of the paper by the style, and thus make a visible record, or even to produce a distinct sound of the armature upon the magnet for reading messages by the ear. The relay is therefore contrived for employing this feeble current to close and open the circuit of a *local battery*, whose current is powerful enough to deliver messages in either form, or even in both forms at once. All which the weak current of the distant battery has to do is to cause the armature A to move toward the magnet M till the top of L *touches* the screw N, and thus closes the circuit of the local battery. When the current ceases, a delicate spring, s, draws L back from contact with N, and breaks the circuit of the local battery.

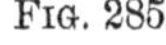
Fig. 285.

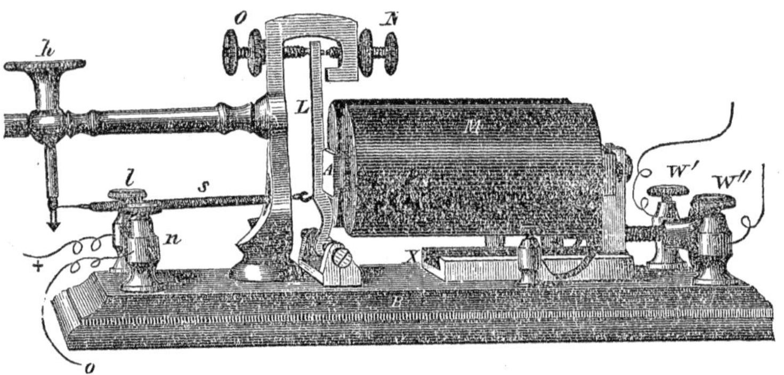

By the adjustments above described, the distance through which L moves, and the force of the spring s, may be made as small as the operator pleases.

506. The Register and the Sounder.—The *second* part of the indicator is either a *register*, or a *sounder*, according as messages are to be addressed to the eye or to the ear. The register (Fig. 282) has been already described; the current of the local battery close at hand has force enough to cause visible indenta-

Fig. 286.

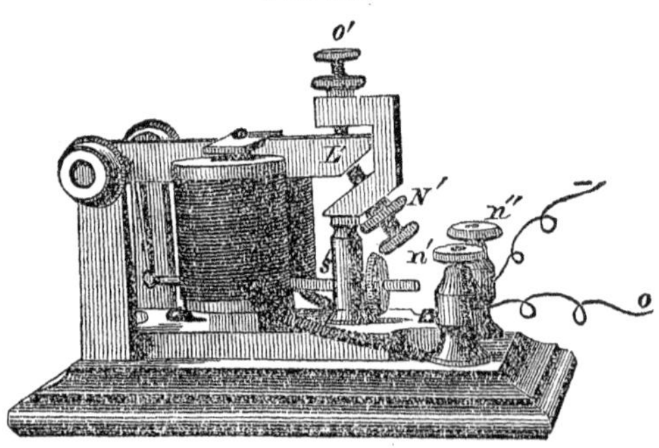

tions in the paper whenever the lever is drawn to the magnet; and this record can be read at any time subsequently. Within a few years a modified form of the register has come into use, and is called the sounder. In this the end of the lever L' (Fig. 286), instead of being furnished with a style, is made to strike against the two screws, *N'*, *O'*. The downward *click* is a little louder than the upward one, and so the beginning and end of each *dot* or *dash* are distinguished from each other. Many operators learn from the first to *read by the ear*, and have never used a register.

Whether a register or a sounder is employed, its coil wire is entirely distinct from the line wire, and belongs only to the local battery. The circuit of this battery may be traced (Figs. 285, 286) from the positive pole through *l*, *L*, *N*, *n*, *o*,. . . .*o*, *n'*, the coil of the sounder, and *n''*, to the negative pole. The binding screws, *l*, *n*, *n'*, *n''*, are connected with their respective levers, or contact screws, by insulated wires concealed in the bases. When *A* is attracted by *M*, *L* touches *N*, and the circuit is closed; when it is withdrawn by *s*, the circuit is opened, because *O* is insulated. Hence the motions of *L* and *L'* are simultaneous.

Since the relay is always in the main circuit, it communicates, by means of the local current, to the operator at whose station it is, all the messages sent between any two stations on the line, including those which he himself sends. Hence, if his own indicator does not operate while he is at work, he knows that his message is not passing over the line, owing to some break in the circuit.

507. Repeaters.—On a well insulated wire the weakness of the current at the distance of a few miles from the battery is mainly due to the *resistance* of the wire. The nature of this resistance is unknown, but it is subject to the same law as the *friction* of a fluid along the interior of a tube; it varies *directly* as the *length*, and *inversely* as the *diameter*. Hence telegraph wires of considerable thickness are employed, and even then, after a certain number of miles, varying according to strength of battery, insulation, etc., the current will not work even a relay. Before reaching that point, therefore, the wire is allowed to pass into the ground, and so complete the circuit.

To pass a message beyond the place at which the current will only work a relay, *N'* and *L'* are made parts of a new circuit called a *repeater*, which is closed and opened simultaneously with the preceding one by the motions of *L'*, just as a local circuit is worked by *L* (Art. 505).

On the 28th of February, 1868, signals were sent through from Cambridge, Mass., to San Francisco, by the employment of thirteen repeaters. The time occupied by the signals in going and returning (making about 7,000 miles) was three-tenths of a second—allowance being made for the coil wires of the electro-magnets through which the current passed.

508. Atlantic Telegraph Cable.—This cable stretches a distance of 3,500 miles, and from the nature of the case is a continuous wire, so that it cannot be advantageously worked by the Morse apparatus. The indicator employed is a sensitive galvanometer needle which is made to oscillate on opposite sides of the zero point by the passage through it of currents in opposite directions. But to reverse the direction of the current throughout the whole length of the cable is a slow process. *For the cable is an immense Leyden jar*, the surface of the copper wire (amounting to 425,000 sq. feet) answering to the *inner* coating, the water of the ocean to the *outer*, and the gutta-percha between the two to the *glass* of an ordinary jar. A current passing into it is therefore detained by electricity of the contrary kind induced in the water, and no effect will be produced at the further end until it is *charged.*

This very circumstance, at first considered a misfortune, is now taken advantage of in a very simple and ingenious manner to facilitate the transmission of signals. The current is allowed to pass into the cable till it is *charged*—then, *without breaking the circuit*, by depressing a key for an instant, a connection is made between it and a wire running out into the sea; that is, between the inner and outer coatings. *This partially discharges it*, and the needle at the other end is deflected. When the key is raised the discharge ceases, the current flows on as before, and the needle is deflected in the opposite direction.

It is said that after this plan was adopted, twenty words could be sent through the cable per minute, whereas only four per minute could be sent before. The greatest speed thus far attained on land wires is believed to have been the transmission in one instance of 1,352 words in thirty minutes between New York and Philadelphia in 1868.

509. Fire-Alarm Telegraph.—Recurring again to the standard (Fig. 280), the end *E* may be so connected with machinery as to cause the striking of a bell in a distant tower whenever the circuit is closed at *n p*. In our large cities boxes are placed at convenient points, each containing a crank, or lever, by which the circuit may be closed and the fire-bell rung. Thus, by previously arranged signals, the locality of the fire is immediately made known at the various engine-houses.

510. Chronograph.—This is used in observatories for recording the passage of stars across the meridian. Imagine the circuit of Fig. 282 to be closed and instantly broken again, by a clock pendulum at the end of every second. As the paper, *R R*, moves uniformly, dots are made on it at *equal distances from each other*, each of which distances, therefore, represents one second. The observer has a key, by which also he closes the circuit for an instant when a certain star passes the meridian. The dot thus made shows, by its situation between the two nearest second dots, at what fraction of the second the transit occurred.

In practice, however, the record is more conveniently made on a large sheet of paper, which is wrapped tightly around a cylinder. The clock-work, which revolves the cylinder, also moves the recording pen in a line parallel to its axis. By these two motions, a spiral ink-line is traced on the paper. At the end of every beat of the observatory clock, the closing of the circuit gives the pen a momentary lateral movement, by which a slight notch is made in the line. A similar notch is made by the touch of the key, when the observer perceives the star on the meridian wire of the telescope. Fig. 287 represents a portion of the sheet after its removal

FIG. 287.

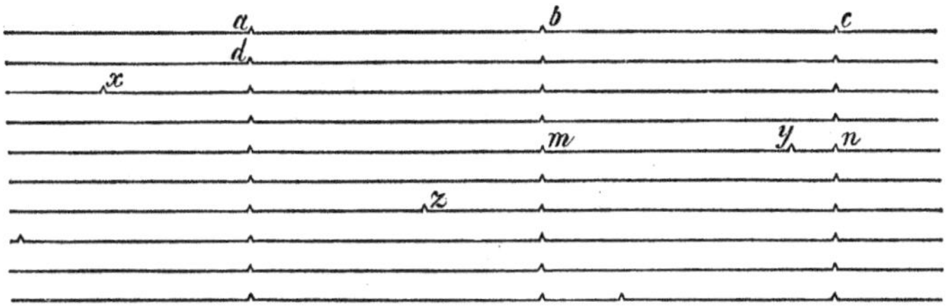

from the cylinder; *a*, *b*, *c*, *d*, &c., are the second marks; *x*, *y*, *z*, &c., are transit records. The ratio $m\,y : m\,n$ shows what fraction of the second $m\,n$ has elapsed when the transit y occurs.

PART VIII.

HEAT.

CHAPTER I.

EXPANSION BY HEAT.—THE THERMOMETER.

511. Nature of Heat.—Heat is another of those agencies which have been regarded as imponderable substances. It was said to emanate in straight lines from the sun and from bodies in combustion, to cause the sensation of warmth when it strikes us, to expand bodies when it enters them, to raise their temperature, &c. But there is abundant reason for believing that heat consists of exceedingly minute and rapid vibrations of ordinary matter and of the ether which fills all space. It is to be regarded as one of the modes of *motion*, which may be caused by any kind of force, and which may be made a measure of that force. Heat affects only one of our senses, that of feeling. Its increase produces the sensation of warmth, and its diminution that of cold.

512. Expansion and Contraction by Heat and Cold.—It is found to be a fact almost without exception, that as bodies are heated they are expanded, and that they contract as they are cooled. It is easy to conceive that the vibratory motion of the several molecules of a body compels them to recede from each other, and to recede the more as the vibration becomes more violent. Although the change in magnitude is generally very small, yet it is rendered visible by special contrivances, and is made the means of measuring temperature.

513. The Thermometer.—This instrument measures the degree of heat, or the *temperature*, of the medium around it, by the expansion and contraction of some substance. The substance commonly employed is mercury. The liquid, being inclosed in a glass bulb, can expand only by rising in the fine bore of the stem, where very small changes of volume are rendered visible. A scale is attached to the stem for reading the degrees of temperature.

The graduation of the thermometer must begin with the fixing of two important points by natural phenomena, the freezing and boiling of water. When the bulb is plunged into powdered ice, the point at which the column settles is the *freezing point* of the thermometer. And if it is placed in pure boiling water under a given atmospheric pressure, the mercury indicates the *boiling point.* Between these two points, namely 32° and 212° F., there must be 180°, and the scale is graduated accordingly. As the bore of the tube is not likely to be exactly equal in all parts, the length of the degrees should vary inversely as the area of the cross-section. This is accomplished by moving a short column of mercury along the different parts and comparing the lengths occupied by it. The degrees in the several parts must vary in the same ratio.

514. Different Systems of Graduation.—There are in use three kinds of thermometer scale, Fahrenheit's, Reaumur's, and the Centigrade. In Fahrenheit's, the freezing point of water is called 32°, and the boiling point 212°; in Reaumur's, the freezing point is called 0°, and the boiling point 80°; in the Centigrade, the freezing point 0°, and the boiling point 100°. In a scientific point of view, the Centigrade is preferable to either of the others, but Fahrenheit's is generally used in this country. The letter F., R., or C., appended to a number of degrees, indicates the scale intended. In this country, F. is understood if no letter is used.

515. To Reduce from one Scale to Another.—Since the zero of Fahrenheit's scale is 32° below the freezing point, while in both of the others it is at the freezing point, 32° must always be subtracted from any temperature according to Fahrenheit, in order to find its relation to the zero of the other scales. Then, since 212° − 32° (= 180°) F. are equal to 80° R., and to 100° C., the formula for changing F. to R. is $\frac{4}{9}$ (F. − 32) = R.; and for changing F. to C., it is $\frac{5}{9}$ (F. − 32) = C. Hence, to change R. to F., we have $\frac{9}{4}$ R. + 32 = F.; and to change C. to F., $\frac{9}{5}$ C. + 32 = F.

Mercury congeals at about −39° F.; therefore, for temperatures lower than that, alcohol is used, which does not congeal at any known temperature.

516. Expansion of Solids.—When the expansion of a solid is considered simply in one dimension, it is called *linear* expansion; in two dimensions only, *superficial* expansion; in all three dimensions, *cubical* expansion.

The linear expansion of a metallic rod is readily made visible by an instrument called the *pyrometer*, which magnifies the motion. The end A of the rod $A\ B$ (Fig. 288) is held in place by a screw. The end B rests against the short arm of the lever C, the longer arm of which bears on the arm D of the long bent lever

Fig. 288.

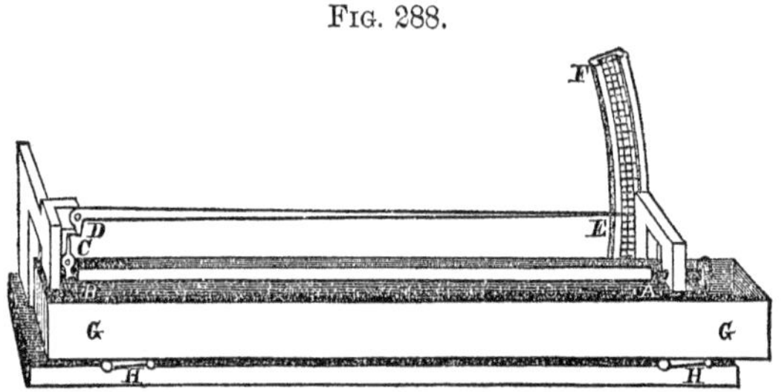

$D\ E$; this serves as an index to the graduated arc $E\ F$. The long metallic dish $G\ G$, being raised on the hinges $H\ H$, so as to enclose the bar $A\ B$, and then filled with hot water, the bar instantly expands, and raises the index along the arc $E\ F$.

517. Coefficient of Expansion.—The *coefficient* of linear expansion of a given substance is the fractional increase of its length, when its temperature is raised *one degree.* But since this increase is generally somewhat greater at higher temperatures, the coefficients of expansion given in tables usually refer to a temperature at or near the freezing point of water. Thus, the coefficient of expansion for silver is 0.00001061; by which is meant that a silver bar one foot long at 32° F. becomes 1.00001061 ft. in length at 33° F.

The coefficient of superficial expansion is *twice,* and that of cubical expansion *three times* as great as the coefficient of linear expansion. For, suppose c to be the coefficient of linear expansion; then if the edge of a cube is 1, and the temperature is raised 1°, the edge becomes $1+c$, and the area of one side becomes $(1+c)^2=1+2c+c^2$, and the volume $(1+c)^3=1+3c+3c^2+c^3$. But as c is very small, the higher powers may be neglected, and the area is $1+2c$, and the volume is $1+3c$; that is, the coefficient of superficial expansion is $2c$, and that of cubical expansion is $3c$, as stated above.

518. The Coefficient of Expansion differs in different Substances.—Copper expands nearly twice as fast as platinum; the ratio of expansion in steel and brass is about as 61 to 100. This ratio is employed in the construction of the compensation pendulum (Art. 171). The same is sometimes used also to render constant the length of the rod with which the base line of a trigonometrical survey is measured.

If two thin slips of metal of different expansibility be soldered together so as to make a slip of double thickness, it will bend one way and the other by changes of temperature. If it is straight at a certain temperature, heating will bend it so as to bring the most expansible metal on the convex side; and cooling will bend it in the opposite direction; and the degree of flexure will be according to the degree of change in temperature. Compensation in clocks and watches is sometimes effected on this plan. If the compound slip has the form of a helix, with the most expansible metal on the inside, heating will begin to uncoil it, and cooling, to coil it closer. A very sensitive thermometer, known as Breguet's thermometer, is constructed on this principle.

519. The Strength of the Thermal Force.—It is found that the force exerted by a body, when expanding by heat or contracting by cold, is equal to the mechanical force necessary to expand or compress the body to the same degree. The force is therefore very great. If the rails were to be fitted tightly end to end on a railroad, they would be forced out of their places by expansion in warm weather, and the track ruined. The tire of a carriage wheel is heated till it is too large, and then put upon the wheel; when cool, it draws together the several parts with great firmness. In repeated instances, the walls of a building, when they have begun to spread by the lateral pressure of an arched roof, have been drawn together by the force of contraction in cooling. A series of iron rods being passed across the building through the upper part of the walls, and broad nuts being screwed upon the ends, the alternate bars are expanded by the heat of lamps, and the nuts tightened. Then, when they cool, they draw the walls toward each other. The remaining bars are then treated in the same manner, and the process is repeated till the walls are restored to their vertical position and secured. For a measure of the force of heat see Art. 555.

520.—Expansion of Liquids.—It has already been noticed that mercury and alcohol expand by heat, and are therefore used in thermometers for measuring temperature. These are the best liquids for such a purpose, because their temperature of congelation is very low.

As liquids have no permanent form, the coefficient of expansion for them is always understood to be that of cubical expansion. There is a practical difficulty in the way of finding the coefficient for liquids, because they must be enclosed in some solid, which also expands by heat. Hence, the *apparent* expansion must be corrected by allowing for the expansion of the inclosing solid, before the coefficient of *absolute* expansion is known.

This fact is illustrated by the following experiment. Fill the bulb and part of the stem of a large thermometer tube with a colored liquid, and then plunge the bulb quickly into hot water; the first effect is, that the liquid *falls*, as if it were cooled; after a moment it begins to rise, and continues to do so till it attains the temperature of the hot water. The first movement is caused by the expansion of the glass, which is heated so as to enlarge its capacity and let down the liquid before the heat has penetrated the latter. It is obvious that what is rendered visible in this case, must always be true when a liquid is heated—namely, that the vessel itself is enlarged, and therefore that the rise of the liquid shows only the difference of the two expansions. Ingenious methods have been devised for obtaining the coefficients of absolute expansion of liquids, and the results are to be found in tables on this subject.

521. Exceptional Case.—There is a very important exception to the general law of expansion by heat and contraction by cold, in the case of water just above the freezing point. If water be cooled down from its boiling point, it continually contracts till it reaches a point somewhat above 39° F., when it begins to expand, and continues to expand till it freezes at 32° F. On the other hand, if water at 32° F. be heated, it contracts till it reaches a point between 39° and 40° F., when it commences to expand. Therefore the density of water is greatest at the point where this change occurs. Different experimenters vary a little as to its exact place, but it is usually called 4° C., or 39.2° F.

The importance of this exception is seen in the fact that ice forms on the *surface* of water, and continues to float until it is again dissolved. As the cold of winter comes on, the upper stratum of a lake grows more dense and sinks; and this process continues till the temperature of the surface reaches 39°, when it is arrested. Below that point the surface grows lighter as it becomes colder, till ice is formed, and shields the water beneath from the severe cold of the air above.

As in solids so in liquids, the thermal force is very great. Suppose mercury to be expanded by raising its temperature one degree, it would require more than 300 pounds to the square inch to compress it to its former volume.

522. Expansion of Gases.—The gases expand by heat more rapidly and more regularly than solids and liquids. The large expansion and contraction of air is made visible by immersing the open end of a large thermometer tube in colored liquid. When the bulb is warmed, bubbles of air are forced out and rise to the top of the liquid; when it is cooled, the air contracts and the liquid rises rapidly in the tube.

The coefficient of expansion for air is about 0.00205, which increases slightly with increase of temperature and of pressure. And most of the gases have coefficients which differ but little from this.

CHAPTER II.

PASSAGE OF HEAT THROUGH SPACE AND MATTER.

523. Heat is Communicated in Several Ways.—1. By *radiation.* Heat is said to be *radiated* when the vibratory motion is transmitted from the source with great swiftness through the ether which fills space. Its velocity is supposed to be the same as that of light. The motion is propagated in straight lines in every direction, and each line is called a *ray* of heat. We feel the rays of heat from the sun or a fire, when no object intervenes between it and ourselves.

2. By *reflection.* When rays of heat, on striking a surface, are thrown back from it, they are said to be *reflected;* and the law of reflection is the same as for sound, namely, the angle of incidence equals the angle of reflection, and they lie on opposite sides of the perpendicular to the surface.

3. By *conduction.* This is the slow progress of the vibratory motion from one atom to another of ordinary matter.

4. By *convection.* This mode of communication takes place only in *fluids.* When the particles are expanded by heat, they are pressed upward by others which are colder and therefore specifically heavier. Heat is thus conveyed from place to place by the motion of the heated matter.

524. Radiation of Heat.—The intensity of heat radiated from a given kind of source, is governed by the three following laws:

1. *The intensity of radiated heat varies as the temperature of the source.*

2. *It varies inversely as the square of the distance.*

3. *It grows less, while the inclination of the rays to the surface of the radiant grows less.*

The truth of these laws is ascertained by a series of careful experiments. But the second may be proved mathematically from the fact of propagation in straight lines, and is true of other emanations, such as sound and light. For the heat, as it advances in every direction from the radiant, is spread over spherical surfaces which increase as the squares of the distances; therefore the intensities must grow less in the same ratio; that is, the intensities vary inversely as the squares of the distances.

The radiating power of a given body depends on the condition of its surface.

If a cubical vessel filled with hot water have one of its vertical sides coated with lamp black, another with mica, a third with tarnished lead, and the fourth with polished silver, and the heat radiated from these several sides be concentrated upon a thermometer bulb, the ratio of radiation will be found nearly as follows

Lamp black,	100
Mica,	80
Tarnished lead,	45
Polished silver,	12

Polished metals generally radiate feebly; and this explains the familiar fact that hot liquids retain their temperature much better in bright metallic vessels than in dark or tarnished ones.

525. Equalization of Temperature.—Radiation is going on continually from all bodies, more rapidly in general from those most heated; and therefore there is a constant tendency toward an equal temperature in all bodies. A system of exchange goes on, by which the hotter bodies grow cool, and the colder ones grow warm, till the temperature of all is the same. But this equality does not check the radiation; it still goes forward, each body imparting to others as much heat as it receives from them.

526. Reflection of Heat.—When rays of heat meet the surface of a body, some of them are *reflected*, passing off at the same angle with the perpendicular on the opposite side. But others pass *into* the body, and are said to be absorbed by it. It is true of waves of heat as of all other kinds of vibration, that when they meet a new surface and are reflected, the angle of incidence equals the angle of reflection, and that their intensity after reflection is weakened.

If a person, when near a fire, holds a sheet of bright tin so as to see the light of the fire reflected by it, he will plainly perceive that heat is reflected also. And if any *sound* is produced by the fire, as the crackling of combustion, or the hissing of steam from wood, the reflection of the sound is likewise heard. This simple experiment proves that waves of sound, of heat, and of light, all follow the same law of reflection.

527. Heat Concentrated by Reflection.—Let two polished reflectors, *M* and *N* (Fig. 289), having the form of concave paraboloids, be placed ten or fifteen feet apart, with their axes in the same straight line, and let a red-hot iron ball be in the focus *A*

FIG. 289.

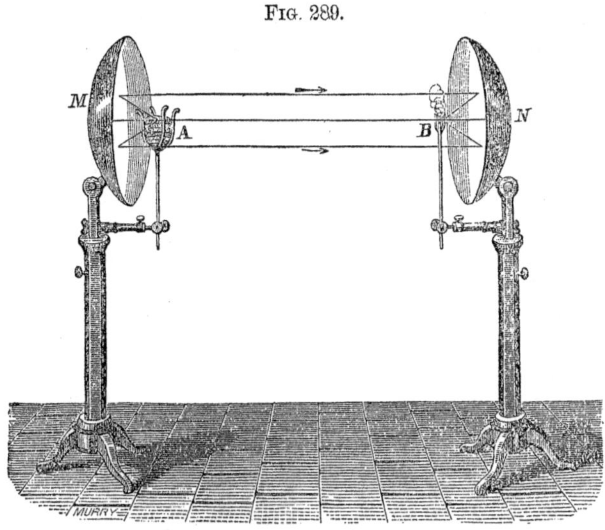

of one, and an inflammable substance, as phosphorus, in the focus *B* of the other; then the latter will be set on fire by the heat of the ball. The rays diverging from *A* to *M* are reflected in parallel lines to *N*, and then converged to *B*.

If, instead of phosphorus, the bulb of a thermometer is put in the focus *B*, a high temperature is of course indicated on the scale. Now remove the hot ball from *A*, and put in its place a lump of ice; then the thermometer at *B* sinks far below the temperature of the room. This last experiment does not prove that *cold* is reflected as well as heat, but confirms what was stated (Art. 525), that all objects radiate to one another till their temperatures are equalized. The ice radiates only a little heat, which is reflected to the thermometer, but the latter radiates much more, which is reflected to the ice, so that the temperature of the thermometer rapidly sinks.

528. Absorption of Heat.—So much of the radiant heat as falls on a body and is not reflected, is absorbed. The absorbing power in a body is found to be in general equal to its radiating power. It is very noticeable that bodies equally exposed to the radiant heat of the sun or a fire, become very unequally heated. A white cloth on the snow, under the sunshine, remains at the surface; a black cloth sinks, because it absorbs heat, and melts the snow beneath it. Polished brass before a fire remains cold; dark, unpolished iron, is soon hot.

529. Conduction of Heat by Solids.—While radiated and reflected heat moves through the empty spaces of the solar system, and through the atmospheres of the planets, with inconceivable velocity, conducted heat, on the contrary, passes through bodies very slowly, and yet at very different rates in different bodies. Those in which heat is conducted most rapidly, are called good conductors, as the common metals; those in which it passes slowly, are called poor conductors, as glass and wood. In general, the bodies which are good conductors of heat, are also good conductors of electricity. Let rods of different metals and other substances, *A*, *B*, *C*, &c. (Fig. 290), all of the same length, be inserted with water-tight joints in the side of a wooden vessel. Then attach by wax a marble under the end of each rod, and fill the vessel with boiling water. The marbles will fall by the melting of the wax, not at the same, but at different times, showing that the heat reaches some of them sooner than others. It will be seen, however, in the chapter on specific heat, that the order in which they fall is not necessarily the order of conducting power.

Fig. 290.

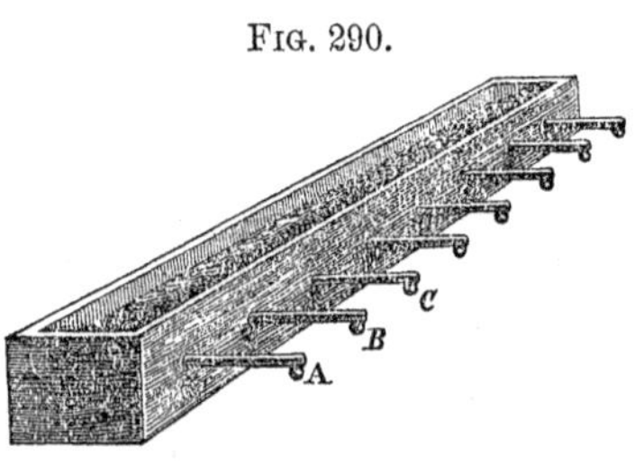

530. Effects of Molecular Arrangement.—Organic substances usually conduct heat poorly; and bodies having a structural arrangement which differs in different directions, are not likely to conduct equally well in all directions. Thus, let two thin plates be cut from the same crystal, one, *A* (Fig. 291), perpendicular, and the other, *B*, parallel to the optic axis. Let a hole be drilled through the centre of each, and after a lamina of wax has been spread over the crystal, let a hot wire be inserted in it. On

Fig. 291.

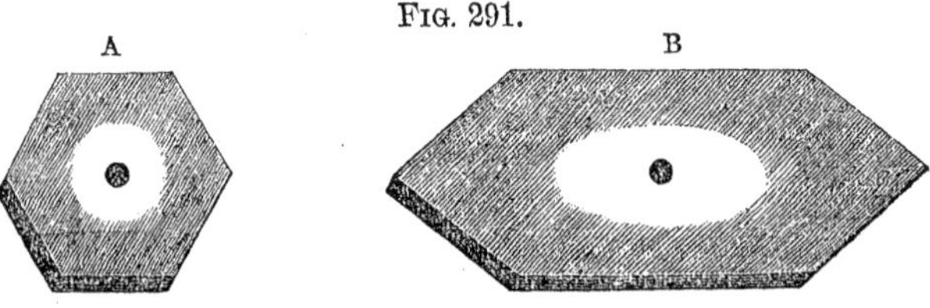

the plate *A*, the melting of the wax will advance in a circle, showing equal conducting power in all directions in the transverse section. In the plate *B*, it will advance in an elliptical form, the major axis being parallel to the optic axis of the crystal, proving the best conduction to be in that direction.

A block of wood cut from one side of the trunk of a tree, conducts most perfectly in the direction of the fiber, and least in a direction which is tangent to the annual rings and perpendicular to the fiber, and in an intermediate degree in the direction of the radius of the rings.

531. Conduction by Fluids.—Fluids, both liquid and gaseous, are in general very poor conductors. Water, for example, can be made to boil at the top of a vessel, while a cake of ice is fastened within it a few inches below the surface. If thermometers are placed at different depths, while the water boils at the top, there is discovered to be a very slight conduction of heat downward. The gases conduct even more imperfectly than liquids.

It will be seen hereafter (Art. 533) that a mass of fluid becomes heated by convection, not by conduction.

532. Illustrations of Difference in Conductive Power.—In a room where all articles are of equal temperature, some feel much colder than others, simply because they conduct the heat from the hand more rapidly; painted wood feels colder than woolen cloth, and marble colder still. If the temperature were higher than that of the blood, then the marble would seem the hottest, and the cloth the coolest, because of the same difference of conduction *to* the hand.

Our clothing does not impart warmth to us, but by its non-conducting property, prevents the vital warmth from being wasted by radiation or conduction. If the air were hotter than our blood, the same clothing would serve to keep us cool.

A pitcher of water can be kept cool much longer in a hot day, if wrapped in a few thicknesses of cloth; for these prevent the heat of the air from being conducted to the water. In the same way ice may be prevented from melting rapidly.

The vibrations of heat, like those of sound, are greatly interrupted in their progress by want of continuity in the material. Any substance is rendered a much poorer conductor by being in the condition of a powder or fiber. Ashes, sand, sawdust, wool, fur, hair, &c., owe much of their non-conducting quality to the innumerable surfaces which heat must meet with in being transmitted through them.

533. Convection of Heat.—Liquids and gases are heated almost entirely by convection. As heat is applied to the sides and bottom of a vessel of water, the heated particles become specifically lighter, and are crowded up by heavier ones which take their place. There is thus a constant circulation going on which tends to equalize the temperature of the whole. This motion is made visible in a glass vessel, by putting into the water some opaque powder of nearly the same density as water. Ascending currents are seen over the part most heated, and descending currents in the parts farthest from the heat, as represented in Fig. 292. The ocean has perpetual currents caused in a similar manner. The hottest portions flow away from the tropical toward the polar latitudes, while at greater depths the cold waters of high latitudes flow back towards the tropics.

Fig. 292.

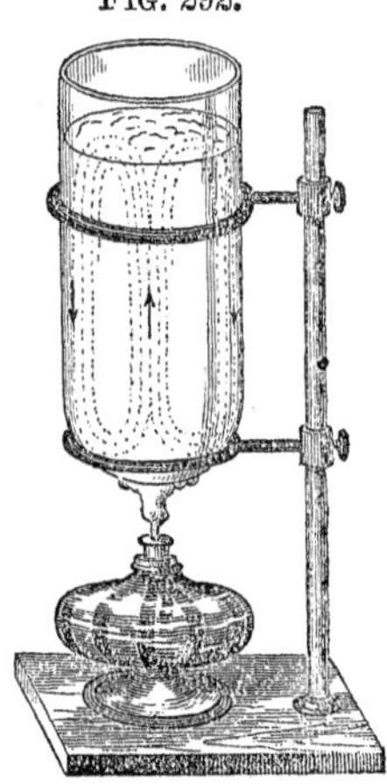

For a like reason, the air is constantly in motion. The atmospheric currents on the earth have been considered in Chapter III of Pneumatics.

534. Diathermancy.—It has already been noticed, that radiant heat passes freely through the atmosphere as well as through vacant space. The air is therefore said to be *diathermal;* it is also transparent, since it permits light to pass freely through it. But there are substances which allow the free transmission of the waves of light, but not those of heat; and there are others through which waves of heat can freely pass, but not those of light.

Water and glass, which are almost perfectly transparent to the faintest light, will not transmit the vibrations of heat unless they are very intense. If an open lamp-flame shines upon a thin film of ice, while nearly the whole of the *light* is transmitted, only 6 per cent. of the *heat* can pass through. On the other hand, rock salt is remarkably diathermal. A plate of it, one-tenth of an inch thick, will transmit 92 per cent. of the heat of a lamp; and if it be coated with lampblack so thick as to stop light completely, the heat is still transmitted with almost no diminution.

If a prism be made of a substance highly diathermal, as rock salt, it is found that heat, as well as light, is *refracted,* being bent from its course less than most of the colors, and falling mostly beyond the red extremity of the visible spectrum, but partially coinciding with that color.

CHAPTER III.

SPECIFIC HEAT.—CHANGES OF CONDITION.—LATENT HEAT.

535. Specific Heat.—The heat which is absorbed by a body is not wholly employed in raising its temperature. While a part of the thermal force which is communicated, throws the atoms into vibration, that is, heats the body, another part performs interior work of some other kind, such as urging the atoms asunder, or forcing them into new arrangements. This latter portion is lost to our sense and to the thermometer, until the body is again cooled, when it re-appears. The relative quantity of the force thus hidden from view is different in different substances. Hence the phrase, *specific heat,* is used to express the amount of heat required to raise a given weight one degree of temperature. The specific heat of water is greater than that of any other substance known, and it is made the standard of comparison.

The *thermal unit* is the amount of heat required to raise the temperature of a pound of water one degree, and is called 1. The specific heat of a few substances is given in the following table, in order to show how greatly they differ.

Water	1.0000	Silver	0.0570
Sulphur	0.2026	Mercury	0.0333
Iron	0.1138	Gold	0.0324
Copper	0.0951	Lead	0.0314

If a pound of water, a pound of iron, and a pound of mercury are each raised one degree in temperature, the water consumes about nine times as much heat as the iron, and thirty times as much as the mercury.

When bodies are cooled, they show the same differences in the quantity of heat which they give off.

It is a benevolent provision in nature, that water, which is extended over so large a portion of the globe, has so great specific heat; for the changes of both heat and cold are by this means greatly moderated.

536. Method of Finding Specific Heat.—The following is one of several methods of finding the specific heat of a substance; it is called the *method of mixtures.* Let a known weight of the substance be heated to a certain temperature, and then plunged into, or mixed with, the same weight of water of a low temperature; after which measure the temperature of the mass. It will thus be known how much one has lost and the other gained in order to reach the common point. If a pound of mercury at the temperature of 132°, be poured into a pound of water at the temperature of 32°, the mass will be found at about 35.25°, the water being heated only 3.25°, while the mercury is cooled 96.75°. Therefore 96.75° : 3.25° : : 1 : 0.0335, which is about the specific heat of mercury.

The specific heat of bodies is in general a little greater as their temperature rises. That of the gases, however, seems to be nearly constant at all temperatures, and under all pressures.

537. Apparent Conduction Affected by Specific Heat.—The conducting power of different substances cannot be correctly compared, without making allowance for their specific heat (Art. 529). For the heat which is communicated to one end of a rod, will reach the other end more slowly, if a great share of it disappears on the way. For instance, at the same distance from the source of heat, wax is melted quicker on a rod of bismuth than on one of iron, though iron is the best conductor, because the specific heat of iron is three times as great as that of bismuth; the heat actually reaches the wax soonest through the iron, but not enough to melt it, because so much is required to raise the iron to a given temperature.

538. Changes of Condition.—Among the most important effects produced by heat, are the changes of condition from solid to liquid and from liquid to gas, or the reverse, according as the temperature of a body is raised or lowered. Increase of heat changes ice to water, and water to steam, and the diminution of heat reverses these effects. A large part of the simple substances, and of compound ones not decomposed by heat, undergo similar changes at some temperature or other; and probably it would be found true of all if the requisite temperature could be reached.

The *melting point* (called also *freezing point*, or *point* of *congelation*) of a substance is the temperature at which it changes from a solid to a liquid or the reverse.

The *boiling point* is the temperature at which it changes from a liquid to a gas or the reverse.

539. Latent Heat.—Whenever a solid becomes a liquid, or a liquid becomes a gas, a large amount of heat disappears, and is said to become *latent*. The thermal force is expended in sundering the atoms, and perhaps in putting them into new relations and combinations, so that there is not the slightest increase of temperature after the change begins till it ends. The force is not *lost*, but is treasured up in the form of *potential energy*, which becomes available whenever a change is made in the opposite direction. Using the force of heat to turn water into steam, is like using the strength of the arm in coiling up a spring, or lifting a weight from the earth. The spring and the weight are each in a condition to perform work. They have potential energy, which can be used at pleasure.

It has been already noticed that much heat disappears in bodies of great specific heat, as their temperature rises. But the amount which becomes latent, while a change of condition takes place, is vastly greater. Let heat be applied at a uniform rate to a mass of water at the temperature of 32°, until it rises to the boiling point, 212°, and note the time occupied. Continuing the same uniform supply, it will require $5\frac{3}{8}$ times as long to change it all into steam. In other words, 180° of heat will raise water from the freezing to the boiling point, and (180° × $5\frac{3}{8}$ =) $967\frac{1}{2}$° are required to change the same into steam, which still remains at the temperature of 212°; the whole of the 967° of thermal force have been consumed in the internal work of re-arranging the atoms.

540. Temperature of Change of Condition.—Different substances change their condition at very different temperatures. Water solidifies at 32° F., mercury at − 39°, tin at 455°, gold at 2016°. Water boils at 212° F., ether at 95°, alcohol at 173°, mercury at 662°. There are some solids, which soften gradually, and pass through a large range of temperature before becoming liquid, as iron and glass. No definite melting point can be given for such substances.

Again, many liquids pass into the gaseous state by a slow and almost insensible process which goes on at the surface. This is called *evaporation;* and it takes place at all temperatures, but more rapidly as the temperature is higher. Even solids evaporate without passing through the liquid form. For example, a thin film of ice on a pavement wastes away in cold weather without melting.

541. Boiling under Pressure.—The boiling point for water is given as 212° F. This means that water boils at that point under the ordinary pressure of the air, and at or near the sea level. At that temperature the steam formed has a *tension* or expansive force equal to the atmospheric pressure. But if the pressure were diminished, water would boil at a lower temperature. On high mountains, boiling water is from 20° to 30° lower in temperature than at the ocean level. And under the receiver of an air pump, as pressure is gradually taken off, water boils at lower and lower points of temperature down to 72°.

The effect of diminished pressure to lower the boiling point is well shown by the following familiar experiment: In a thin glass flask, boil a little water, and after removing it from the fire, cork and invert the flask. The steam which is formed will soon press so strongly upon the water as to stop the boiling. When this happens, pour a little cold water upon the flask; the water within will immediately commence boiling violently, because the vapor is condensed and the pressure removed. This effect may be reproduced several times before the water in the flask is too cool to boil in a vacuum.

542. Freezing Produced by Melting.—Since a great amount of heat disappears in a substance as it passes from the solid to the liquid state, the loss thus occasioned may produce freezing in a contiguous body. When salt and powdered ice are mixed, their union causes liquefaction. And if this mixture is surrounded by bad conductors, and a tin vessel containing some liquid be placed in the midst of it, the latter is frozen by the abstraction of heat from it, by the melting of the ice and salt. In this way ice creams and similar luxuries are easily prepared in hot as well as in cold weather.

543. Freezing by Evaporation.—In like manner, freezing by evaporation is explained. Put a little water in a shallow dish of thin glass, and set it on a slender wire-support under the receiver of an air pump. Beneath the wire-support place a broad dish containing sulphuric acid. When the air is exhausted, the water in a few moments is found frozen. As the pressure of the air is taken off, evaporation proceeds with increased rapidity, and the requisite heat for this change of condition can be taken only from the dish of water. But the atmosphere of vapor retards the process by its pressure; hence the sulphuric acid is placed in the receiver, so as to seize upon the vapor as fast as formed, and thus render the vacuum more complete. The water is frozen by giving up its heat to become latent in the vapor, so rapidly formed; but when this vapor becomes liquid again in combining with the acid, the same heat reappears in raising the temperature of the acid.

Thin cakes of ice may sometimes be procured, even in the hottest climates, by the evaporation of water in broad shallow pans under the open sky, where radiation by night aids in reducing the temperature. The pans should be so situated as to receive the least possible heat by conduction.

544. Spheroidal Condition.—When a little water is placed in a red-hot metallic cup, instead of boiling violently, and disappearing in a moment, as might be expected, it rolls about quietly in the shape of an oblate spheroid, and wastes very slowly. So drops of water, falling on the horizontal surface of a very hot stove, are not thrown off in steam and spray with a loud hissing sound, as they are when the stove is only moderately heated, but roll over the surface in balls, slowly diminishing in size till they disappear.

In such cases, the water is said to be in the *spheroidal state.* Not being in contact with the metal, it assumes the shape of an oblate spheroid, in obedience to its own molecular attractions and the force of gravity, as small masses of mercury do on a table. The reason why the water does not touch the hot metal is, that the heat causes a coat of vapor to be instantly formed about the drop, on which it rests as on an elastic cushion; and as the vapor is a poor conductor of heat, further evaporation proceeds very slowly. It is easily seen that the spheroid does not touch the metal, by so arranging the experiment that a beam of light may shine horizontally upon the drop, and cast its shadow completely separated from that of the hot plate below it, as in Fig. 293.

Fig. 293.

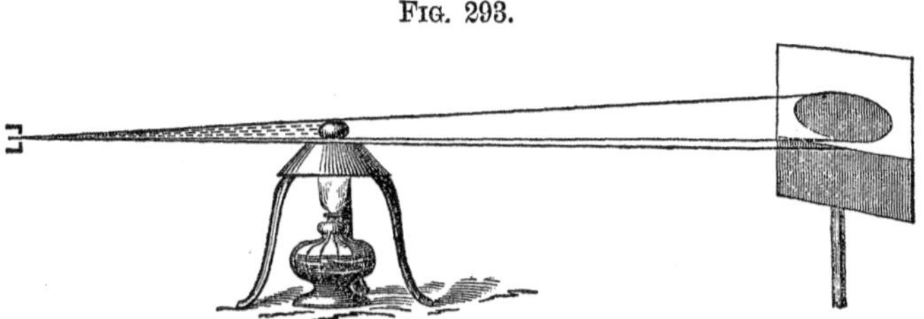

If the heated surface is cooling, the temperature may become so low that the drop at length touches it, when in an instant violent ebullition takes place, and the water quickly disappears in vapor.

CHAPTER IV.

STEAM.—THE STEAM-ENGINE.—MECHANICAL EQUIVALENT OF HEAT.

545. Thermal Force in Steam.—It has been already noticed that while water is heated, and especially while it is converted into steam by boiling, the heat apparently lost is so much *force* treasured up ready for use, as truly as when strength is expended in lifting great weights, which by their descent can do the work desired. In modern engineering, the force of steam is employed more extensively, and for more varied purposes, than any other. Every steam-engine is a machine for transforming the internal motion of heated steam into some of the visible forms of motion.

546. Tension of Steam.—When steam is formed by boiling water in the open air, its tension is equal to that of the air, and therefore ordinarily about fifteen pounds to the square inch. But when it is formed in a tight vessel, so that it cannot expand, as the temperature of the water is raised the tension is increased in a much greater ratio; because the same steam has greater tension at a higher temperature, and besides this, new steam is continually added. The following table gives the temperature for successive atmospheres of tension:

Atmospheres.	Degrees of Temperature.	Atmospheres.	Degrees of Temperature.
1	212	11	367
2	251	12	374
3	275	13	381
4	294	14	387
5	307	15	393
6	320	16	399
7	332	17	404
8	342	18	409
9	351	19	414
10	359	20	418

It is seen by the above table that thirty-nine degrees of heat are needed to add the second atmosphere of tension, and that the number diminishes constantly, so that only four degrees are required to add the twentieth atmosphere.

In the formation of ordinary steam at 212°, one cubic inch of water is expanded to about 1,700 cubic inches of steam, or nearly a cubic foot. At higher temperatures, the volume diminishes nearly as fast as the temperature increases.

547. The Steam-engines of Savery and Newcomen.—The only steam-engines that were at all successful before the great improvements made by Watt, were the engine of Savery and that of Newcomen. No other purpose was proposed by either than that of removing water from mines.

In the engine of Savery, steam was made to raise water by acting on it directly, and not through the intervention of machinery. First, the steam in a vessel was condensed by cold water flowing over the outside, and the atmosphere raised water into the exhausted vessel by its pressure. In the next place, steam was let into the vessel, and by its tension forced the water out, and raised it still higher. The water raised by each part of this operation was prevented from returning by a valve, as in a forcing pump.

Newcomen employed steam in a very different way, namely, as a power to work a common pump. The pump rod was attached to one end of a working beam, and to the other end of the same was attached the rod of the steam piston, which moved steam-tight in a cylinder. The end of the beam next to the pump was made heavy enough to keep the steam piston at the top of the cylinder, when no force was applied. The space beneath the piston being filled with steam, a little cold water was injected, the steam condensed, and the piston forced down by the weight of the air on the top of it. Then, as soon as steam was admitted again below, though having no greater tension than the atmosphere, the piston was drawn up by the weight of the opposite end of the beam. Since the water was raised directly by the weight of the atmosphere, after the steam had given it opportunity to act, this invention of Newcomen was called the *atmospheric engine.*

But in neither of these methods was steam used economically as a power. The movements in both cases were sluggish, and a large part of the force was wasted, because the steam was compelled to act upon a cold surface, which condensed it before its work was done.

548. The Steam-engine of Watt.—Steam did not give promise of being essentially useful as a power till Watt, in the year 1760, made a change in the atmospheric engine, which prevented the great waste of force. Newcomen introduced the cold water which was to condense the steam into the steam cylinder itself; and the cylinder must be cooled to a temperature below 100°, else there would be steam of low tension to retard the descent of the piston. But when the piston was to be raised, the cylinder must be heated again to 212°, in order that the admitted steam might balance the pressure of the air.

In the engine of Watt, the steam is condensed in a separate vessel called the *condenser.* The steam cylinder is thus kept at the uniform temperature of the steam. In the first form which he gave to his engine, he so far copied the atmospheric engine as to allow the piston, after being pressed down by steam, to be raised again by the load on the opposite end of the great beam, while the steam circulates freely below and above the piston. This was called the *single-acting* engine, and might be successfully used for the only use to which any steam-engine was as yet applied, namely, pumping water from mines. But he almost immediately introduced the change by which the whole force of the steam was brought to act on the upper and the under side of the piston. It thus became *double-acting*, and the steam force was no longer intermittent.

549. The Double-acting Engine.—Let *S* (Fig. 294) be the steam cylinder, *P* the piston, *A* the piston rod, passing with steam-tight joint through the top of the cylinder, *C* the condenser, kept cold by the water of the cistern *G*, *B* the steam pipe from the boiler, *K* the eduction pipe, which opens into the valve chest at *O*, *D D* the D-valve, *E F* the openings from the valve-chest into the cylinder. As the D-valve is situated in the figure, the steam can pass through *B* and *E* into the cylinder below the piston, while the steam above the piston can escape by *F* through *O* and *K* to the condenser, where it is condensed as fast as it enters; so that in an instant the space above the piston is a vacuum, while the whole force of the steam is exerted on the under side. The piston is therefore driven upward without any force to oppose it. But before it reaches the top, the D-valve, moved by the machinery, begins to descend, and shut off the steam from *E* and admit it to *F*, and, on the other hand, to shut *F* from the eduction pipe *O*, and open *E* to the same. The steam will then press on the top of the piston, and there will be a vacuum below it, so that the piston descends with the whole force of the steam, and without resistance. To render the condensation more sudden, a little cold water is thrown into the condenser at each stroke through the pipe *H*.

FIG. 294.

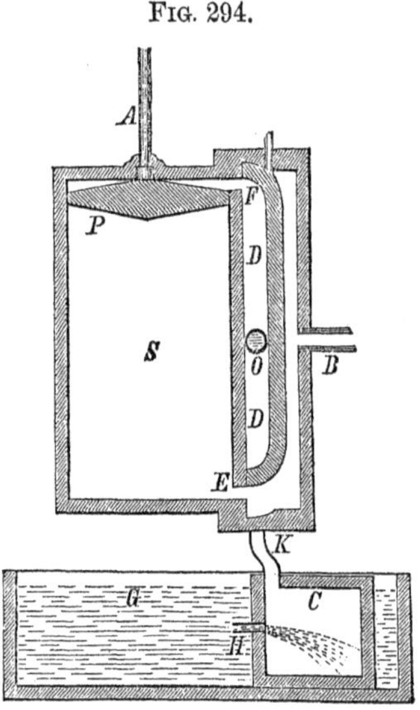

550. The Low-pressure Engine.—The principle of the low-pressure engine is illustrated by the figure and description of the preceding article. But the condensing apparatus of this kind of engine requires many other parts, most of which are presented in Fig. 295. *C* is the steam cylinder; *R* the rod connecting its piston with the end of the working beam, not represented; *A* the steam pipe and throttle valve; *B B* the D-valve; *D D* the eduction pipe, leading from the valve chest to the condenser *E*; *G G* the cold water surrounding the condenser; *F* the air-pump, which keeps the condenser clear of air, steam, and water of condensation; *I* the hot well, in which the water of condensation is deposited by the air-pump; *K* the hot-water pump, which forces the water

Fig. 295.

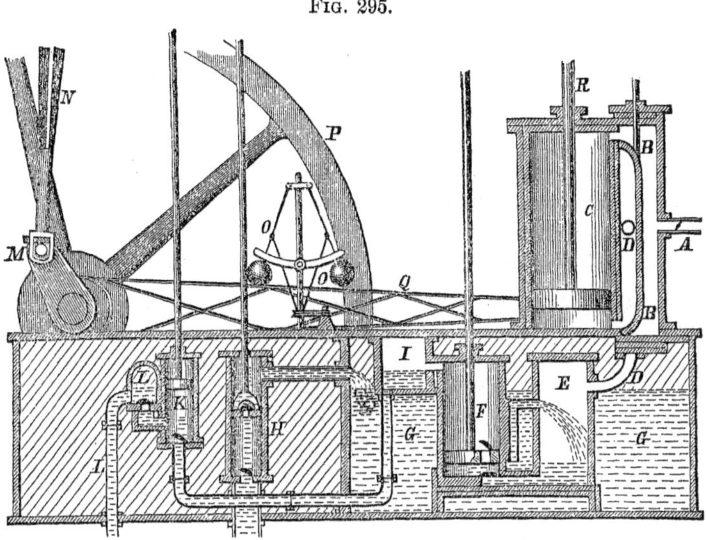

in the hot well through *L* to the boiler; *H* the cold-water pump, by which water is brought to the cistern *G G*; the rods of all the pumps, *F*, *K*, and *H*, are moved by the working beam; *P* the fly-wheel; *M* the crank of the same, *N* the connecting-rod, by which the working beam conveys motion to the fly-wheel; *Q* the excentric rod, by which the D-valve is moved; *O O* the governor, which regulates the throttle valve in the steam pipe *A*.

551. Steam-Valves.—What are commonly called valves in steam machinery are not strictly such, since they are not opened by the pressure of a fluid in one direction, and closed by a pressure in the opposite direction. On the contrary, they are opened and shut by the action of an eccentric cam on the principal axis. The *puppet valve* is the frustum of a cone, fitting into a conical socket, and opens the pipe by being raised. The *sliding valve* does not rise, but slides over the aperture. The *rotary valve*, like the common stop-cock, is a cylinder lying across the pipe, and having an aperture through it, so that by a quarter revolution it opens or shuts the pipe. The *throttle valve*, like the damper of a stove-pipe, is a partition in the pipe, turning on an axis, so as to lie crosswise or lengthwise. The *D-valve*, so called from its form, is a sliding valve which takes the place of four valves in the earlier engines, connecting both the top and the bottom of the steam cylinder with the boiler and with the condenser.

There is much economy of fuel and saving of wear in the machinery, arising from the proper adjustment of the valves. If the steam enters the cylinder during the whole length of a stroke of the piston, its motion is *accelerated;* and is therefore swiftest at the instant before being stopped; thus the machinery receives a violent shock. If the valve is adjusted to *cut off* the steam when the piston has made one-third or one-half of its stroke, the diminishing tension may exert about force enough, during the remaining part, to keep up a uniform motion. The *cut-off*, however, should be regulated in each engine, according to friction and other obstructions.

552. High-Pressure Engine.—The engine of Watt, already described, is properly called a *low-pressure* engine, because the steam, having a vacuum on the opposite side of the piston, works at the least possible tension. For many purposes, especially those of locomotion, it is advantageous to dispense with the large weight and bulk of machinery necessary for condensation, and do the work with steam of a higher tension. In Fig. 295, if the condenser, cistern, and all the pumps are removed, then the steam is discharged from E and F at each stroke into the air. Therefore the steam in that part of the cylinder which is open to the air, will have a tension of 15 lbs. per inch; and, consequently, the steam on the opposite side of the piston must have a tension 15 lbs. per inch greater than before, in order to do the same work.

553. Applications of Steam Power.—For more than half a century, the only use of the steam-engine was to work the water pumps of the English mines. But the genius of Watt has rendered it available for nearly every purpose which requires the use of machinery. Every description of machine, for the heaviest and the lightest operation, may have a steam-engine for its prime mover. Near the beginning of the present century, it began to be used for locomotion on land and water; and at the present day, both traveling and the transportation of merchandise are principally accomplished by means of steam.

554. Estimation of Steam Power.—It is customary to express the power of a steam-engine by comparing it with the number of horses whose strength it equals. In making this comparison, Watt took as a measure of *one horse-power*, the ability to raise 2,000,000 lbs. through the height of *one foot* in an hour; or 2,000,000 *foot-pounds* per hour. It is obviously immaterial what the respective factors for feet, and for pounds, are, if the product only equals 2,000,000. For example, 2,000 lbs. through 1,000 feet, or 5,000 lbs. 400 feet per hour, &c., is equal to one horse-power. It is found that the available force of *one cubic foot* of water, when changed to steam, is about equal to 2,000,000 foot-pounds; that is, to one horse-power. Hence, an engine of fifty horse-power is one which can change fifty cubic feet of water into steam in one hour.

555. Mechanical Equivalent of Heat.—In all cases in which mechanical force produces heat, and again in all those in which heat produces visible motion, careful experiment proves that heat and mechanical force may each be made a measure of the other. Forces of any kind may be compared, by observing the weights which they will lift through a given distance. The *mechanical equivalent of heat* (commonly called, from the name of an English experimenter, Joule's equivalent) is given in the following statement:

The force required to heat one pound of water one degree F., is equal to that which would lift 772 pounds the distance of one foot, or is equal to 772 foot-pounds.

The force requisite to raise one pound of water 1° F., is sometimes called the *thermal unit* (Art. 535), and all forces may be brought to this as a standard of comparison. Thus, one horse-power (2,000,000 foot-pounds per hour) is 2,590 thermal units per hour, or about 43 per minute.

Since a force of 772 foot-pounds is expended in heating a pound of water 1° F., therefore to heat the same from 32° to 212° requires a force of 138,960 foot-pounds; and to change the same pound of water into steam of atmospheric tension requires an additional force of 746,900 foot-pounds (Art. 539).

CHAPTER V.

TEMPERATURE OF THE ATMOSPHERE.—MOISTURE OF THE ATMOSPHERE.—DRAFT AND VENTILATION.

556. Manner in which the Air is Warmed.—The space through which the earth moves around the sun is intensely cold, probably 75° below zero. And the one or two hundred miles of height occupied by the atmosphere is too cold for animal or vegetable life, except the lowest stratum, three or four miles in thickness. This portion receives its heat mainly by convection. The radiated heat of the sun passes through the air, warming it but little, and on reaching the earth is partly absorbed by it. The air lying in contact with the earth, and thus becoming warmed, grows lighter and rises, while colder portions descend and are warmed in their turn. So long as the sun is shining on a given region of the earth, this circulation is going on continually. But the heated air which rises is expanded by diminished pressure, and thus cooled. Hence the circulation is limited to a very few miles next to the earth.

557. Limit of Perpetual Frost.—At a moderate elevation, even in the hottest climate, the temperature of the air is always as low as the freezing point. Hence the permanent snow on the higher mountains in all climates. The limit at the equator is about three miles high, and with many local exceptions it descends each way to the polar regions, where it is very near the earth. The descent is more rapid in the temperate than in the torrid or frigid zones.

558. Isothermal Lines.—These are imaginary lines on each hemisphere, through all those points whose mean annual temperature is the same. At the equator, the mean temperature is about 82°, and it decreases each way toward the poles, but not equally on all meridians. Hence the isothermal lines deviate widely from parallels of latitude. Their irregularities are due to the difference between land and water, in absorbing and communicating heat, to the various elevations of land, especially ranges of mountains, to ocean currents, &c. In the northern hemisphere, the isothermal lines, in passing westward round the earth, generally descend toward the equator in crossing the oceans, and ascend again in crossing the continents. For example, the isothermal of 50°, which passes through China on the parallel of 44°, ascends in crossing the eastern continent, and strikes Brussels, lat. 51°; and then on the Atlantic, descends to Boston, lat. 42°, whence it once more ascends to the N. W. coast of America. The lowest mean temperature in the northern hemisphere is not far from zero, but it is not situated at the north pole. Instead of this, there are *two* poles of greatest cold, one on the eastern continent, the other on the western, near 20° from the geographical pole. There are indications, also, of two south poles of maximum cold.

559. Moisture of the Atmosphere.—By the heat of the sun all the waters of the earth form above them an atmosphere of vapor, or invisible moisture, having more or less extent and tension, according to several circumstances. Even ice and snow, at the lowest temperatures, throw off some vapor. A gaseous body diffuses itself by its force of tension, whether another gas occupies the same space or not; that is, the particles of one do not exert a perceptible attraction or repulsion on those of the other, but each is a vacuum to the other, except so far as it *obstructs* its movements. Therefore, at a given temperature, there can exist an atmosphere of *vapor* of the same height and tension, whether there is an atmosphere of *oxygen* and *nitrogen* or not. Vapor, then, is not strictly *suspended* in the air, or *dissolved* by it, but exists independently. And yet it is by no means always true that there is actually the same tension of vapor as there would be if it existed alone, because of the time required for the formation of vapor, on account of mechanical obstruction presented by the air; whereas, if no air existed, the vapor would form almost instantly. It is on the same account that water will boil in a vacuum at 72°, but under the pressure of the air must be heated to 212°.

560. Temperature and Tension of Vapor.—The degree of tension of vapor forming without obstruction, depends on its temperature, but varies far more rapidly; increasing pretty nearly in a geometrical ratio, while the heat increases arithmetically. Hence, if vapor should receive its full increment of tension, while the thermometer rises 10 degrees from 80° to 90°, a vastly greater quantity would be added than when it rises 10 degrees from 40° to 50°. On the contrary, if vapor is at its full tension in each case, much more water will be precipitated in cooling from 90° to 80° than from 50° to 40°.

561. Dew-point.—This is the temperature at which vapor, in a given case, is precipitated into water in some of its forms. If there was no air, the dew-point would always be the same as the existing temperature; since lowering the temperature in the least degree would require a diminished tension or quantity of vapor, some must therefore be condensed into water. But in the air the tension may not be at its full height, and therefore the temperature may need to be reduced several degrees before precipitation will take place. A comparison of the temperature with the dew-point is one of the methods employed for measuring the humidity of the air.

562. Measure of Vapor.—The measure of the vapor existing at a given time, is expressed by two numbers, one indicating its *tension*,—*i. e.*, the height of the column of mercury which it will sustain; the other, *humidity*,—*i. e.*, its quantity per cent., as compared with the greatest possible amount at that temperature. Thus, tension = 0.6, humidity = 83, signifies that the quantity of vapor is sufficient to support six-tenths of an inch of mercury, and is 83 hundredths of the quantity which *could* exist at that temperature. The greatest tension possible at zero, is 0.04; at the freezing point, 0.18; at 80°, 1.0. At the lowest natural temperatures, the maximum tension is doubled every 12° or 14°; at the highest, every 21° or 22°.

563. Hygrometers.—This is the name usually given to instruments intended for measuring the moisture of the air. But the one most used of late years is called the *psychrometer*, which gives indication of the amount of moisture by the degree of *cold* produced in evaporation; for evaporation is more rapid, and therefore the cold occasioned by it the greater, according as the air is drier. The psychrometer consists of two thermometers, one having its bulb covered with muslin, and moistened before the observation. The wet-bulb thermometer will ordinarily indicate a lower temperature than the dry-bulb; if, in a given case, they read alike, the humidity is 100. The instrument is accompanied by tables, giving tension and humidity for any observation.

564. Precipitations of Moisture.—Whenever the air is cooled below the dew-point, a part of the vapor is deposited in the liquid or solid form. The precipitations occur under various conditions, and receive the following names: dew, frost, fog, cloud, rain, mist, hail, sleet, and snow.

565. Dew.—Frost.—The deposition called *dew* takes place on the surface of bodies, by which the air is cooled below its dew-point. It is at first in the form of very small drops, which unite and enlarge as the process goes on. Dew is formed in the evening or night, when the surfaces of bodies exposed to the sky become cold by radiation. As soon as their temperature has descended to the dew-point, the stratum of air contiguous to them deposits moisture, and continues to do so more and more as the cold increases.

Of two bodies in the same situation, that will receive most dew which radiates most rapidly. Many vegetable leaves are good radiators, and receive much dew. Polished metal is a poor radiator, and ordinarily has no dew deposited on it.

Sometimes, however, good radiators have little dew, because they are so situated as to obtain heat nearly as fast as they radiate it. Dew is rarely formed on a bed of sand, though it is a good radiator, because the upper surface gets heat by conduction from the mass below. Dew is not formed on water, because the upper stratum sinks and gives place to warmer ones.

Bodies most exposed to the open sky, other things being equal, have most dew precipitated on them. This is owing to the fact, that in such circumstances, they have no return of heat either by reflection or radiation. If a body radiates its heat to a building, a tree, or a cloud, it also gets some in return, both reflected and radiated. Hence, little dew is to be expected in a cloudy night, or on objects surrounded by high trees and buildings.

Wind is unfavorable to the formation of dew, because it mingles the strata, and prevents the same mass from resting long enough on the cold body to be cooled down to the dew-point.

When the radiating body is cooled below the freezing point, the water deposited takes the solid form in fine crystals, and is called *frost*. Frost will often be found on the best radiators, or those exposed to the open sky, when only dew is found elsewhere.

566. Fog.—This form of precipitation consists of very small globules of water sustained in the lower strata of the air. Fog occurs most frequently over low grounds and bodies of water, where the humidity is likely to be great. If air thus humid mixes with air cooled by neighboring land, even of less humidity, there will probably be more vapor than can exist at the intermediate temperature, for the reason mentioned in Art. 560. The case may be illustrated thus. Let two masses of air of equal volumes be mixed, the temperature of one being 40°, the other 60°; and each containing vapor at the highest tension. Then the mixture will have the mean temperature of 50°, and the vapor of the mixture will also be the *arithmetical* mean between that of the two masses. But, according to the law (Art. 560), the vapor can only have a tension which is nearly a *geometrical* mean between the two, and that is necessarily lower than the *arithmetical* mean; hence the excess must be precipitated. If 8 lbs. of vapor were in one volume and 18 lbs. in the other, an equal volume of the mixture would have $\frac{1}{2}(8 + 18) = 13$ lbs. of moisture; but at the mean temperature of 50°, only $\sqrt{8 \times 18} = 12$ lbs. could exist as vapor; therefore *one pound* must be precipitated. And even if one of the masses had a humidity somewhat below 100, still some precipitation is likely to take place.

567. Cloud.—The same as fog, except at a greater elevation. Air rising from heated places on the earth, and carrying vapor with it, is likely to meet with masses much colder than itself, and depositions of moisture are therefore likely to take place. Mountain-tops are often capped with clouds, when all around is clear. This happens when lower and warmer strata are driven over them, and thus cooled below the dew-point. The same air, as it continues down the other side, takes up its vapor again, and is as transparent as it was before ascending. A person on the summit perceives a chilly fog driving by him, but the fog was an invisible vapor a few minutes before reaching him, and returns to the same condition soon after leaving him. The cloud *rests* on the mountain; but all the particles which compose it are swiftly *crossing over*. Clouds are often above the limit of perpetual frost; they then consist of crystals of ice.

568. Classification of Clouds.—The aspects of clouds are various, and depend in some measure at least on the circumstances of their formation. The usual classification is the following:

1. *Cirrus.*—This cloud is fibrous in its appearance, like *hair* or flax, sometimes straight, sometimes bent, and frequently at one end is gathered into a confused heap of fibers. The cirrus is high, and often consists of frozen particles, even in summer.

2. *Cumulus.*—This consists of compact rounded heaps, which often resemble mountain-tops covered with snow. This form of cloud is confined mostly to the summer season; it usually begins to form after the sun rises, and to disappear before it sets, and is rarely seen far from land. The cumulus is generally not so high as the cirrus.

3. *Stratus.*—*Sheets* or stripes of cloud, sometimes overspreading the whole sky, or as a fog covering the surface of the earth or water. The stratus is the most common, and usually lies lowest in the air.

4, 5, 6. *Cirro-cumulus, cirro-stratus, cumulo-stratus.*—Intermediate or combined forms.

7. *Nimbus.*—A cloud, which forms so fast as to fall in rain or snow, is called by this name.

569. Rain, Mist.—Whether the precipitated moisture has the form of cloud or rain, depends on the rapidity with which precipitation takes place. If currents of air are in rapid motion, if the temperature of masses, brought into contact by this motion, are widely different, and if their humidity is at a high point, the vapor will be precipitated so rapidly, that the globules will touch each other, and unite into larger drops, which cannot be sustained. Globules of fog and cloud, however, are specifically as heavy as drops of rain; but they are sustained by the slightest upward movements of the air, because they have a great surface compared with their weight. A globule whose diameter is 100 times less than that of a drop of rain, meets with 100 times more obstruction in descending, since the weight is diminished a million times $(\frac{1}{100})^3$, and the surface only ten thousand times $(\frac{1}{100})^2$. So the dust of even heavy minerals is sustained in the air for some time, when the same substances, in the form of sand, or coarse gravel, fall instantly.

Mist is fine rain; the drops are barely large enough to make their way slowly to the earth.

570. Hail, Sleet, Snow.—When the air in which rapid precipitation occurs, is so cold as to freeze the drops, hail is produced. As hailstones are not usually in the spherical form when they reach the earth, it is supposed that they are continually receiving irregular accretions in their descent through the vapor of the air. Hail-storms are most frequent and violent in those regions where hot and cold bodies of air are most easily mixed. Such mixtures are rarely formed in the torrid zone, since there the cold air is at a great elevation; in the frigid zone, no hot air exists at any height; but in the temperate climates, the heated air of the torrid, and the intensely cold winds of the frigid zone, may be much more easily brought together; and accordingly, in the temperate zones it is that hail-storms chiefly occur. Even in these climates, they are not frequent except on plains and in valleys contiguous to mountains which are covered with snow during the summer. The slopes of the mountain sides give direction to currents of air, so that masses of different temperature are readily mingled together.

Sleet is frozen mist, that is, it consists of very small hailstones.

Snow consists of the small crystals of frozen cloud, united in flakes. Like all transparent substances, when in a pulverized state, it owes its whiteness to innumerable reflecting surfaces. A cloud, when the sun shines upon it, is for the same reason intensely white.

571. Theories of Precipitation.—It is probable that clouds and rain are caused not only by the mixing of air of different temperatures, but also by the changes which take place in the condition of the air as it ascends.

In the lower strata, the air is about one degree colder for every 300 feet of elevation. If, therefore, a mass of air is transferred from the surface of the earth to a height in the atmosphere, it will be cooled to the temperature of the stratum which it reaches; not principally by giving off its heat, but by *expanding*, and thus having its own heat reduced by being diffused through a larger space. Now, if the rising mass was saturated with moisture, this moisture would begin at once to be precipitated by the cooling which it undergoes in consequence of expansion. If, instead of being saturated, its dew-point is a certain number of degrees below its temperature, it must ascend far enough to be cooled to the dew-point, before precipitation of its moisture will take place. Suppose, for instance, the temperature at the earth is 70°, and the dew-point is 65°; then after the warm air has risen 1500 feet (5×300

ft.), it will become 5° cooler, and contain all the moisture which is possible at that temperature. At that point precipitation begins, and forms the base of a cloud. The clouds, called *cumulus*, which are seen forming during many summer forenoons, are the precipitations of columns rising from warm spots of earth so high that they are cooled below their dew-point. But the movement and the precipitation do not stop here; for, as moisture is precipitated, its latent heat is given off in large quantities, which elevates the temperature of the mass, and causes it to rise still higher, and precipitate still more of its moisture. As it becomes rarer, it spreads laterally, and causes the cumulus often to assume the overhanging form which distinguishes that species of cloud.

572. Cyclones.—The late Mr. Redfield investigated with great success the phenomena of violent storms, especially of *Atlantic hurricanes*, and showed that they are generally, if not always, great whirlwinds, called *cyclones*. They usually take their rise in the equatorial region eastward of the West India Islands; they rotate on a vertical axis, advancing slowly to the northwest, until they approach the coast of the United States near the latitude of 30°, and then gradually veer to the northeast, running nearly parallel to the American coast, and finally spend themselves in the northern Atlantic. Their rotary motion is always in one direction, namely, from the east through the north to the west, or against the sun. This motion is also far more violent, especially in the central parts of the storm, than the progressive motion. The rotary motion may amount to 50 or 100 miles per hour, while the forward motion of the storm is not more than 15 or 20 miles.

In the southern hemisphere also, cyclones occur, having a progressive and a rotary motion, both symmetrical with those of the northern cyclones. On the axis they revolve *with* the sun, not against it; and they first advance toward the southwest, and gradually veer toward the southeast, as they recede from the equator.

573. Draught of Flues.—The effect of the sun's heat in causing circulation of the air has been already considered (Art. 289–293). Similar movements on a limited scale are produced whenever a portion of the air is heated by artificial means. Thus, the air of a chimney is made lighter by a fire beneath it, than a column of the outer air extending to the same height. It is therefore pressed upward by the heavier external air, which descends and moves toward the place of heat. The difference of weight in the two columns is greater, and therefore the draught stronger, if the chimney is high, provided the supply of heat is sufficient to maintain the requisite temperature. Chimneys are frequently built one or two hundred feet high for the uses of manufactories. The high fireplaces and large flues of former times were unfavorable for draught, both because much cold air could mingle with that which was heated, and because there was room for external air to descend by the side of the ascending column. For good draught, no air should be allowed to enter the flue except that which has passed through the fire.

574. Ventilation of Apartments.—The air of an apartment, as it becomes vitiated by respiration, may generally be removed, and fresh air substituted, by taking advantage of the same inequality of weight in air-columns, which has been mentioned. If opportunity is given for the warm impure air to escape from the top of a room, and for external air to take its place, there will be a constant movement through the room, as in the flue of a chimney, though at a slower rate. If the external air is cold, the weight of the columns differs more, and therefore the ventilation is more easily effected. But in cold weather, the air, before being admitted to the room, is warmed by passing through the air-chambers of a furnace. When there is a chimney-flue in the wall of a room, with a current of hot air ascending in it, the ventilation is best accomplished by admitting the air into the flue at the upper part of the room; since it will then be removed with the velocity of the hot-air current.

The tendency of the air of a warm room to pass out near the top, while a new supply enters at the lower part, is shown by holding the flame of a candle at the top, and then at the bottom, of a door which is opened a little distance. The flame bends outward at the top and inward at the bottom.

The impure air of a large audience-room is sometimes removed by a mechanical contrivance, as, for instance, a fan-wheel placed above an opening at the top, and driven by steam.

The ventilation of mines is accomplished sometimes by a fire built under a shaft, fresh air being supplied by another shaft, and sometimes by a fan-wheel at the top of the shaft. If there happen to be two shafts which open to the surface at very different elevations, ventilation may be effected by the inequality of temperature which is likely to exist within the earth and above it. Let *M M* (Fig. 296) be the vertical section of a mine through two shafts *A* and *B*, which open at different heights to the surface of the earth. If the external air is of the same temperature as the air within the earth, then the column *A* in the longer shaft has the same weight as *B* and *C* together, measured upward to the same level. In that case, which is likely to occur in spring and fall, there is no circulation without the use of other means. But in summer the air C is warmer than A and B; therefore A is heavier than $B + C$. Hence there is a current of air down A and up B. In winter, C is colder than air within the earth; therefore $B + C$ are together heavier than A, and the current sets in the opposite direction, down B and up A.

FIG. 296.

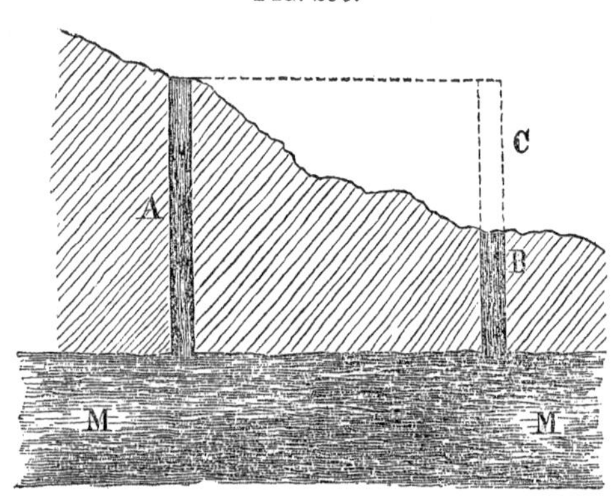

575. Sources of Heat.—*The sun,* although nearly a hundred millions of miles from the earth, is the source of nearly all the heat existing at its surface. The interior of the earth, except a thickness of forty or fifty miles next to the surface, is believed to be in a condition of heat so intense that all the materials composing it are in the melted state. But the earth's crust is so poor a conductor that only an insensible fraction of all this heat reaches the surface.

Mechanical operations are usually attended by a development of heat. For example, if a broad surface of iron were made to revolve, rubbing against another surface, nearly all the force expended in overcoming the friction would appear as heat, a comparatively small part being conveyed through the air as sound. The cutting tool employed in turning an iron shaft has been known to generate heat enough to raise a large quantity of cold water to the boiling point, and to keep it boiling for an indefinite time. It is a fact familiar to all, that violent friction of bodies against each other will set combustibles on fire. The axles of railroad cars are made red-hot if not duly oiled; boats are set on fire by the rope drawn swiftly over the edge by a whale after he is harpooned; a stream of sparks flies from the emory wheel when steel is polished, &c. Condensation and percussion, as well as friction, and all sudden applications of force, cause sensible heat. Indeed, wherever the full equivalent of any force is not obtained in some other form, the deficiency may be detected in the heat which is developed.

Chemical action is another very common source of heat. Combustion is the effect of violent chemical attraction between atoms of different natures, when both light and heat are manifested. If the union goes on slowly, as in the rusting of iron, the amount of heat is the same, but it is diffused as fast as developed. The molecular forces, expended in most cases of chemical combination, as measured by their heating effects, are enormously great.

The warmth produced by the vital processes in plants and animals is supposed by many physicists to be caused by chemical action. In breathing the air, some of its oxygen is consumed, which becomes united with the blood. This process is in some respects analogous to a slow combustion, by which heat is evolved in the animal system.

PART IX.

LIGHT.

CHAPTER I.

MOTION AND INTENSITY OF LIGHT.

576. Definitions.—Light is supposed to consist of exceedingly minute and rapid vibrations in a medium or ether which fills space; which vibrations, on reaching the retina of the eye, cause vision, as the vibrations of the air cause hearing, when they impinge on the tympanum of the ear, and as thermal vibrations produce a sensation of warmth, when they fall on the skin.

Bodies, which of themselves are able to produce vibrations in the ether surrounding them, are said to *emit* light, and are called *self-luminous,* or simply *luminous ;* those, which only *reflect* light, are called *non-luminous.* Most bodies are of the latter class. A *ray* of light is a line, along which light is propagated; a *beam* is made up of many parallel rays; a *pencil* is composed of rays either diverging or converging; and is not unfrequently applied to those which are parallel.

A substance, through which light is transmitted, is called a *medium ;* if objects are clearly seen through the medium, it is called *transparent ;* if seen faintly, *semi-transparent ;* if light is discerned through a medium, but not the objects from which it comes, it is called *translucent ;* substances which transmit no light are called *opaque.*

577. Light Moves in Straight Lines.—So long as the medium continues uniform, the line of each ray is perfectly straight. For an object cannot be seen through a bent tube; and if three disks have each a small aperture through it, a ray cannot pass through the three, except when they are exactly in a straight line. The shadow which is projected through space from an opaque body proves the same thing; for the edges of the shadow, taken in the direction of the rays, are all straight lines.

From every point of a luminous surface light emanates in all possible directions, when not prevented by the interposition of an opaque body. Thus, a candle is seen by night at the distance of one or two miles; and within that limit, no space so small as the pupil of the eye is destitute of rays from the candle. A point from which light emanates is called a *radiant*. If light from a radiant falls perpendicularly on a circular disk, the pencil is a cone; if on a square disk, it is a square pyramid, &c., the illuminated surface in each case being the base, and the radiant the vertex.

578. The Velocity of Light.—It has been ascertained by several independent methods, that light moves at the rate of about 192,500 *miles per second.*

One method is by means of the eclipses of Jupiter's satellites. The planet Jupiter is attended by four moons which revolve about it in short periods. These small bodies are observed, by the telescope, to undergo frequent eclipses by falling into the shadow which the planet casts in a direction opposite to the sun. The exact moment when the satellite passes into the shadow, or comes out of it, is calculated by astronomers. But sometimes the earth and Jupiter are on the same side, and sometimes on opposite sides of the sun; consequently, the earth is, in the former case, the whole diameter of its orbit, or about one hundred and ninety millions of miles nearer to Jupiter than in the latter. Now it is found by observation, that an eclipse of one of the satellites is seen about sixteen minutes and a half sooner when the earth is nearest to Jupiter, than when it is most remote from it, and consequently, the light must occupy this time in passing through the diameter of the earth's orbit, and must therefore travel at the rate of about 192,000 miles per second.

Another method of estimating the velocity of light, wholly independent of the preceding, is derived from what is called the *aberration of the fixed stars.* The apparent place of a fixed star is altered by the motion of its light being combined with the motion of the earth in its orbit. The place of a luminous object is determined by the direction in which its light meets the eye. But the direction of the impulse of light on the eye is modified by the motion of the observer himself, and the object appears forward of its true place. The stars, for this reason, appear slightly displaced in the direction in which the earth is moving; and the velocity of the earth being known, that of light may be computed in the same manner as we determine one component, when the angles and the other component are known.

The velocity of light has been determined also by direct experiment, in a manner somewhat analogous to that employed by Wheatstone for ascertaining the velocity of electricity.

579. Loss of Intensity by Distance.—The *intensity* of light varies *inversely as the square of the distance.* In Fig. 297, suppose light to radiate from S, through the rectangle $A\ C$, and fall on $E\ G$, parallel to $A\ C$. As $S\ A\ E$, $S\ B\ F$, &c., are straight lines, the triangles, $S\ A\ B$, $S\ E\ F$, are similar, as also the rectangles, $A\ C$, $E\ G$; therefore, $A\ C : E\ G :: A\ B^2 : E\ F^2 :: S\ A^2 : S\ E^2$. But the same quantity of light, being diffused over $A\ C$ and $E\ G$, will be more intense, as the surface is smaller. Hence, the intensity of light at E : intensity at $A :: A\ C : E\ G :: S\ A^2 : S\ E^2$, which proves the proposition. This demonstration is applicable to every kind of emanation in straight lines from a point.

FIG. 297.

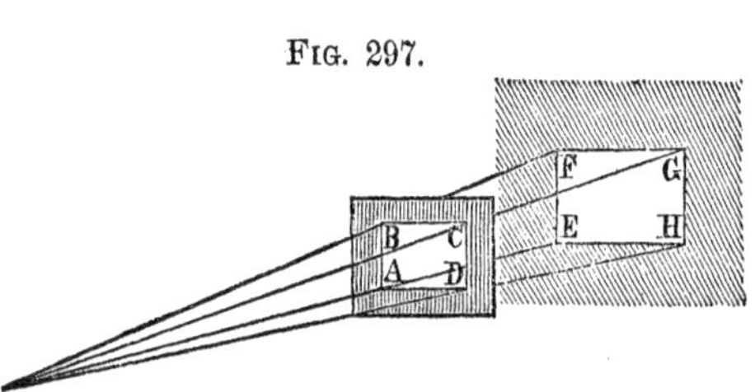

580. Brightness the Same at all Distances.—The *brightness* of an object is the quantity of light which it sheds, as compared with the apparent area from which it comes. Now the *quantity* (or intensity), as has just been shown, varies inversely as the square of the distance. The apparent *area* of a given surface also diminishes in the same ratio, as we recede from it. Hence, the brightness is constant. For illustration, if we remove to *three* times the distance from a luminous body, we receive into the eye nine times less light, but the body also appears nine times smaller, so that the relation of light to apparent area remains the same.

581. Loss of Intensity by Absorption.—In a uniform medium, while the distance increases *arithmetically*, the intensity diminishes *geometrically*. Imagine the medium to be divided by parallel planes into strata of equal thickness; and suppose the first stratum to diminish the intensity by $\frac{1}{n}$ of the whole. Then the intensity of the light which reaches the second stratum is $1 - \frac{1}{n} = \frac{n-1}{n}$. But on account of the uniformity of the medium, every stratum produces the same effect, that is, it transmits to the next, $\frac{n-1}{n}$ of that which falls upon it. Therefore, $\frac{n-1}{n}$

of $\frac{n-1}{n}$, or $\frac{(n-1)^2}{n^2}$, leaves the second stratum, $\frac{(n-1)^3}{n^3}$, the third, and so on, in a geometrical series. For example, if a piece of colored glass is $1\frac{1}{4}$ inch thick, and each quarter of an inch absorbs $\frac{3}{5}$ of the light which falls upon it, then about *one-hundredth* of what enters the first surface will escape from the last. For $\left(\frac{2}{5}\right)^5 = .01$ nearly.

582. Photometers.—These are instruments designed for the measurement of the relative intensities of light. We cannot determine by the eye alone how many times more intense one light is than another, though we can judge with tolerable accuracy when two surfaces are *equally* illuminated. Photometers are, therefore, generally constructed on the plan of determining the ratio of intensities of two lights, by means of our ability to decide when they illuminate two surfaces equally. It is sufficient to mention Rumford's method *by shadows.* Let the two unequal lights be so placed that the two shadows of an opaque body cast by them shall fall side by side on a white screen. If one shadow appears more luminous than the other, remove to a greater distance the light which illuminates it (or bring the other nearer), until the shadows appear of the same degree of illumination. Then measure the distances from the lights to the screen, and the intensities of the lights will be *directly as the squares of the distances.* For the light at the greater distance, since it illuminates the screen equally with the other, must gain as much by intensity as it loses by distance; that is, in the ratio of the square of the distance.

583. Shadows.—When a luminous body shines on one which is opaque, the space beyond the latter, from which the light is excluded, is called a *shadow.* The same word, as commonly used, denotes only the *section* of a shadow made by a surface which crosses it. Shadows are either *total* or *partial.* If tangents are drawn on all the corresponding sides of the two bodies, the space inclosed by them beyond the opaque body is the total shadow; if other tangents are drawn, crossing each other between the bodies, the space between the total shadow and the latter system of tangents is the partial shadow, or *penumbra.* In case the bodies are spheres, as in Fig. 298, the total shadow will be a cylinder, or conical frustum, each of infinite length, or a complete cone, according to the relative size of the spheres. But, in every case, the penumbra and inclosed total shadow will form an increasing frustum. It is obvious that the shade of the penumbra grows gradually deeper from the outer surface to the total shadow within it.

Every shadow cast by the sun has a penumbra bordering it, which gives to the shadow an ill-defined edge; and the more re-

FIG. 298.

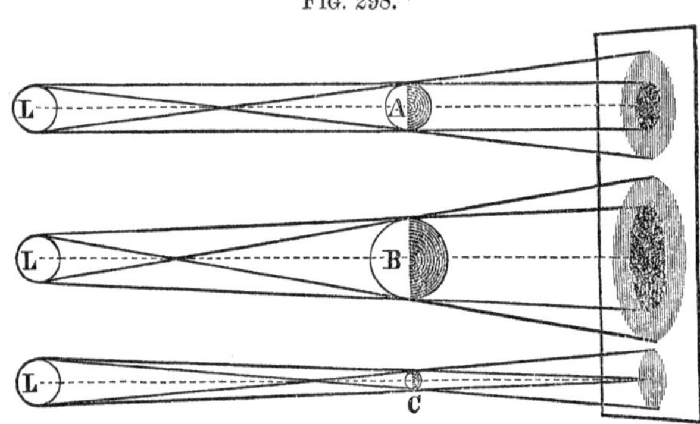

mote the sectional shadow is from the opaque body which casts it, the broader will be the partial shadow on the edge.

CHAPTER II.

REFLECTION OF LIGHT.

584. Radiant and Specular Reflection.—Light is said to be *reflected* when, on meeting a surface, it is turned back into the same medium. In ordinary cases of reflection, the light is diffused in all directions, and it is by means of the light thus scattered from a body that it becomes visible, when it sheds no light of its own. This is called *radiant reflection.* It is produced by unpolished surfaces. But when a surface is highly polished, a beam of light falling on it is reflected in some particular direction; and, if the eye is placed in this reflected beam, it is not the reflecting surface which is seen, but the original object, apparently in a new position. This is called *specular reflection.* It is, however, generally accompanied by some degree of radiant reflection, since the reflector itself is commonly visible in all directions. Ordinary mirrors are not suitable for accurate experiments on reflection, because light is modified by the glass through which it passes. The *speculum* is therefore used, which is a reflector made of solid metal, and accurately ground to any required form, either *plane, convex*, or *concave.* The word *mirror* is, however, much used in optics for every kind of reflector.

Optical experiments are usually performed on a beam of light admitted through an aperture into a darkened room; the direction of the beam being regulated by an adjustable mirror placed outside. An instrument consisting of a plane speculum moved by a clock, in such a manner that the reflected sunbeam shall remain stationary at all hours of the day, is called a *heliostat.*

585. The Law of Reflection.—When a ray of light is incident on a mirror, the angle between it and a perpendicular to the surface at the point of incidence, is called the *angle of incidence;* and the angle between the reflected ray and the same perpendicular, is called the *angle of reflection.* The law of reflection found to be universally true is the following:

The angles of incidence and reflection are on opposite sides of the perpendicular, and are equal to each other.

This is well shown by attaching a small mirror to the centre of a graduated semicircle perpendicular to its plane. Let *M D N* (Fig. 299) be the semicircle, graduated from *D* both ways to *M* and *N*, and mounted so that it can be revolved on its centre, and clamped in any position. Let the small mirror be at *C*, with its plane perpendicular to *C D*; then a ray from the heliostat, as *A C*, passing the edge at a particular degree, will be seen after reflection to pass the corresponding degree in the other quadrant. By revolving the semicircle, any angle of incidence may be tried, and the two rays are always found to be in the same plane with *C D*, and equally inclined to it.

FIG. 299.

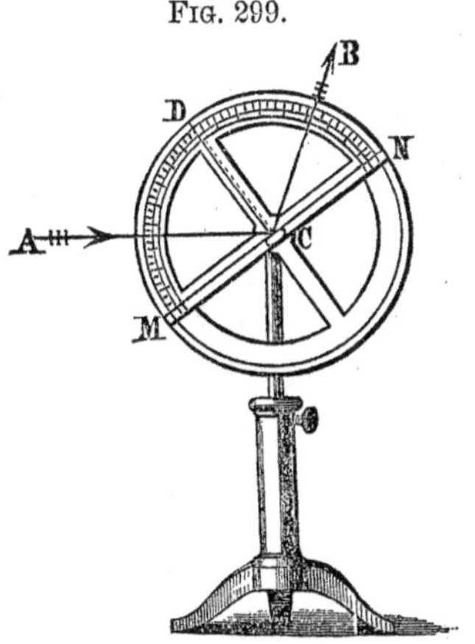

As the mirror revolves, the reflected ray revolves twice as fast.

For *A C D* is increased or diminished by the angle through which the mirror turns; therefore *D C B* is also increased or diminished by the same; hence *A C B*, the angle between the two rays, is increased or diminished by the sum of both, or twice the same angle.

It follows from the law of reflection, that a ray which falls on a mirror perpendicularly, retraces its own path after reflection. It is obvious, also, that the complements of the angles of incidence and reflection are equal, i. e. $A\,C\,M = B\,C\,N$. The law of reflection is applicable to curved as well as to plane mirrors; the radius of curvature at any point being the perpendicular with which the incident and reflected rays make equal angles.

Radiant reflection forms no exception to the foregoing law, though the incident rays are in one and the same direction, and the reflected rays are scattered every way. For the minute cavities and prominences which constitute the roughness of the general surface are bounded by small surfaces lying at all inclinations; and each one reflecting the rays which meet it in accordance with the law, those rays are necessarily thrown off in all possible directions.

586. Inclination of Rays to each other not altered by the Plane Mirror.—

1. Rays which *diverge* before reflection, diverge at the same angle after reflection.

Let $M N$ (Fig. 300) be a plane mirror, and $A B$, $A C$, any two rays of light falling upon it from the radiant A, and reflected in the lines $B E$, $C G$. Draw the perpendicular $A P$, and produce it indefinitely, as to F, behind the mirror; also produce the reflected rays back of the mirror. Let $Q R$ be perpendicular to the mirror at the point B; it is therefore parallel to $A F$, and the plane passing through $A F$ and $Q R$, is that which includes the ray $A B$, $B E$. Therefore, $E B$, when produced back of the mirror, intersects $A P$ produced. Let F be the point of intersection. $B A F = A B Q$, and $A F B = E B Q$; but $A B Q = E B Q$ (Art. 585); $\therefore B A F = A F B$, and $A B = F B$. If P and B be joined, $P B$ being in the plane $M N$ is perpendicular to $A F$, and therefore bisects it. Hence, the reflected ray meets the perpendicular $A F$ as far behind the mirror, as the incident ray does in front. In the same way it may be proved that $A C = C F$, and that $C G$, when produced back of the mirror, meets $A F$ at the same point F.

Fig. 300

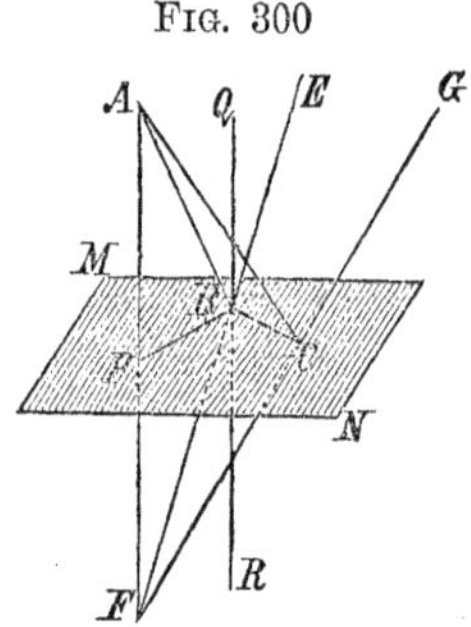

Now, since the triangles $A C B$ and $F C B$, have their sides respectively equal, their angles are equal also; hence $B A C = B F C$. Therefore any two rays diverge at the same angle after reflection as they did before reflection.

Since the reflected rays seem to emanate from F, that point is called the *apparent radiant; A* is the *real radiant.*

2. Rays which *converge* before reflection, converge at the same angle after reflection. Let $E B$, $G C$, be incident rays converging toward F, and let $B A$, $C A$, be the reflected rays. It may be

proved as before, that A and F are in the same perpendicular, $A F$, and equidistant from P, and that $E F G = B A C$.

The point F, to which the incident rays were converging, is called the *virtual focus;* A is the *real focus.*

3. Rays which are *parallel* before reflection are parallel after reflection.

It has been proved in case 1, that F, the intersection of the reflected rays, is as far behind the mirror, as A, the intersection of incident rays, is before it. Now, if the incident rays are parallel, A is at an infinite distance from the mirror. Therefore F is at an infinite distance behind it, and the reflected rays are parallel.

In all cases, therefore, rays reflected by a plane mirror retain the same inclination to each other which they had before reflection.

587. Spherical Mirrors.—A *spherical mirror* is one which forms a part of the surface of a sphere, and is either convex or concave. The *axis* of such a mirror is that radius of the sphere which passes through the middle of the mirror. In the practical use of spherical mirrors, it is found that the light must strike the surface very nearly at right angles; hence, in the following statements, the mirror is supposed to be a very small part of the whole spherical surface, and the rays nearly coincident with the axis.

It is sufficient to trace the course of the rays on one side of the axis, since, on account of the symmetry of the mirror around the axis, the same effect is produced on every side.

588. Converging Effect of a Concave Mirror.—

1. *Parallel* rays are converged to the *middle* point between the centre and surface, which is therefore called the *focus of parallel rays*, or the *principal focus.* Let $R A$, $L E$ (Fig. 301), be parallel rays incident upon the concave mirror $A B$, whose centre of concavity is

FIG. 301.

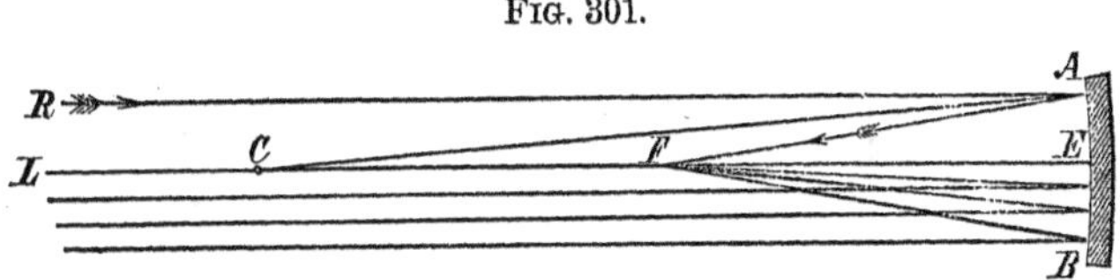

C. The ray $L E$, passing through C, and therefore perpendicular to the mirror at E, is reflected directly back. Join $C A$, and make $C A F = R A C$; then $R A$ is reflected in the line $A F$, and the two reflected rays meet at F. $R A C = A C F$, $\therefore A C F = F A C$, and $A F = C F$; and as A and E are very near together, $E F = F C$; that is, the focus of parallel rays is at the middle point between C and E.

2. *Diverging* rays, falling on a given concave mirror, are reflected *converging, parallel, or less diverging,* according to the degree of divergency in the original pencil. Let C (Fig. 302) be the centre of concavity, and F the focus of parallel rays. Then,

FIG. 302.

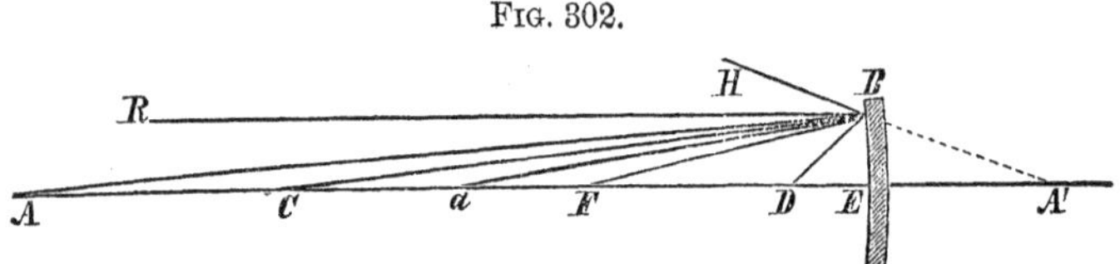

rays diverging from any point, A, beyond C, will be converged to some point, a, between C and F, since the angles of incidence and reflection are less than those for parallel rays. Rays diverging from C are reflected back to C; those from points between C and F, as a, are converged to points beyond C, as A; those diverging from F become parallel; and those from points between F and the mirror, as D, *diverge* after reflection, but at a less angle than before, and seem to flow from A'. To prove, in the last case, that the angle of divergence, A', after reflection, is less than the angle D, the divergence before reflection, observe that the angle A' is less than the exterior angle $H\ B\ C$; but $H\ B\ C = D\ B\ C$ (Art. 585), which is less than $D\ B\ R$, which is equal to $A'\ D\ B$; much more, then, is A' less than $A'\ D\ B$.

3. *Converging* rays are made to *converge more.* The rays $H\ B$, $A\ E$, converging to A', are reflected to D, nearer the mirror than F is. And it has been shown that the angle D is larger than A', hence the convergency is increased.

From the three foregoing cases, it appears that the *concave* mirror always tends to produce *convergency;* since, when it does not actually produce it, it diminishes divergency.

589. Conjugate Foci.—When light radiates from A, it is reflected to a; when it radiates from a, it meets at A. Any two such interchangeable points are called *conjugate foci.* If the radius of the mirror and the distance of one focus from the mirror are given, the distance of its conjugate focus may be determined. Let the radius $= r$; the distance $A\ E = m$; and $a\ E = n$. As the angle $A\ B\ a$ is bisected by $B\ C$, $A\ B : a\ B :: A\ C : a\ C$; that is, since $B\ E$ is very small, $A\ E : a\ E :: A\ C : a\ C$, or, $m : n :: m - r : r - n$.

$$\therefore \quad m = \frac{n\ r}{2\ n - r}; \text{ and } n = \frac{m\ r}{2\ m - r}.$$

If A is not on the axis of the mirror, as in Fig. 303, let a line be drawn through A and C, meeting the mirror in E; this is called a *secondary* axis, and the light radiating from A will be reflected to a on the same secondary axis, for $A\ E$ is perpendicular to the mirror, and will be reflected directly back; and if $A\ E$ and $C\ E$ are given, $a\ E$ may be found as before.

Fig. 303.

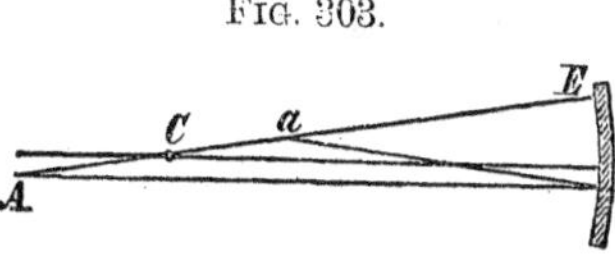

590. Diverging Effect of a Convex Mirror.—

1. *Parallel* rays are reflected diverging from the *middle point* between the centre and surface. Let C (Fig. 304) be the centre of convexity of the mirror $M\ N$, and draw the radii, $C\ M$, $C\ D$,

Fig. 304.

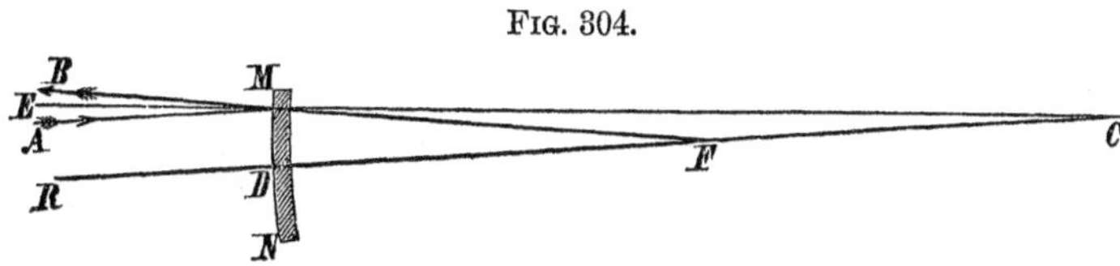

producing them in front of the mirror; these are perpendicular to the surface. The ray $R\ D$ will be reflected back; $A\ M$ will be reflected in $M\ B$, making $B\ M\ E = A\ M\ E$. Produce the reflected ray back of the mirror, and it will meet the axis in F, midway from C to D; for $F\ C\ M = A\ M\ E$, and $F\ M\ C = B\ M\ E$; therefore the triangle $F\ C\ M$ is isosceles, and $C\ F = F\ M$, and as M is very near D, $C\ F = F\ D$. Hence the rays, after reflection, diverge as if they radiated from a point in the middle of $C\ D$, which is the apparent radiant.

2. *Diverging* rays have their divergency increased. Let $A\ D$, $A\ M$ (Fig. 305), be the diverging rays; $D\ A$, $M\ B$, the reflected

Fig. 305.

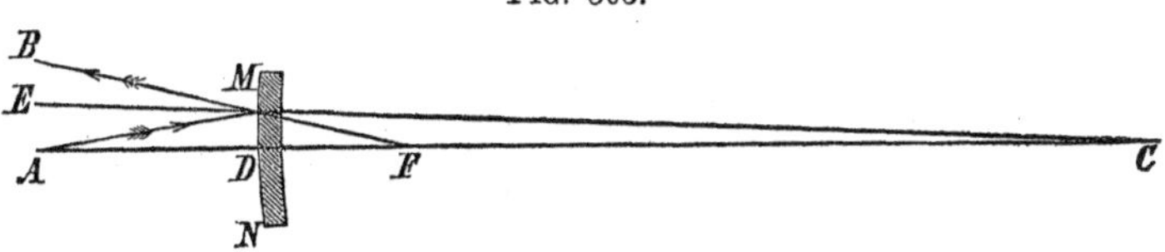

rays; these when produced meet at F, which is the apparent radiant. $M\ A\ F$ is the divergency of the incident rays, and $A\ F\ B$ of the reflected rays. Now the exterior angle, $A\ F\ B$, is greater than $C\ M\ F$, or $B\ M\ E$, or $A\ M\ E$. But $A\ M\ E$, being exterior, is greater than $M\ A\ F$; much more, then, is $A\ F\ B$ greater than $M\ A\ F$.

3. *Convergent* rays are at least rendered less convergent, and

may become parallel or divergent, according to the degree of previous convergency. The two first effects are shown by Figs. 304 and 305, reversing the order of the rays. And it is easy to perceive that rays converging to *C*, will diverge from *C* after reflection; if to a point more distant than *C*, they will diverge afterward from a point between *C* and *F* (Fig. 304), and *vice versa*.

The general effect, therefore, of a *convex* mirror, is to produce *divergency*.

A and *F* (Fig. 305) are called conjugate foci, being interchangeable points; for rays from *A* move after reflection as though from *F*, and rays converging to *F* are by reflection converged to *A*. Conjugate foci, in the case of the convex mirror, are in the same axis either principal or secondary, as they are in the concave mirror, and for the same reason, viz., that every axis is perpendicular to the surface.

591. Images by Reflection.—An optical image consists of a collection of focal points, from which light either really or apparently radiates. When rays are converged to a focus they do not stop, but cross, and diverge again, as if originally emanating from the focal point. A collection of such points, arranged in order, constitutes a *real image*. When rays are reflected diverging, they proceed *as though* they emanated from a point behind the mirror. A collection of such imaginary radiants forms an *apparent* or *virtual image*. The images formed by *plane* and *convex* mirrors are always apparent; those formed by *concave* mirrors may be of either kind.

592. Images by a Plane Mirror.—When an object is before a plane mirror, its image is at the *same distance behind* it, of the *same magnitude*, and *equally inclined* to it. Let *M N* (Fig. 306) be a plane mirror, and *A B* an object before it, and let the position of the object be such that the reflected rays may enter the eye placed at *H*. From *A* and *B* let fall upon the plane of the mirror the perpendiculars *A E*, *B G*, and produce them, making $E a = A E$, and $G b = B G$. Now, since the rays from *A* will, after reflection, radiate as if from *a* (Art. 586), and those from *B*, as if from *b*, and the same of all other points, therefore the image and object are equally distant from the mirror. *A C*, *a c*, parallel to the mirror, are equal; as $B G = b G$, and $A E = a E$, therefore, by subtraction, $B C = b c$; also the right angles at *C* and *c* are equal. There-

Fig. 306.

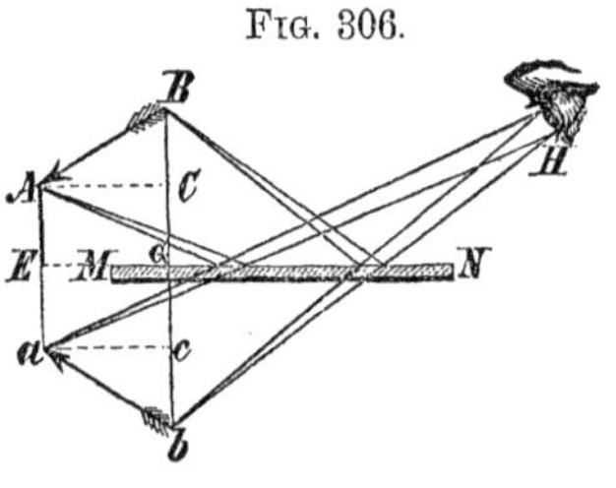

fore $A\ B = a\ b$, and $B\ A\ C = b\ a\ c$; that is, the object and image are of equal size, and equally inclined to the mirror.

It appears from the demonstration, that the object and its image are comprehended between the same perpendiculars to the plane of the mirror.

The object and image obviously have to each other twice the inclination that each has to the mirror. Hence, in a mirror inclined 45° to the horizon, a horizontal surface appears vertical, and one which is vertical appears horizontal.

593. Symmetry of Object and Image.—All the three dimensions of the object and image are respectively *equal*, as shown above, but one of them is *inverted* in position, namely, that dimension which is perpendicular to the mirror. Hence, a person and his image face in opposite directions; and trees seen in a lake have their tops downward. Those dimensions which are parallel to the mirror are not inverted. In consequence of the inversion of *one* dimension alone, the object and its image are not *similar*, but *symmetrical* forms; and one could not coincide with the other if brought to occupy the same space. The image of a *right* hand is a *left* hand, and all relations of right and left are reversed. It is for this reason that a printed page, seen in a mirror, is like the type with which it was printed.

594. The Length of Mirror Requisite for Seeing an Object.—If an object is parallel to a mirror, the length of mirror occupied by the image is to the length of the object as the reflected ray to the sum of the incident and reflected rays. Let $A\ B$ (Fig. 307) be the length of the object, $C\ D$ that of the image, and $F\ G$ that of the space occupied on the mirror; then, by similar triangles, $F\ G : C\ D :: E\ F : E\ C$. But $C\ D = A\ B$, and $C\ F = A\ F$; $\therefore F\ G : A\ B :: E\ F : A\ F + F\ E$. If the eye is brought nearer the mirror, the space on the mirror occupied by the image is diminished, because $E\ F$ has to $A\ F + F\ E$ a less ratio than before. The same effect is produced by removing the object further from the mirror. The length of mirror necessary for a person to see himself is equal to half his height, because in that case, $E\ F : A\ F + F\ E :: 1 : 2$, which ratio will not be altered by change of distance.

FIG. 307.

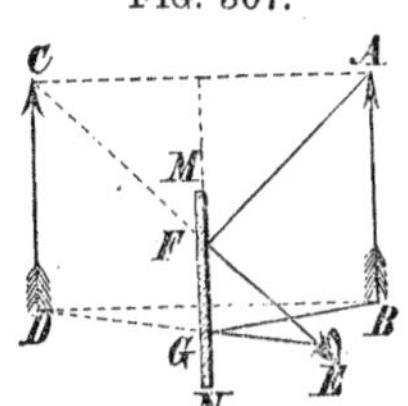

595. Displacement of Image by Two Reflections.—If an image is seen by light reflected from two mirrors in a plane perpendicular to their common section, its angular deviation from the object is equal to twice the inclination of the mirrors. Let $A\,B\,C\,D$ (Fig. 308) be two plane mirrors inclined at the angle $A\,G\,C$. If an eye at H sees the star S in the direction O, the angle $S\,H\,O = 2\,A\,G\,C$. For $H\,B\,G = A\,B\,S = G\,B\,D$; $\therefore H\,B\,D = 2\,G\,B\,D$. In like manner, $B\,D\,O = 2\,B\,D\,C$. But $S\,H\,O = B\,D\,O - H\,B\,D = 2\,B\,D\,C - 2\,G\,B\,D = 2\,B\,G\,D$; $\therefore S\,H\,O = 2\,B\,G\,D$.

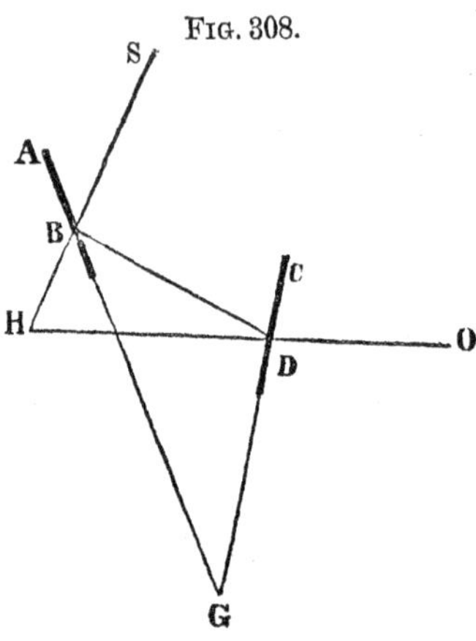

FIG. 308.

This principle is employed in the construction of *Hadley's quadrant*, and the *sextant*, used at sea for measuring angular distances. The angles measured are twice as great as the arc passed over by the index which carries the revolving mirror; hence, in the quadrant, an arc of 45° is graduated into 90°; and, in the sextant, an arc of 60° is graduated into 120°.

596. Multiplied Images by Two Mirrors.—

1. *Parallel Mirrors*. The series of images is *infinite* in number, and arranged *in a straight line*, perpendicular to the mirrors. The object E between the parallel mirrors, $A\,B$, $C\,D$ (Fig. 309),

FIG. 309.

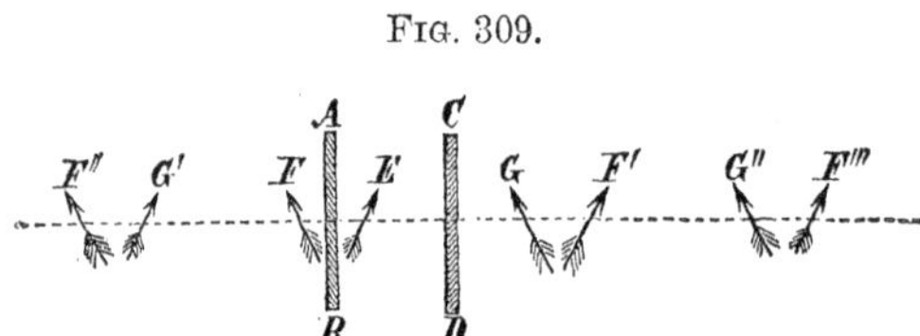

has an image at F, as far behind $A\,B$ as E is in front of it, and between the same perpendiculars. The rays reflected by $A\,B$ diverge, as though they emanated from F; hence, F may be regarded as an object before $C\,D$, whose image is at F', as far behind it. Again, F' may be considered as an object before $A\,B$, and so on indefinitely. Another series exists in the same line, by beginning with G, the first image behind $C\,D$. As light is absorbed and scattered by each reflection, these images grow fainter, and at length disappear. Articles of jewelry are sometimes apparently multiplied and extended over a large surface, by lining the cases with parallel mirrors.

The multiplied images of a small bright object, sometimes seen in a looking-glass, are produced by repeated reflections between the front and the silvered covering on the back side. At each internal impact on the first surface some light escapes, and shows us an image, while another portion is reflected to the back, and thence forward again. The image of a lamp viewed very obliquely in a mirror is sometimes repeated eight or ten times; and a planet, or bright star, when seen in a looking-glass, will be accompanied by three or four faint images, caused in the same way.

2. *Inclined Mirrors.* In this case, the images are *limited* in number, and arranged *in the circumference of a circle*, whose centre is in the line of common section of the planes of the mirrors, and whose radius is the distance of the object from that line. Let $A\,B$, $A\,C$ (Fig. 310) be the mirrors, and E the object. Draw $E\,G$ perpendicular to $A\,B$, and make $E\,F = F\,G$, then will G be the first image: in the same way, find I, the image of G by $A\,C$; K, the image of I; and V, that of K. Then begin with the mirror $A\,C$, and find, as before, M, O, P, Q, the successive images by the two mirrors. No image of V or Q can be formed, because they are behind both mirrors. All these images are in the circumference of a circle, whose radius is $E\,A$; for $E\,F$, $F\,A$,

FIG. 310.

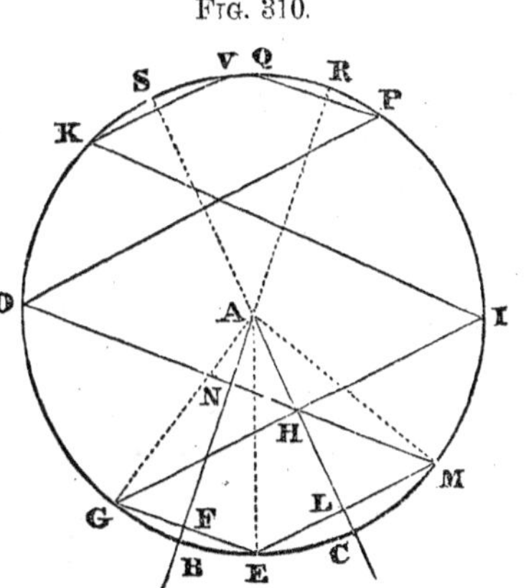

and angle at F, are respectively equal to $G\,F$, $F\,A$, and angle at F; $\therefore E\,A = G\,A$; and in the same way it may be proved, that $E\,A = A\,M$, $A\,I$, &c. If the edges at A be separated, making the inclination of the mirrors less and less, the number of images will increase, and the circumference will approach a straight line, so that ultimately we shall have the case described in (1), in which the mirrors are parallel.

597. Path of the Pencil by which each Image is Seen.— Fig. 311 will assist to understand how each image is seen by a pencil of light which passes back and forth between the mirrors, until it reaches the eye. If the eye is at O, and the object at Q, and its images at A, B, C, D, each image is of course seen by a pencil which comes from the mirror to the eye, as if it originated in that image. Therefore, draw a line from any image as D, to the eye, and from its intersection with the mirror draw a line to the preceding image; from the intersection of that line with the other mirror, a line to the image next preceding, and so on back to Q; the whole path of the pencil will then be traced. Thus, $A\ O$ being joined, and $Q\ a$ drawn to the intersection a, the image A is seen by the ray $Q\ a$, $a\ O$. In like manner, B is seen by $Q\ b$, $b\ c$, $c\ O$; C, by $Q\ d$, $d\ e$, $e\ f$, $f\ O$; and D, by $Q\ g$, $g\ h$, $h\ i$, $i\ k$, $k\ O$.

Fig. 311.

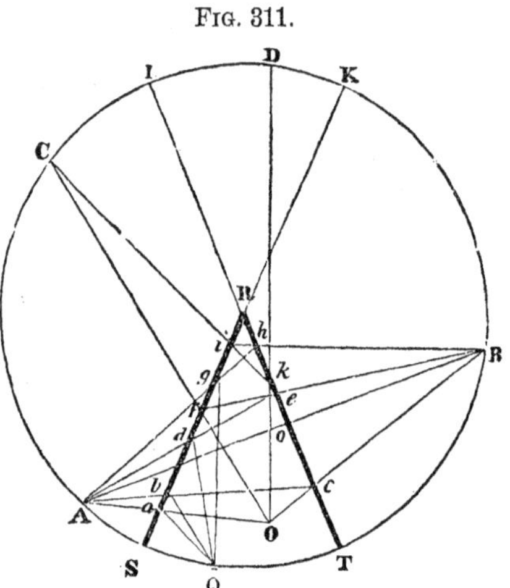

598. The Kaleidoscope.—This instrument, when carefully constructed, beautifully exhibits the phenomenon of multiplied reflection by inclined mirrors. It consists of a tube containing two long, narrow, metallic mirrors, inclined at a suitable angle; and is used by placing the objects (fragments of colored glass, etc.) at one end, and applying the eye to the other. In order that there may be perfect symmetry in the figure made up of the objects and their successive images, the angle of the mirrors should be of such size, that it can be exactly contained an even number of times in 360°. The best inclination is 30°; and the field of view is then composed of 12 sectors. It is also essential, that the small objects forming the picture, should lie at the least possible distance beyond the mirrors. To insert three mirrors instead of two, as is often done, only serves to confuse the picture, and mar its beauty.

599. Images by the Concave Mirror.—The concave mirror forms various images, either *real* or *apparent*, either *greater* or *less* than the object, either *erect* or *inverted*, according to the place of the object.

1. The object *between the mirror and its principal focus.* By Art. 588 (2), rays which diverge from a point between the mirror and its principal focus, continue to diverge after reflection, but in a less degree. Let C be the centre, and F the principal focus of the mirror $M\ N$ (Fig. 312), and $A\ B$ the object. Draw the axes,

C A, *C B*, and produce them behind the mirror. The pencil from *A* will be reflected to the eye at *H*, radiating as from *a*, in the same axis; likewise, those from *B*, as from *b*. Therefore, the

FIG. 312.

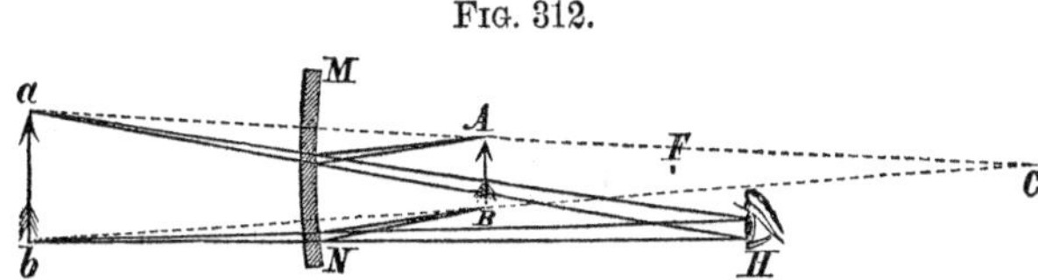

image is *apparent*, since rays do not actually flow from it; *erect*, as the axes do not cross each other between the object and image; *enlarged*, because it subtends the angle of the axes at a greater distance than the object does. As the object approaches, and finally reaches the principal focus, the reflected rays approach parallelism, and the image departs from the mirror, till it is at an infinite distance, and is viewed as a heavenly body.

2. Object *between the principal focus and the centre*. As soon as the object passes the principal focus, the rays of each pencil begin to converge; and each radiant of the object has its conjugate focus in the same axis beyond the centre (Art. 589). For example, the pencil *A d g* (Fig. 313) is converged to *a* in the axis *A C a*, and *B D G* to *b*, in the axis *B C b*. Therefore, the image of *A B* is *a b* beyond the centre; and if an observer is beyond *a b*, the rays, after crossing at the image, will reach him, as though they

FIG. 313.

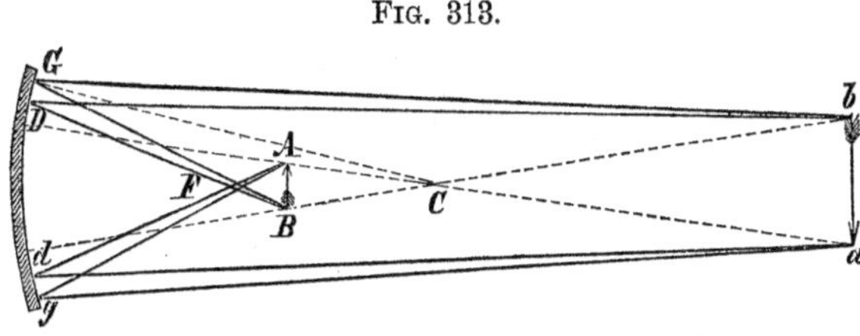

originated in *a b*; or if a screen is placed at *a b*, the light which is collected in the focal points will be thrown in all directions by radiant reflection from the screen. Hence, the image is *real;* it is also *inverted*, because the axes cross between the conjugate foci; and it is *enlarged*, since it subtends the angle of the axes at a greater distance than the object does. That *b C* is greater than *B C*, is proved by joining *C G*, which bisects the angle *B G b*, and therefore divides *B b* so that *B C* : *C b* : : *B G* : *G b*. When the object reaches the centre, the image is there also, but inverted in position, since rays which proceed from one side of *C*, are reflected to the other side of it.

3. Object *beyond the centre.* This is the reverse of (2), the conjugate foci having changed places; *a b*, therefore, being the object, *A B* is its image, *real, inverted, diminished.* As the object removes to infinity, the image proceeds only to the principal focus *F.*

600. Illustrated by Experiment.—These cases are shown experimentally by placing a lamp close to the mirror, and then carrying it along the axis to a considerable distance away. While the lamp moves from the mirror to the principal focus, its image behind the mirror recedes from its surface to infinity; we may then regard it as being either at an infinite distance behind, or an infinite distance in front, since the rays of every pencil are parallel. After the lamp passes the principal focus, the image appears in the air at a great distance in front, and of great size, and they both reach the centre together, where they pass each other; and, as the lamp is carried to great distances, the image, growing less and less, approaches the principal focus, and is there reduced to its smallest size. The only part of the infinite line of the axis before and behind, in which no image can appear, is the small distance between the mirror and its principal focus.

If a person looks at *himself*, so long as he is between the mirror and the principal focus, he sees his image behind the mirror and enlarged. But when he is between the principal focus and centre, the image is *real*, and behind him; the converging rays of the pencils, however, enter his eyes, and give an indistinct view of his image as if at the mirror. When he reaches the centre, the pupil of the eye is seen covering the entire mirror, because rays from the centre are perpendicular, and return to it from all parts of the surface. Beyond the centre, he sees the real image in the air before him, distinct and inverted.

601. Images by the Convex Mirror.—The convex mirror affords no variety of cases, because diverging rays, which fall upon

Fig. 314.

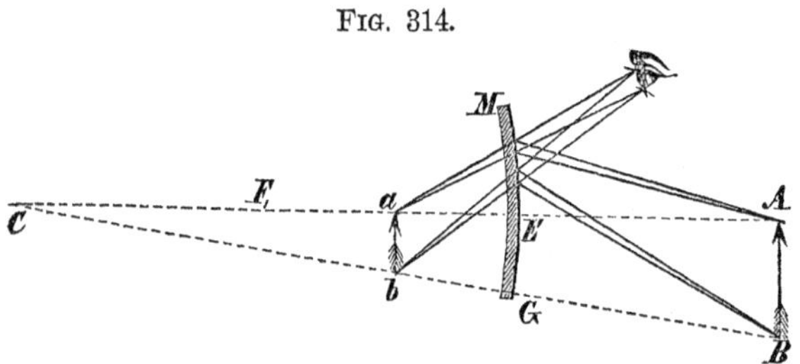

it, are made to diverge still more by reflection. In Fig. 314, the pencil from *A* is reflected, as if radiating from *a* in the same axis

A C, and that from *B*, as from *b* in the axis *B C*; and these apparent radiants are always nearer the surface than the middle point between it and *C* (Art. 590). The image is therefore *apparent;* it is *erect*, since the axes do not cross between the object and image; and it is *diminished*, as it subtends the angle of the axes at a less distance than the object.

602. Caustics by Reflection.—These are luminous curved surfaces, formed by the intersections of rays reflected from a hemispherical concave mirror. The name *caustic* is given from the circumstance that *heat*, as well as light, is concentrated in the focal points which compose it. *B A D* (Fig. 315), represents a section of the mirror, and *B F D* of the caustic; the point *F*, where all the sections of the caustic through the axis meet each other, is called the *cusp*. When the incident rays are parallel, as in the figure, the cusp is at the principal focus, that is, the middle point between *A* and *C*. The rays near the axis *R A*, after reflection meet at the cusp (Art. 588); but those a little more distant cross them, and meet the axis a little further toward *A*. And the more distant the incident ray from the axis, the further from the centre does the reflected ray meet the axis. Thus each ray intersects all the previous ones, and this series of intersections constitutes the curve, *B F*. The curve is luminous, because it consists of the foci of the successive pencils reflected from the arc *A B*.

FIG. 315.

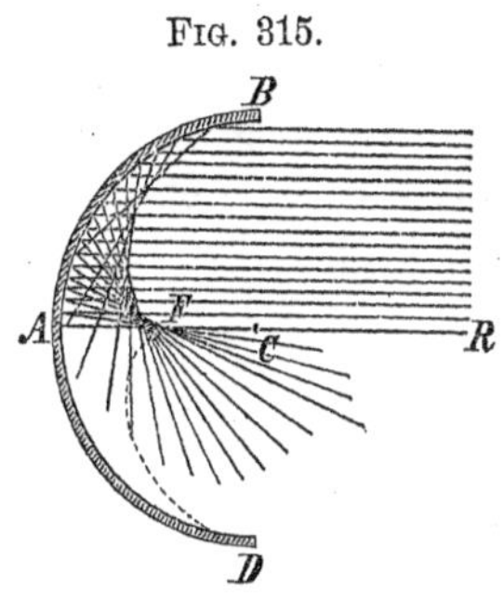

If the incident rays, instead of being parallel, diverge from a lamp near by, the form of the caustic is a little altered, and the cusp is nearer the centre. This case may be seen on the surface of milk, the light of the lamp being reflected by the edge of the bowl which contains it.

If parallel or divergent light falls on a convex hemispherical mirror, there will be *apparent* caustics behind the mirror; that is, the light will be reflected as if it radiated from points arranged in such curves.

603. Spherical Aberration of Mirrors.—It has already been mentioned (Art. 587), that the statements in this chapter relating to focal points and images, as produced by spherical mirrors, are true only when the mirror is a very small part of the whole spherical surface. In Art. 602 we have seen the effect of using a large part of the spherical surface—viz., the rays neither converge *to*, nor diverge *from* a single point, but a series of points arranged in a curve. This general effect is called the *spherical aberration* of a mirror; since the deviation of the rays is due to the spherical curvature. The deviation, as we have seen, is quite apparent in a hemisphere, or any considerable portion of one; but it exists in some degree in any spherical mirror, unless infinitely small compared with the hemisphere.

But there are curves which will reflect without aberration. Let a concave mirror be ground to the form of a paraboloid, and rays parallel to its axis will be converged to the focus without aberration. For, at any point on such a mirror, a line parallel to the axis, and a line drawn to the focus, make equal angles with the tangent, and therefore, equal angles with the perpendicular to the surface. And rays, parallel to the axis of a convex paraboloid, will diverge as if from its focus, on the same account. Again, if a radiant is placed at the focus of a concave parabolic mirror, the reflected rays will be parallel to the axis, and will illuminate at a great distance in that direction. Such a mirror, with a lamp in its focus, is placed in front of the locomotive engine to light the track, and has been much used in light-houses. If a concave mirror is ellipsoidal, light emanating from one focus is collected without aberration to the other, because lines from the foci to any point of the curve make equal angles with the tangent at that point.

Since heat is reflected according to the same law as light, a concave mirror is a burning-glass. When it faces the sun, the light and heat are both collected in a small image of the sun at the principal focus. And, if no heat were lost by the reflection, the intensity at the focus would be to that of the direct rays, as the area of the mirror to the area of the sun's image. Burning mirrors have sometimes been constructed on a large scale, by giving a concave arrangement to a great number of plane mirrors.

CHAPTER III.

REFRACTION OF LIGHT.

604. Division of the Incident Beam.—When light falls on an *opaque* body, we have noticed that it is arrested, and a shadow formed beyond. Of the light thus arrested, a portion is reflected, and another portion lost, which is said to be absorbed by the body. When light meets a *transparent* body, a part is still reflected, and a small portion absorbed, but, in general, the greater part is transmitted. The ratio of intensities in the reflected and transmitted beams varies with the angle of incidence, but little being reflected at small angles of incidence, and almost the whole at angles near 90°.

605. Refraction.—The transmitted beam suffers important changes, one of which is a change in *direction*. This change is called *refraction*, and takes place at the surface of a new medium. In Fig. 316, *A C*, incident upon *R S*, the surface of a different medium, is turned at *C* into another line, as *C E*, which is called the *refracted ray*. The angle *E C Q*, between the refracted ray and the perpendicular is called the *angle of refraction;* the angle *G C E*, between the directions of the incident and the refracted rays, is the *angle of deviation.*

FIG. 316.

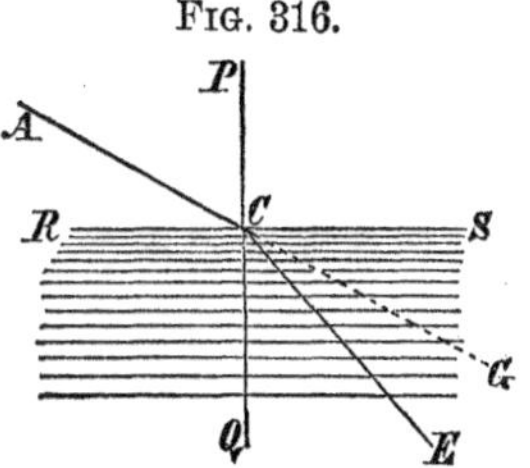

It is a general fact, to which there are but few exceptions, that a ray of light in passing out of a rarer into a denser medium is refracted *toward* the perpendicular to the surface; and in passing out of a denser into a rarer medium, it is refracted *from* the perpendicular. But the chemical constitution of bodies sometimes affects their refracting power. Some inflammable bodies, as sulphur, amber, and certain oils, have a great refracting power in comparison with other bodies; and in a given instance, a ray of light in passing out of one of these substances into another of greater density may be turned from the perpendicular instead of toward it. In the optical use of the words, therefore, *denser* is understood to mean, *of greater refractive power;* and *rarer* signifies, *of less refractive power.* In Fig. 316, the medium below *R S* is of greater refractive power than that above.

We see an example of refraction in the bent appearance of an oar in the water, the light which comes to the eye from the part immersed is bent *from* the perpendicular as it passes from water into air, and causes it to appear higher than its true place. In the same manner, the bottom of a river appears elevated, and diminishes the apparent depth of the stream. Let a small object be placed in the bottom of a bowl, and let the eye be withdrawn till the object is hidden from view by the edge of the bowl. If now the bowl be filled up with water, the object is no longer concealed, for the light, as it emerges from the water, is bent away from the perpendicular, and brought low enough to enter the eye.

606. Law of Refraction.—The law which is found to hold true in all cases of common refraction is this:

The angles of incidence and refraction are on opposite sides of the perpendicular to the surface, and, for any given media, the sines of the angles have a constant ratio for all inclinations.

For example, in Fig. 317, if $A\ C$ is refracted to E, then $a\ C$ will be refracted to e, so that $A\ D : E\ F :: a\ d : e\ f$; and if the rays pass out in a contrary direction, the ratio is also constant, being the reciprocal of the former, viz., $E\ F : A\ D :: e\ f : a\ d$.

Fig. 317.

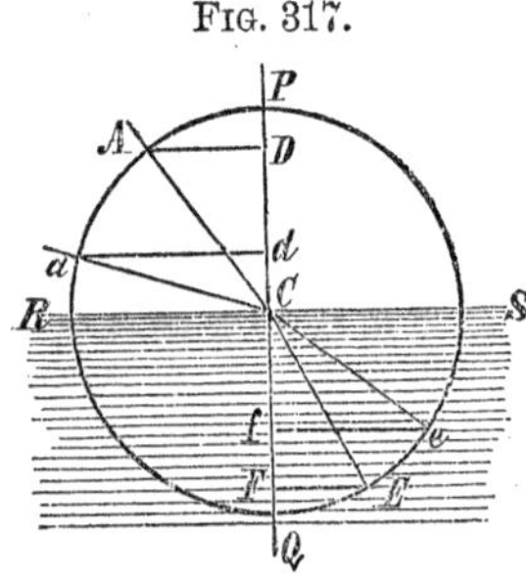

A ray perpendicular to the surface, passing in either direction, is not refracted; for, according to the law, if the sine of one angle is zero, the sine of the other must be zero also. Whichever way light passes, when air is one of the media, suppose the sine of the smaller angle, i. e. the angle in the denser medium, to be 1, then the sine of the larger angle for water is 1.336; and for crown glass, it is about 1.5. The number, in each case, expresses the constant ratio of the sines for the given media, and is called the *index of refraction,* and is employed as the measure of refractive power. The following table gives the index of refraction for a few substances:

Chromate of lead, . . .	2.974	Amber, ...	1.547
Red silver ore, ...	2.564	Crown glass, ...	1.530
Diamond, ...	2.439	Oil of olives, ...	1.470
Phosphorus, ...	2.224	Alum, ...	1.457
Sulphur, ...	2.148	Fluor spar,.	1.434
Flint glass,.	1.830	Mineral acids, ...	1.410
Sapphire, ...	1.800	Alcohol, ...	1.372
Sulphuret of carbon, . .	1.768	Water, ...	1.336
Oil of cassia, ...	1.641	Ice, ...	1.309
Quartz, ...	1.548	Tabasheer,.	1.111

Fig. 318.

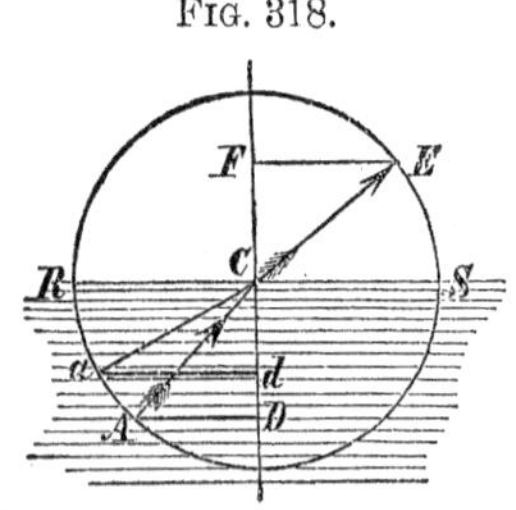

607. Limit of Transmission from a Denser to a Rarer Medium.—As a consequence of the law of refraction, there is a limit beyond which a ray cannot escape from a denser medium. Let $A\ C$ (Fig. 318) be the ray incident upon the rarer medium $R\ E\ S$. It will be refracted from the perpendicular $D\ F$ into the direction $C\ E$, so that $A\ D$ is to

E F in a constant ratio (Art. 606). If the angle *A C D* be increased, the angle *F C E* must also increase till its sine equals *C S.* Make *a d* : *C S* : : *A D* : *E F*; then *a C* is the limit of the incident rays which can emerge. For if *a C D* is enlarged, its sine is increased, and therefore the sine of refraction must increase; but this is impossible, since it is already equal to the radius *C S.* Hence it follows, that whenever the angle of incidence is greater than that at which the sine of the angle of refraction becomes equal to radius, the ray cannot be refracted consistently with the constant ratio of the sines.

This is proved also by experiment; the emerging ray increases its angle of refraction till it at length ceases to pass out. Beyond that limit all the incident rays are *reflected* from the inner surface of the denser medium; and this reflection is more perfect than any external reflection, and is called *total reflection.* If $n =$ the index of refraction, the limit at which refraction ceases and total reflection begins is found by the proportion, $n : 1 : :$ rad. : sine of the limit. If the refractive power is greater, the limit is smaller; for, by the above proportion, since the means are constant, n varies inversely as sine of limit. For *water,* it is 48° 28′; for *crown glass,* 40° 49′; for *diamond,* 24° 12′.

608. Transmission through Plane Surfaces.—

1. A medium bounded by *parallel* planes. In this case the *incident* and *emergent* rays are *parallel.* Let *D E* (Fig 319) enter the medium *A B b a* at *E,* and leave it at *F,* and let *P Q, R S* be the perpendiculars at *E, F.* The first angle of refraction *Q E F,* and the second angle of incidence, *E F R,* are equal, being alternate; therefore, *D E P* = *S F G,* since their sines have a constant ratio to those of *Q E F, E F R.* Hence, if the incident rays are produced to *C* and *H,* the angles of deviation are equal; but *D E F* is supplement to the first angle of deviation, and *E F G* of the second. Therefore, as *D E F* and *E F G* are alternate and equal, *D E* is parallel to *F G.*

Fig. 319.

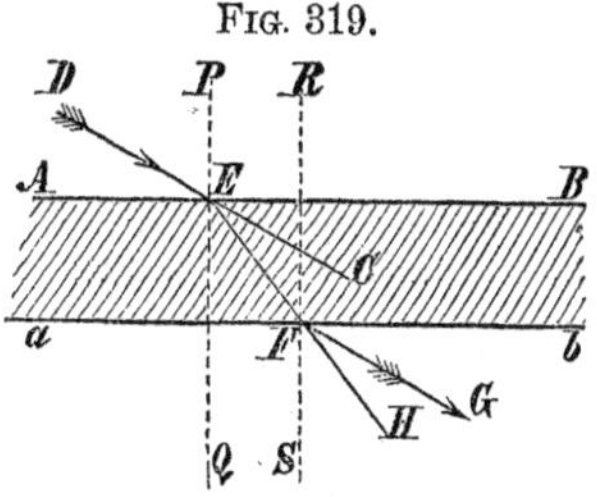

Fig. 320.

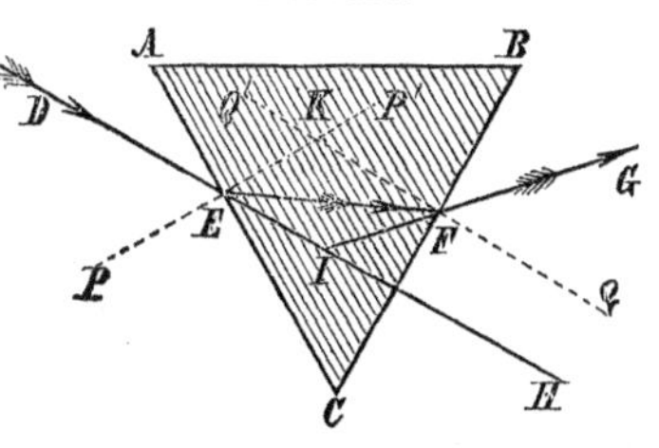

2. A medium bounded by *inclined* planes, called a *prism.* The transmitted ray is turned *from the refracting angle.* Let

$A\ B\ C$ (Fig. 320) be that section of a glass prism which is perpendicular to its axis, and $A\ C$, $B\ C$, the inclined sides of it, through which the light is transmitted. C is called the *refracting angle*, and $A\ B$ the *base*. As prisms are usually constructed and mounted, either A, B, or C may be the refracting angle; but it is not essential that any of the faces should meet at an edge, as the effect on the light depends only on the inclination. In ordinary directions of the ray, the two refractions, one on entering, the other on leaving the prism, conspire to increase the deviation of the ray from its original direction. $D\ E$ is first bent *toward* $E\ K$, making the deviation $H\ E\ F$; at F, it is turned *from* $F\ Q$, making a second deviation, $E\ F\ I$, the same way. The sum of the two deviations, $I\ E\ F + E\ F\ I = G\ I\ H$, the total deviation away from the refracting angle, C.

To an eye at G, the radiant D is seen in the direction $G\ F\ I$.

609. The Multiplying Glass.—A piece of glass ground with one side plane, and the other in any number of plane facets on a convex surface, is called a *multiplying glass*. Each facet, along with the opposite plane surface, forms a prism; and if a radiant A is placed in the axis (a perpendicular through the centre of the plane surface), the pencils, falling on the several facets, will be turned from the edge, and may by two refractions at the opposite surfaces be brought to an eye placed also in the axis, and thus as many images will be seen as there are facets. Fig. 321 exhibits the effect of seven such facets.

FIG. 321.

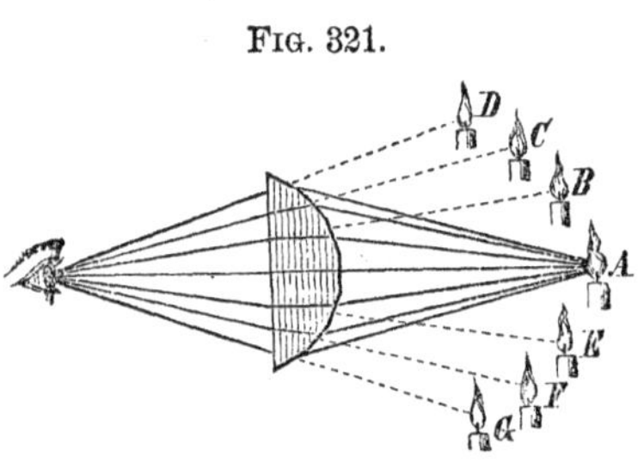

610. Prism used for Measuring Refractive Power.—The following theorem may be used for determining the refractive power of a substance, after first forming it into a prism of small angle:

If the angle of deviation be divided by the refracting angle of the prism, and the quotient be added to unity, the sum is the index of refraction.

In proving this, it is assumed that all the angles are very small, so that they vary as their sines. Let $n =$ the index of refraction, then (Fig. 320),

$K\ E\ I\ (=D\ E\ P) : K\ E\ F :: n : 1$; $\therefore\ F\ E\ I : K\ E\ F :: n-1 : 1$;
also $K\ F\ I\ (=G\ F\ Q) : K\ F\ E :: n : 1$; $\therefore\ E\ F\ I : K\ F\ E :: n-1 : 1$;

$$\therefore F E I + E F I : K E F + K F E :: n - 1 : 1;$$
$$\therefore F I H : P' K F :: n - 1 : 1.$$

But $P' K F = A C B$, each being the supplement of $E K F$. Therefore, $F I H : A C B :: n - 1 : 1$;

$$\therefore \quad n - 1 = \frac{F I H}{A C B}; \therefore n = \frac{F I H}{A C B} + 1.$$

Now, in crown glass, $\frac{F I H}{A C B}$ is found by trial to be very near $\frac{1}{2}$; $\therefore n = 1.5$ nearly.

In order to find the index of refraction for any solid substance, grind it into a prism whose sides are nearly parallel, and carefully measure their inclination. Then measure the displacement of a distant object seen through it at right angles to its surface. For example, the faces of a transparent mineral incline 1° 10′; and when held before the eye, it displaces a distant object 50′; ∴ the index of refraction $= 1 + \frac{5}{7} = 1.714$.

611. Light through one Surface.—

1. *Plane Surface.* When *parallel* rays pass into another medium through a plane surface, they remain parallel. For the perpendiculars being parallel, the angles of incidence are equal, and therefore the angles of refraction are equal also, and the refracted rays parallel. But a pencil of *diverging* rays is made to diverge *less*, when it *enters a denser medium.* For the outer rays make the largest angles of incidence, and are therefore most refracted toward the perpendiculars, and thus toward parallelism with each other. And when *diverging* rays *enter a rarer medium*, they diverge *more;* because the outside rays make the largest angles of incidence, and therefore the largest angles of refraction, by which means they spread more from each other.

Fig. 322.

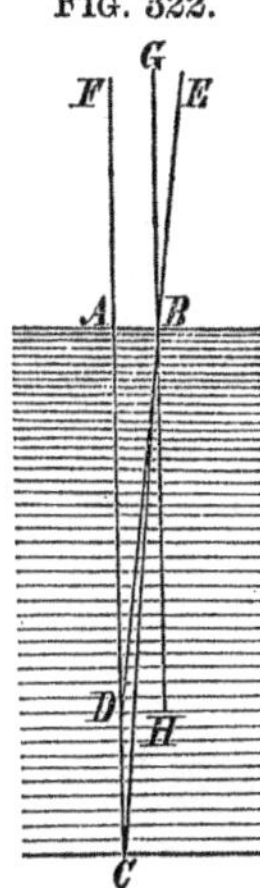

The last case is illustrated when we look perpendicularly into water, and see its depth apparently diminished by about one-fourth of the whole. Let $A B$ (Fig. 322) be the surface, and C a point at the bottom, from which a pencil comes to the eye. Let $C F$, the axis of the pencil, be perpendicular to $A B$, and $C B E$ an oblique ray of the pencil. The angle $C = C B H =$ angle of incidence; and $A D B = G B E =$ angle of refraction. Now, in the triangle $B D C$, $B C : B D$ ($:: A C : A D$ nearly) $::$ sin D : sin C :: sine of refraction : sine of incidence :: 1.34 : 1. Hence the apparent depth is one-fourth less than the real depth. The apparent depth of water may be diminished much more than this by looking into it obliquely.

2. *Convex surface of the denser. A convex surface tends to converge rays.* Let *C'* (Fig. 323) be the centre of convexity, and *C' D, C' C,* two radii produced. As rays are bent *toward* the per-

Fig. 323.

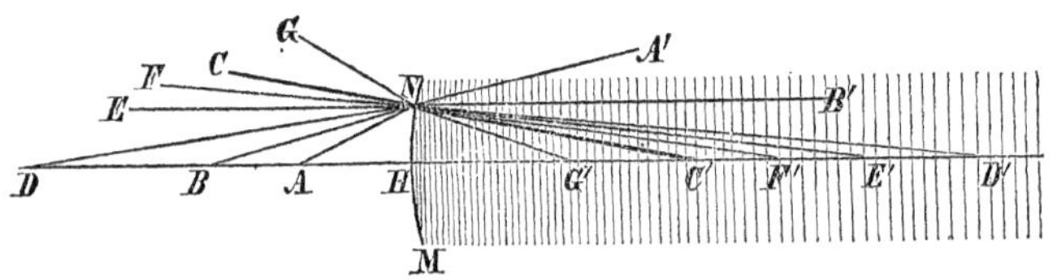

pendiculars in entering a denser medium, and as the perpendiculars themselves converge to *C'*, the general effect of such a surface is to produce convergency. The pencil, *A H, A N,* is merely made less divergent, *H D', N A'*; *B H, B N* become parallel, *H D', N B'*; *D H, D N,* convergent to *D'*; the parallel rays, *D H, E N,* convergent to *E'*; the convergent pencil, *D H, F N,* more convergent to *F'*; but *D H, C N,* which converge equally with the radii, are not changed; and *D H, G N,* which converge more than the radii, converge less than before, to *G'*. The two last cases, which are exceptions to the general effect, rarely occur in the practical use of lenses.

If we trace in the opposite direction the rays, *A', B', D',* &c., comparing each with *D' D,* we find, in this case also, that the convex surface tends to converge the rays, by bending them *from* *C' D, C' C.*

3. *Concave surface of the denser. A concave surface tends to diverge rays.* Let *C C', C D* (Fig. 324), be the radii of concavity produced. As the radii *diverge* in the direction in which the light

Fig. 324.

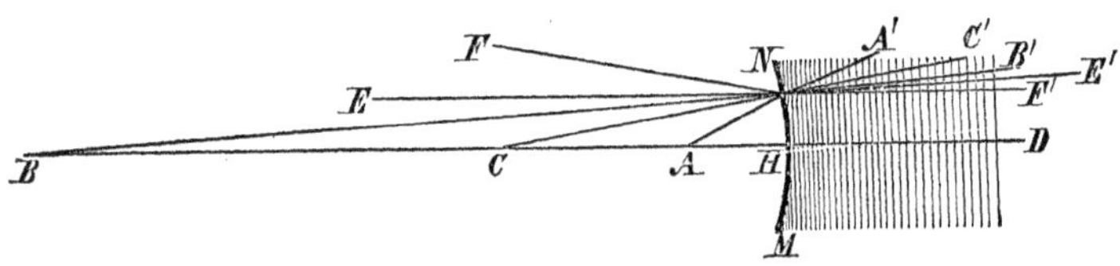

moves, the rays, being bent *toward* them, will generally be made to diverge also. Hence, parallel rays, *B H, E N,* are diverged, *H D, N E'*; and diverging rays, *B H, B N,* are diverged more, *H D, N B'*. If, however, rays diverge as much as the radii, or more, they proceed in the same direction, or diverge less, a case which rarely occurs.

If the rays are traced in the opposite direction, the tendency in general to produce divergency appears from the fact that the perpendiculars are now *converging* lines, and the rays are refracted *from* them.

612. Lenses.—A *lens* is a circular piece of glass, whose surfaces are plane or spherical, and the spherical surface either convex or concave. The usual varieties are shown in Fig. 325.

FIG. 325.

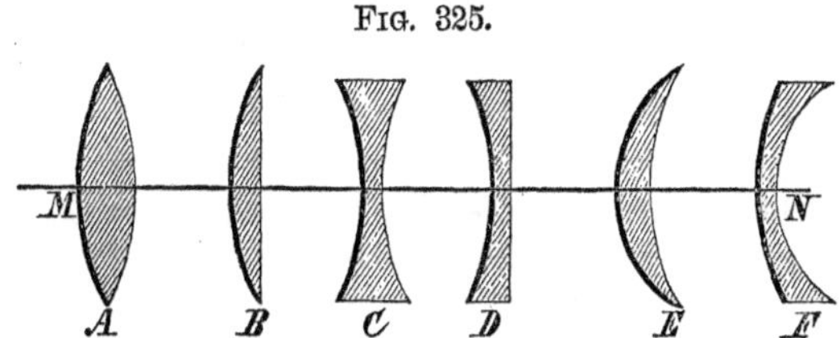

A *double convex lens* (A) consists of two spherical segments, either equally or unequally convex, having a common base.

A *plano-convex lens* (B) is a lens having one of its sides convex and the other plane, being simply a segment of a sphere.

A *double concave lens* (C) is a solid bounded by two concave spherical surfaces, which may be either equally or unequally concave.

A *plano-concave lens* (D) is a lens one of whose surfaces is plane and the other concave.

A *meniscus* (E) is a lens one of whose surfaces is convex and the other concave, but the concavity being less than the convexity, it takes the form of a crescent, and has the effect of a convex lens whose convexity is equal to the difference between the sphericities of the two sides.

A *concavo-convex lens* (F) is a lens one of whose surfaces is convex and the other concave, the concavity exceeding the convexity, and the lens being therefore equivalent to a concave lens whose concavity is equal to the difference between the sphericities of the two sides.

A line ($M\,N$) passing through a lens, perpendicular to its opposite surfaces, is called the *axis*. The axis usually, though not necessarily, passes through the centre of the figure.

613. General Effect of the Convex Lens.—Whether double-convex or plano-convex, its general effect is to converge light. It has been shown (Art. 611) that the convex surface of a denser medium tends to converge rays, whichever way they pass through it. Therefore, if *E* (Fig. 326) is a radiant, while *E C′ C* follows

FIG. 326.

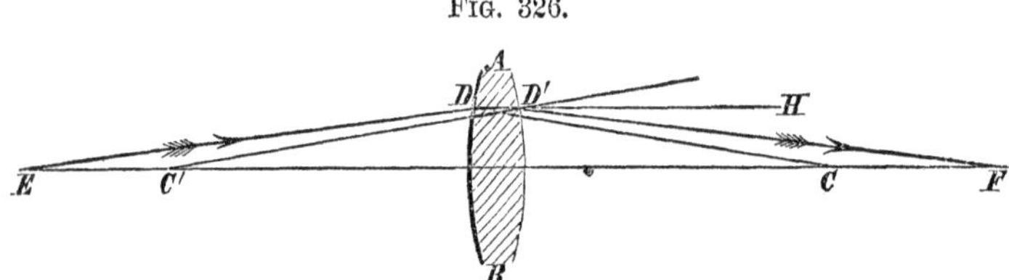

the axis without change of direction, the oblique ray *E D* is first refracted *toward D C*, and then *from C′ D′* produced, and both actions conspire to converge it to the axis. The rays are represented as meeting in the focus *F*. Whether the rays are *actually* converged, depends on their previous relation to each other. If the lens is *plano-convex*, the plane surface has usually but little effect in converging the light; but by Art. 611 it may be shown that its action will usually conspire with that of the convex surface.

614. General Effect of the Concave Lens.—This lens, whether double-concave or plano-concave, tends to produce *divergency.* This is evident from what has been shown in Art. 611. The ray *E D* (Fig. 327), in entering the denser medium, is first

FIG. 327.

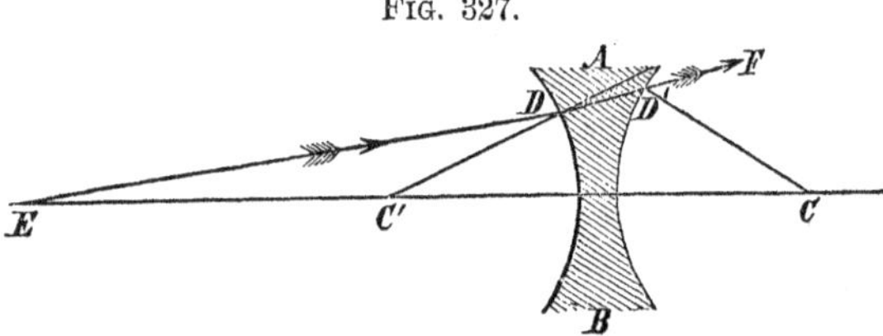

refracted *toward C′ D* produced, and on leaving the medium at *D′*, is refracted *from D′ C*; and is thus twice refracted from the ray *E C*, which being in the axis, is not refracted at all. If the lens is *plano-concave*, the effect of the plane surface may, or may not, conspire with that of the concave surface.

615. The Optic Centre of a Lens.—Within every lens there is a point called the *optic centre,* so situated that the incident and emergent portions of every ray which passes through it are parallel to each other. Let *C, C′* (Fig. 328), be the centres of the two surfaces of the lens; draw the axis *C C′*, also any oblique radius $C A$, and $C' B$ parallel to it; then join $A B$; the point E, in which $A B$ intersects the axis, is the optic centre, and $R A$ the incident, and $B R'$ the emergent portion of the ray passing through A and B, are parallel to each other. For the angles $E A C$, and $E B C'$, are equal, being alternate, and therefore the ray is refracted at A and B equally and in opposite directions, making $R A$ and $B R'$ parallel, as proved in Art. 608, 1. But the point E is the same, whatever may be the points A and B, to which the parallel radii, $C A$, $C' B$, are drawn. For, since the triangles, $E A C$, $E B C'$, are similar, $C A : C' B :: C E : C' E$; $\therefore C A + C' B : C' B :: C E + C' E : C' E$; and as the three first terms are constant, the fourth, $C' E$, is constant also, and E is a fixed point.

Fig. 328.

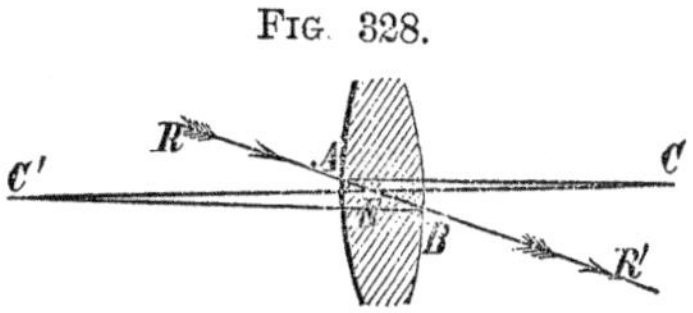

When the lens is thin, and the rays are nearly parallel to its axis, the ray $R A B R'$ may be considered a straight line; and it forms the axis of the pencil of light which passes through the lens in that direction.

616. Conjugate Foci.—If the rays from R (Fig. 329) are collected at F, then rays emanating from F will be returned to R; and the two points are called *conjugate foci*. Their relative distances from the lens may be determined when the radii of

Fig. 329.

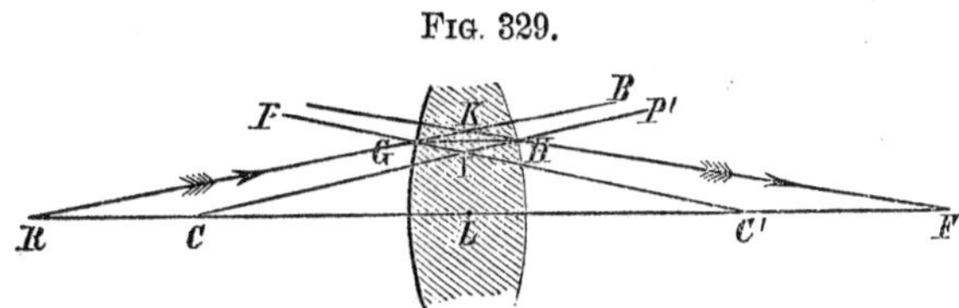

the surfaces and the index of refraction are known. Let n be the index of refraction, and assume, what is practically true, that the angles of incidence and refraction are so small that their ratio is the same as the ratio of their sines. Then

$$R G P (= K G I) : I G H :: n : 1;$$
$$\therefore \quad K G H : I G H :: n - 1 : 1;$$
in like manner $$K H G : I H G :: n - 1 : 1;$$
$$\therefore K G H + K H G : I G H + I H G :: n - 1 : 1.$$
But $$K G H + K H G = B K F = R + F;$$
and $$I G H + I H G = G I C = C + C';$$

24

naming the acute angles at R, C, C', F, by those letters respectively

$$\therefore R + F : C + C' :: n - 1 : 1.$$

Now, the lens being thin, and the angles R, C, C', and F very small, the same perpendicular to the axis, at L, the centre of the lens, may be considered as subtending all those angles. Hence, each angle is as the reciprocal of its distance from L. Let $R\,L = p$; $F\,L = q$; $C\,L = r$; and $C'\,L = r'$. Then the equation above becomes

$$\frac{1}{p} + \frac{1}{q} : \frac{1}{r} + \frac{1}{r'} :: n - 1 : 1;$$

which expresses in general the relation of the conjugate foci. To adapt it to crown-glass, call $n = \frac{3}{2}$, and we have

$$\frac{1}{p} + \frac{1}{q} : \frac{1}{r} + \frac{1}{r'} :: 1 : 2.$$

617. To find the Principal Focus.—The radiant from which parallel rays come is at an infinite distance. Therefore, making $p = \infty$, and the distance of the principal focus $= F$, we have $\frac{1}{p} = 0$, and

$$\frac{1}{F} : \frac{1}{r} + \frac{1}{r'} :: n - 1 : 1; \text{ or,}$$

$$F = \frac{r\,r'}{(n-1)\,(r + r')}; \text{ for crown-glass this is}$$

$$F = \frac{2\,r\,r'}{r + r'}.$$

If the curvatures are equal, $F = \frac{r}{2\,(n-1)}$; for crown-glass this becomes $F = r$; that is, the principal focus of a double convex lens of crown-glass, having equal curvatures, is at the centre of convexity.

The foregoing formulæ are readily adapted to the other forms of lens. When a surface is plane, its radius is infinite, and $\frac{1}{r}$, or $\frac{1}{r'} = 0$. When concave, its centre is thrown upon the same side as the surface, and its radius is to be called negative. And if the focal distance, as given by the formula, becomes negative, it is understood to be on the same side as the radiant; that is, the focus is a virtual radiant.

618. Images by the Convex Lens.—The *convex* lens forms a variety of images, whose character and position depend on the place of the object. If it is at the *principal focus*, the rays of every pencil pass out parallel, and seem to come from an infinite distance. If the object is *nearer* than the principal focus, the emergent rays of each pencil diverge less than the incident rays, and therefore seem to radiate from points further back; the image is therefore *apparent*. Let *M N* (Fig. 330) be the object nearer than the principal focus, *F*. Then the pencil from *M* will, after refrac-

Fig. 330.

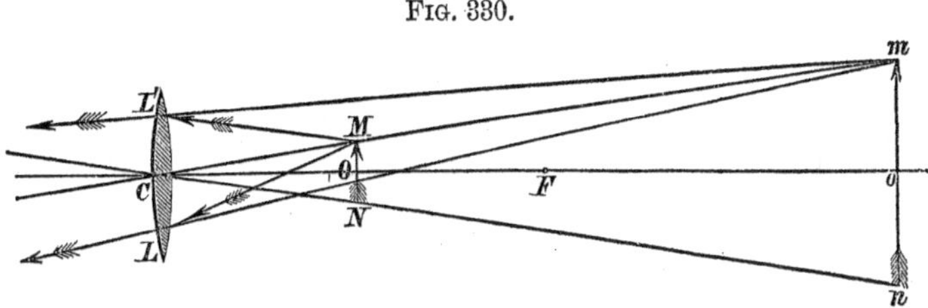

tion, diverge as from *m* in *C M* produced, and so of every point; hence *m n* is the image. It is *erect*, because the axes of the pencils do not cross between the object and image; and it is *enlarged*, because it subtends the angle *M C N* at a greater distance than the object.

But if the object is *further* from the lens than the principal focus, the rays of each pencil converge to a point in the axis of that pencil produced through the lens; and thus light is collected in focal points, which consequently become actual radiants. The last case is illustrated by Fig. 331, in which *M N* is the object, and

Fig. 331.

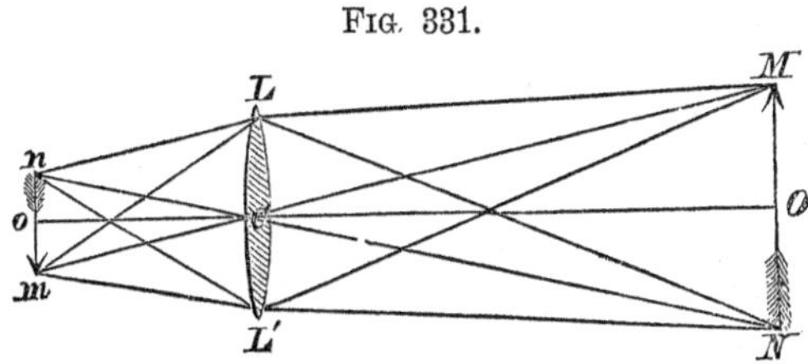

m n the image. A cone of rays from *O* covers the lens *L L*, and is converged again into the axis at *o*, the conjugate focus of *O*, and there cross, and proceed as from a radiant. The cone of rays from *M* is converged to *m* in the axis *M m* of that cone, which is a straight line through the optic centre (Art. 615); and so from every point of the object. Though the rays of every radiant *converge* from the lens to the conjugate focus of that radiant, yet the axes of the pencils *diverge* from each other, having all crossed at the optic centre. The image is therefore *inverted*, as are all real images, in whatever way produced.

The formula for conjugate foci shows that if p is increased, q is diminished; therefore the further $M\ N$ is removed from the lens, the nearer $m\ n$ approaches to it; but the nearest position is the principal focus, which it reaches when the object is at an infinite distance. As the object and image subtend equal angles at the optic centre, and are parallel, or nearly parallel with each other, their *diameters* are proportioned to their *distances* from the lens. But the area of the lens has no effect on the size of the image, since change of area does not alter the relation of the axes, but only the size of the luminous cones, and thus the quantity of light in each pencil.

619. Images by the Concave Lens.—As the rays of each pencil are diverged more after passing through the lens than before, the image is *apparent*, and is situated between the lens and the object. Let $M\ N$ (Fig. 332) be the object; the cone of rays from N will, after refraction, diverge more, as from n, in the same axis

FIG. 332.

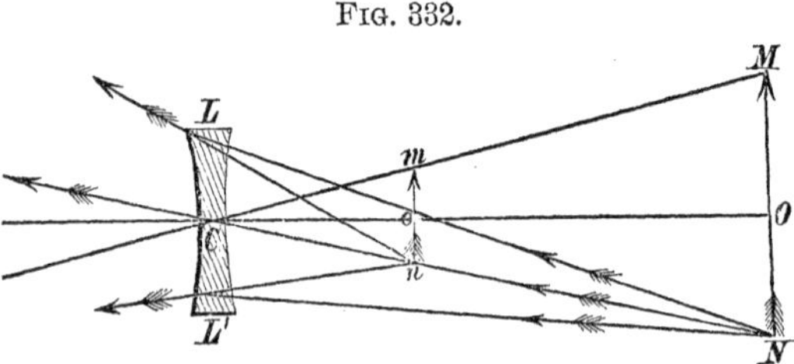

$C\ N$; and all other pencils will be affected in a similar manner, and form an apparent image $m\ n$. It will be *erect*, since the axes do not cross between, and *diminished*, being nearer the angle C, which is subtended by both object and image.

It is noticeable that the *concave mirror* and the *convex lens* are analogous in their effects, forming images on both sides, both real and apparent, both erect and inverted, both larger and smaller than the object; while the *convex mirror* and the *concave lens* also resemble each other, producing images always on one side, always apparent, always erect, always smaller than the object.

620. Caustics by Refraction.—If the convex surface of a lens is a considerable part of a hemisphere, the rays more distant from the axis will be so much more refracted than others, as to cross them and meet the axis at nearer points, thus forming caustics by refraction. Fig. 333 shows this effect in the case of parallel rays; those near the axis inter-

FIG. 333.

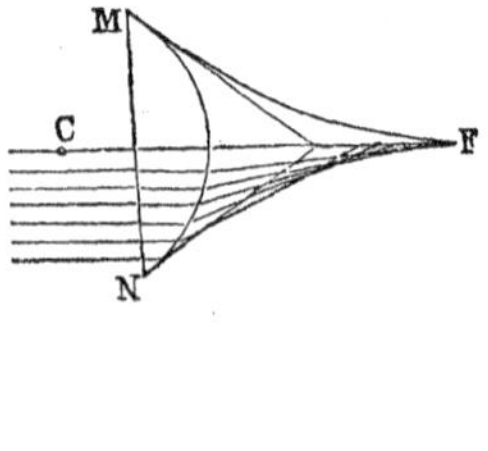

secting it at the principal focus F, and the intersections of remoter rays being nearer and nearer to the lens, so that the whole converging pencil assumes a form resembling a cone with concave sides.

621. Spherical Aberration of a Lens.—The production of caustics is an extreme case of what is called spherical aberration. Unless the lens is of small angular breadth, a pencil whose rays originated in one point of an object is not converged accurately to one point of the image, but the outer rays are refracted too much, and make their focus nearer the lens than that of the central rays, as represented in Fig. 334. If F is the focus of the central rays, and F' of the extreme ones, other rays of the same beam are collected in intermediate points, and $F\,F'$ is called the *longitudinal spherical aberration;* and $G\,H$, the breadth covered by the pencil at the focus of central rays, is called the *lateral spherical aberration.*

FIG. 334.

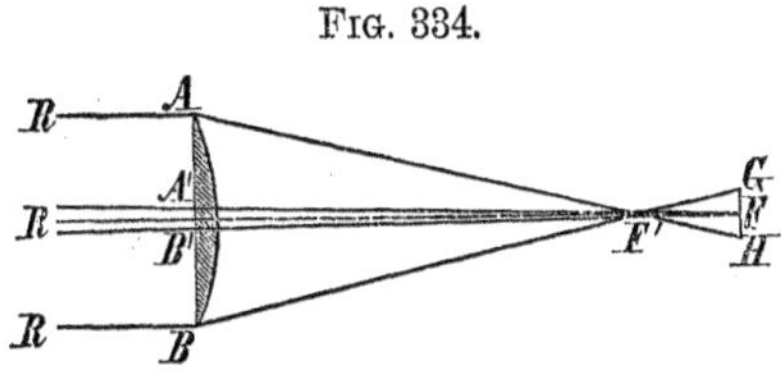

Such a lens cannot form a distinct image of any object; because perfect distinctness requires that all rays from any one point of the object should be collected to one point in the image. If, for example, the beam whose outside rays are $R\,A$, $R\,B$, comes from a point of the moon's disk, that point will not be perfectly represented by F, because a part of its light covers the circle, whose diameter is $G\,H$, thus overlapping the space representing adjacent points of the moon. And if that point had been on the edge of the moon's disk, F could not be a point of a well-defined edge of the image, since a part of the light would be spread over the distance $F\,G$ outside of it, and destroy the distinctness of its outline.

622. Remedy for Spherical Aberration.—As spherical lenses refract too much those rays which pass through the outer parts, it is obvious that, to destroy aberration, a lens is required whose curvature diminishes toward the edges. Accordingly, forms for *ellipsoidal* lenses have been calculated, which in theory will completely remove this species of aberration. But no curved solids can be so accurately ground as those whose curvature is uniform in all planes, that is, the spherical. Hence, in practice it is found better to *reduce* the aberration as much as possible by spherical lenses, than to attempt an entire *removal* of it by other forms which cannot be well made.

In a plano-convex lens, whose plane surface is toward the object, the spherical aberration is 4.5; that is (Fig. 334), $FF' = 4.5$ times the thickness of the lens. But the same lens, with its convex side toward the object, is far better, its aberration being only 1.17. In a double convex lens of equal curvatures, the aberration is 1.67; if the radii of curvature are as 1 : 6, and the most convex side is toward the object, the aberration is only 1.07. By placing two plano-convex lenses near each other, the aberration may be still more reduced.

623. Atmospheric Refraction.—The atmosphere may be regarded as a transparent spherical shell, whose density increases from its upper surface to the earth. The radii of the earth produced are the perpendiculars of all the laminæ of the air; and rays of light coming from the vacuum beyond, if oblique, are bent *gradually* toward these perpendiculars; and therefore heavenly bodies appear more elevated than they really are. The greatest elevation by refraction takes place at the horizon, where it is about half a degree.

624. Mirage.—This phenomenon, called also *looming*, consists of the formation of one or more images of a distant object, caused by horizontal strata of air of very different densities. Ships at sea are sometimes seen when beyond the horizon, and their images occasionally assume distorted forms, contracted or elongated in a vertical direction. These effects are generally ascribed to *extraordinary refraction* in horizontal strata, whose difference of density is unusually great. But many cases of mirage seem to be instances of *total reflection* from a highly rarefied stratum resting on the earth. These occur frequently on extended sandy plains, as those of Egypt. When the surface becomes heated, distant villages, on more elevated ground, are seen accompanied by their images inverted below them, as in water. As the traveler advances, what appeared to be an expanse of water retires before him. By placing alcohol upon water in a glass vessel, and allowing them time to mingle a little at their common surface, the phenomena of mirage may be artificially represented.

CHAPTER IV.

DECOMPOSITION AND DISPERSION OF LIGHT.

625. The Prismatic Spectrum.—Another change which light suffers in passing into a new medium, is called *decomposition*, or the separation of light into colors. For this purpose, the glass prism is generally employed. It is so mounted on a jointed stand, that it can be placed in any desired position across the beam from the heliostat. The beam, as already noticed, is bent away from the refracting angle, both in entering and leaving the prism, and deviates several degrees from its former direction. If the light is admitted through a narrow aperture, *F* (Fig. 335), and

FIG. 335.

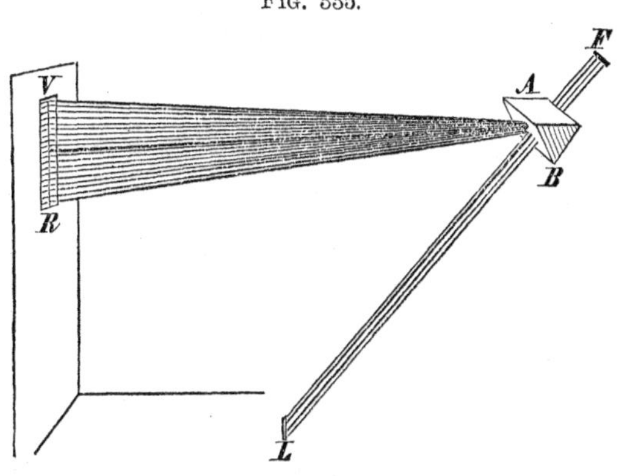

the axis of the prism is placed parallel to the length of the aperture, the light no longer falls, as before, in a narrow line, *L*, but is extended into a band of colors, *R V*, whose length is in a plane at right angles to the axis of the prism. This is called the prismatic spectrum. Its colors are usually regarded as seven in number—*red, orange, yellow, green, blue, indigo, violet.* The red is invariably nearest to the original direction of the beam, and the violet the most remote; and it is because the elements of white light are unequally refrangible, that they become separated, by transmission through a refracting body. The spectrum is properly regarded as consisting of innumerable shades of color. Instead of Newton's division into *seven* colors, many choose to consider all the varieties of tint as caused by the combination of *three* primitive colors, *red, yellow,* and *blue,* varying in their proportions throughout the entire spectrum. The number *seven*, as perhaps any other particular number, must be regarded as arbitrary.

The spectrum contains other elements besides those which affect the eye. If a thermometer bulb be moved along the spectrum, it is found that the greatest heat lies outside of the visible spectrum at the red extremity; hence heat is less refrangible than light. On the other hand, the chemical or actinic rays are more refrangible than the luminous rays, and fall at and beyond the violet end of the spectrum.

Light from other sources is also susceptible of decomposition by the prism; but the spectrum, though resembling that of the sun, usually differs in the proportion of the colors.

626. The Individual Colors of the Spectrum cannot be Decomposed by Refraction.—Some of the colors of the spectrum are called *simple* colors, namely, *red, yellow*, and *blue*, while the others are generally regarded as *compound* colors; for orange may be formed of red and yellow, green of yellow and blue, while indigo and violet are mixtures of blue and red in different proportions. It is nevertheless true, that none of the colors of the spectrum can be decomposed by refraction. For if the spectrum formed by the prism *A* be allowed to fall on the screen *E D* (Fig. 336), and one color of it, green for example, be let through the

FIG. 336.

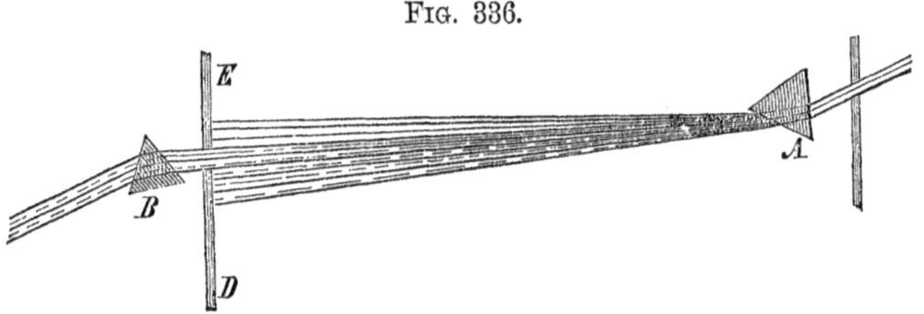

screen, and received on a second prism, *B*, it is still refracted as before, but all its rays remain together and of the same color. The same is true of every color of the spectrum. Therefore, so far as refrangibility is concerned, all the colors of the spectrum are alike simple.

627. Colors of the Spectrum Recombined.—It may be shown, in several ways, that if all the colors of the spectrum be combined, they will *reproduce white light*. One method is by transmitting the beam successively through two prisms whose refracting angles are on opposite sides. By the first prism, the colors are separated at a certain angle of deviation, and then fall on the second, which tends to produce the same deviation in the opposite direction, by which means all the colors are brought upon the same ground, and the illuminated spot is white as if no prism had been interposed. Or the colors may be received on a series of small plane mirrors, which admit of such adjustment as to reflect all the beams upon one spot. Or finally, the several colors can, by different methods, be passed so rapidly before the eye that their visual impressions shall be united in one; in which case the illuminated surface appears white.

Fig. 337.

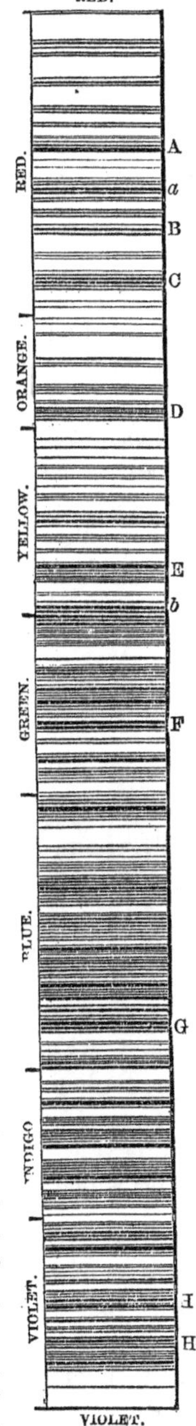

628. Complementary Colors.—If certain colors of the spectrum are combined in a compound color, and the others in another, these two are called *complementary colors*, because, when united, they will produce white. For example, if *green, blue,* and *yellow* are combined, they will produce green, differing slightly from that of the spectrum; the remaining colors, *red, orange, indigo,* and *violet,* compose a kind of purple, unlike any color of the spectrum. But these particular shades of *green* and *purple,* if mingled, will make perfectly white light, and are therefore complementary colors.

629. Fixed Dark Lines of the Spectrum.—Let the aperture through which the sunbeam enters be made exceedingly narrow, and let the prism be of uniform density, and then let the refracted pencil pass immediately through a small telescope, and thence into the eye, and there appears a phenomenon of great interest—the *dark lines,* or the *Fraunhofer lines,* as they are often called from the name of their discoverer. These lines, an imperfect view of which is presented in Fig. 337, are unequal in breadth, in darkness, and in distance from each other, and so fine and crowded in many parts that the whole number cannot be counted. Fraunhofer himself described between 500 and 600, among which a few of the most prominent are marked by letters, and used in measuring refractive power. At least as many as six thousand are now known and mapped, so that any one of them may be identified. They are parallel to each other, and perpendicular to the length of the spectrum. When the pencil passes through a succession of prisms, all bending it the same way, the spectrum becomes more dilated, and more lines are seen. The instrument fitted up as above described, either with one prism or a series of prisms, is called a *spectroscope*.

630. Bright Lines in the Spectrum of Flame.—If the spectroscope be used for the examination of the flame of different substances in combustion, the spectrum is found to consist of certain bright lines, differing in color and number, according to the substance under examination. Thus, the spectrum of sodium flame, besides showing other fainter lines, consists mainly of two conspicuous *yellow* lines, very close together, so as ordinarily to appear as one. The flame of carbon shows two distinct lines, one of which is green, the other indigo. In this respect every substance differs from every other, and each may be as readily distinguished by the lines which compose its spectrum as by any other property. The lines of some substances are very numerous; as, for example, iron, whose spectrum lines amount to four or five hundred.

But a *solid* or *liquid* substance, when raised to a red or white heat, without passing into the gaseous state and producing flame, forms a continuous spectrum, having neither bright nor dark lines.

631. The Spectrum of a Heated Solid or Liquid Shining through Flame.—The condition of a spectrum is entirely changed when the light from a heated solid or liquid substance shines through the flame of a burning gas. The *bright* lines instantly become *dark* lines. The flame seems to absorb just those rays, and only those, which are like the rays emitted by itself. As an example, the spectrum of sodium flame consists of a bright double yellow line, and a few fine luminous lines of other colors. If now iron at an intense white heat shines through this flame, the whole spectrum becomes luminous, except the very lines which were before bright; these are now dark.

632. Composition of the Sun's Surface.—A great number of the dark lines of the solar spectrum are identical in position with lines in the spectrum of terrestrial substances. The spectroscope can be attached to the eye-piece of a telescope, so as to bring half the breadth of the solar spectrum side by side with half the breadth of the spectrum of the flame of some substance; and their lines can thus be compared with each other on the divisions of the same scale. When this is done, there is found, with regard to several substances, an identity of position and relative breadth and intensity so exact that it is impossible to regard the agreement as accidental. The double line *D* of the sunbeam, is the prominent line of sodium. So all the numerous lines of potassium, iron, and several other simple substances, exactly coincide with the dark lines of the spectrum of sunlight.

The foregoing facts seem to indicate that the photosphere of the sun consists of the flame of many substances, among which are some such as belong to the earth, namely, sodium, potassium, iron, &c.; and that the luminous liquid matter beneath the photosphere shines through it, and changes all the bright lines to dark ones.

633. Dispersion of Light.—Decomposition of light refers to the *fact* of a separation of colors; *dispersion*, rather to the *measure* or *degree* of that separation. The *dispersive power* of a medium indicates the amount of separation which it produces, compared with the amount of refraction. For example, if a substance, in refracting a beam of light 1° 51′ from its course, separates the violet from the red by 4′, then its dispersive power is $\frac{4}{111} = .036$. The following table gives the dispersive power of a few substances much used in optics:

	Dispersive power.		Dispersive power.
Oil of Cassia, . .	0.139	Plate-Glass, . . .	0.032
Sulphuret of Carbon,	0.130	Sulphuric Acid, . .	0.031
Oil of Bitter Almonds,	0.079	Alcohol, ...	0.029
Flint-Glass, . . .	0.052	Rock-Crystal, . .	0.026
Muriatic Acid, . .	0.043	Blue Sapphire, . .	0.026
Diamond, ...	0.038	Fluor-spar, . . .	0.022
Crown-Glass, . . .	0.036		

The discovery that different substances produce different degrees of dispersion, is due to Dollond, who soon applied it to the removal of a serious difficulty in the construction of optical instruments.

634. Chromatic Aberration of Lenses.—This is a deviation of light from a focal point, occasioned by the different refrangibility of the colors. If the surface of a lens be covered, except a narrow ring near the edge, and a sunbeam be transmitted through the ring, the chromatic aberration becomes very apparent; for the most refrangible color, violet, comes to its focus nearest, and then the other colors in order, the focus of red being most remote. Since the distinctness of an image depends on the accurate meeting of rays of the same pencil in one point, it is clear that discoloration and indistinctness are caused by the separation of colors.

635. Achromatism.—In order to refract light, and still keep the colors united, it is necessary that, after the beam has been refracted, and thus separated, a substance of greater dispersive power should be used, which may bring the colors together again, by refracting the beam only a part of the distance back to its original direction. For instance, suppose two prisms, one of crown-glass and one of flint-glass, each ground to such a refracting angle as to separate the violet from the red ray by 4′. In order for this, the crown-glass, whose dispersive power is .036, must refract the beam 1° 51′; for $\frac{4'}{1^\circ\ 51'} = .036$; and the flint-glass, whose dispersive power is .052, must refract only 1° 17′; for $\frac{4'}{1^\circ\ 17'} = .052$. Place these two prisms together, base to edge, as in Fig. 338, *C* being the crown-glass and *F* the flint-glass. Then

Fig. 338.

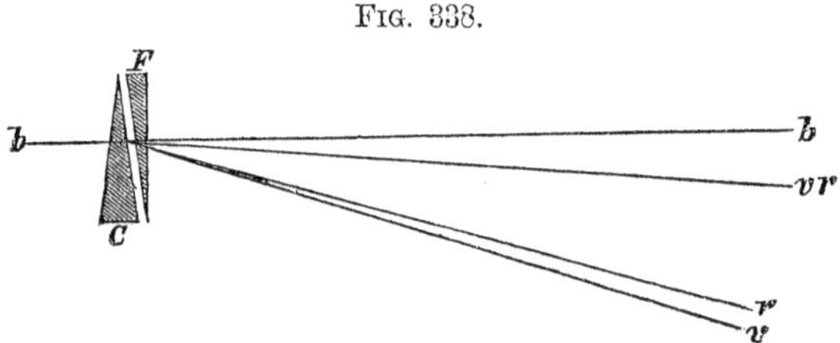

C will refract the beam, *b b*, downward 1° 51′, and the violet, *v*, 4′ more than the red, *r*; *F* will refract this decomposed beam upward 1° 17′, and the violet 4′ more than the red, which will just bring them together at *v r*. Thus the colors are united again, and yet the beam is refracted downward 1° 51′ − 1° 17′ = 34′, from its original direction.

636. Achromatic Lens.—If two prisms can thus produce achromatism, the same may be effected by lenses; for a convex lens of crown-glass may converge the rays of a pencil, and then a

Fig. 339.

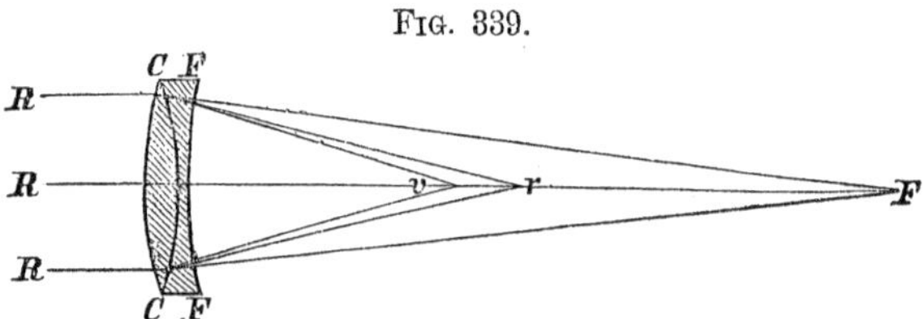

concave lens of flint-glass may diminish that convergency sufficiently to unite the colors. A lens thus constructed of two lenses of different materials and opposite curvatures, so adapted as to produce an image free from chromatic aberration, is called an *achromatic lens.* Fig. 339 shows such a combination. The convex lens of crown-glass alone would gather the rays into a series of colored foci from v to r; the concave flint-glass lens refracts them partly back again, and collects all the colors at one point, F.

637. Colors not Dispersed Proportionally.—It is assumed in the foregoing discussion, that when the red and violet are united, all the intermediate colors will be united also. It is found that this is not strictly true, but that different substances separate two given colors of the spectrum by intervals which have different ratios to the whole length of the spectrum. This departure from a constant ratio in the distances of the several colors, as dispersed by different media, is called the *irrationality of dispersion.* In consequence of it there will exist some slight discoloration in the image, after uniting the extreme colors. It is found better in practice to fit the curvatures of the lenses, for uniting those rays which most powerfully affect the eye.

CHAPTER V.

RAINBOW AND HALO.

638. The Rainbow.—This phenomenon, when exhibited most perfectly, consists of two colored circular arches, projected on falling rain, on which the sun is shining from the opposite part of the heavens. They are called the *inner* or *primary* bow, and the *outer* or *secondary* bow. Each contains all the colors of the spectrum, arranged in contrary order; in the primary, red is outermost; in the secondary, violet is outermost. The primary bow is narrower and brighter than the secondary, and, when of unusual brightness, is accompanied by *supernumerary* bows, as they are called; that is, narrow red arches just within it, or overlapping the violet; sometimes three or four supernumeraries can be traced for a short distance. The common centre of the bows is in a line drawn from the sun through the eye of the spectator.

639. Action of a Transparent Sphere on Light.—It will aid in understanding the manner in which the bow is formed, to notice the experiments which first led to a correct theory respecting it. Let a hollow sphere of glass be filled with water and placed in the sunlight, and then let the directions and conditions of the most luminous pencils which emerge from it be observed. Fig. 340 exhibits the general result. The entire hemisphere exposed to the rays is of course penetrated by them; but a narrow pencil, *S A*, about 60° distant from *S C*, the axis of the drop, that is, the ray which passes through its centre, is converged to *B*, where some light escapes, but a large part is reflected to *D*; at that point a division occurs again, and the emerging pencil, *D E*, consists of decomposed light, each color of which can be seen at a great distance. The part reflected at *D* divides again near *F*, but the emergent portion diverges, and has not the intensity of *D E*. But if a pencil, which strikes the drop at *a* about 10° outside of *S A*, be traced, we find that a part is reflected near *B*, a part of that again reflected at *d*, and then, on reaching the point *F*, the emerging portion is not only decomposed, but retains its intensity to a great distance. The pencil *D E*, which has been *once* reflected at *B*, is the one concerned in the production of the primary bow. The pencil *F G*, which leaves the drop at *F*, after it has been *twice* reflected, is one of those which constitute the secondary bow.

FIG. 340.

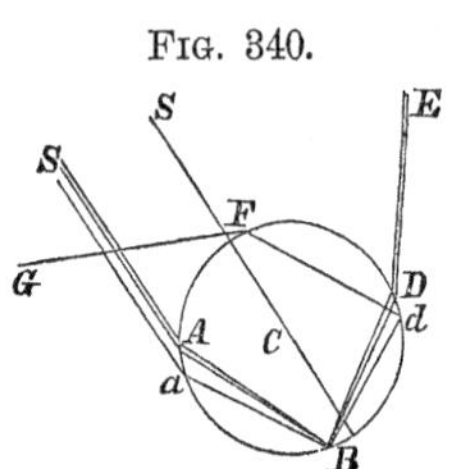

Most of the light which enters a drop of rain, and leaves it again, either not reflected at all, or reflected one or more times, is scattered in various directions, and brings to the eye of an observer no impression of intense light. It is only such rays as are reflected and transmitted in circumstances to be contiguous to each other, and to continue parallel after leaving the drop, which can produce the bright colors of the rainbow.

640. Course of Rays in the Primary Bow.—Let *f z p q* (Fig. 341) be the section of a drop of rain, *f p* a diameter, *a b*, *c d*, &c., parallel rays of the sun's light, falling upon the drop. Now *y f*, a ray coinciding with the diameter, suffers no refraction; and *a b*, a ray near to *y f*, is refracted very little toward the radius, so as to meet the remoter surface of the drop about half as far from the axis as when it entered; but the rays which lie further from *y f*, making greater angles with the radius, are more and more refracted as they are further removed from the diameter.

FIG. 341.

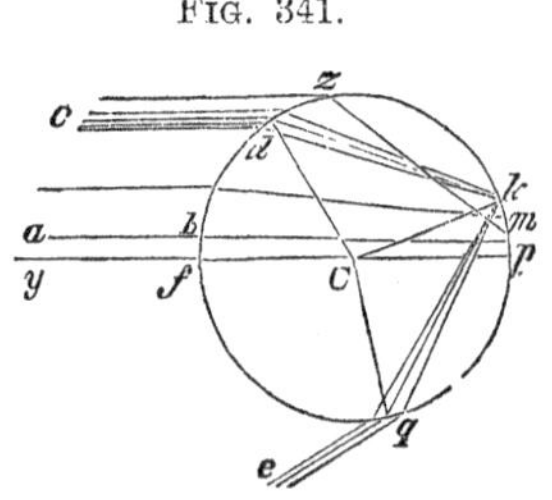

And it is found, by a simple calculation on the course of the rays, that those which enter beyond the limit of about 60°, cross more or less of those entering nearer the axis, the furthest one of all at 90° being refracted almost to *p*. Hence all the rays falling on the quadrant *f z*, meet the circumference within the arc *k p*. But when a varying quantity is approaching its limit, or is beginning to depart from it, its changes are nearly insensible. Thus, a large number of rays near *c d*, on both sides of it, meet very near *k*, the limit of the arc *p k*. Consequently, many more rays are reflected from that point than from any other in the arc. Now were these rays to return in the same lines, they would emerge parallel in the lines near *c d*; but if, instead of returning back in the quadrant *f z*, they are reflected on the other side of the radius, they make the same angles with the radius, and therefore with each other, as the incident rays do, and consequently meet the curve at the same inclination on the other side of the axis, and emerge parallel. Hence it appears that there is a particular point in the section of the drop on the back side, where the rays of the sun's light *accumulate*, and then diverge, so that, on emerging, those of a given color form a compact pencil of parallel rays. It is found by calculation that the angle which the incident and emergent rays make with each other—that is, the angle included by *c d* and *e q* produced—is, for the *red* rays, 42° 2′, and for the *violet* rays, 40° 17′, and for other colors, between these limits. Calculation shows, also, that these are the *greatest* deviations possible for rays once reflected; since all rays on the quadrant *f z*, whether nearer or further than the pencil *c d*, at 60°, emerge with smaller deviations.

641. Course of Rays in the Secondary Bow.—There is also an accumulation at a certain limit, for the light which emerges after suffering *two* reflections. If, as before, we calculate the course of the rays which fall on the quadrant *f z*, we shall find that those, *d e* (Fig. 342), which enter at about 71° or 72°, from the axis *f p*, after crossing each other in the drop at *a*, are reflected at *h* into parallel lines, and, consequently, after a second reflection at *m*, have their relations to each other and the radii exactly reversed. Hence, they cross a second time at *b*, and emerge parallel at *q*. Such a pencil, entering above the axis, will, on emerging, ascend and cross its own path, outside of the drop, the violet rays intersecting *d e* at an angle of 54° 9′, the red at 50° 59′, and the other colors

FIG. 342.

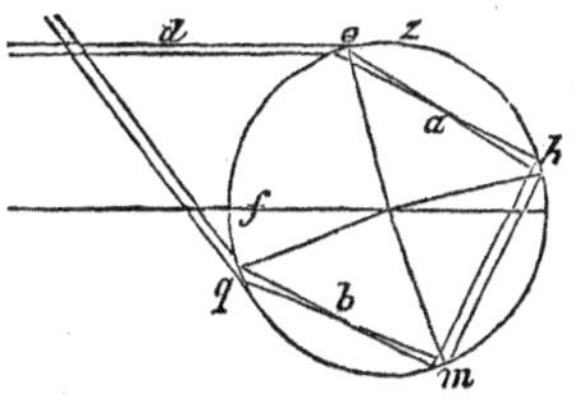

in order between. That the emergent pencil may *descend* to the observer, the incident pencil must enter *below* the axis, and come out above it. These rays, entering at the distance of 71° and 72° from the axis, are the only ones which, after *two* reflections, emerge compact and parallel, and give a bright color at a great distance. All rays which enter nearer the axis, and also those which enter more remote, make, after two reflections, larger angles of deviation, and also diverge from each other.

642. Axis of the Bows.—Let *A B D G I* (Fig. 343) represent the path of the pencil of red light in the primary bow. If *A B* and *I G* are produced to meet in *K*, the angle *K* is the deviation, 42° 2′, of the incident and emergent red rays. Suppose the spectator at *I*, and let a line from the sun be drawn through his position to *T*; it is sensibly parallel to *A B*, and therefore the angles *I* and *K* are equal. As *T* is opposite to the sun, the red

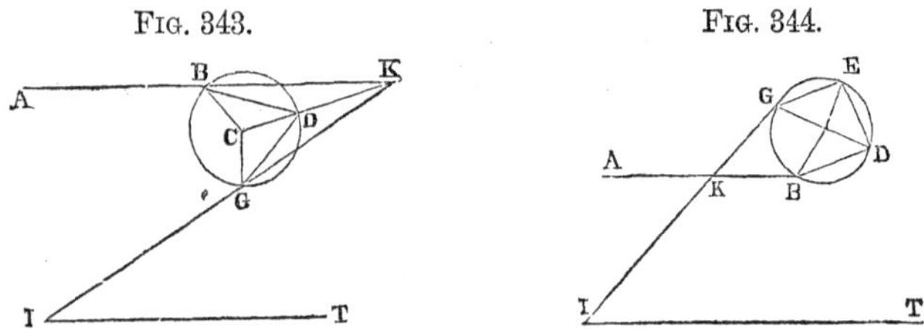

FIG. 343. FIG. 344.

color is seen at the distance of 42° 2′, on the sky, from the point *T*; and so the angular distance of each color from *T* equals the angle which the ray of that color makes with the incident ray. In like manner, in the secondary bow, if *I T* (Fig. 344) be drawn through the sun and the eye of the observer, it is parallel to *A B*, and the angular distance of the colored ray from *T* is equal to *K*, the deviation of the incident and emergent rays. *I T* is called the *axis* of the bows, for a reason which is explained in the next article.

643. Circular Form of the Bows.—Let *S O C* (Fig. 345) be a straight line passing from the sun, through the observer's place at *O*, to the opposite point of the sky; and let *V O*, *R O* be the extreme rays, which after *one* reflection bring colors to the eye at *O*, and *R′ O*, *V′O*, those which exhibit colors after *two* reflections; then (according to Arts. 640, 641), *V O C* = 40° 17′, *R O C* = 42° 2′, *R′ O C* = 50° 59′, *V′O C* = 54° 9′. Now, if we suppose the whole system of lines, *S V′O*, *S V O*, to revolve about *S O C*, as an axis, the relations of the rays to the drops, and to each other will not be at all changed; and the same colors will describe the same lines, whatever positions those lines may occupy in the revolution. The emergent rays, therefore, all describe the surfaces of cones, whose common vertex is in the eye at O; and

Fig. 345.

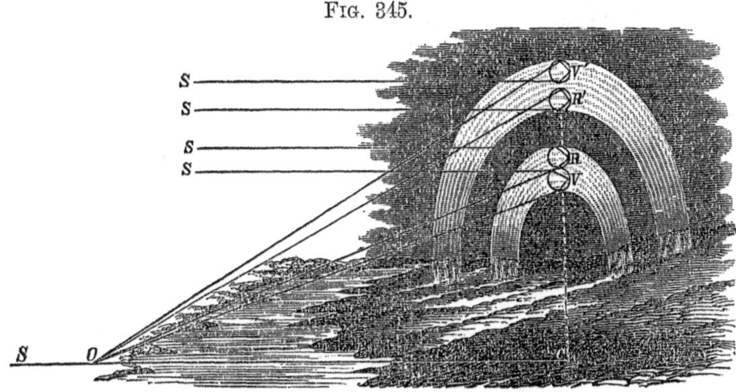

the colors, as seen on the cloud, are the circumferences of their bases.

In a given position of the observer, the extent of the arches depends on the elevation of the sun. When on the horizon, the bows are semicircles; but less as the sun is higher, because their centre is depressed as much below the horizon as the sun is elevated above it. If rain is near, however, the lower parts of the bows may sometimes be seen projected on the landscape as arcs of ellipses, parabolas, or hyperbolas; for the surface of the earth cuts the axis of the cones obliquely. From the top of a mountain, the bows have been seen as entire circles.

644. Colors of the Two Bows in Reversed Order.— The reason for the inversion of colors in the two bows may be seen in the fact that, in the primary bow, the rays which descend to the observer's eye, must emerge from the lower or inner quadrant of the drop, and be bent *upward* (*outward*) from the radius produced; while, in the secondary, they must emerge from the upper or outer quadrant, and be bent from the radius *downward.* The ray $V\,O$ (Fig. 345), of the primary, being supposed a violet ray, is the most refrangible, and therefore all other rays from that drop fall below it, and fail to reach the eye. To bring other colors to O, drops must be selected higher up; hence, violet is the color seen nearest the axis. In the secondary bow, if $V'\,O$ is the violet ray, the other colors, being bent in a less degree from the radius of the drop, lie above $V'\,O$; and therefore, in order that other colors may reach O, they must emerge from lower drops, i. e. drops nearer the axis. Hence, violet is the outer color of the secondary bow.

645. Rainbows, the Colored Borders of Illuminated Segments of the Sky.—The primary bow is to be regarded as the *outer edge* of that part of the sky from which rays can come to the eye after suffering but *one* reflection in drops of rain; and the secondary bow is the *inner edge* of that part from which light, after being *twice* reflected, can reach the eye.

It is found by calculation, that in case of one reflection, the incident and emergent rays can make no inclinations with each other greater than 42° 2′ for red light, and 40° 17′ for violet; but the inclinations may be less in any degree down to 0°. Therefore, all light, once reflected, comes to the eye from *within* the primary bow.

But the angles, 50° 59′ and 54° 9′, are, by calculation, the *least* deviations of red and violet light from the incident rays after *two* reflections. But the deviations may be greater than these limits up to 180°. Therefore rays twice reflected can come to the eye from any part of the sky, except between the secondary bow and its centre.

It appears, then, that from the zone lying between the two bows, no light, reflected by drops internally, either once or twice, can possibly reach the eye. Observation confirms these statements; when the bows are bright, the rain within the primary is more luminous than elsewhere; and outside of the secondary bow, there is more illumination than between the two bows, where the cloud is perceptibly darkest.

646. The Tertiary Bow.—A *tertiary* bow, or a bow formed by light three times reflected in drops of rain, is on the same side of the sky with the sun, and distant about 40° 40′ from it. The incident rays, which form it, enter the drops about 77° from their axis, and emerge on the back side. But this order of bow is so very faint from repeated reflections, and so unfavorably situated, that it is very rarely seen.

647. The Common Halo.—This, as usually seen, is a white or colored circle of about 22° radius, formed around the sun or moon. It might, without impropriety, be termed the *frost-bow*, since it is known to be formed by light refracted by crystals of ice suspended in the air. It is formed when the sun or moon shines through an atmosphere somewhat hazy. About the sun it is a white ring, with its inner edge red, and somewhat sharply defined, while its outer edge is colorless, and gradually shades off into the light of the sky. Around the moon it differs only in showing little or no color on the inner edge.

648. How Caused.—The phenomenon is produced by light

passing through crystals of ice, having sides inclined to each other at an angle of 60°. Let the eye be at E (Fig. 346), and the sun in the direction $E\ S$. Let $S\ A$, $S\ B$, &c., be rays striking upon such crystals as may happen to lie in a position to refract the light toward $S\ E$ as an axis. Each crystal turns the ray from the refracting edge on entering; and again, on leaving, it is bent still more, and the emergent pencil is decomposed. The color, which comes from each one to the eye E, depends on its angular distance from $E\ S$, and the position of its refracting angle. The angle of deviation for A is $E\ A\ D = S\ E\ A$; for B, it is $S\ E\ B$, and so on. It is found by calculation, that the least deviation for red light is 21° 45′; the least for orange must be a little greater, because it is a little more refrangible, and so on for the colors in order. The greatest deviation for the rays generally is about 43° 13′. All light, therefore, which can be transmitted by such crystals must come to the observer from points somewhere between these two limits, 21° 45′ and 43° 13′ from the sun. But by far the greater part of it, as ascertained by calculation, passes through near the least limit.

Fig. 346.

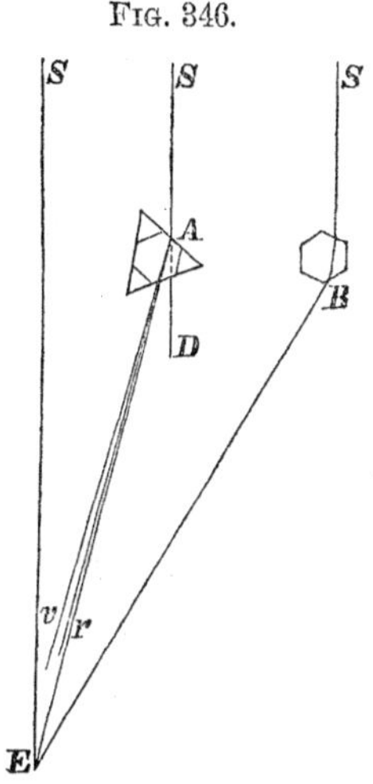

649. Its Circular Form.—What takes place on one side of $E\ S$ may occur on every side; or, in other words, we may suppose the figure revolved about $E\ S$ as an axis, and then the transmitted light will appear in a ring about the sun S. The inner edge of the ring is red, since that color deviates least; just outside of the red the orange mingles with it; beyond that are the red, orange, and yellow combined; and so on, till, at the minimum angle for violet, all the colors will exist (though not in equal proportions), and the violet will be scarcely distinguishable from white. Beyond this narrow colored band the halo is white, growing more and more faint, so that its outer limit is not discernible at all.

650. The Halo, a Bright Border of an Illuminated Zone.—As in the rainbow, so in the halo, the visible band of colors is only the border of a large illuminated space on the sky. The ordinary halo, therefore, is the bright inner border of a zone, which is more than 20° wide. The whole zone, except the inner edge, is too faint to be generally noticed, though it is perceptibly more luminous than the space between the halo and the luminary.

651. Frequency of the Halo.—The halo is less brilliant and beautiful, but far more frequent, than the rainbow. Scarcely a week passes during the whole year in which the phenomenon does not occur. In summer the crystals are three or four miles high, above the limit of perpetual frost. As the rainbow is sometimes seen in dew-drops on the ground, so the frost-bow, just after sunrise, has been noticed in the crystals which fringe the grass.

652. The Mock Sun.—The mock sun, or sun-dog, is a short arc of the halo, occasionally seen at 22° distance, on the right and left of the sun, when near the horizon. The crystals, which are concerned in producing the mock sun, are supposed to have the form of *spiculæ*, or six-sided *needles*, whose alternate sides are inclined to each other at an angle of 60°; these, if suspended in the air in a vertical position, could refract the light only in directions nearly horizontal, and therefore present only the right and left sides of the halo.

In high latitudes, other and complex forms of halo are frequent, depending for their formation on the prevalence of crystals of other angles than 60°. [See Appendix for calculations of the angular radius of rainbows and halo.]

CHAPTER VI.

COLOR, BY REFLECTION, INFLECTION, STRIATION OF SURFACE, AND THIN PLATES.

653. Natural Colors of Bodies.—The colors which bodies exhibit, when seen in ordinary white light, are owing to the fact that they decompose light by absorbing or transmitting some colors and reflecting the others. We say that a body *has* a certain color, whereas it only *reflects* that color; a flower is called red, because it reflects only or principally red light; another yellow, because it reflects yellow light, &c. A white surface is one which reflects all colors in their due proportion; and such a surface, placed in the spectrum, assumes each color perfectly, since it is capable of reflecting all. A substance which reflects no light, or but very little, is black. What peculiarity of constitution that is which causes a substance to reflect a certain color, and to absorb others, is unknown.

Very few objects have a color which exactly corresponds to any color of the spectrum. This is found to result from the fact that most bodies, while they reflect some one color chiefly, reflect the others in some degree. A red flower reflects the red light abundantly, and perhaps some rays of all the other colors with the red. Hence there may be as many shades of red as there can be different proportions of other colors intermingled with it. The same is true of each color of the spectrum. Thus there is an infinite variety of tints in natural objects. These facts are readily established by using the prism to decompose the light which bodies reflect.

654. Inflection, or Diffraction of Light.—This phenomenon consists of delicate *colored fringes bordering the edges of shadows*, when the light comes from a luminous point or line.

For the purpose of experiments on this subject, a beam of light is admitted into a dark room, through a very small aperture, as a pin-hole made in sheet-lead; or, what is better, a convex lens is placed in the window-shutter, which brings the rays to a focus, and affords a divergent pencil of light. If we introduce into this pencil any opaque body, as a knife-blade, for example, and observe the shadow which it casts on a white screen, we shall observe on both sides of the shadow *fringes of colored light*, the different colors succeeding each other in the order of the spectrum, from violet to red. Three or four series can usually be discerned, the one nearest to the shadow being the most complete and distinct, and the remoter ones having fewer and fainter colors. The phenomenon is independent of the density or thickness of the body which casts the shadow. The light, in passing by the edge or back of a knife, by a block of marble or a bubble of air in glass, is in each case affected in the same way. But if the body is very narrow, as, for example, a fine wire, a modification arises from the light which passes the opposite side; for now fringes appear *within* the shadow, and at a certain distance of the screen the central line of the shadow is the most luminous part of it.

655. Breadth of Fringe varies with the Color.—If, in the foregoing experiments, we use light of one color alone instead of white light, then the fringes are only of that color, separated from each other by lines which are comparatively dark; and, on measuring the breadths and distances of fringes of different colors, those of *red* light are found to be widest, those of *violet* narrowest, and the other colors have breadths according to their order. This explains why, in the case of white light, the several colors appear in a series, with the red outermost; for each element of the white light forms its own system of fringes, but the systems do not coincide—the wider ones project beyond the narrower, and thus become separately visible.

If the screen is moved further from the body, the distance of a given color from the edge of the shadow becomes greater, but not in proportion to the distance of the screen from the body; which proves that the color is not propagated in a straight line, but in a curve. These curves are found to be *hyperbolas*, having their concavity on the side next the shadow, and are in fact a species of caustics.

656. Light through Small Apertures.—The phenomena of inflection are exhibited in a more interesting manner when we view with a magnifying glass a pencil of light after it has passed through a small aperture. For instance, in the cone already described as radiating from the focus of a lens in a dark room, let a plate of lead be interposed, having a pin-hole pierced through it, and let the slender pencil of light which passes through the pin-hole fall on the magnifier. The aperture will be seen as a luminous circle surrounded by several rings, each consisting of a prismatic series. These are, in truth, the fringes formed by the edge of the circular puncture, but they are modified by the circumstance that the opposite edges are so near to each other. If, now, the plate be removed, and another interposed having *two* pin-holes, within one-eighth of an inch of each other, besides the colored rings round each, there is the additional phenomenon of long lines crossing the space between the apertures; the lines are nearly straight, and alternately luminous and dark, and varying in color, according to their distance from the central one. These lines are wholly due to the overlapping of two pencils of light, for on covering one of the apertures they entirely disappear. By combining circular apertures and narrow slits in various patterns in the screen of lead, very brilliant and beautiful effects are produced.

657. Why Inflection is not always noticed in looking by the Edges of Bodies.—It must be understood that light is *always* inflected when it passes by the edges of bodies; but that it is rarely observed, because, as light comes from various sources at once, the colors of each pencil are overlapped and reduced to whiteness by those of all the others. By using care to admit into the eye only isolated pencils of light, some cases of inflection may be observed which require no apparatus. If a person standing at some distance from a window holds close to his eye a book or other object having a straight edge, and passes it along so as to come into apparent coincidence with the sash-bars of the window, he will notice, when the edge of the book and the bar are very nearly in a range, that the latter is bordered with colors, the violet extremity of the spectrum being on the side of the bar nearest to the book, and the red extremity on the other side. Again, the effect produced when light passes through a narrow aperture may be seen by looking at a distant lamp through the space between the bars of a pocket-rule, or between any two straight edges brought almost into contact. On each side of the lamp are seen several images of it, growing fainter with increased distance, and finely colored. An experiment still more interesting is to look at a distant lamp through the net-work of a bird's feather. There are several series of colored images, having a fixed arrangement in relation to the disposition of the minute apertures in the feather; for the system of images revolves just as the feather itself is revolved.

658. Striated Surfaces.—If the surface of any substance is ruled with fine parallel grooves, 2000 or more to the inch, it will reflect bright colors when placed in the sunbeam. *Mother-of-pearl* and many kinds of sea-shell exhibit colors on account of delicate striæ on their surface. These are the edges of thin laminæ which compose the shell, and which crop out on the surface in fine and nearly parallel lines. It may be known that the color arises from such a cause, if, when the substance is impressed on fine cement, its colors are communicated to the cement. Indeed, it was in this way that Dr. Wollaston accidentally discovered the true cause of such colors. The changeable hues in the plumage of some birds, and the wings of some insects, are owing to a striated structure of their surfaces. But the metals can be made to furnish the most brilliant spectra, by stamping them with steel dies, which have been first ruled by a diamond with lines from 2000 to 10,000 per inch, and then hardened. Gilt buttons and other articles for dress are sometimes prepared in this manner, and are called *iris ornaments.* The color in a given case depends on the distance between the grooves, and the obliquity of the beam of light. Hence, the same surface, uniformly striated, may reflect all the colors, and every color many times, by a mere change in its inclination to the beam of light.

659. Thin Laminæ.—Any transparent substance, when reduced in thickness to a few millionths of an inch, reflects brilliant colors, which vary with every change of thickness. Examples are seen in the thin laminæ of air occupying cracks in glass and ice, and the interstices between plates of mica, also in thin films of oil on water, and alcohol on glass, but most remarkably in soapy water blown into very thin bubbles.

If a lens of slight convexity is laid on a plane lens, and the two are pressed together by a screw, and viewed by reflected light, rings of color are seen arranged around the point of contact. The rings of least diameter are broadest and most brilliant, and each one contains the colors of the spectrum in their order, from violet on the inner edge to red on the outer. But the larger rings not only become narrower and paler, but contain fewer colors; yet the succession is always in the same order as above. Increased pressure causes the rings to dilate, while new ones start up at the centre, and enlarge also, until the centre becomes black, after which no new rings are formed. These are commonly called Newton's rings, because Sir Isaac Newton first investigated their phenomena.

660. Ratio of Thicknesses for Successive Rings.—A given color appears in a circle around the point of contact, because equal thicknesses are thus arranged. If the diameters of the successive rings of any one color be carefully measured, their squares are found to be as the odd numbers, 1, 3, 5, 7; and hence the thicknesses of the laminæ of air at the repetitions of the same color are as the same numbers. For, let Fig. 347 represent a section of the spherical and plane surfaces in contact at a. Let $a\,b$, $a\,d$, be the radii of two rings at their brightest points. Suppose $a\,i$, perpendicular to $m\,n$, to be produced till it meets the opposite point of the circle of which $a\,g$ is an arc, and call that point f; then $a\,f$ is the diameter of the sphere of which the lens is a segment. Let $b\,e$, $d\,g$, be parallel to $a\,i$, and $e\,h$, $g\,i$, to $m\,n$, then we have

FIG. 347.

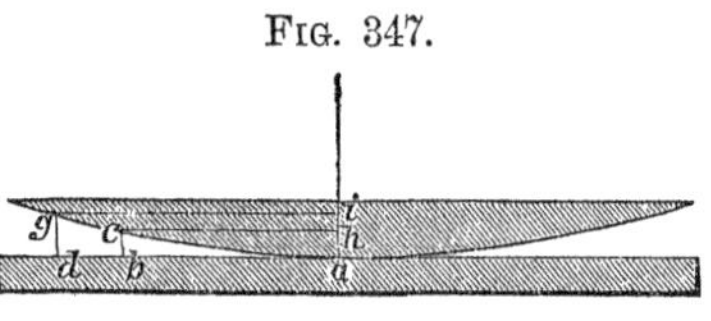

$$(e\,h)^2 : (g\,i)^2 :: a\,h \times h\,f : a\,i \times i\,f.$$

But the distances between the two lenses being exceedingly small in comparison with the diameter of the sphere, $h\,f$ and $i\,f$ may be taken as equal to $a\,f$, whence, by substitution,

$$(e\,h)^2 : (g\,i)^2 :: a\,h \times a\,f : a\,i \times a\,f :: a\,h : a\,i :: b\,e : d\,g.$$

Therefore the thicknesses of successive rings are as the odd numbers.

661. Thickness of Laminæ for Newton's Rings.—The absolute thickness, $b\,e$, $d\,g$, &c., can also be obtained, $a\,f$ being known, since

$$a\,f : a\,e :: a\,e : a\,h \text{ or } b\,e;$$

for in so short arcs the chord may be considered equal to the sine, that is, the radius of the ring. When air is between the lenses, all the rings range between the thickness of *half a millionth* of an inch and 72 *millionths;* if water is used, the limits are $\frac{3}{8}$ *of a millionth* and 58 *millionths.* Below the smaller limit the medium appears black, or no color is reflected; above the highest limit the medium appears white, all colors being reflected together. When water is substituted for air, all the rings contract in diameter, indicating that a particular order of color requires less thickness of water than of air; the thicknesses for different media are found to be in the inverse ratio of the indices of refraction.

662. Relation of Rings by Reflection and by Transmission.—If the eye is placed beyond the lenses, the transmitted light also is seen to be arranged in very faint rings, the brightest portions being at the same thicknesses as the darkest ones by reflection; and these thicknesses are as the even numbers, 2, 4, 6, &c. The centre, when black by reflection, is white by transmission, and where red appears on one side, blue is seen on the other; and, in like manner, each color by reflection answers to its *complementary color* by transmission.

663. Newton's Rings by a Monochromatic Lamp.—The number of reflected rings seen in common light is not usually greater than from *five* to *ten.* The number is thus small, because as the outer rings grow narrower by a more rapid separation of the surfaces, the different colors overlap each other, and produce whiteness. But if a light of only one color falls on the lenses, the number may be multiplied to several hundreds; the rings are alternately of that color and black, growing more and more narrow at greater distances, till they can be traced only by a microscope. A good light for such a purpose is the flame of an alcohol lamp, whose wick has been soaked in strong brine, and dried.

CHAPTER VII.

DOUBLE REFRACTION AND POLARIZATION.

664. Double Refraction.—There are many transparent substances, particularly those of a crystalline structure, which, instead of refracting a beam of light in the ordinary mode, *divide it into two beams.* This effect is called *double refraction,* and substances which produce it are called *doubly-refracting substances.*

This phenomenon was first observed in a crystal of carbonate of lime, denominated *Iceland spar.* It is bounded by six rhomboidal faces, whose inclinations to each other are either 105° 5′, or 74° 55′. There are two opposite solid angles, *A* and *X* (Fig. 348), each of which is formed by the meeting of three obtuse plane angles; and when the edges of the crystal are equal, the diagonal *A X* is equally inclined to the edges which it meets, as *A B*, *A C*, and *A D*; *A X* is called the *axis* of the crystal. But every other line in the crystal parallel to *A X* is also an axis, because the crystal may be conceived to be divided into any number of similar crystals, each having its own axis; the axis is therefore a *direction* rather than a line. If a thick crystal of spar be laid on a line of writing, it appears as *two* lines, one of which seems not only thrown aside from the other, but brought a little nearer to the eye. Therefore every ray of light, in passing through, is divided into two rays, which come to the eye in different directions. The double refraction may also be seen by letting a very slender sunbeam, *R r* (Fig. 349), fall on the crystal; as it enters it takes two directions, *r O*, and *r E*, which on passing out describe the lines *O O′*, *E E′*, parallel to the incident beam, *R r*. One of these rays, *O O′*, is called the *ordinary* ray, because it is always refracted according to the ordinary law of refraction (Art. 606); that is, it remains in the plane of incidence, and the sines of incidence and refraction have a constant ratio to each other at all inclinations. The other, *E E′*, is called the *extraordinary* ray, because in some positions it departs from this law of refraction in one or both particulars.

Fig. 348.

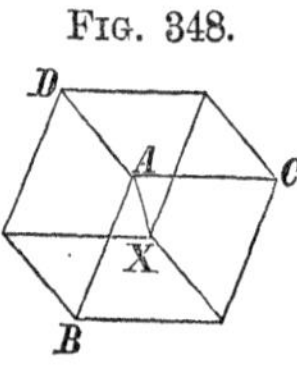

Fig. 349.

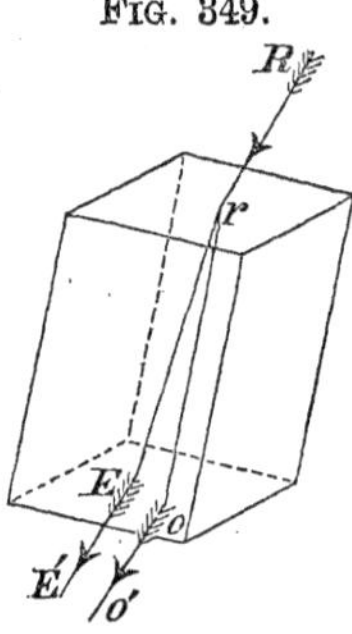

The property of double refraction belongs to a large number of crystals, and also to some animal substances, as hair, quills, &c.; and it may be produced artificially in glass by heat or pressure.

665. Optical Relations of the Axis.—The axis of Iceland spar has been defined with reference to *form*; but it is also the axis with respect to its optical relations, for in the direction of that line a ray is never doubly refracted, while it is doubly refracted in all other directions.

Every plane which includes the axis of a crystal is called a *principal section*. In every principal section the extraordinary ray conforms to one part of the law of refraction, but not to the other; it remains in the plane of incidence, but does not preserve a constant ratio of sines at different inclinations.

In a plane at right angles to the axis, the extraordinary ray conforms to both parts of the law; but in all planes besides this and the principal sections, it conforms to neither part.

Crystals of a *positive* axis, are those in which the extraordinary ray has a *larger* index of refraction than the ordinary ray; crystals of a *negative* axis are those in which the index of the extraordinary ray is less than that of the ordinary ray. Iceland spar is a crystal of negative axis.

Some crystals have two axes of double refraction; that is, there are two directions in which light may be transmitted without being doubly refracted. A few crystals have more than two axes.

666. Polarization of Light.—This name is given to a change which may be produced in light, such that it has different properties on different sides. *Common light,* as, for instance, a direct sunbeam, has the same relation to space on all sides. If it falls on a piece of glass at a given angle, it will suffer reflection equally well in every plane, as we turn the glass round, and so of refraction, or any change we may attempt. But if a beam were so changed in its character that it could be reflected upward, but could not be reflected to the right, it would be called, not *common,* but *polarized* light.

667. Polarization by Reflection.—Let two tubes, *M N* and *N P* (Fig. 350), be fitted together in such a manner that one can

FIG. 350.

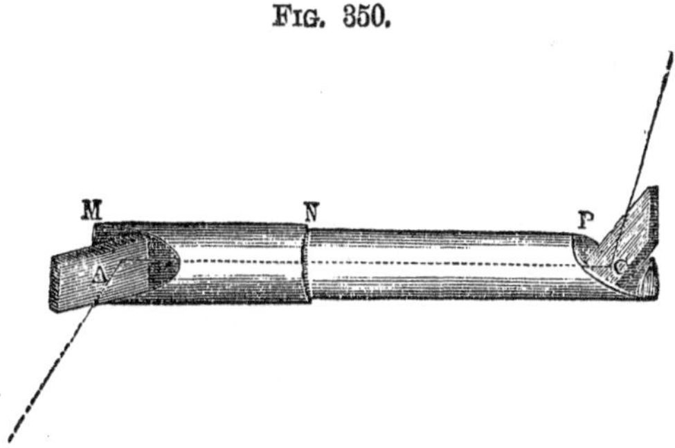

be revolved upon the other; and to the end of each let there be attached a plate of dark-colored glass, *A* and *C*, capable of reflecting only from the first surface. These plates are hinged so as to be adjusted at any angle with the axis of the tube. Let the plane of each glass incline to the axis of the tube at an angle of 33°, and let the beam *R A* make an incidence of 57°, the complement of 33°, on *A*; then it will, after reflection, pass along the axis of the tube, and make the same angle of incidence on *C*. If now the tube *NP* be revolved, the second reflected ray will vary its intensity, according to the angle between the two planes of incidence on *A* and *C*. The beam *A C* is *polarized light;* the glass *A*, which has produced the polarization, is called the *polarizing plate;* the glass *C*, which shows, by the effects of its revolution, that *A C* is polarized, is the *analyzing plate;* and the whole instrument, constructed as here represented, or in any other manner for the same purpose, is called a *polariscope*.

668. Changes of Intensity Described.—The changes in the ray *C E* are as follows: When the tube *N P* is placed so that the plane of incidence on *C* is coincident with the former plane of incidence, *R A C*, whether *C E* is reflected forward or backward in that plane, the intensity at *E* will be the same as if *A C* had been a beam of common light. If *N P* is revolved, *E* will begin to grow fainter, and reach its minimum of intensity when the planes *R A C* and *A C E* are at right angles, which is the position indicated in the figure. Continuing the revolution, we find the intensity increasing through the second quadrant of revolution, and reaching its maximum, when the two planes of incidence again coincide, 180° from the first position. The next half revolution repeats these changes in the same order.

669. The Polarizing Angle.—The angle of 57° is called the polarizing angle for glass, not because glass will not polarize at other angles of incidence, but because at all other angles it polarizes the light in a less degree; and this is indicated by the fact that, in revolving the analyzing plate, there is less change of intensity, and the light at *E* does not become so faint. Different substances have different polarizing angles, and these are found to be so connected with the degree of refractive power, that by a knowledge of the index of refraction for any substance, its polarizing angle can be calculated, and *vice versa*. Hence the refractive power of opaque bodies may be determined. No substance *entirely* polarizes the light incident upon it, even at the angle of polarization. Complete polarization of the ray *A C* would be indicated by the entire extinction of *C E*, at two opposite points of its revolution. On the other hand, every substance polarizes, in *some* degree, the light which it reflects. The polarization produced by reflection from the metals is very slight.

670. Polarization by a Bundle of Plates.—Light may also be polarized by *transmission* through a bundle of laminæ of a transparent substance, at an angle of incidence equal to its polarizing angle. Let a pile of twenty or thirty plates of transparent glass, no matter how thin, be placed in the same position as the reflector *A*, in Fig. 350, and a beam of light be transmitted through them in a direction toward *C*. In entering and leaving the bundle *A*, situated as in the figure, the angles of incidence and refraction are in a horizontal plane. When *C* is revolved, the beam undergoes the same changes as before, with this difference, that the places of greatest and least intensity will be reversed. If the light is reflected from *C* in the same plane in which it was refracted by *A*, its intensity is least, and it is greatest when reflected in a plane at right angles to it, as at *E* in the figure.

671. Polarization by Crystals.—The third and most perfect method of polarizing light, is by *transmission through certain crystals*. *Some* crystals polarize the transmitted light by *absorption;* and *every* doubly-refracting crystal polarizes both the ordinary and the extraordinary ray. If a thin plate be cut from a crystal of tourmaline, by planes parallel to its axis, the beam transmitted through it is polarized, and, when received on the analyzing plate, will alternately become bright and faint, as the tube of the analyzer is revolved. And if a beam is passed through a doubly-refracting crystal, and the two parts fall on the analyzing plate, they will come to their points of greatest and least brightness at alternate quadrants; indeed, when one ray is brightest, the other is entirely extinguished. Therefore the two rays which emerge from a doubly-refracting crystal are polarized *completely*, and in planes at right angles with each other.

672. Every Polarizer an Analyzer.—We have seen that light is polarized by reflection from glass at an incidence of 57°, and analyzed by another plate at the same angle of incidence. This is but an instance of what is always true, that every method of polarizing light may be used to analyze, i. e., to test its polarization. Hence, a bundle of thin plates of glass may take the place of the analyzer *C*, as well as of the polarizer *A*. For, on turning it round, though the transmitted beam remains in the same place, yet it will, at the alternate quadrants, brighten to its maximum and fade to its minimum of intensity.

So, again, if light has passed through a tourmaline, and is received on a second whose crystalline axis is parallel to that of the former, the ray will proceed through that also; but if the second is turned in its own plane, the transmitted ray grows faint, and nearly disappears at the moment when the two axes are at

90° of inclination, and this alternation continues at each 90° of the whole revolution.

Finally, place a double-refractor at each end of the polariscope, and let a beam pass through them and fall on a screen. The first crystal will polarize each ray, and the second will doubly refract and also analyze each, exhibiting a very interesting series of changes. In general, four rays will emerge from the second crystal, producing four luminous spots on the screen. But, on revolving the tube, not only do the rays commence a revolution round each other, but two of them increase in brightness, and the other two at the same time diminish as fast, till two alone are visible, at their greatest intensity. At the end of the second quadrant, the spots before invisible are at their maximum of brightness, and the others are extinguished. This alternation continues as long as the crystal is revolved. In the middle of each quadrant the four are of equal brightness.

673. Color by Polarized Light.—The phenomena of *color* produced by polarized light are beautiful, and of great interest.

Let a very thin plate of some doubly-refracting crystal be placed perpendicularly across the axis of the polariscope (Fig. 351), and let the analyzed ray, *C E*, fall on a screen. When the principal section of the crystal, *D H*, coincides with the first plane of reflection, *R A C*, or is perpendicular to it, all the phenomena are the same as if no crystal was interposed. But let the film be revolved in its own plane till *D H* makes 45° with the plane *R A C*; then, instead of the dark spot at *E*, a brilliant color appears. That color may be any tint of the spectrum, according to the thickness of the interposed film. If now the revolution of the crystal is continued, the color fades out at the end of the next 45°, reappears at 90°, and so on. But if the crystal be so placed as to give color, and the analyzing plate be revolved, a different series presents itself. The color observed at *E*, during the first 45°, gradually fades, and during the next 45° its *complement* appears and brightens to its maximum. The original color is restored at 180°, and the complementary color at 270°.

FIG. 351.

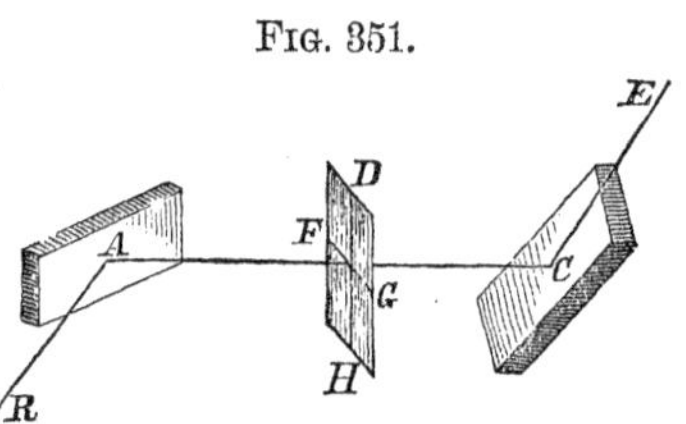

The most interesting form of this experiment is seen when the light is polarized and analyzed by means of double-refractors; since the polarization is more perfect, and the two pairs of oppositely polarized rays are on the screen at once. When two of the images are of a certain color, the other two have the complementary color.

674. Systems of Colored Rings.—Systems of irised bands and rings may also be produced by the polariscope. Let a plate be cut from a doubly-refracting crystal of one axis by planes perpendicular to that axis; and place it between the polarizer and analyzer. If now a pencil of sufficient divergency is transmitted, a system of colored circles will be formed, resembling Newton's rings between lenses. If a polariscope is formed of two tourmalines, and the crystal laid between them, and the whole combination, less than half an inch thick, is brought close to the eye, the pencil of light will consist of rays of various obliquity, and the rings may be seen beautifully projected on the sky. Or the ring systems may be projected on a screen by a polariscope furnished with concentrating lenses. Fig. 352 presents the system as seen through Iceland spar when the planes of reflection in the polari-

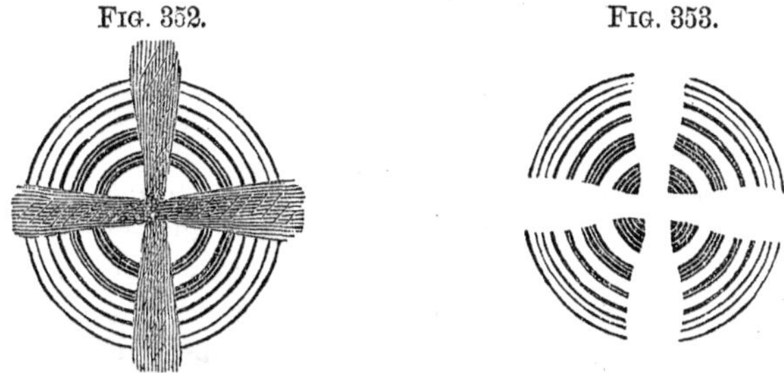

FIG. 352. FIG. 353.

scope are at right angles. Two dark diameters cross the system and interrupt the rings. If the planes of reflection are coincident, the system is in every respect complementary to the other (Fig. 353). The colors of the rings are all reversed, and the crossing bands are white. If double-refractors of two axes are used instead of the spar, compound systems are shown, of various forms and great beauty.

CHAPTER VIII.

NATURE OF LIGHT.—WAVE THEORY.

675. The Wave Theory.—Light has sometimes been regarded as consisting of *material particles* emanating from luminous bodies. But this, called the *corpuscular* or *emission theory*, has mostly yielded to the *undulatory* or *wave theory*, which supposes light to consist of vibrations in a medium. This medium, called the *luminiferous ether*, is imagined to exist throughout all space, and to be of such rarity as to pervade all other matter. It is supposed also to be elastic in a very high degree, so that undulations excited in it are transmitted with great velocity. If radiant heat consists of undulations of the same ether, they perhaps differ from those of light only in being slower. For it is a familiar fact, that when the heat of a body is increased, a point is at length reached at which the body becomes luminous; that is, the vibrations then affect the sense of sight as well as that of feeling. Moreover, the rays of heat are somewhat less refrangible than those of light (Art. 625), from which it is inferred that its vibrations are slower.

676. Postulates of the Wave Theory.—

1. *The waves are propagated through the ether at the rate of 192,500 miles per second.*

As this is the known velocity of light, it must be the rate at which the waves are transmitted.

2. *The atoms of the ether vibrate at right angles to the line of the ray in all possible directions.*

It was at first assumed that the luminous vibrations, like the vibrations of sound, are *longitudinal*, that is, back and forth in the line of the ray; but the discoveries in polarization require that the vibrations of light should be assumed to be *transverse*, that is, in a plane perpendicular to the line of the ray; and, moreover, that in that plane the vibrations are in every possible direction within an inconceivably short space of time. Thus, if a person is looking at a star in the zenith, we must consider each atom of the ether between the star and his eye as vibrating across the vertical in all horizontal directions, north and south, east and west, and in innumerable lines between these.

3. *Different colors are caused by different rates of vibration.*

Red is caused by the *slowest* vibrations, and *violet* by the *quickest*, and other colors by intermediate rates. White light is to the eye what harmony is to the ear, the resultant effect of several rates of vibration combined. There are slower vibrations of the ether than those of red light, and quicker ones than those of violet light, but they are not adapted to affect the vision. The former affect the sense of feeling as *heat*, the latter produce chemical effects, and are called *actinic* rays.

4. *The ether within bodies is less elastic than in free space.*

This is inferred from the fact that light moves with less velocity in passing through bodies than in free space; the greater the refractive power of a body, the slower does light move within it. And in some bodies of crystalline structure, it happens that the velocity is different in different directions, so that the elasticity of the ether within them must be regarded as varying with the direction.

677. Reflection and Refraction on the Wave Theory.—When the waves of light reach the surface of a new medium, the ether within it being generally in a different state of elasticity, a system of waves will be propagated backward in the former medium, and another onward in the new medium. The reflected system will make the same angle with the perpendicular as the incident system, analogous to the reflection of waves of water and of sound. But the system which enters the medium will change its direction according to its velocity in the medium; and the velocity depends on the elasticity of the ether. In media of greater refractive power, the elasticity is considered to be less than in those of less refractive power; and the waves are therefore propagated more slowly in the former than in the latter. Let *A B*, *d D* (Fig. 354), represent the parallel waves of a beam falling on *M N*, the surface of a denser medium. The side of the wave which enters first at *D*, advances more slowly than the side still moving in the rarer medium. Suppose *D* to reach *c*, while *d* is going to *C*; then the wave, now wholly within the medium, lies in the position *C c*, and advances in a line perpendicular to *C c*, so long as it continues in the medium. Thus the light is refracted *toward* the perpendicular *G H*, in entering a denser medium. In a similar manner, it is shown that *E F C c*, in entering a rarer medium, is refracted *from* the perpendicular, since the side *C d* emerges first and then gains velocity over the side *c D*.

Fig. 354.

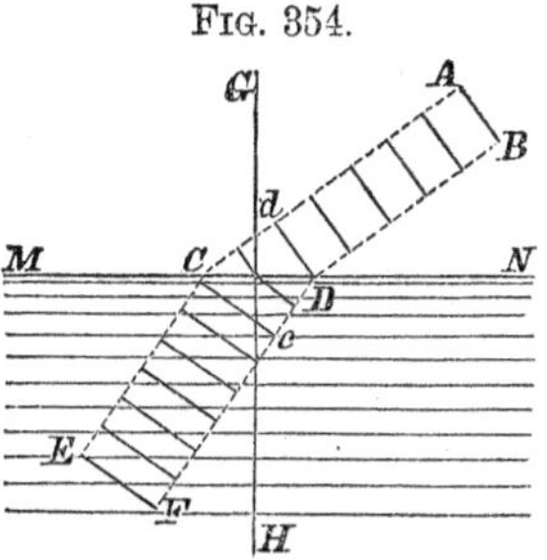

678. Refraction on the Emission Theory.—It appears, therefore, that the wave theory requires us to suppose light to move more slowly in denser media. But, in the emission theory, it is necessary to suppose it to move more swiftly. For the bending of the path of a particle of light toward the perpendicular must be attributed to the attraction exerted by the medium on the particle. Suppose, then, that a particle of light moves along *B A* (Fig. 355), and enters a denser medium. Let the velocity, *B A*, be resolved into *B E*, *E A*; the latter will be increased by the attraction of the medium; the former will not be changed. Make *A G* = *B E* or *D A*, and *A F* greater than *A E*; then *A C* represents the direction and velocity of the ray after entering the medium. But as AF is greater than EA, while $AG = DA$, $\therefore AC$ is greater than BA. On the other hand, if a ray, CA, is entering a rarer medium, the attraction of the denser draws it *backward*, and renders the component AE less than FA; and hence the velocity AB, in the rarer medium, is less than CA, the velocity in the denser. The two theories are thus in conflict on the question whether light *gains* or *loses* velocity in entering a more refractive medium. Several direct tests have been applied in order to determine this; and they all agree in proving that light moves more slowly in substances which have greater refractive power.

FIG. 355.

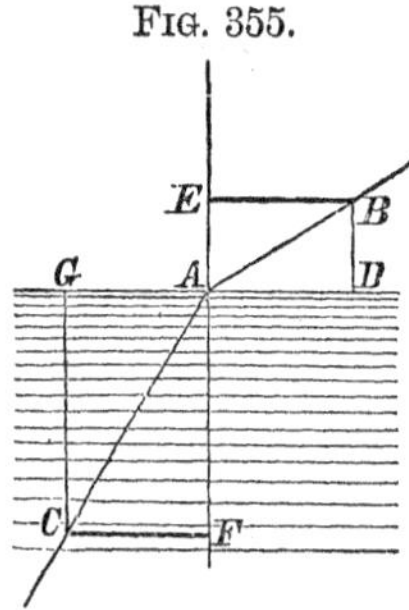

679. Interference.—Many interesting phenomena are explained on the principle of interference of waves. As two systems of water-waves may increase or diminish their height by being combined, and as sounds, when blended, may produce various results, and even destroy each other, so may two pencils of light either augment or diminish each other's brightness, and even produce darkness.

Any one may try for himself the following experiment: Prick two very small holes, quite near each other, through paper, and holding the paper close to one eye, look through both holes at any small bright spot, such as occurs in a crack of glass when the sun shines upon it; then will the bright spot be seen striped across with parallel black lines, which will be further apart as the holes are closer together. The two pencils of light, through the two apertures, overlap on the retina of the eye, and cause bright and dark lines by interference. Where like phases meet, the lines are bright; where opposite phases meet, there is *no* light, and the lines are black.

But there are other forms of experiment by which the exact length of wave for each color may be determined.

680. Interference by Thin Plates.—Let light of any one color, as yellow, fall on the lenses which exhibit Newton's rings. A system of waves is reflected from the first surface of the thin stratum of air which lies between the lenses, and another system from the second, and these two come to the eye together. Suppose, at a given point, the thickness of air is such that the reflected waves of the second system meet those of the first, phase for phase, in exact concert; at that point is seen a brighter yellow than if there was but one reflecting surface. But, at another point, the thickness of the air may be such that the two systems disagree by half a wave, bringing opposite phases together; in which case all motion is destroyed, and the point is black. The former is one point of a yellow circle, the latter of a black circle, each around the point of contact. It is obvious that at the smallest bright ring, the reflected waves from the second surface must be just *one* wave-length behind those from the first; at the second ring, *two* waves behind, &c.; and, in general, luminous circles appear where the two systems differ by an exact number of whole waves, and dark circles where they differ by half a wave, or any whole number and a half. The exact measurement of the thicknesses of air at any point (Art. 661), has led to the determination of the length of waves of each color.

681. Change of Color alters the Size of the Rings.—If orange or red light is used instead of yellow, the rings are a little enlarged, being formed where the lamina of air is a little thicker, and therefore the waves for those colors are longer; but if green, blue, indigo, and violet are each tried separately, the rings grow smaller in each case; and it is inferred that the lengths of waves are less in the same order, and in the ratio of the thicknesses.

The reason becomes obvious why, in white light, the rings are few in number, and consist of a series of different colors, without any black circles between. As rings of different colors are of different sizes when separate, so when all colors are used together they will be arranged side by side, and some will be likely to fall where the black circles between others would occur. Again, as all the rings grow narrower at greater distances, because the thickness of the lamina increases faster, they crowd upon and overlap each other, and produce white light. Hence, a full prismatic series occurs only near the centre, and after *five* or *ten* repetitions, growing less and less perfect, white light covers the whole surface.

682. Interference by Two Mirrors.—If two plane reflectors, inclined at a very obtuse angle, receive light from a minute radiant, and reflect it to one spot on a screen, the reflected pencils will interfere, and produce bright and dark lines. Suppose light of one color, as violet, flows from a radiant point A (Fig. 356); let mirrors $B\ C$ and $B\ D$ reflect it to the screen $K\ L$. F and E may be so selected that the ray $A\ F + F\ G$ equals the ray $A\ E + E\ G$. Then G will be luminous, because the two paths being equal, the same phase of wave in each ray will occur at the point G. But if H be so situated that $A\ f + f\ H$ differs *half a violet wave* from $A\ e + e\ H$, then H will be a dark point, because opposite phases meet there. A similar point, I, will lie on the other side of G. Again, there are two points, K and L, one on each side of G, to each of which the whole path of light by one mirror will exceed

FIG. 356.

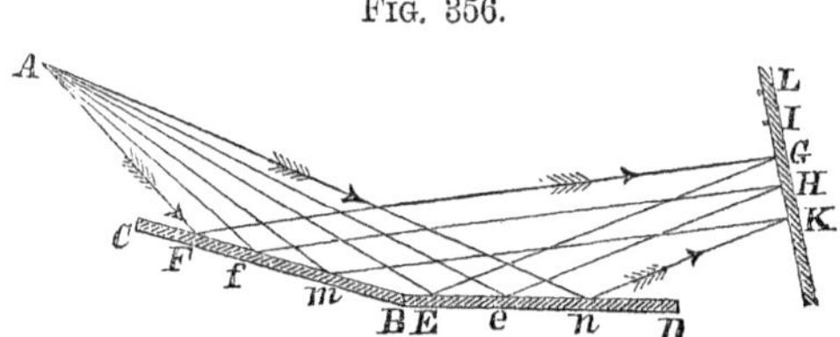

the whole by the other by just *one* violet wave; those points are bright. Thus, there is a series of bright and dark points on the screen; or rather a series of bright and dark hyperbolic *lines*, of which these points are sections. Other colors will give bands separated a little further, indicating longer waves. And white light, producing all these results at once, will give a repetition of the prismatic series.

683. Interference by Inflection.—One of the forms of inflection is explained as follows: Through an opaque screen, $A\ B$ (Fig. 357), let there be a very narrow aperture, $c\ d$, by which is admitted the beam of light, $e\ f\ g\ h$, of some one color, and emanating from a single point. That part of the aperture near d may be regarded as a luminous centre, from which emanate waves in all directions, and the same is true of the other part of the aperture near c. Let i be a point on one side of the beam, so situated that the distances $d\ i$ and $c\ i$ shall differ by half a wave of the color employed; then, as opposite phases meet there, i will be a dark point. Let j be a point still further removed from the beam, where $c\ j - d\ j$ equals the length of a wave, then j will be luminous, since like phases meet in that point. This alternation will be repeated a few times till the luminous points become crowded and feeble. If the aperture is made narrower, the intervals $h\ i$, $i\ j$, &c., will increase, as they obviously must, in order to preserve $c\ i - d\ i$ equal to a half wave, and $c\ j - d\ j$ equal to a wave. Violet light produces the narrowest lines, red the widest, and white light the prismatic series, for the same reason as in Newton's rings. If $A\ c$, the left side of the screen, is entirely removed, so that light passes only one edge, d, the fringes will still exist, though somewhat modified.

FIG. 357.

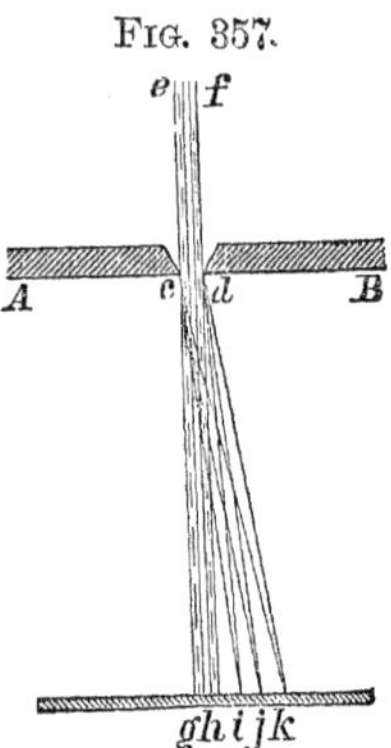

684. Length and Number of Luminous Waves.—The other cases of inflection, and the phenomena of *striation*, as well as the *supernumerary* rainbows, are fully accounted for on the principle of interference. The careful measurements which have been made in nearly all these instances, have led, by so many independent methods, to the accurate determination of the length of a wave of each color. When the length of wave of any color is known, the number of vibrations per second is readily obtained by dividing the velocity of light by the length of the wave. The remarkable results of these investigations are given in the following table:

Colors.	Length of a wave in decimals of an inch.	Number of vibrations per second.
Extreme red, . .	.0000266	458,000,000,000,000
Red, ...	.0000256	477,000,000,000,000
Orange, ...	.0000240	506,000,000,000,000
Yellow, ...	.0000227	535,000,000,000,000
Green, ...	.0000211	577,000,000,000,000
Blue, ...	.0000196	622,000,000,000,000
Indigo, ...	.0000185	658,000,000,000,000
Violet, ...	.0000174	699,000,000,000,000
Extreme violet, . .	.0000167	727,000,000,000,000
Mean, ...	.0000225	541,000,000,000,000

685. Change of Vibrations in Polarized Light.—It has been stated (Art. 676) that the vibrations of the ether, in the case of common light, must be supposed to be *transverse in all directions*. But, instead of this, we may conceive, what is mechanically equivalent to it, that the vibrations are made in *two* transverse directions at right angles to each other. Thus, in the descent of light from a star in the zenith, we may suppose each atom of the ether to vibrate in the two transverse lines, one north and south, and the other east and west; because every motion oblique to these can be resolved into two components, one on each of these two. Or any other two lines, perpendicular to each other in the same plane, may be assumed as the directions of vibration.

This being the nature of common light, it is easy to state what is meant by polarized light. It is that in which the vibrations are performed in only *one* of the transverse directions. For example, in the ray of star-light just supposed, if all the easterly and westerly vibrations, and all the easterly and westerly components of the oblique vibrations, were destroyed, then no motions would remain except in the north and south direction, and the light of that star would be polarized. It is, of course, immaterial what particular transverse motion is cut off, provided all the motion at right angles to it is retained.

686. Polarizing and Analyzing by Reflection.—When light is reflected, those vibrations of the ray which are *in* the plane of incidence are generally weakened in a greater or less degree, while those which are *perpendicular to* the same plane are not affected. How much the vibrations are weakened depends on the elasticity of the ether within the medium, and on the angle of incidence. But reflection of light rarely if ever takes place without diminishing the amplitude of those vibrations which are in the plane of incidence; so that a reflected ray is always polarized, at least, in a slight degree.

It will now be readily understood how the analyzing plate (Fig. 350) proves the light to be polarized. Suppose the reflectors *A* and *C* are so perfect polarizers that vibration in the plane of incidence is *entirely* destroyed. Along *R A* the particles of ether vibrate across it both horizontally and vertically; and as the plane of incidence *R A C* is horizontal, the atoms along *A C* will vibrate only vertically, because the horizontal vibrations, being in the plane of incidence, are destroyed. Now let *C* be placed so as to reflect horizontally; the light will not be weakened by this reflection, because there are no horizontal vibrations to be destroyed. But let *C* be turned so as to reflect vertically, for instance, upward; now there can be no reflection, since all the vibrations left in *A C* are in the vertical plane, which is the plane of incidence; and they are destroyed. For the same reason that reflection at *A* extinguished all *horizontal* motions in the atoms of ether, the reflection at *C* extinguishes all *vertical* motions; hence there is *no* motion beyond *C*.

687. Polarizing by Transmission through a Bundle of Plates.—At each of the surfaces some reflection occurs, so that all vibrations in the plane of incidence at length disappear from the reflected ray, even though the laminæ are not *perfect* polarizers; while all vibrations perpendicular to this plane are preserved. Hence the reverse must be true of the transmitted ray; it will retain the vibrations, so far as they coincide with the plane of incidence, and lose them, so far as they are perpendicular to it. Thus the two sets of rectangular vibrations are separated from each other; one exists in the reflected ray, the other in the transmitted ray. The two rays are therefore polarized in planes at right angles to each other.

688. Polarizing by Absorption.—A tourmaline absorbs, or in some way extinguishes the vibrations, so far as they are perpendicular to its crystalline axis, but leaves all motion which is parallel to its axis unimpaired. It is at once apparent why a second tourmaline analyzes; for if its axis is parallel to that of the first, the same vibrations which could pass the one, could pass the other also; but if the two axes are at right angles, the same system of vibrations which could pass the first, because parallel to its axis, will be absorbed by the second, because perpendicular to its axis.

689. Polarizing by Double Refraction.—In doubly-refracting crystals, the ether possesses different degrees of elasticity in different directions; hence, so far as vibrations lie in *one* plane, they may be more retarded in their progress, and in a plane at right angles to that they may be less retarded, and the degree of refraction depends on the amount of retardation (Art. 677). Thus the two systems become separated, and emerge at different places. Each ray is of course polarized, having vibrations in only one direction; and the two planes of polarization are at right angles to each other.

690. Different Kinds of Polarization.—Since the discovery was made that the etherial atoms may by certain methods be thrown into circular movements, and by others into vibrations in an ellipse with the axis in a fixed direction, the polarization already described has been called *plane polarization,* since the atoms vibrate in a plane. *Circular polarization* is that in which the atoms revolve in circles; and *elliptical polarization* denotes a state of vibration in ellipses, whose major axes are confined to one plane.

CHAPTER IX.

VISION.

691. Image by Light through an Aperture.—If light from an external object pass through a small opening of any shape in the wall of a dark room, it will form an ill-defined inverted image on the opposite wall. Imagine a minute square orifice, through which the light enters and falls on a screen several feet distant. A pencil of light, in the shape of a square pyramid, emanating from the highest point of the object, passes through the aperture, and forms a luminous square near the bottom of the screen. From an adjacent point another pencil, crossing the first at the aperture, forms another square, overlapping and nearly coinciding with the former. Thus every point of the object is represented by its square on the screen; and as the pencils all cross at the aperture, the image formed is every way inverted. It is also indistinct, because the squares overlap, and the light of contiguous points is mingled together. If the orifice is smaller, the image is less luminous, but more distinct, because the pencils which form it overlap in a less degree. If the hole is circular, or triangular, or of irregular form, there is no change in the appearance of the image, which is now composed of small circles, or triangles, or irregular figures, whose shape is completely lost by overlapping.

692. Effect of a Convex Lens at the Aperture.—The image will become distinct, and more luminous also, if the aperture be enlarged to a diameter of two or three inches, and then covered by a convex lens of the proper curvature. The image will be *distinct*, because the rays from each point of the object are converged to a point again, and *luminous*, in proportion as the lens has a larger area than the aperture before employed. This is a real, and therefore an inverted image (Art. 618). A *scioptic ball* is a sphere containing a lens, and so fitted in a socket that it can be turned in any direction, and thus bring into the room the images of different parts of the landscape. The *camera obscura* is a darkened room furnished with a scioptic ball and adjustable screen for producing distinct pictures of external objects.

Instead of connecting the lens with the wall of a room, it is frequently attached to a portable box or case, within which the image is formed. The *Daguerreotype*, or *photograph*, is the image produced by the convex lens, and rendered permanent by the chemical action of light on a surface properly prepared. The lens for photographic purposes needs to be achromatic, and corrected, also, as far as possible, for spherical aberration.

693. The Eye.—The eye is a camera obscura in miniature; we find here the darkened room, the aperture, the convex lens, and the screen, with inverted images of external objects painted on it. A horizontal section of the eye is represented in Fig. 358.

The optical apparatus of the eye, and the spherical case which incloses it, constitute what is called the *eye-ball*. The case itself, except about a sixth part of it in front, is a strong white substance, called, on account of its hardness, the *sclerotic coat, S, S* (Fig. 358). In the front, this opaque coat changes to a perfectly transparent covering, called the *cornea, C, C,* which is a little more convex than the sclerotic coat. The increased convexity of the cornea may be felt by laying the finger gently on the eye-lid when closed, and then rolling the eye one way and the other.

Fig. 358.

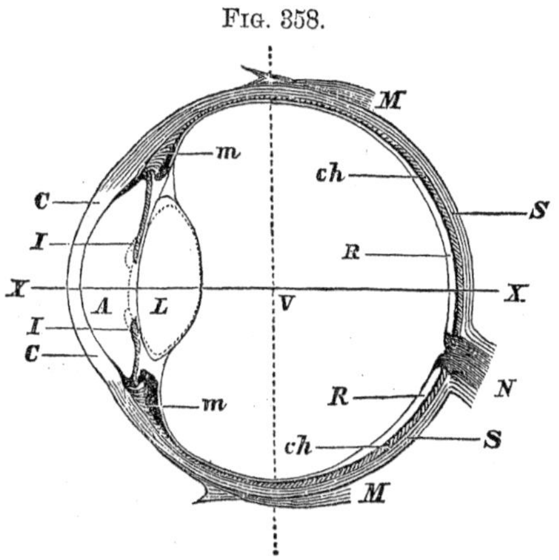

The bony socket, which contains the eye, is of pyramidal form, its vertex being some distance behind the eye-ball; room is thus afforded for the mechanism which gives it motion. This cavity, except the hemisphere in front occupied by the eye itself, is filled up with fatty matter and with the six muscles by which the eye-ball is revolved in all directions.

694. The Interior of the Eye.—Behind the cornea is a fluid, *A*, called the *aqueous humor*. In the back part of this fluid lies the *iris*, *I*, *I*, an opaque membrane, having in the centre of it a circular aperture, the *pupil*, through which the light enters. The iris is the colored part of the eye; the back side of it is black. Directly back of the aqueous humor and iris, is a flexible double convex lens, *L*, called the *crystalline lens*, or *crystalline humor*, having the greatest convexity on the back side. The large space back of the crystalline is occupied by the vitreous humor, *V*, a semi-liquid, of jelly-like consistency. Next to the vitreous humor succeed those inner coatings of the eye, which are most immediately concerned in vision. First in order is the *retina*, *R*, *R*, on which the light paints the inverted pictures of external objects. The fibres of the optic nerve, which enter the ball at *N*, are spread all over the retina, and convey the impressions produced there to the brain. Outside of the retina is the *choroid coat*, *c h*, *c h*, covered with a black pigment, which serves to absorb all the light so soon as it has passed through the retina and left its impressions. The choroid is inclosed by the sclerotic already described. The nerve-fibres, which are spread over the interior of the retina, are gathered into a compact bundle about one-tenth of an inch in diameter, which passes out through the three coatings at the back part of the ball, about fifteen degrees from the axis, *X X*, on the side toward the other eye. *M*, *M* represent two of the muscles, where they are attached to the eye-ball.

695. Vision.—The index of refraction for the cornea, and the aqueous and vitreous humors, is just about the same as that for water; for the crystalline lens, the index is a little greater. The light, therefore, which comes from without, is converged principally on entering the cornea, and this convergency is a little increased both on entering and leaving the crystalline. If the convergency is just sufficient to bring the rays of each pencil to a focus on the retina, then the images are perfectly formed, and there is distinct vision. To prevent the reflection of rays back and forth within the chamber of the eye, its walls are made perfectly black throughout by a pigment which lines the choroid, the ciliary processes, and the back of the iris. Telescopes and other optical instruments are painted black in the interior for a similar purpose.

The cornea is prevented from producing spherical aberration by the form of a prolate spheroid which is given to its surface, and the crystalline, by a gradual increase of density from its edge to its centre.

696. Adaptations.—By the prominence of the cornea rays of considerable obliquity are converged into the pupil, so that the eye, without being turned, has a range of vision more or less perfect, through an angle of about 150°.

The quantity of light admitted into the eye is regulated by the size of the pupil. The iris, composed of a system of circular and radial muscles, expands or contracts the pupil according to the intensity of the light. These changes are involuntary; a person may see them in his own eyes by shading them, and again letting a strong light fall upon them, while he is before a mirror.

The pupils in the eyes of animals have different forms according to their habits; in the eyes of those which graze, the pupil is elongated horizontally, and in the eyes of beasts and birds of prey, it is elongated vertically.

The eyes of animals are adapted, in respect to their refractive power, to the medium which surrounds them. Animals which inhabit the water have eyes which refract much more than those of land animals. The human eye being fitted for seeing in air, is unfit for distinct vision in water, since its refractive power is nearly the same as that of water, and therefore a pencil of parallel rays from water entering the eye would scarcely be converged at all. The effect is the same as if the cornea were deprived of all its convexity.

697. Accommodation to Diminished Distance.—It has been shown (Art. 618), that as an object approaches a lens, its image moves away, and the reverse. Therefore in the eye there must be some change in order to prevent this, and keep the image distinct on the retina while the object varies its distance. In a state of rest, the eye converges to the retina only the pencils of *parallel* rays, that is, those which come from objects at great distances. Rays from near objects diverge so much that, while the eye is at rest, it cannot sufficiently converge them so that they will meet on the retina; but each conical pencil is cut off before reaching its focus, and all the points of the object are represented by overlapping circles, causing an indistinct image. The change in the eye, which fits it for seeing near objects distinctly, is called *accommodation.* This is effected by increasing the convexity of the crystalline lens, principally the front surface. The *ciliary muscle, m, m,* surrounds the crystalline, and is attached to the sclerotic coat just on the circle where it changes into the cornea. This muscle is connected with the edge of the crystalline by the circular ligament which surrounds the latter and holds it in place. When the muscle contracts, it relaxes the ligament, and the crystalline, by its own elastic force, begins to assume a more convex form, as represented by the dotted line. The eye is then accommodated for the vision of objects more or less near, according to the degree of change in the lens. On the other hand, when the ciliary muscle relaxes, the ligament again draws upon the lens to flatten it, and adapt it for the view of distant objects. In Fig. 359 these two conditions of the crystalline are more distinctly shown. The dotted line exhibits the shape of the lens when accommodated for seeing near objects. Accompanying this action of the ciliary muscle is that of the iris, which diminishes the pupil for near objects, so as to exclude the outer and more divergent rays. The dotted lines in front of the iris represent its situation when pushed forward by the crystalline accommodated for near objects.

FIG. 359.

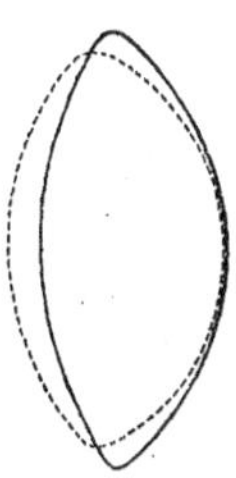

698. Long-Sightedness.—As life advances, the crystalline becomes harder and less elastic. It therefore assumes a less convex form when the ligament is relaxed, and cannot be accommodated to so short distances as in earlier years; and at length it remains so flattened in shape that only very distant objects can be seen distinctly. The eye is then said to be long-sighted, and requires a convex lens to be placed before it, to compensate for insufficient convexity in the crystalline.

There are, however, cases of long-sightedness in early life. Such instances are found to be the result of an oblate form of the eye-ball, as shown in Fig. 360; it is too short from front to back to furnish room for the convergency of the pencils, and they are cut off by the retina before reaching their focal points. In order to bring the distinct image forward upon the retina, convex glasses are needed in such cases, just as for the eyes of most people when advanced in life. As the term long-sightedness is now applied to this abnormal condition of the eye, the effect of age upon the sight is more properly called old-sightedness.

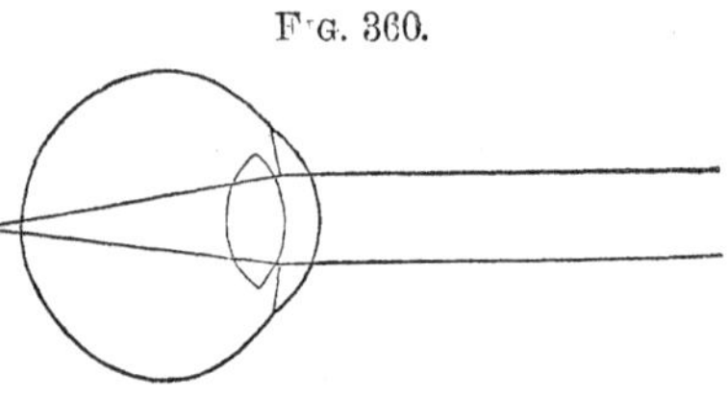
F'G. 360.

699. Short-Sightedness.—The eyes of the short-sighted have a form the reverse of that just described; the eye-ball is elongated from cornea to retina (Fig. 361), resembling a prolate spheroid, so that rays parallel, or nearly so, are converged to a point before reaching the retina, and after crossing, fall on it in a circle; and the image, made up of overlapping circles instead of points, is indistinct. If this elongation of the eyeball is extreme, an object must be brought very near, in order that its image may move back to the retina, and distinct vision be produced. This inconvenience is remedied by the use of concave lenses, which increase the divergency of the rays before they enter the eye, and thus throw their focal points further back.

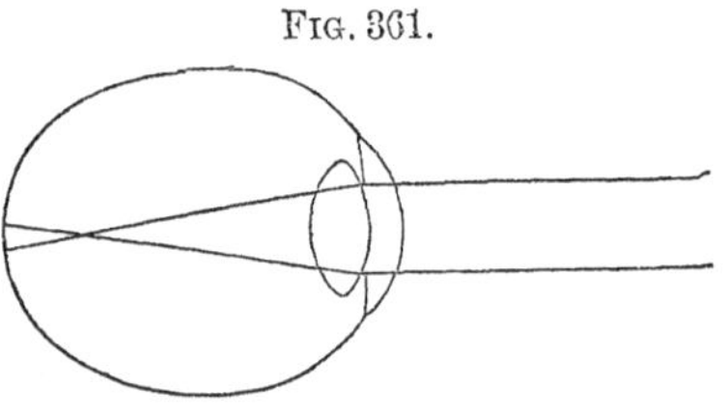
FIG. 361.

In the normal condition of the eyes in early life, the nearest limit of distinct vision is about *five* inches. This limit slowly increases with advance of life, but much more slowly in some cases than others, till it is at an indefinitely great distance. The near limit of distinct vision for the short-sighted varies from *five* down to *two* inches, according to the degree of elongation in the eye-ball.

700. Why an Object is Seen Erect and Single.—The image on the retina is *inverted;* and that is the very reason why the object is seen erect; the image is not the thing *seen*, but *that by means of which we see.* The impression produced at any point on the retina is referred outward in a straight line through a point near the centre of the lens, to something external as its cause; and therefore that is judged to be highest without us which makes its image lowest on the retina, and the reverse.

An object appears as *one*, though we see it by means of *two* images; but this is only one of many instances in which we have learned by experience to refer two or more sensations to one thing as the cause. Provided the images fall on parts of the retina, which in our ordinary vision *correspond* with each other, then by experience we refer both impressions to one object; but if we press one eye aside, the image falls in a new place in relation to the other, and the object seems double.

701. Indirect Vision.—The Blind Point.—To obtain a clear and satisfactory view of an object, the axes of both eyes are turned directly upon it, in which case each image is at the centre of the retina. But when the light from an object is exceedingly faint, it is better seen by *indirect vision*, that is, by looking to a point a little on one side, and especially by changing the direction of the eyes from moment to moment, so that the image may fall in various places *near* the centre of the retina. Many heavenly bodies are plainly discerned by indirect vision, which are too faint to be seen by direct vision.

In the description of the eye it was stated that the retina, as well as the choroid and the sclerotic, is perforated to allow the optic nerve to pass through. At that place there is no vision, and it is called the *blind point*. In each eye it is situated about 15° from the centre of the retina toward the other eye. Let a person close his right eye, and with the left look at a small but conspicuous object, and then slowly turn the eye away from it toward the right; presently the object will entirely disappear, and as he looks still further to the right, it will after a moment reappear, and continue in sight till the axis of the eye is turned 70° or 80° from it. The same experiment may be tried with the right eye in the opposite direction. The reason why people do not generally notice the fact till it is pointed out, is that an object cannot disappear to both eyes at once, nor to either eye alone, when directed to the object.

702. Continuance of Impressions.—The impression which a visible object makes upon the retina continues about one-eighth or one-ninth of a second; so that if the object is removed for that length of time, and then occupies its place again, the vision is uninterrupted. A coal of fire whirled round a centre at the rate of eight or nine times per second, appears in all parts of the circumference at once. When riding in the cars, one sometimes gets a faint but apparently an uninterrupted view of the landscape beyond a board fence, by means of successive glimpses seen through the cracks between the upright boards. Two pictures, on opposite sides of a disk, are brought into view together, as parts of one and the same picture, by whirling the disk rapidly on one of its diameters. Such an instrument is called a *thaumatrope.* The *phantasmascope* is constructed on the same principle. Several pictures are painted in the sectors of a circular disk, representing the same object in a series of positions. These are viewed in a mirror through holes in the disk, as it revolves quickly in its own plane. Each glimpse which is caught whenever a hole comes before the eye, presents the object in a new attitude; and all these views are in such rapid succession that they appear like one object going through the series of movements.

703. Accidental Colors.—There are impressions on the retina of another kind, which are produced by intense lights; they continue longer, and are in respect to color unlike the objects which cause them. They are commonly called *accidental colors.* If a particular part of the retina is for some time affected by the image of a bright colored object, and then the eyes are shut, or turned upon a white surface, the *form* appears to remain, but the *color* is complementary to that of the object; and its continuance is for a few seconds or several minutes, according to the vividness of the impression. This is the cause of the *green* appearance of the sky between clouds of brilliant *red* in the morning or evening.

704. Estimate of the Distance of Bodies.—

1. If objects are near, we judge of relative distance by the *inclination of the optic axes* to each other. The greater that inclination is, or, which is the same thing, the greater the change of direction in an object, as it is viewed by one eye and then by the other, the nearer it is. If objects are *very* near, we can with one eye alone judge of their distance by the degree of effort required to accommodate the eye to that distance.

2. If objects are known, we estimate their distance by the *visual angle* which they fill, having by experience learned to associate together their distance and their *apparent*, that is, their *angular* size.

3. Our judgment of distant objects is influenced by their *clearness* or *obscurity.* Mountains, and other features of a landscape, if seen for the first time when the air is remarkably pure, are estimated by us nearer than they really are; and the reverse, if the air is unusually hazy.

4. Our estimate of distance is more correct when *many objects intervene.* Hence it is that we are able to place that part of the sky which is near the horizon further from us than that which is over our heads. The apparent sky is not a hemisphere, but a flattened semi-ellipsoid.

705. Magnitude and Distance Associated.—Our judgments of distance and of magnitude are closely associated. If objects are known, we estimate their distance by their visual angle, as has been stated; but if unknown, we must first acquire our notion of their distance by some other means, and then their visual angle gives us a definite impression as to their size. And if our judgment of distance is erroneous, a corresponding error attaches to our estimate of their magnitude. An insect crawling slowly on the window, if by mistake it is supposed to be some rods beyond the window, will appear like a bird flying in the air. The moon near the horizon seems larger than above us, because we are able to locate it at a greater distance.

706. Binocular Vision.—The Stereoscope.—If objects are placed quite near us, we obtain simultaneously *two views*, which are essentially different from each other—one with one eye, and one with the other. By the right eye more of the right side, and less of the left side, is seen, than by the left eye. Also, objects in the foreground fall further to the left compared with distant objects, when seen with the right eye than when seen with the left. And we associate with these combined views the form and extent of a body, or group of bodies, particularly in respect to distance of parts from us. It is, then, by means of *vision with two eyes*, or *binocular vision*, that we are enabled to get accurate perceptions of prominence or depression of surface, reckoned in the visual direction. A picture offers no such advantage, since all its parts are on one surface, at a common distance from the eyes. But, if two perspective views of an object should be prepared, differing as those views do, which are seen by the two eyes, and if the right eye could then see only the right-hand view, and the left eye only the left-hand view, and if, furthermore, these two views could be made to appear on one and the same ground, the vision would then be the same as is obtained of the real object by both eyes. This is effected by the *stereoscope*. *Two* photographic views are taken, in directions which make a small angle with each other, and these views are seen at once by the two eyes respectively, through a pair of half-lenses, placed with their thin edges toward each other, so as to turn the visual pencils away from each other, as though they emanated from one object. An appearance of *relief* and *reality* is thus given to superficial pictures, precisely like that obtained from viewing the objects themselves.

CHAPTER X.

OPTICAL INSTRUMENTS.

707. The Camera Lucida.—This is a four-sided prism, so contrived as to form an apparent image at a surface on which that image may be copied, the surface and image being both visible at the same time. It has the form represented by the section in Fig. 362; $A = 90°$, $C = 135°$; B and D, of any convenient size, their sum of course $= 135°$. A pencil of light from the object M, falling perpendicularly on $A\,D$, proceeds on, and makes, with $D\,C$, an angle equal to the complement of D. After suffering total reflection at G, and again at H, its direction $H\,E$ is perpendicular to $M\,F$. For, produce $M\,F$ and $E\,H$, till they intersect in I; then, since $C = 135°$, $C\,G\,H + C\,H\,G = 45°$; but $I\,G\,H = 2\,C\,G\,H$, and $I\,H\,G = 2\,C\,H\,G$; $\therefore\ I\,G\,H + I\,H\,G = 90°$; $\therefore\ I = 90°$. Therefore $H\,E$ emerges at right angles to $A\,B$, and is not refracted. Now, if the pupil of the eye be brought over the edge B, so that, while $E\,H$ enters, there may also enter a pencil from the surface at M', then both the surface M' and the object M will be seen coinciding with each other, and the hand may therefore sketch M on the surface at M'. The reason for two reflections of the light is, that the inversion produced by one reflection may be restored by the second.

Fig. 362.

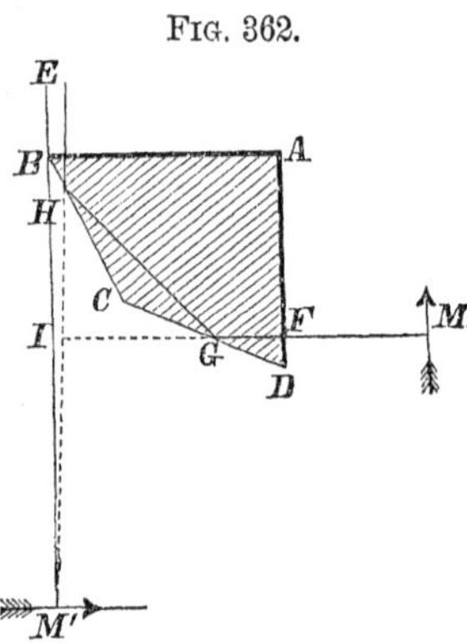

One of the most useful applications of the camera lucida is in connection with the compound microscope, where it is employed in copying with exactness the forms of natural objects, too small to be at all visible to the naked eye.

708. The Microscope.—This is an instrument for *viewing minute* objects. The nearer an object is brought to the eye, the larger is the angle which it fills, and therefore the more perfect is the view, provided the rays of each pencil are converged to a point on the retina. But if the object is nearer than the limit of distinct vision, the eye is unable to produce sufficient convergency. If the letters of a book are brought close to the eye, they become blurred and wholly illegible. But let a pin-hole be pricked through paper, and interposed between the eye and the letters, and, though faint, they are *distinct* and *much enlarged.* The *distinctness* is owing to the fact that the outer rays, which are most divergent, are excluded, and the eye is able to converge the few central rays of each pencil to a focus. The letters appear *magnified,* because they are so near, and fill a large angle. The microscope utilizes these excluded rays, and renders the image not only large and distinct, but luminous.

709. The Single Microscope.—The single microscope is merely a convex lens. It aids the eye in converging the rays, which come from a very near object, so that a distinct and luminous image may be formed on the retina. The lens may be regarded as a part of the eye, and the diameter of an object is magnified in the ratio of the limit of distinct vision to the focal distance of the lens. Taking *five* inches as the limit of distinct vision, if the principal focal distance is *one-fourth* of an inch, then we may consider the object twenty times nearer the eye than in viewing it without a lens, and therefore magnified twenty times in diameter, or 400 times in area. Now glass lenses are made whose focal length is not more than $\frac{1}{50}$ inch, and whose magnifying power, therefore, is $5 : \frac{1}{50}, = 250$ in diameter, or 62,500 in area.

Though the focal distance of a lens may be made as small as we please, yet a practical limit to the magnifying power is very soon reached.

1. The *field of view,* that is, the extent of surface which can be seen at once, diminishes as the power is increased.

2. Spherical aberration increases rapidly, because the outer rays are very divergent. Hence the necessity of diminishing the aperture of the lens, in order to exclude the most divergent rays.

3. It is more difficult to illuminate the object as the focal length of the lens becomes less; and this difficulty becomes a greater evil on account of the necessity of diminishing the aperture in order to reduce the spherical aberration.

Magnifying glasses are single microscopes of low power, such as are used by watchmakers. Lenses of still lower power and several inches in diameter are used for viewing pictures.

710. The Compound Microscope.—It is so called because it consists of two parts, an object-glass, by which a real and magnified image is formed, and an eye-glass, by which that image is again magnified. Its general principle may be explained by Fig. 363, in which *a b* is the small object, *c d* the object-glass, and *e f* the eye-glass. Let *a b* be a little beyond the principal focus of *c d*, and then the image *g h* will be real, on the opposite side of *c d*, and larger than *a b*. Now apply *e f* as a single microscope for viewing *g h*, as though it were an object of comparatively large size. Let *g h* be at the principal focus of *e f*, so that the rays of each pencil shall be parallel; they will, therefore, come to the eye at *k*, from an *apparent* image on the same side as the *real* one, *g h*; and the extreme pencils, *e k*, *f k*, if produced backward, will include the image between them, *e k f* being the angle which it fills.

Fig. 363.

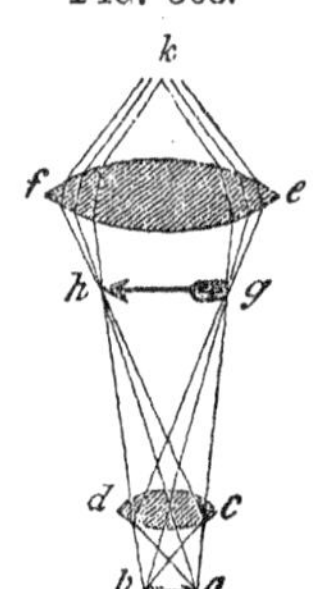

711. The Magnifying Power.—The *magnifying power* of the compound microscope is estimated by compounding two ratios; first, the distance of the image from the object-glass, to the distance of the object from the same; and secondly, the limit of distinct vision to the distance of the image from the eye-glass. For the image itself is enlarged in the first ratio (Art. 618); and the eye-glass enlarges that image in the second ratio (Art. 709). The advantage of this form over the single microscope is not so much that a great magnifying power is obtained, as that a given magnifying power is accompanied by a larger field of view.

712. Modern Improvements.—Great improvements have been made in the compound microscope, principally by combining lenses in such a manner as greatly to reduce the chromatic and spherical aberrations. The object-glass generally consists of one, two, or three achromatic pairs of lenses. The eye-piece usually contains two plano-convex lenses, a combination which is found to be the most favorable for diminishing the spherical aberration, and for enlarging the field of view. For convenience, the direction of the rays is, in many instruments, changed from a vertical to a horizontal direction, by total reflection in a right-angled prism. In Fig. 364, *A* is the object; *B*, *C*, and *C'*, achromatic plano-convex lenses, the plano-concave part being of flint-glass, the double-convex part of crown-glass, and the two parts fitted and cemented together; *D* the right-angled prism; *E* the field-glass, so called because it enlarges the field of view, by bending the outer pencils so that they come within the limit of the eye-glass *G*; *G* the eye-glass,

Fig. 364.

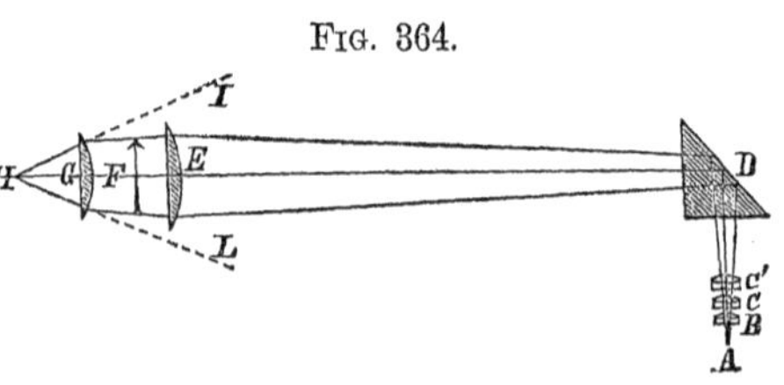

converging the pencils to the eye at H, while the rays of each pencil diverge a little, as from the magnified image back of G. The image seen by the eye at H fills the angle $I H L$.

713. Microscopes for Projecting Images.—For the purpose of forming magnified images on a screen, to be viewed by an audience, the microscope is modified in its arrangements. One form for projecting transparencies, whether paintings or photographs, is called the *magic lantern*. Another form, especially adapted for the exhibition of small objects in natural history, is the *solar microscope*.

Such instruments are valuable as means of instruction and entertainment, but they are of no use for investigation and discovery.

714. The Magic Lantern.—It consists of a box, represented in Fig. 365, containing a lamp, and having openings so arranged as to permit the air to pass freely through it, without letting light escape. In front of the lamp is a tube containing a concentrating lens, C, the painting on glass, B, and the lens, A, for producing

FIG. 365.

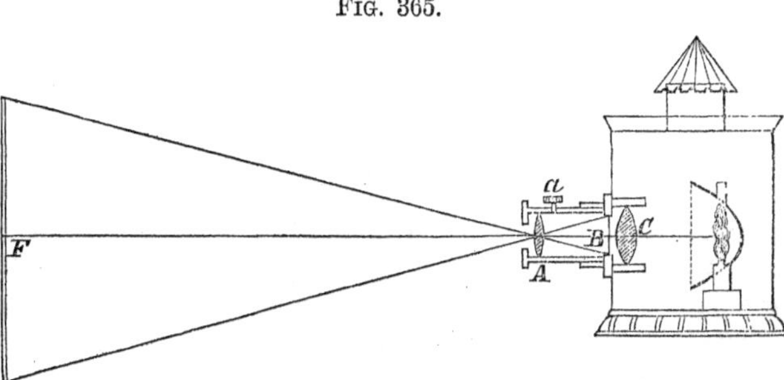

the image; back of the lamp may be a concave mirror for reflecting additional light on the lens C. The transparency B is a painting on glass, and the strong light which falls on it proceeds through the lens A, as from an original object brilliantly colored. It is a little further from A than its principal focus, and therefore the rays from any point are converged to the conjugate focus in a real image, F, on a distant screen. This image is of course inverted relatively to the object, and therefore, if the picture B is inverted, F will be erect. The lens may be placed at various distances from B by the adjusting screw a, so as to give the greatest distinctness to the image at any given distance of the screen. According to Art. 618, the diam. of B : diam. of F : : $A B$: $A F$; and therefore, theoretically, the image may be as large as we please.

But spherical aberration will increase rapidly as the image is enlarged, and even if this evil could be remedied, the want of light would render the image too faint to be well seen; for the illumination is as much less than that of the painting as the area is greater. Two magic lanterns placed side by side, may throw different images on the same ground, so as to produce the effect called *dissolving views.*

715. The Solar Microscope.—This does not differ in principle from the magic lantern. For illumination the solar or electric light is employed, and images are formed, not of artificial paintings, but of small natural objects. The lens *A* (Fig. 366),

FIG. 366.

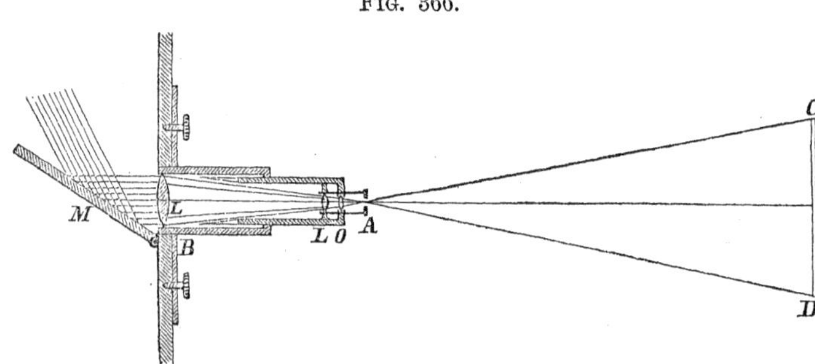

which forms the image, is fixed in the end of a tube, *A B*, and at the other end is a mirror, *M*, which can be turned on a hinge to incline at any angle with the tube. This apparatus is attached to a window-shutter, the mirror on the outside, and the tube within. By adjusting screws the mirror is inclined so as to reflect the sunbeam along the tube, where it is concentrated by lenses, *L L*, upon the object, *O*. Just beyond the object is the lens *A*, of very small aperture, by which the image *CD* is formed. If the sunbeam is large, and the screen at a sufficient distance, the images of objects may be plainly seen when magnified millions of times in area. Spherical aberration, however, is considerable; and this prevents the instrument from being of service for investigation.

716. The Telescope.—The telescope aids in *viewing distant* bodies. An image of the distant body is first formed in the principal focus of a convex lens or a concave mirror; and then a microscope is employed to magnify that image as though it were a small body. The image is much more luminous than that formed in the eye, when looking at the heavenly body, because there is concentrated in the former the large beam of light which falls upon the lens or mirror, while the latter is formed by the slender pencil only which enters the pupil of the eye. If the image in a telescope is formed by a lens, the instrument is called a *refracting telescope;* but if by a mirror, a *reflecting telescope.*

717. The Astronomical Telescope.—This is the most simple of the refracting telescopes, consisting of a lens to form an image of the heavenly body, and a single microscope for magnifying that image. The former is called the *object-glass,* the latter the *eye-glass.* The image is of course at the principal focus of the object-glass, and the eye-glass is placed at its own focal distance beyond the image, in order that the rays of each pencil may emerge parallel; therefore the two lenses are separated from each other by the sum of their focal distances. The lines marked *A*, *A'*, *A''* (Fig. 367), represent the cylinder of rays which flow

FIG. 367.

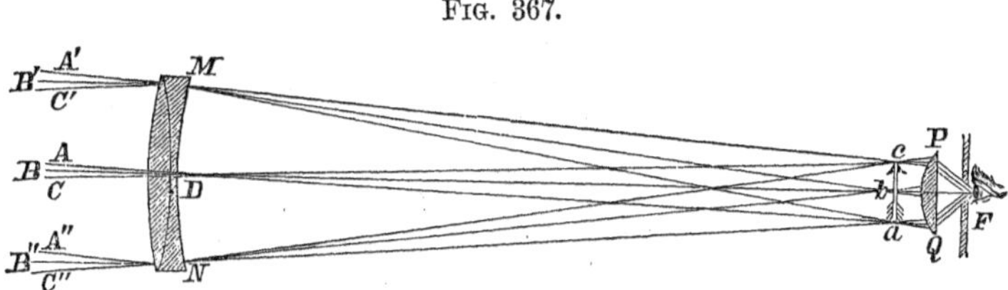

from the highest point of the object, and which cover the whole object-glass, *M N.* All these rays are collected at *a*, the lowest point of the image, the axis of the pencil, *A a*, being a straight line (Art. 615). After crossing at *a*, they are received on the lower edge of the eye-glass, *P Q*, by which they are made parallel, but the entire pencil is bent toward the axis of the lenses, and meets it at *F.* The beam, *B*, *B'*, *B''*, coming from the centre of the object, forms the centre, *b*, of the image; and *C*, *C'*, *C''*, from the lowest point of the object, forms the top, *c*, of the image. In a similar manner each point of the image is formed by the concentrated rays which emanate from a corresponding point in the object. These innumerable pencils, after diverging from their focal points in the image, are turned toward the axis by passing through the eye-piece, while the rays of each become parallel. At *F* there is a diaphragm having an aperture, at which the eye is placed.

718. The Powers of the Telescope.—The *magnifying power* of the astronomical telescope is expressed by the ratio of the *focal distance of the object-glass to that of the eye-glass.* For (Fig.

367) the object, as seen by the naked eye, fills the angle $A D C$, between the axes of its extreme pencils. But, since the axes cross each other in straight lines at the optic centre of the lens, $A D C = a D c$. Therefore, to an eye placed at the object-glass, the image, $a c$, appears just as large as the object; while at the eye-glass it appears as much larger in diameter as the distance is less.

The *illuminating power* is important for objects which shed a very feeble light on account of their immense distance. This power depends on the size of the beam, that is, on the aperture of the object-glass.

The *defining power* is the power of giving a clear and sharply defined image, without which both the other powers are useless. And it is the power of producing a well-defined image which limits both of the other powers. For every attempt to increase the magnifying power by giving a large ratio to the focal lengths of the object-glass and the eye-glass, or to increase the illuminating power by enlarging the object-glass, increases the difficulties in the way of getting a perfect image. These difficulties are three—the spherical aberration (Art. 621), the chromatic aberration (Art. 634), and unequal densities in the glass. The third difficulty is a very serious one, especially in large lenses. Very few good object-glasses have been made so large as fifteen inches in diameter.

719. Manner of Mounting.—The *equatorial mounting* of large telescopes is quite essential for accuracy of observation or measurement. When the magnifying power is great, the diurnal motion is very perceptible, and the body quickly leaves the field of view. To prevent this, the telescope is so mounted as to revolve on an axis parallel to the earth's axis, and then by means of a clock it has a motion communicated to it, by which it exactly keeps up with the apparent motion of a heavenly body. Another axis, at right angles with the former, allows the telescope to be directed to a point at any distance north or south of the celestial equator.

Astronomical telescopes, when of portable size, are usually mounted upon a tripod stand, and admit of motion on a horizontal and a vertical axis.

720. The Terrestrial Telescope.—In order to secure simplicity, and thus the highest excellence, in the astronomical telescope, the image is allowed to be *inverted*, which circumstance is of no importance in viewing heavenly bodies. But, for terrestrial objects, it would be a serious inconvenience; and, therefore, a *terrestrial telescope*, or *spy-glass*, has additional lenses for the purpose of forming a second image, inverted, compared with the first, and, therefore, erect, compared with the object. In Fig. 368, m, m, m, represent a pencil of rays from the top of a distant object, and *n, n, n,* from the bottom; *A B,* the object-glass; *m n,* the first image; *C D,* the first eye-glass, which converges the pencils of

FIG. 368.

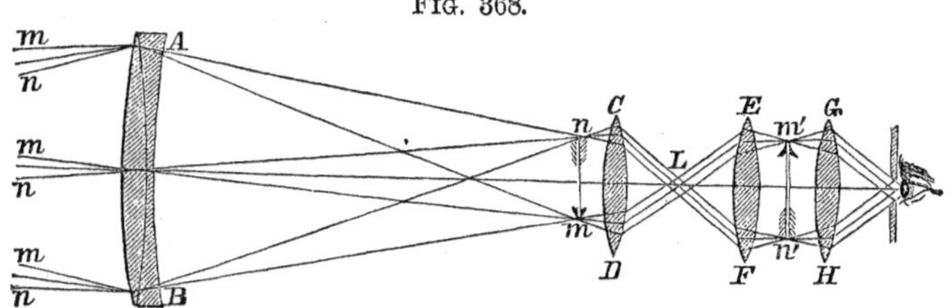

parallel rays to *L.* Instead of placing the eye at *L,* the pencils are allowed to cross and fall on the second eye-glass, *E F,* by which the rays of each pencil are converged to a point in the second image, *m' n',* which is viewed by the third eye-glass, *G H.* The second and third lenses are commonly of equal focal length, and add nothing to the magnifying power.

Such instruments are usually of a portable size, and hence the aberrations are corrected with comparative ease, by the methods already described. The spy-glass, for convenient transportation, is made of a series of tubes, which slide together in a very compact form.

721. Galileo's Telescope.—This was the first form of telescope, having been invented by Galileo, whose name it therefore bears. It differs from the common astronomical telescope in having for the eye-glass a *concave* instead of a convex lens, which receives the rays at such a distance from the focus to which they tend, as to render them parallel. Thus, the rays, *M, M, M* (Fig. 369), from the top of the object, are converged by the object-glass,

FIG. 369.

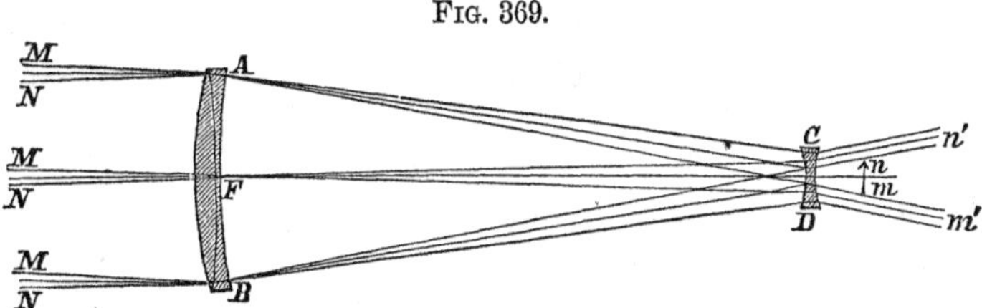

A B, toward *m,* in the image; and the pencil, *N, N, N,* from the bottom of the object, is converged toward *n*; but the concave lens *C D* is interposed at such a point as to render these converging rays parallel, and in this way they come to the eye situated behind the lens. But, though the *rays* converge before they reach the concave lens, the *pencils* diverge, having crossed at *F*; therefore, in passing the concave lens, they are made to diverge more, and will enter the eye as if they had crossed at a much nearer point than F. The angle between these extreme pencils is the angle which the object appears to fill; and the magnifying power is in the ratio of this angle to the angle $MFN = mFn$; and that equals the ratio of the focal distance of AB to the focal distance of CD. The object appears *erect* in the Galilean telescope, since the pencil, which comes from the top of the object, *appears* to come from the top of the virtual image; thus, the parts of the object and image are similarly situated. It is obvious that, since the pencils diverge, only the central ones, within the size of the pupil, can enter the eye. This circumstance exceedingly limits the field of view, and unfits the instrument for telescopic use. It is employed for opera-glasses, having a power usually of only *two* or *three* in diameter.

722. The Gregorian Telescope.—This is the most frequent form of reflecting telescope, and receives its name from the inventor, Dr. Gregory, of Scotland. The light from a heavenly body, entering the open tube (Fig. 370), is received on the large concave

FIG. 370.

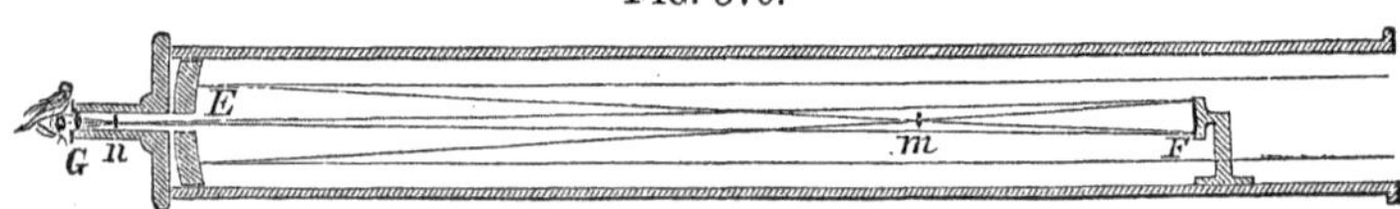

speculum, E, which forms an inverted image, m, at the principal focus; the rays of each pencil crossing there next meet the small concave mirror F, which forms an erect image, n, at the conjugate focus, beyond the speculum, the centre of the latter being perforated to let the light pass through. The eye-glass, G, magnifies this image. To avoid confusion, only two rays are drawn in the figure, and those belong to the central pencil. Rays from the *top* of the object would enter the tube inclining slightly downward, and be reflected to the bottom of m, and again to top of n. Rays from the *bottom* would ascend, and be reflected to the top of the first image, and to the bottom of the second.

723. The Herschelian Telescope.—Sir William Herschel modified the Gregorian by dispensing with the small reflector F, and inclining the large speculum E, so as to form the image near the edge of the tube, where the eye-glass is attached. Thus, the observer is situated with his back to the object. The speculum of Herschel's telescope was about four feet in diameter, and weighed more than 2,000 pounds, and its focal length was forty feet. The Earl of Rosse has since constructed a Herschelian telescope having an aperture of *six* feet, and a focal length of *fifty* feet.

APPENDIX.

APPLICATIONS OF THE CALCULUS.

I. Fall of Bodies.

1. Differential Equations for Force and Motion.—These are three in number, as follows:

$$1. \quad v = \frac{d\,s}{d\,t}.$$

$$2. \quad f = \frac{d\,v}{d\,t} = \frac{d^2\,s}{d\,t^2}.$$

$$3. \; f\,d\,s = v\,d\,v.$$

These equations are readily derived from the elementary principles of mechanics. In Art. 6 we have $v = \frac{s}{t}$. Reducing the numerator and denominator to infinitesimals, v remains finite, and the equation becomes $v = \frac{d\,s}{d\,t}$; which is Equation 1st. Therefore, if the space described by a body is regarded as a function of the time, the *first* differential coefficient expresses the *velocity.*

Again (Art. 12), $f = \frac{v}{t}$, where f represents a *constant* force. Making velocity and time infinitely small, we get the intensity of the momentary force, $f = \frac{d\,v}{d\,t}$. But, by Equation 1st, $v = \frac{d\,s}{d\,t}$; $\therefore f = \frac{d^2\,s}{d\,t^2}$; which is Equation 2d. Hence we learn that the *first* differential coefficient of the *velocity* as a function of the time, or the *second* differential coefficient of the *space* as a function of the time, expresses the *force.*

Equation 3d is obtained by multiplying the 1st and 2d crosswise, and removing the common denominator.

We proceed to apply these equations to the preparation of formulæ for falling bodies.

2. Bodies falling through Small Distances near the Earth's Surface.—In this case, let the accelerating force, which is considered *constant*, be called g. Then, by Eq. 2, $g = \frac{dv}{dt}$, $\therefore dv = g\,dt$. Integrating, we have $v = gt + C$. But, since $v = 0$ when $t = 0$, $\therefore v = gt$, and $t = \frac{v}{g}$, as in formulas 5, 6, Art. 28.

Again, substituting gt for v in Eq. 1, $ds = gt\,dt$; and by integration, $s = \frac{1}{2}gt^2 + C$; but $C = 0$, for the same reason as before; $\therefore s = \frac{1}{2}gt^2$, and $t = \sqrt{\frac{2s}{g}}$, as in formulas 1, 2, Art. 28.

Once more, equating the two foregoing values of t, we have $v = \sqrt{2gs}$, and $s = \frac{v^2}{2g}$, as in formulas 3, 4, Art. 28.

If, in the equation, $s = \frac{1}{2}gt^2$, v be substituted for gt, we have $s = \frac{1}{2}vt$, or $vt = 2s$; that is, the acquired velocity multiplied by the time of fall gives a space twice as great as that fallen through (Art. 25).

3. Bodies falling through Great Distances, so that Gravity is Variable, according to the Law in Art. 16.—

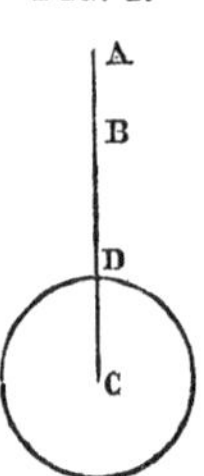

FIG. 1.

Suppose a body to fall from A to B (Fig. 1), toward the centre C. Let $AC = a$; $BC = x$; $DC = r$, the radius of the earth.

The force f at B, is found by the principle, Art. 16,

$$x^2 : r^2 :: g : f = gr^2\frac{1}{x^2} = gr^2x^{-2}.$$

4. To find the Acquired Velocity.—Substitute gr^2x^{-2} for f, and $a - x$ for s, in Equation 3d, and we have $gr^2x^{-2} \,.\, d(a - x) = v\,dv$; $\therefore$ by integration $\frac{1}{2}v^2 = \int - gr^2x^{-2}\,dx = gr^2x^{-1} + C$. But $v = 0$, when $x = a$; $\therefore C = - gr^2a^{-1}$; and

$$\frac{1}{2}v^2 = gr^2x^{-1} - gr^2a^{-1};$$

$$\therefore \quad v^2 = \frac{2gr^2(a - x)}{ax};$$

$$\therefore \quad v = \left\{\frac{2gr^2(a - x)}{ax}\right\}^{\frac{1}{2}}$$

This is the general formula for the acquired velocity. If the body falls to the earth, $x = r$, and the formula becomes

$$v = \left\{\frac{2gr(a - r)}{a}\right\}^{\frac{1}{2}}$$

Again, if the body falls to the earth through so small a space that $\frac{r}{a}$ may be regarded as a unit, the formula reduces to

$$v = \{2g(a-r)\}^{\frac{1}{2}} = (2gs)^{\frac{1}{2}};$$

the same as obtained by other methods.

If a body falls to the earth from an infinite distance, it does not acquire an infinite velocity. For then, as we may put a for $a-r$,

$$v = \left\{\frac{2gr\,.\,a}{a}\right\}^{\frac{1}{2}} = (2gr)^{\frac{1}{2}} =$$

$$(2\,.\,32\tfrac{1}{6}\,.\,3956\,.\,5280)^{\frac{1}{2}} \text{ feet} = 6.95 \text{ miles.}$$

Therefore, the greatest possible velocity acquired in falling to the earth is less than *seven miles;* and a body projected upward with that velocity would never return.

5. To find the Time of Falling.—From equation first we obtain $dt = \frac{ds}{v}$; in this, substitute $d(a-x)$ for ds, and $\frac{\{2gr^2(a-x)\}^{\frac{1}{2}}}{(ax)^{\frac{1}{2}}}$ for v, as found in the preceding article; then

$$dt = \frac{(ax)^{\frac{1}{2}}\,.\,d(a-x)}{\{2gr^2(a-x)\}^{\frac{1}{2}}} = \left(\frac{a}{2gr^2}\right)^{\frac{1}{2}} \cdot \frac{-x^{\frac{1}{2}}\,dx}{(a-x)^{\frac{1}{2}}};$$

$\therefore$ by integration $t = \left(\frac{a}{2gr^2}\right)^{\frac{1}{2}} \cdot \int -x^{\frac{1}{2}}\,dx\,(a-x)^{-\frac{1}{2}}.$

By the formula in the calculus for reducing the index of x we obtain

$$\int -x^{\frac{1}{2}}dx\,(a-x)^{-\frac{1}{2}} = (ax-x^2)^{\frac{1}{2}} - \frac{a}{2}\,vers^{-1}\left(\frac{2x}{a}\right) + C.$$

Now, when $t = 0$, $x = a$; $\therefore C = \frac{a\pi}{2}$;

hence, $t = \left(\frac{a}{2gr^2}\right)^{\frac{1}{2}}\left\{(ax-x^2)^{\frac{1}{2}} - \frac{a}{2}\,vers^{-1}\left(\frac{2x}{a}\right) + \frac{a\pi}{2}\right\}.$

6. Bodies falling within the Earth (supposed to be of uniform density), where Gravity Varies as the Distance from the Centre.—

Fig. 2.

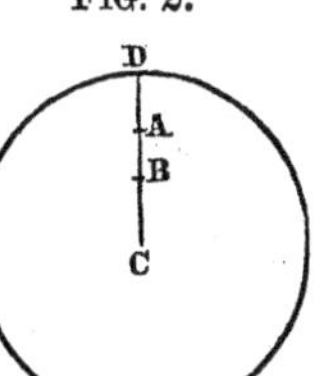

Suppose a body to fall from A to B (Fig. 2); and let $DC = r$, $AC = a$, and $BC = x$. Then

$$r : x :: g : f = \frac{g}{r}x = \text{force at B.}$$

To find the velocity acquired.—By Eq. 3d,

$$v\,d\,v = f\,d\,s; \therefore v\,d\,v = \frac{g}{r}x\,.\,d\,(a - x) = -\frac{g\,x\,d\,x}{r};$$

$$\therefore \tfrac{1}{2}\,v^2 = -\frac{g\,x^2}{2\,r} + C; \text{ but } v = 0 \text{ when } x = a;$$

$$\therefore C = \frac{g\,a^2}{2\,r}, \text{ and } \tfrac{1}{2}\,v^2 = \frac{g\,(a^2 - x^2)}{2\,r}; \therefore v = \left\{\frac{g}{r}\,(a^2 - x^2)\right\}^{\frac{1}{2}}.$$

If the body falls from the surface to the centre, $x = 0$, and this formula becomes $v = (g\,r)^{\frac{1}{2}} = (32\tfrac{1}{6} \times 3956 \times 5280)^{\frac{1}{2}} = 25{,}904$ feet per second.

To find the time of falling.—By Equation 1st, and substitutions, we obtain $d\,t = \frac{d\,s}{v} = \frac{d\,(a - x)}{v} = -\frac{d\,x}{v} = \frac{-\,d\,x}{\left\{\frac{g}{r}\,(a^2 - x^2)\right\}^{\frac{1}{2}}}$

$$= \left(\frac{r}{g}\right)^{\frac{1}{2}} \times \frac{-\,d\,x}{(a^2 - x^2)^{\frac{1}{2}}}; \therefore t = \left(\frac{r}{g}\right)^{\frac{1}{2}} \int \frac{-\,dx}{(a^2 - x^2)^{\frac{1}{2}}} = \left(\frac{r}{g}\right)^{\frac{1}{2}} \cos^{-1}\frac{x}{a} + C.$$

When $t = 0, x = a, \frac{x}{a} = 1$, and the arc, whose cosine is $1 = 0$;

$$\therefore C = 0. \quad \therefore t = \left(\frac{r}{g}\right)^{\frac{1}{2}} \times \cos^{-1}\frac{x}{a}.$$

If the body falls to the centre, $x = 0$, and $t = \left(\frac{r}{g}\right)^{\frac{1}{2}} \times \frac{\pi}{2}$; in which a does not appear at all; so that the time of falling to the centre from any point within the surface is the same; and equals $\left(\frac{3956 \times 5280}{32\frac{1}{6}}\right)^{\frac{1}{2}} \times 1.570796$ in seconds, or 21m. 5.8s.

II. Centre of Gravity.

7. Principle of Moments.—In order to apply the processes of the calculus to the determination of the centre of gravity, the principle is used, which was proved (Art. 78), that if every particle of a body be multiplied by its distance from a plane, and the sum of the products be divided by the sum of the particles, the quotient is the distance of the common centre from the same plane. The product of any particle or body by its distance from the plane, is called its *moment* with respect to that plane.

8. General Formulæ.—Let $B\,A\,C$ (Fig. 3) be any symmetrical curve, having $A\,X$ for its axis of abscissas, and $A\,Y$, at right angles to it, for its axis of ordinates. It is obvious that the centre of gravity of the line $B\,A\,C$, of the area $B\,A\,C$, of the solid of revolution around the axis $A\,X$, and of the surface of the same solid, are all situated on $A\,X$, on account of the symmetry of the figure. It is proposed to find the formula for the distance of the centre from $A\,Y$, in each of these cases. Let G in every instance represent the distance of the general centre of gravity from the axis $A\,Y$, or the plane $A\,Y$, at right angles to $A\,X$. The distance G would plainly be the same for the *half* figure $B\,A\,D$, as for the whole $B\,A\,C$; expressions may therefore be obtained for either, according to convenience.

FIG. 3.

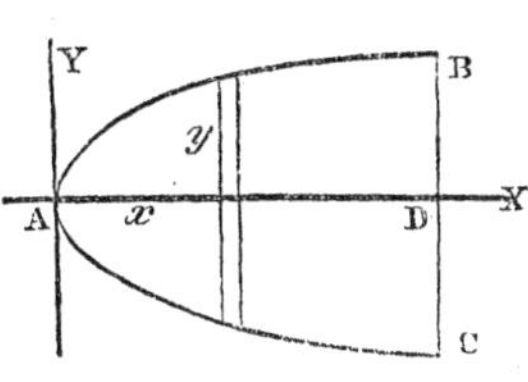

1. *The line A B.*—Let x be the abscissa, and y the ordinate; then $(d\,x^2 + d\,y^2)^{\frac{1}{2}}$ is the differential of the line $A\,B$. For brevity, let $s =$ the line, and $d\,s$ its differential. If we now multiply this differential by its distance from $A\,Y$, $x\,d\,s$ is the moment of a minute portion of the line; and the integral of it, $\int x\,d\,s$, is the moment of the whole. Dividing this by the line itself, i. e. by s, we have $\frac{\int x\,d\,s}{s}$ for the distance G.

2. *The area B A D.*—The differential of the area is $y\,d\,x$; the differential of its moment is $x\,y\,d\,x$; hence the moment itself is $\int x\,y\,d\,x$; and the distance $G = \frac{\int x\,y\,d\,x}{\text{area}}$.

3. *The solid of revolution.*—The differential of the solid, generated by the revolution of $A\,B$ on $A\,X$, is $\pi\,y^2 d\,x$; the differential of its moment is $\pi\,x\,y^2 d\,x$; and the moment is $\int \pi\,x\,y^2 d\,x$; hence the distance $G = \frac{\int \pi\,x\,y^2 d\,x}{\text{solid}}$.

4. *The surface of revolution.*—The differential of the surface is $2\,\pi\,y\,d\,s$; the differential of its moment is $2\,\pi\,x\,y\,d\,s$; and therefore the moment is $\int 2\,\pi\,x\,y\,d\,s$; and the distance $G = \frac{\int 2\,\pi\,x\,y\,d\,s}{\text{surface}}$.

9. Application of Formulæ.—We proceed to determine the centre of gravity in a few cases by the aid of these formulæ:

1. *A straight line.*—Imagine the line placed on $A\,X$, with one extremity at the origin A. The moment of a minute part of it is $x\,d\,x$, and that of the whole is $\int x\,d\,x$, while the length of the whole is x; $\therefore\ G = \frac{\int x\,dx}{x} = \frac{\frac{1}{2}\,x^2 + C}{x} = \frac{1}{2}\,x$, as it evidently should

be. In all the cases considered here, $C = 0$, because the function vanishes when x does.

2. *The arc of a circle.*—By formula 1st we have $G = \frac{\int x\,d\,s}{s}$. but $d\,s = (d\,x^2 + d\,y^2)^{\frac{1}{2}}$; by the equation of the circle, $y^2 = 2\,a\,x - x^2$;

$$\therefore y\,d\,y = (a - x)\,d\,x; \therefore d\,y^2 = \frac{(a-x)^2\,d\,x^2}{y^2} = \frac{(a-x)^2\,d\,x^2}{2\,a\,x - x^2};$$

$$\therefore (d\,x^2 + d\,y^2)^{\frac{1}{2}} = \frac{a\,d\,x}{(2\,a\,x - x^2)^{\frac{1}{2}}};$$

$$\therefore \frac{\int x\,d\,s}{s} = \int \frac{x}{s} \times \frac{a\,d\,x}{(2\,a\,x - x^2)^{\frac{1}{2}}} = \frac{a}{s}\int \frac{x\,d\,x}{(2\,a\,x - x^2)^{\frac{1}{2}}} = \frac{a}{s}\left\{\text{vers}^{-1}\,x\right.$$

$$\left. -(2\,a\,x - x^2)^{\frac{1}{2}}\right\} = \frac{a}{s}(s - y) = a - \frac{a\,y}{s} = a - \frac{a\,c}{t},$$ if the arc is doubled and called t, and c (chord) put for $2\,y$. As $a - \frac{a\,c}{t}$ is the distance from the origin A, and a = radius of the arc; $\therefore$ the distance from the centre of the circle to the centre of gravity of the arc, is $\frac{a\,c}{t}$, which is a fourth proportional to the *arc*, the *chord*, and the *radius*.

When the arc is a semi-circumference, $c = 2\,a$, and $t = \pi\,a$; $\therefore$ the distance of the centre of gravity of a semi-circumference from the centre of the circle is $\frac{2\,a}{\pi}$.

3. *The area of a circular sector.*—Suppose the given sector to be divided into an infinite number of sectors; then each may be considered a triangle, and its centre of gravity therefore distant from the centre of the circle by the line $\frac{2\,a}{3}$. Hence the centres of gravity of all the sectors lie in a circular arc, whose radius is $\frac{2\,a}{3}$; so that the centre of gravity of the whole sector coincides with the centre of gravity of that arc. The distance of the centre of gravity of the arc from the centre of the circle, by the preceding case, is $\frac{2}{3}a \times \frac{2}{3}c \div \frac{2}{3}t = \frac{2\,a\,c}{3\,t}$, which is therefore the distance of the centre of gravity of the sector from the centre of the circle. When the sector is a semicircle the distance becomes $\frac{2\,a \times 2\,a}{3\,\pi\,a}$ $= \frac{4\,a}{3\,\pi}$

4. *The area of a parabola.*—The equation of the curve is

$$y^2 = p\,x, \text{ or } y = p^{\frac{1}{2}}\,x^{\frac{1}{2}};$$

therefore the formula 2 for moment,

$$\int x\,y\,d\,x = \int p^{\frac{1}{2}}\,x^{\frac{3}{2}}\,d\,x = \tfrac{2}{5}p^{\frac{1}{2}}\,x^{\frac{5}{2}}\,(+C = 0);$$

but the area of the half parabola $= \frac{2}{3}\,p^{\frac{1}{2}}\,x^{\frac{3}{2}}$;

$$\therefore\ G = \tfrac{2}{5}\,p^{\frac{1}{2}}\,x^{\frac{5}{2}} \div \tfrac{2}{3}\,p^{\frac{1}{2}}\,x^{\frac{3}{2}} = \tfrac{3}{5}\,x.$$

To find the distance of the centre of gravity of the semi-parabola from the axis $A\,X$, proceed as follows: The differential of the area, as before, equals $y\,d\,x$; and the distance of its centre from $A\,X$ is $\frac{1}{2}\,y$. Therefore its moment with respect to $A\,X$ is $\frac{1}{2}\,y^2\,d\,x = \frac{1}{2}\,p\,x\,d\,x$; and the moment of the whole is $\int \frac{1}{2}\,p\,x\,d\,x = \frac{1}{4}\,p\,x^2$; ∴ the distance of the centre from

$$A\,X = \tfrac{1}{4}\,p\,x^2 \div \tfrac{2}{3}\,p^{\frac{1}{2}}\,x^{\frac{3}{2}} = \tfrac{3}{8}\,p^{\frac{1}{2}}\,x^{\frac{1}{2}} = \tfrac{3}{8}\,y.$$

5. *The area of a circular segment.*—The equation of the circle is, $y = (2\,a\,x - x^2)^{\frac{1}{2}}$. Therefore (formula 2),

$$\int x\,y\,d\,x = \int x\,(2\,a\,x - x^2)^{\frac{1}{2}}\,d\,x.$$

Add and subtract $a\,(2\,a\,x - x^2)^{\frac{1}{2}}\,d\,x$, and it becomes

$$\int a\,(2\,a\,x - x^2)^{\frac{1}{2}}\,d\,x - \int (a - x)\,(2\,a\,x - x^2)^{\frac{1}{2}}\,d\,x =$$

$$a\int y\,d\,x - \frac{(2\,a\,x - x^2)^{\frac{3}{2}}\,(a - x)\,d\,x}{\frac{3}{2}\,(2\,a - 2\,x)\,d\,x} = a\,.\,\text{area } A\,B\,D - \tfrac{1}{3}\,(2\,a\,x - x^2)^{\frac{3}{2}}.$$

$$\therefore\ G = a - \frac{(2\,a\,x - x^2)^{\frac{3}{2}}}{3 \text{ area } A\,B\,D}.$$

When $x = a$, $G = a - \dfrac{4\,a}{3\,\pi}$; and the distance of the centre of gravity of a semicircle from the centre of the circle $= \dfrac{4\,a}{3\,\pi}$. When $x = 2\,a$, $G = a$, as it plainly should be.

6. *A spherical segment.*—The equation of the circle is $y^2 = 2\,a\,x - x^2$. Therefore (formula 3),

$$\int \pi x y^2 d\,x = \int \pi\,x\,d\,x\,(2ax - x^2) = \int 2a\pi x^2 dx - \int \pi x^3 dx = \tfrac{2}{3}a\pi x^3 - \tfrac{1}{4}\pi x^4;$$

$$\therefore\ G = \frac{\frac{2}{3}\,a\,\pi\,x^3 - \frac{1}{4}\,\pi\,x^4}{a\,\pi\,x^2 - \frac{1}{3}\,\pi\,x^3} = \frac{8\,a\,x - 3\,x^2}{12\,a - 4\,x}.$$

When $x = a$, $G = \frac{5}{8}\,a$; that is, the centre of gravity of a hemisphere is $\frac{5}{8}$ of radius from the surface, or $\frac{3}{8}$ of radius from the centre of the sphere. If $x = 2\,a$, $G = a$.

7. *A right cone.*—In this case $A\,B$ (Fig. 3), is a straight line, and its equation is $y = a\,x$, where a is any constant.

$\therefore\ y^2 = a^2x^2;\ \therefore\ \int \pi x y^2 dx = \int \pi a^2 x^3 dx = \frac{\pi}{4} a^2 x^4;\ \therefore\ G = \frac{\frac{1}{4}\pi a^2 x^4}{\frac{1}{3}\pi a^2 x^3} = \frac{3}{4}x.$

Hence the centre of gravity of a cone is three-fourths of the axis from the vertex. See Art. 75.

8. *The convex surface of a right cone.*—The equation is

$$y = a x;\ \therefore d y^2 = a^2 d x^2;\ \text{and}\ (d x^2 + d y^2)^{\frac{1}{2}} = (a^2 + 1)^{\frac{1}{2}} d x.$$

Therefore (formula 4),

$\int 2 \pi x y ds = \int 2\pi x y (dx^2 + dy^2)^{\frac{1}{2}} = \int 2\pi a x^2 (a^2+1)^{\frac{1}{2}} dx = \frac{2}{3}\pi a x^3 (a^2+1)^{\frac{1}{2}}$ = the moment of the surface. The surface itself,

$$= \pi y (x^2 + y^2)^{\frac{1}{2}} = \pi a x^2 (a^2 + 1)^{\frac{1}{2}}.\quad \therefore G = \frac{\frac{2}{3}\pi a x^3 (a^2 + 1)^{\frac{1}{2}}}{\pi a x^2 (a^2 + 1)^{\frac{1}{2}}} = \frac{2}{3} x.$$

The centre of gravity of the convex surface of a right cone is on the axis, at a distance equal to two-thirds of its length from the vertex.

III. Centre of Oscillation.

9. To find the Moment of Inertia of a Body for any given Axis.—To render the formula $l = \frac{S(m r^2)}{M k}$ suitable to the application of the calculus, we have simply to substitute the sign of integration for S, and $d\,M$ for m, and we have

$$l = \frac{\int r^2 d M}{M k}. \qquad (1)$$

It is useful to know how to find the *moment of inertia* with respect to any axis by means of the *known moment* with respect to *another* axis parallel to it and passing through the centre of gravity of the body.

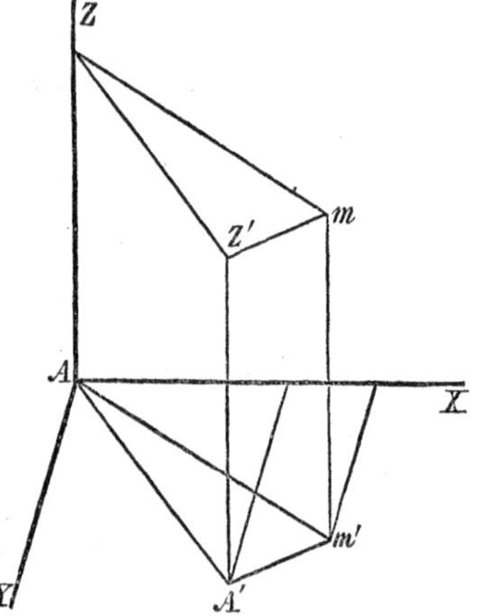

Fig. 4.

Let $A\,Z$ (Fig. 4) be the axis passing through the centre of gravity of the body for which the moment of inertia is $\int r^2 dM$, and let $A'\,Z'$ be the axis parallel to it, for which the moment of inertia, $\int r'^2\, d\,M$ of the same mass M, is to be determined. For every particle m of the body the corresponding value of $A\,m'$ is $r^2 = x^2 + y^2$. In like man-

ner, if we denote the co-ordinates of A' by a and β, and the distance between the axes by a, we shall have $a^2 = a^2 + \beta^2$. Now the distance of the particle m from $A'\,Z'$ is $r'^2 = (x - a)^2 + (y - \beta)^2 = x^2 + y^2 + a^2 + \beta^2 - 2\,a\,x - 2\,\beta\,y = r^2 + a^2 - 2\,a\,x - 2\,\beta\,y$; $\therefore$ $\int r'^2\,d\,M = \int r^2\,d\,M + a^2 \int d\,M - 2\,a \int x\,d\,M - 2\,\beta \int y\,d\,M = a^2\,M + \int r^2\,d\,M$, ... (2)
since $A\,Z$ passes through the centre of gravity of the body. Hence, *the moment of inertia of a body with respect to any axis is equal to the moment of inertia with respect to a parallel axis through the centre of gravity, plus the mass of the body multiplied by the square of the distance between the two axes.*

Put $C =$ the moment of inertia with respect to an axis through the centre of gravity; then the distance from the axis of suspension to the centre of oscillation, the axes being parallel, will be

$$l = \frac{C + a^2\,M}{M\,k} \qquad (3)$$

10. Examples.—

1. Find the centre of oscillation of a slender rod or straight line suspended at any point.

Let a and b be the lengths on opposite sides of the axis of suspension, then by (1)

$$l = \frac{\int r^2\,d\,M}{M\,k} = \frac{\int r^2\,d\,r}{(a + b)\frac{1}{2}\,(a - b)} = \frac{2\,(a^3 + b^3)}{3\,(a^2 - b^2)} = \frac{2\,(a^2 - a\,b + b^2)}{3\,(a - b)}$$

between the limits $r = + a$ and $r = - b$.

If the rod is suspended at its extremity, $b = 0$, and $l = \frac{2}{3}\,a$. If it is suspended at its middle point, $a = b$ and $l = \infty$.

2. Find the centre of oscillation of an isosceles triangle vibrating about an axis in its own plane passing through its vertex.

Put b and h for the base and altitude of the triangle; then by

(1), $l = \dfrac{\int_0^h r^2 \cdot \frac{b}{h}\,r\,d\,r}{\frac{1}{2}\,b\,h \cdot \frac{2}{3}\,h} = \frac{3}{4}\,h.$

If the axis of suspension coincides with the base of the triangle, then $l = \dfrac{\int_0^h r^2 \cdot \frac{b}{h}\,(h - r)\,d\,r}{\frac{1}{2}\,b\,h \cdot \frac{1}{3}\,h} = \dfrac{h}{2}.$

3. Find the centre of oscillation of a circle vibrating about an axis in its own plane.

$$C = \int r^2\,d\,M = 2 \int x^2\,y\,d\,x = 2 \int x^2\,(R^2 - x^2)^{\frac{1}{2}}\,d\,x =$$
$$- x\,\frac{(R^2 - x^2)^{\frac{3}{2}}}{2} + \frac{R}{2} \int (R^2 - x^2)^{\frac{1}{2}}\,d\,x.$$

Taking this integral between $x = -r$ and $x = +r$, we have

$$C = \frac{R^2}{2} \cdot \frac{\pi R^2}{2} = \frac{\pi R^4}{4}.$$

Substituting this value of C in (3) we have

$$l = \frac{\frac{\pi R^4}{4} + a^2 \pi R^2}{a \pi R^2} = a + \frac{R^2}{4a}.$$

4. Find the centre of oscillation of a circle vibrating about an axis perpendicular to it.

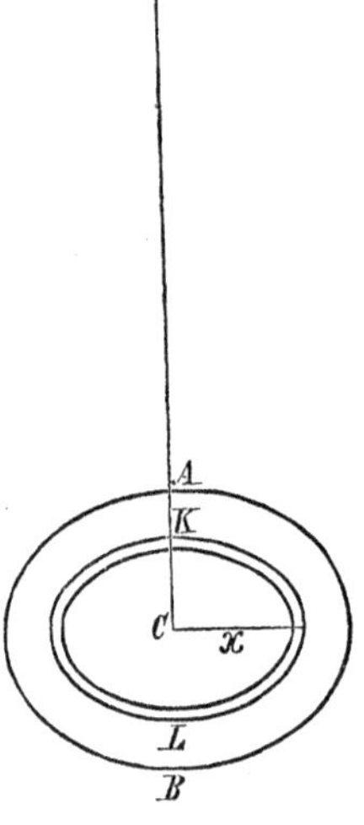

FIG. 5.

Let $K L$ (Fig. 5) be an elementary ring whose radius is x and whose breadth is $d x$; then

$$d M = 2 \pi x d x, \text{ and } C = \int_0^R x^2 \cdot 2 \pi x d x$$

$$= \frac{\pi R^4}{2}; \therefore l = \frac{\frac{\pi R^4}{2} + a^2 \pi R^2}{\pi R^2 a} = a + \frac{R^2}{2a}.$$

As $a + \frac{R^2}{2a}$ is greater than $a + \frac{R^2}{4a}$, a circular pendulum will vibrate faster when the axis of suspension is in its plane, than when it is perpendicular to it.

IV. CENTRE OF HYDROSTATIC PRESSURE.

11. General Formula.—Let the surface pressed upon be plane and vertical; and let the water level be the plane of reference. Suppose the surface to have a symmetrical form with reference to a vertical axis, x, whose ordinate is y (Fig. 6). A horizontal element of the surface is $2 y d x$, and (since the pressure varies as the depth) the pressure on that element $2 x y d x$. Hence the whole pressure to the depth x is $\int 2 x y d x = 2 \int x y d x$. The moment of the pressure on the element of surface is $2 x^2 y d x$; and the sum of all the moments to the same depth is $\int 2 x^2 y d x = 2 \int x^2 y d x$. Therefore, putting p for the depth of the centre of pressure, $p = \frac{\int x^2 y d x}{\int x y d x}$.

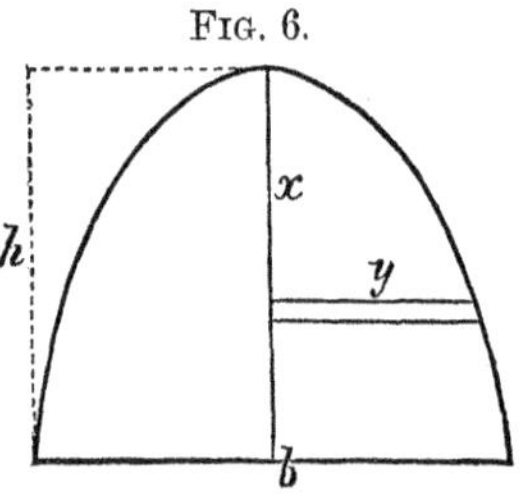

FIG. 6.

12. Examples.

1. *A rectangle.*—Let its height $= h$, and its base $= b$; then $2y$ everywhere equals b, and a horizontal element at the depth x is $b\,d\,x$, the pressure on it is $b\,x\,d\,x$, and the moment of that pressure is $b\,x^2\,d\,x$; $\therefore$ the depth of the centre of pressure $p = \frac{\int b\,x^2\,d\,x}{\int b\,x\,d\,x} = \frac{\frac{1}{3}\,b\,x^3 + c}{\frac{1}{2}\,b\,x^2 + c'}$. Since the pressure and area is each zero, when x is zero, c and c' both disappear, and $p = \frac{2}{3}\,x$, which for the whole surface becomes $p = \frac{2}{3}\,h$. That is, the centre of pressure on a vertical rectangular surface reaching to the water level, is two-thirds of the distance from the middle of the upper side to the middle of the lower.

2. *A triangle whose vertex is at the surface of the water, and its base horizontal.*—Let the triangle be isosceles, its height $= h$, and its base $= b$; then $h : b :: x : 2\,y = \frac{b}{h}\,x$. Therefore $p = \frac{\int \frac{b}{h}\,x^3\,d\,x}{\int \frac{b}{h}\,x^2\,d\,x}$ $= \frac{\frac{1}{4}\,x^4}{\frac{1}{3}\,x^3} = \frac{3}{4}\,x$; and for the whole height, $\frac{3}{4}\,h$.

If the triangle is not isosceles, it may be easily shown that the centre of pressure is on the line joining the vertex and the middle of the base, at a distance from the vertex equal to three-fourths of the length of that line.

3. *A triangle whose base is at the water level.*—Then $h : b :: h - x : 2\,y = b - \frac{b}{h}\,x$. Therefore the pressure is $\int -\,b\,x\,d\,x - \int -\,\frac{b}{h}\,x^2\,d\,x$, because $d\,x$ is negative. The moment of the pressure is $\int -\,b\,x^2\,d\,x - \int -\frac{b}{h}\,x^3\,d\,x$.

Therefore $p = \frac{-\int b\,x^2\,d\,x + \int \frac{b}{h}\,x^3\,d\,x}{-\int b\,x\,d\,x + \int \frac{b}{h}\,x^2\,d\,x} = \frac{-\frac{1}{3}\,x^3 + \frac{1}{4\,h}x^4}{-\frac{1}{2}\,x^2 + \frac{1}{3\,h}\,x^3} =$ $\frac{4\,h\,x^3 - 3\,x^4}{6\,h\,x^2 - 4\,x^3} = \frac{4\,h\,x - 3\,x^2}{6\,h - 4x}$; and, when $x = h$, this becomes $\frac{1}{2}\,h$. In general, the centre of pressure is at the middle of the line joining the vertex and the middle of the base.

4. *A parabola whose vertex is at the surface.*—As $y = p^{\frac{1}{2}}\,x^{\frac{1}{2}}$, therefore $p = \frac{\int x^2\,p^{\frac{1}{2}}\,x^{\frac{1}{2}}\,d\,x}{\int x\,p^{\frac{1}{2}}\,x^{\frac{1}{2}}\,d\,x} = \frac{\int x^{\frac{5}{2}}\,d\,x}{\int x^{\frac{3}{2}}\,d\,x} = \frac{\frac{2}{7}\,x^{\frac{7}{2}}}{\frac{2}{5}\,x^{\frac{5}{2}}} = \frac{5}{7}x$; or $\frac{5}{7}\,h$, for the whole area.

5. *A parabola whose base is at the surface.*—As $h - x$ is the depth of an element, $d\,x$ is negative. $p = \dfrac{-\int (h-x)^2 x^{\frac{1}{2}}\,d\,x}{-\int (h-x)\cdot x^{\frac{1}{2}}\,d\,x} =$

$$\frac{\int (h^2 x^{\frac{1}{2}}\,d\,x - 2\,h\,x^{\frac{3}{2}}\,d\,x + x^{\frac{5}{2}}\,d\,x)}{\int (h\,x^{\frac{1}{2}}\,d\,x - x^{\frac{3}{2}}\,d\,x)} = \frac{\frac{2}{3}h^2 x^{\frac{3}{2}} - \frac{4}{5}h\,x^{\frac{5}{2}} + \frac{2}{7}x^{\frac{7}{2}}}{\frac{2}{3}h\,x^{\frac{3}{2}} - \frac{2}{5}x^{\frac{5}{2}}} =$$

$\dfrac{\frac{1}{3}h^2 - \frac{2}{5}h\,x + \frac{1}{7}x^2}{\frac{1}{3}h - \frac{1}{5}x}$; and when $x = h$, the expression becomes

$$\frac{\frac{1}{3}h^2 - \frac{2}{5}h^2 + \frac{1}{7}h^2}{\frac{1}{3}h - \frac{1}{5}h} = \frac{4}{7}h.$$

V. Angular Radius of the Primary and Secondary Rainbow and the Halo.

13. The Primary Rainbow.—Since the primary bow is formed by those rays which, on emerging after one reflection, make the largest angle with the incident rays, proceed to find what angle of incidence will cause the largest deviation of the emerging rays.

In Fig. 7, let x = angle of incidence; y = angle of refraction; z = angle of deviation; n = index of refraction. Then, in the quadrilateral $B\,D\,G\,K$, $D\,B\,K = D\,G\,K = x - y$; angle at $D = 360 - 2\,y$; $\therefore K = z = 4\,y - 2\,x$;

Fig. 7.

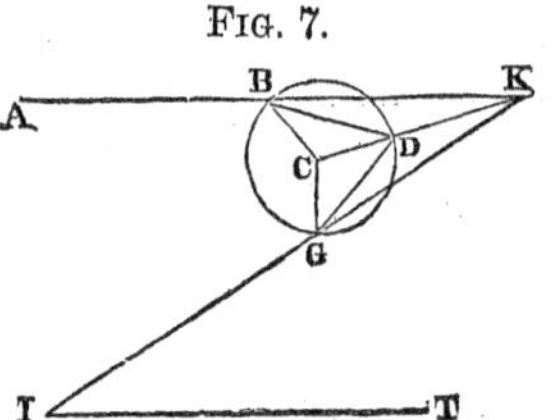

$$\therefore \frac{d\,z}{d\,x} = \frac{4\,d\,y}{d\,x} - 2 = 0.$$

But $\qquad \sin x = n \sin y;$

$$\therefore \cos x\,d\,x = n \cos y\,d\,y, \text{ and } \frac{d\,y}{d\,x} = \frac{\cos x}{n \cos y}.$$

By substitution, $\dfrac{4 \cos x}{n \cos y} = 2.$

$$\therefore 2 \cos x = n \cos y; \text{ and } 4 \cos^2 x = n^2 \cos^2 y.$$

But $\qquad \sin^2 x = n^2 \sin^2 y;$

$$\therefore 3 \cos^2 x + 1 = n^2; \text{ since } \sin^2 + \cos^2 = 1.$$

$$\therefore \cos x = \sqrt{\frac{n^2 - 1}{3}}$$

If 1.33 and 1.55, the values of n for extreme red and violet, be used in this formula, we obtain x, and therefore y and z, for the limiting angles of the primary bow.

14. The Secondary Bow.—To find the angle of minimum deviation. Using the same notation as before, we have in the pentagon $G\,E\,D\,B\,K$ (Fig. 8), $G = B = 180 - x + y$; $E = D = 2\,y$; $\therefore K = z = 180 + 2x - 6\,y$;

FIG. 8.

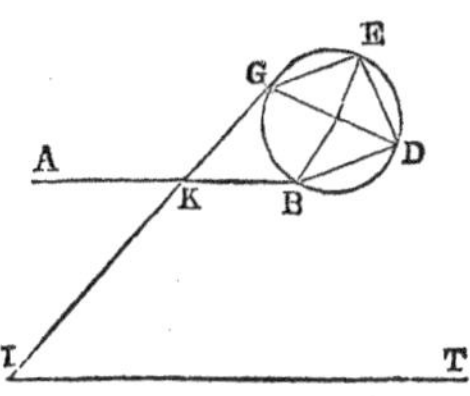

$$\therefore \frac{d\,z}{d\,x} = 2 - \frac{6\,d\,y}{d\,x} = 0.$$

$$\therefore \frac{6 \cos x}{n \cos y} = 2; \text{ and } 3 \cos x = n \cos y;$$

$$\therefore 9 \cos^2 x = n^2 \cos^2 y;$$

but $\quad \sin^2 x = n^2 \sin^2 y$;

$$\therefore 8 \cos^2 x + 1 = n^2;$$

$$\therefore \cos x = \sqrt{\frac{n^2 - 1}{8}};$$

which, as before, will furnish z for each limiting color of the secondary bow.

15. The Common Halo.—Let $D\,E$ (Fig. 9) be the ray from the sun, and $F\,G$ the emergent ray. Let $D\,E\,p = x$; $K\,E\,F = y$; $K\,F\,E = x'$; $G\,F\,p = y'$; $I = z = x - y + y' - x'$. Now, $y + x' = p'\,K\,F - C = 60°$.

FIG. 9.

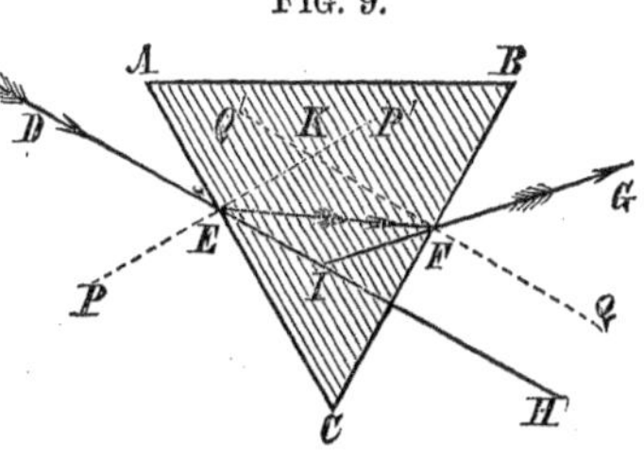

$$\therefore z = x + y' - C.$$

$\sin x = n \sin y$,

and $\sin y' = n \sin x'$;

$$\therefore x = \sin^{-1} (n \sin y),$$

and $\quad y' = \sin^{-1} (n \sin x') = \sin^{-1} \{n \sin C - y\}$.

By substitution,

$z = \sin^{-1} (n \sin y) + \sin^{-1} \{n \sin (C - y)\} - C$. Therefore z is a function of y; and, by differentiating, we have

$$\frac{d\,z}{d\,y} = \frac{n \cos y}{\sqrt{1 - n^2 \sin^2 y}} - \frac{n \cos (C - y)}{\sqrt{1 - n^2 \sin^2 (C - y)}} = 0.$$

$$\therefore \frac{n^2 \cos^2 y}{1 - n^2 \sin^2 y} = \frac{n^2 \cos^2 (C - y)}{1 - n^2 \sin^2 (C - y)};$$

$$\therefore \frac{1 - \sin^2 y}{1 - n^2 \sin^2 y} = \frac{1 - \sin^2 (C - y)}{1 - n^2 \sin^2 (C - y)};$$

$$\therefore (n^2 - 1) \sin^2 y = (n^2 - 1) \sin^2 (C - y);$$

$$\therefore y = C - y, \text{ and } y = \tfrac{1}{2}\,C;$$

$$\text{and } x' = \tfrac{1}{2}\,C.$$

Hence, the minimum deviation occurs when the ray within the crystal is equally inclined to the sides. Knowing n, the index of refraction for ice, x, and its equal, y', can be obtained, and then z, the deviation required.

www.ingramcontent.com/pod-product-compliance
Lightning Source LLC
LaVergne TN
LVHW021130110826
845150LV00005B/983